Emerging Converter Topologies and Control for Grid Connected Photovoltaic Systems

Emerging Converter Topologies and Control for Grid Connected Photovoltaic Systems

Editors

Dmitri Vinnikov
Samir Kouro
Yongheng Yang

MDPI • Basel • Beijing • Wuhan • Barcelona • Belgrade • Manchester • Tokyo • Cluj • Tianjin

Editors

Dmitri Vinnikov
Department of Electrical Power
Engineering and Mechatronics,
Tallinn University of Technology
Estonia

Samir Kouro
Electronics Engineering
Department, Universidad
Técnica Federico Santa María
Chile

Yongheng Yang
Department of Energy
Technology, Aalborg University
Denmark

Editorial Office
MDPI
St. Alban-Anlage 66
4052 Basel, Switzerland

This is a reprint of articles from the Special Issue published online in the open access journal *Energies* (ISSN 1996-1073) (available at: https://www.mdpi.com/journal/energies/special_issues/Emerging_Converter_Photovoltaic_Systems).

For citation purposes, cite each article independently as indicated on the article page online and as indicated below:

LastName, A.A.; LastName, B.B.; LastName, C.C. Article Title. *Journal Name* **Year**, *Volume Number*, Page Range.

ISBN 978-3-03943-909-6 (Hbk)
ISBN 978-3-03943-910-2 (PDF)

Contents

About the Editors

Dmitri Vinnikov received his Dipl.Eng., M.Sc., and Dr.Sc.techn. degrees in electrical engineering from Tallinn University of Technology, Tallinn, Estonia, in 1999, 2001, and 2005, respectively. He is currently the Head of the Power Electronics Group, Department of Electrical Power Engineering and Mechatronics, Tallinn University of Technology (Estonia) and a Visiting Professor at the Institute of Industrial Electronics and Electrical Engineering, Riga Technical University (Latvia). He is the CTO and co-founder of Ubik Solutions LLC – an Estonian start-up company dedicated to innovative and smart power electronics for renewable energy systems. He has authored more than 200 published papers on power converter design and development and is the holder of numerous patents and utility models in this field. His research interests include the applied design of power electronic converters and control systems, renewable energy conversion systems, impedance-source power converters, and implementation of wide bandgap power semiconductors. Dr. Vinnikov has directed over 25 National and International R&D projects, he is one of the founders and leading researchers of ZEBE – Estonian Centre of Excellence for zero energy and resource-efficient smart buildings and districts. He has authored or co-authored two books, five monographs, and one book chapter. D. Vinnikov is a Chair of the IES/PELS Joint Societies Chapter of the IEEE Estonia Section. He leads the Inverters/Rectifiers subcommittee of the Power Electronics Technical Committee (PETC) of the IEEE-IES and is a member of the Student and Young Professionals Activity Committee of the IEEE-IES. Dr. Vinnikov received the Best Young Scientist and Scientist of the Year awards of Tallinn University of Technology in 2010 and 2017, correspondingly. In 2014, he received the Estonian National Research Award in Technical Sciences for his contribution to innovative power electronic systems for renewable energy applications. In 2016, he was awarded a Doctor Honoris Causa from Chernihiv National University of Technology (Ukraine).

Samir Kouro received the M.Sc. and Ph.D. degrees in electronics engineering from the Universidad Tecnica Federico Santa Maria (UTFSM), Valparaíso, Chile, in 2004 and 2008, respectively. He currently is an academic at the Department of Electronic Engineering and serves as Director of Innovation and Technology Transfer of UTFSM. From 2009 to 2011, he was a Postdoctoral Fellow at the Department of Electrical and Computer Engineering, Ryerson University, Toronto, Canada. His research interests include power electronics, renewable energy conversion systems (photovoltaic and wind), and automotive applications. Dr. Kouro is a Principal Investigator of the Solar Energy Research Center (SERC-Chile), and Deputy Director of the Advanced Center of Electrical and Electronics Engineering (AC3E), both of which are R&D centers of excellence in Chile. He has directed 39 R&D projects, co-authored one book, 10 book chapters, 4 patents, and over 200 refereed journal and conference papers. He currently serves as the President of the IEEE PELS Chapter Chile and is a Board Member of the Solar Energy and Energy Innovation Committee of the Chilean Government. Dr. Kouro was included in the Clarivate Analytics 2018 Highly Cited Researcher List, and received the 2018 IEEE-AIE Outstanding Engineer Award, the 2016 IEEE Industrial Electronics Bimal K. Bose Award for Industrial Electronics Applications in Energy Systems, the 2015 IEEE Industrial Electronics Society J. David Irwin Early Career Award, and the 2012 IEEE Power Electronics Society Richard M. Bass Outstanding Young Power Electronics Engineer Award. He has also received 4 best paper awards in IEEE journals in the years 2019, 2016, 2012, and 2008.

Yongheng Yang received his B.Eng. degree from Northwestern Polytechnical University, China, in 2009, and his Ph.D. degree from Aalborg University, Denmark, in 2014. He was a postgraduate student at Southeast University, China, from 2009 to 2011. In 2013, he spent three months as a Visiting Scholar at Texas A&M University, USA. Currently, he is an Associate Professor with the Department of Energy Technology, Aalborg University, where he is also the Vice Program Leader for the research program on photovoltaic systems. Dr. Yang is the Chair of the IEEE Denmark Section. He is now an Associate Editor for several prestigious IEEE Transactions/Journals. He is a Deputy Editor of the IET Renewable Power Generation for Solar Photovoltaic Systems. He was the recipient of the 2018 IET Renewable Power Generation Premium Award and was an Outstanding Reviewer for the IEEE TRANSACTIONS ON POWER ELECTRONICS in 2018. His current research includes grid-integration of photovoltaic systems and multi-energy vectors with an emphasis on the power converter design, control, and reliability.

Preface to "Emerging Converter Topologies and Control for Grid Connected Photovoltaic Systems"

Continuous cost reduction of photovoltaic (PV) systems and the rise of power auctions resulted in the establishment of PV power not only as a green energy source but also as a cost-effective solution to the electricity generation market. Various commercial solutions for grid-connected PV systems are available at any power level, ranging from multi-megawatt utility-scale solar farms to sub-kilowatt residential PV installations. Compared to utility-scale systems, the feasibility of small-scale residential PV installations is still limited by the existing technologies that have not yet properly addressed issues like operation in weak grids, opaque and partial shading, etc. New market drivers such as warranty improvement to match the PV module lifespan, operation voltage range extension for application flexibility, and embedded energy storage for load shifting have again put small-scale PV systems in the spotlight. This Special Issue collects the latest developments in the field of power electronic converter topologies, control, design, and optimization for better energy yield, power conversion efficiency, reliability, and longer lifetime of the small-scale PV systems. This Special Issue will serve as a reference and update for academics, researchers, and practicing engineers to inspire new research and developments that pave the way for next-generation PV systems for residential and small commercial applications. Enjoy reading.

Dmitri Vinnikov, Samir Kouro, Yongheng Yang
Editors

Article

Compact Single-Stage Micro-Inverter with Advanced Control Schemes for Photovoltaic Systems

Yoon-Geol Choi [1], Hyeon-Seok Lee [1], Bongkoo Kang [1], Su-Chang Lee [2] and Sang-Jin Yoon [3,*]

[1] Department of Electrical Engineering, Pohang University of Science and Technology, Pohang 37673, Korea; ygchoi@postech.ac.kr (Y.-G.C.); hsasdf@postech.ac.kr (H.-S.L.); bkkang@postech.ac.kr (B.K.)
[2] LG Electronics Co., Ltd., Energy Business Center, Gumi 39368, Korea; suchang.lee@lge.com
[3] Department of Electrical Engineering, Korea Polytechnics, Gumi 39257, Korea
* Correspondence: sjyoon@kopo.ac.kr; Tel.: +82-54-468-5248

Received: 18 March 2019; Accepted: 27 March 2019; Published: 31 March 2019

Abstract: This paper proposes a grid-connected single-stage micro-inverter with low cost, small size, and high efficiency to drive a 320 W class photovoltaic panel. This micro-inverter has a new and advanced topology that consists of an interleaved boost converter, a full-bridge converter, and a voltage doubler. Variable switching frequency and advanced burst control schemes were devised and implemented. A 320 W prototype micro-inverter was very compact and slim with 60-mm width, 310-mm length, and 30-mm height. In evaluations, the proposed micro-inverter achieved CEC weighted efficiency of 95.55%, MPPT efficiency >95% over the entire load range, and THD 2.65% at the rated power. The proposed micro-inverter is well suited for photovoltaic micro-inverter applications that require low cost, small size, high efficiency, and low noise.

Keywords: single stage micro-inverter; burst control; variable frequency control; maximum power-point tracking

---

1. Introduction

The photovoltaic (PV) generation is emerging as a future energy system because of its installation convenience, no-noise, infinite, and eco-friendly characteristics [1–4]. It is classified into the centralized power system and the distributed power system depending on the scale of solar power generation [5]. The centralized power system has a simple circuit structure with PV strings as the input energy source, but it has a disadvantage that the power generation is considerably lowered when some panels of the PV string are shaded. On the other hand, in the distributed power system, the optimal power extraction is possible because the maximum power point tracking (MPPT) control can be applied to each PV panel with a micro-inverter connected. So, it can minimize the loss of power generation caused by the shading effect. However, one micro-inverter is required for each PV panel, so implementation of this strategy is expensive. Therefore, many attempts have recently been made to lower the cost of micro-inverters.

In general, considering the cost, micro-inverters have been designed to use circuit architectures with a flyback converter [6–10], which provides galvanic isolation with fewer switches than other designs. Although the flyback converter has the advantage of circuit simplicity and low cost, the design must use a transformer with a high turns ratio to achieve a high voltage-conversion ratio from low dc voltage on a single PV panel. In the transformer, the high turns ratio causes a large leakage inductance which increases the stress on semiconductor switches. Moreover, due to low utilization of the transformer, this topology is most suitable for low-power applications <200 W. Recently, multi-phase interleaved technology has been applied to solar power generation from PV panels that output ≥320 W, but this technology requires large and expensive components.

This paper proposes a low-cost, slim, single-stage micro-inverter to drive a 320-W-class PV panel. The proposed micro-inverter has an interleaved structure based on the boost half-bridge (BHB) converter [11] with a cascaded voltage doubler. The interleaved BHB has an inversely-coupled inductor for the voltage step-up operation. The coupled inductor can reduce input ripple current and can be reduced in size. The voltage doubler increases the ac output voltage from the interleaved converter. Therefore, the transformer can have a lower turns ratio in the interleaved BHB than in a flyback converter and can be reduced in size. In the proposed micro-inverter, semiconductor switches achieve turn-on zero-voltage-switching (ZVS) and turn-off zero-current-switching (ZCS) by exploiting the resonance between the leakage inductance of the transformer and output capacitors of the voltage doubler, without additional components.

This paper also presents two advanced control algorithms. First, a variable switching frequency control scheme was implemented to reduce total harmonic distortion (THD) by reducing output ripple current. Then an advanced burst control scheme was implemented to improve power-conversion efficiency at light loads. By distributing output current temporally at light loads, input ripple voltage can be reduced. Therefore, the size of decoupling capacitors is reduced and MPPT efficiency is improved compared with the conventional burst control [12,13]. Section 2 describes the circuit structure and operating principles of the proposed micro-inverter, Section 3 gives the proposed control schemes, Section 4 shows experimental results using a 320-W prototype micro-inverter, and Section 5 concludes the paper.

2. Circuit Structure and Operating Principles of the Proposed Micro-inverter

The proposed micro-inverter (Figure 1) consists of an interleaved boost converter, a full-bridge converter, and a voltage doubler. The portion that is composed of the interleaved boost and full-bridge converters is based on a boost half-bridge topology. The interleaved boost converter consists of an inversely-coupled inductor L_B, four switches S_1–S_4, and a storage capacitor C_S. The full-bridge converter consists of a transformer T_1 and the same four switches S_1–S_4 as the interleaved boost converter. The voltage doubler has four switches S_5–S_8 and two capacitors C_1 and C_2.

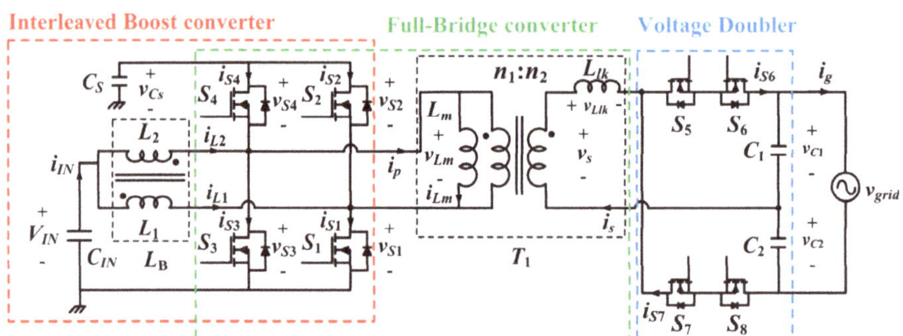

Figure 1. The circuit structure of the proposed micro-inverter.

In the proposed interleaved boost converter, two inductors L_1 and L_2 form L_B (Figure 2) by using a single magnetic core instead of two separate magnetic cores used in the conventional interleaved boost converter [14]. L_B has a turns ratio of 1:1; L_1 and L_2 each have self-inductance L. The mutual inductance M between L_1 and L_2 is represented as:

$$M = kL, \ (k < 0), \tag{1}$$

where k is the coupling coefficient. The voltage drops of L_1 and L_2 are given, respectively, by

$$v_1 = L\frac{di_1}{dt} - M\frac{di_2}{dt}, \tag{2}$$

and

$$v_2 = L\frac{di_2}{dt} - M\frac{di_1}{dt}. \tag{3}$$

Using Equations (2) and (3) and $v_m = -Md(i_1 + i_2)/dt$ yields

$$v_1 - v_m = (L + M)\frac{di_1}{dt} \tag{4}$$

and

$$v_2 - v_m = (L + M)\frac{di_2}{dt}. \tag{5}$$

S_1, the body diode of S_2, and L_1 form one boost power stage. S_3, the body diode of S_4 and L_2 form the other boost power stage. The two boost power stages form an interleaved boost converter and two outputs operate out of phase. When S_1 or S_3 is turned on, voltage v_{IN} is applied to L_1 or L_2, respectively. When S_1 or S_3 is turned off, voltage $v_{IN} - v_{Cs}$ is applied to L_1 or L_2, respectively. The energy accumulated during the on-state for each boost power stage is transferred into C_S. There are four cases of the voltage v_1 of L_1 and the voltage v_2 of L_2 depending on the states of S_1 and S_3. Using Equations (4) and (5), the equivalent inductance for each case is obtained (Table 1). $M < 0$ in Equation (1), so Table 1 demonstrates that appropriate design of the inversely coupled inductor can reduce the input ripple current of the micro-inverter [15].

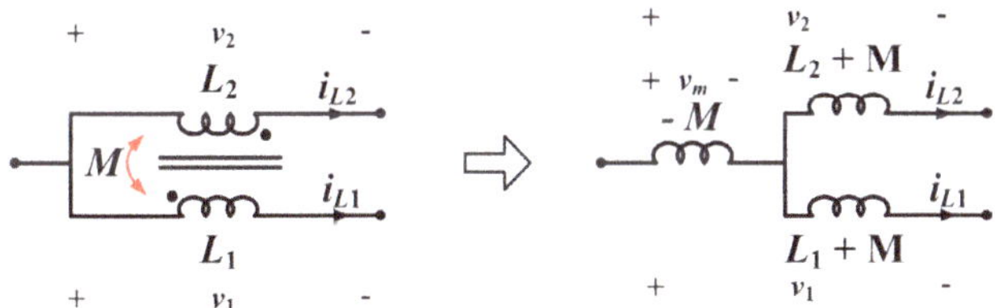

Figure 2. The equivalent circuit of the inversely coupled inductor L_B.

Table 1. Equivalent inductances in the interleaved boost converter.

Symbol	Value	Condition
L_{eq1}	$\frac{L^2 - M^2}{L + DM/1 - D}$	$v_1 = v_{IN}, v_2 = v_{IN} - v_{Cs}$
L_{eq2}	$L + M$	$v_1 = v_2 = v_{IN}$
L_{eq3}	$L + M$	$v_1 = v_2 = v_{IN} - v_{Cs}$
L_{eq4}	$\frac{L^2 - M^2}{L + (1-D)M/D}$	$v_1 = v_{IN} - v_{Cs}, v_2 = v_{IN}$

The full-bridge converter shares four switches S_1–S_4 with the interleaved boost converter, and its input power comes from C_S. The leakage inductance L_{lk} of T_1 and capacitors C_1 and C_2 in the voltage doubler form an LC resonant circuit. The LC resonant current flows through the primary and secondary sides of T_1 with turns ratio $n_1:n_2$. This current causes the body diode of each switch to conduct before the turn-on gate signal is applied, thus achieving zero-voltage-switching (ZVS) for S_1–S_4.

In the voltage doubler, S_5–S_8 rectify current on the secondary side of T_1. When grid voltage is positive, both S_5 and S_8 are turned on, and both S_6 and S_7 act as diodes. When grid voltage is negative, both S_6 and S_7 are turned on, and both S_5 and S_8 act as diodes. The energy transferred to the voltage doubler through T_1 is stored in C_1 and C_2. C_1 and C_2 are connected in series, and the output voltage of the micro-inverter is the sum of the voltage v_{C1} of C_1 and the voltage v_{C2} of C_2.

In the proposed micro-inverter, variable-switching-frequency control is used, and the output voltage is a sinusoidal grid voltage. However, for simplicity, the analysis is based on the assumption that the micro-inverter generates a constant output voltage with a fixed switching frequency at a certain point in the analysis. In addition, the electrical losses of all components are ignored, and the following conditions are assumed: $2\pi\sqrt{L_{lk}(C_1 + C_2)} > DT_s$ and $n^2 L_m >> L_{lk}$, where L_m is the magnetizing inductance and T_s is the switching period. The operation cycle S_1–S_4 is the same regardless of the polarity of the grid voltage, so the analysis considers only positive grid voltage.

The operating waveforms (Figure 3) of the proposed micro-inverter depend on the duty ratio D. First, operational states are analyzed for $D \leq 0.5$ (Figures 3a and 4).

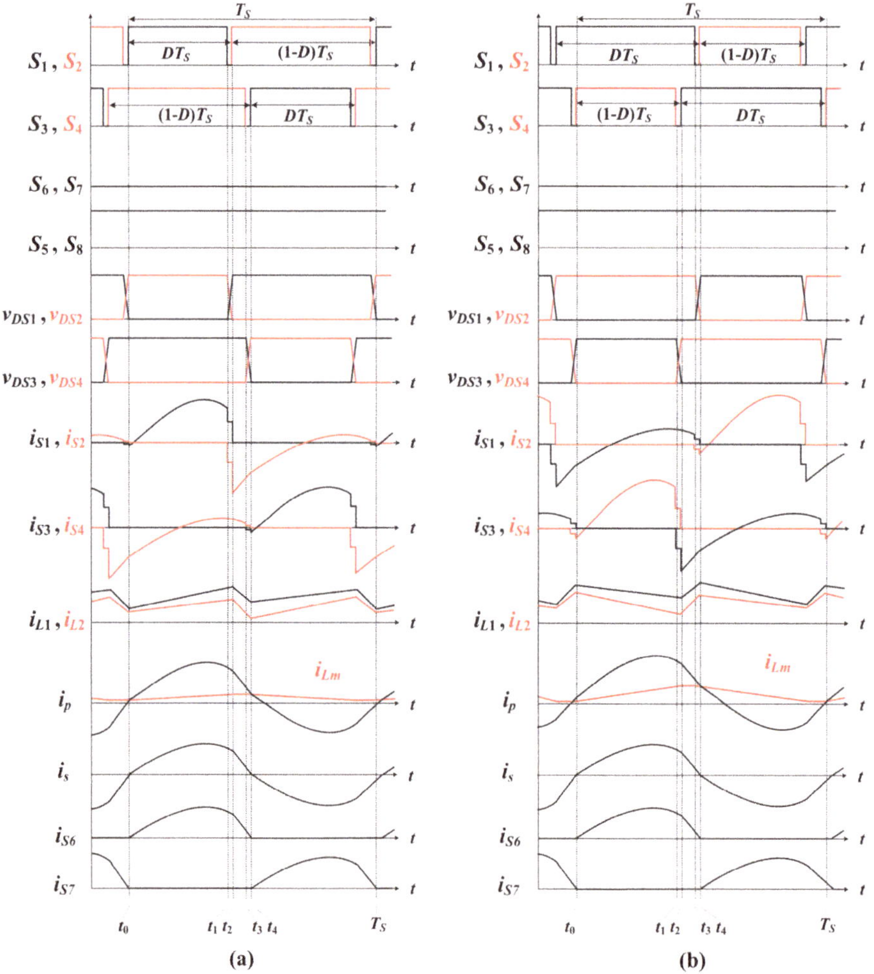

Figure 3. Operating waveforms of the proposed micro-inverter for (**a**) $D \leq 0.5$ and (**b**) $D > 0.5$.

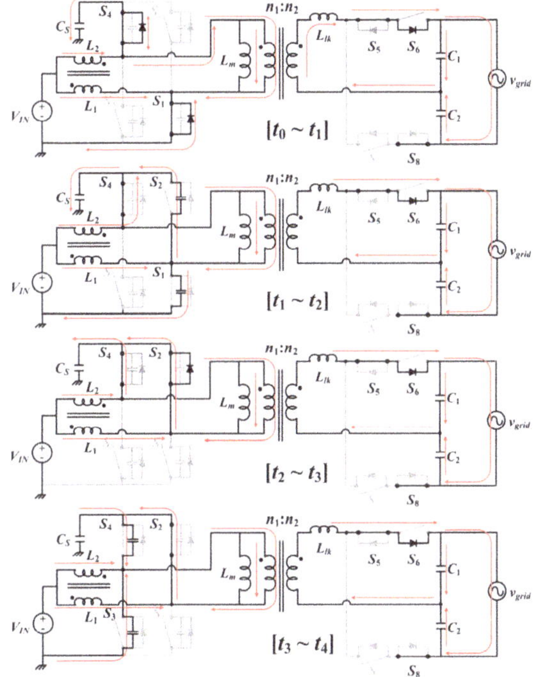

Figure 4. Operating modes when $D \leq 0.5$.

State 1 $(t_0–t_1)$: At $t = t_0$, S_1 is turned on, $v_{DS1} = 0$, and $i_{SW1} < 0$. S_4 remains in the turn-on state, and both S_2 and S_3 remain in the turn-off state. For T_1, the voltage v_{Lm} across L_m is equal to v_{Cs}, and the secondary voltage v_s proportional to the turns ratio $n_1{:}n_2$ is generated on the secondary side of T_1. The magnetizing current i_{Lm} is increased and is given by:

$$i_{Lm}(t) = i_{Lm}(t_0) + \frac{v_{Cs}}{L_m}(t - t_0). \tag{6}$$

Resonance is generated by L_{lk} on the secondary side of T_1 and capacitors C_1 and C_2, and the state equation is given by

$$L_{lk}\frac{di_s}{dt} = nv_{Lm} - v_{C1}, \tag{7}$$

$$i_s = C_1\frac{dv_{C1}}{dt} - C_2\frac{dv_{C2}}{dt} = (C_1 + C_2)\frac{dv_{C1}}{dt}. \tag{8}$$

Using Equations (7) and (8), the secondary current i_s of T_1 is obtained as

$$i_s(t) = \frac{nv_{Lm} - v_{C1}}{Z_r}\sin[\omega_r(t - t_0)], \tag{9}$$

where

$$Z_r = \sqrt{\frac{L_{lk}}{C_1 + C_2}} \tag{10}$$

is the resonant impedance and

$$\omega_r = \frac{1}{\sqrt{L_{lk}(C_1 + C_2)}} \tag{11}$$

is the resonant angular frequency.

From Equations (6) and (9), the primary current i_p of T_1 is obtained as

$$i_p(t) = i_{Lm}(t_0) + \frac{v_{Cs}}{L_m}(t - t_0) + \frac{n^2 v_{Lm} - n v_{C1}}{Z_r} \sin[\omega_r(t - t_0)]. \tag{12}$$

From Table 1, the currents i_{L1} and i_{L2} of the coupled inductor are obtained as

$$i_{L1}(t) = i_{L1}(t_0) + \frac{v_{IN}}{L_{eq1}}(t - t_0), \quad i_{L2}(t) = i_{L2}(t_0) + \frac{v_{IN} - v_{Cs}}{L_{eq1}}(t - t_0). \tag{13}$$

State 2 (t_1–t_2): At $t = t_1$, S_1 is turned off and S_4 remains in the turn-on state. Both S_2 and S_3 remain in the turn-off state. This interval is a dead time to prevent shoot-through before S_2 is turned on. During this state, the drain-source voltage of S_1 increases from 0 V to v_{Cs} and that of S_2 decreases from v_{Cs} to 0 V by charging and discharging parallel capacitance across each switch, respectively.

State 3 (t_2–t_3): At $t = t_2$, S_2 is turned on, $v_{DS2} = 0$, and $i_{SW2} < 0$. S_4 remains in the turn-on state, and both S_1 and S_3 remain in the turn-off state. For T_1, the voltage v_{Lm} across L_m is 0 V and the voltage v_{lk} across L_{lk} is $-v_{C1}$. The amplitude of i_{Lm} remains unchanged during state 3 as:

$$i_{Lm}(t) = i_{Lm}(t_2) = i_{Lm}(t_0) + \frac{v_{Cs}}{L_m}(t_2 - t_0). \tag{14}$$

i_s begins to decrease because the energy stored in L_{lk} is transferred to C_1, and is given by

$$i_s(t) \cong i_s(t_2) - \frac{v_{C1}}{L_{lk}}(t - t_2) = \frac{n v_{Lm} - v_{C1}}{Z_r} \sin[\omega_r(t_2 - t_0)] - \frac{v_{C1}}{L_{lk}}(t - t_2). \tag{15}$$

From Equations (14) and (15), i_p is obtained as

$$i_p(t) = i_{Lm}(t_0) + \frac{v_{Cs}}{L_m}(t_2 - t_0) + \frac{n^2 v_{Lm} - n v_{C1}}{Z_r} \sin[\omega_r(t_2 - t_0)] - \frac{n v_{C1}}{L_{lk}}(t - t_2). \tag{16}$$

From Table 1, i_{L1} and i_{L2} are obtained as

$$i_{L1}(t) = i_{L1}(t_2) + \frac{v_{IN}}{L_{eq3}}(t - t_2), \quad i_{L2}(t) = i_{L2}(t_2) + \frac{v_{IN} - v_{Cs}}{L_{eq3}}(t - t_2). \tag{17}$$

State 4 (t_3–t_4): At $t = t_3$, S_4 is turned off and S_2 remains in the turn-on state. Both S_1 and S_3 remain in the turn-off state. This time interval is a dead time to prevent shoot-through before S_3 is turned on. During this state, the drain-source voltage of S_4 increases from 0 V to v_{Cs} and that of S_3 decreases from v_{Cs} to 0 V.

The proposed micro-inverter has an interleaved structure, so both the operating principle of the next half cycle for $D \leq 0.5$ and the operating principle for $D > 0.5$ are the same as the above analysis except for the switches used. Thus, further analysis for the others is not given.

The voltage gain G_v of the proposed micro-inverter is twice the product of the boost converter voltage gain and the full bridge converter voltage gain:

$$G_v = \frac{V_{grid}}{V_{IN}} = 2 \cdot \frac{1}{1 - D} \cdot 2nD = \frac{4nD}{1 - D}. \tag{18}$$

3. The Proposed Control Schemes

The main controller (Figure 5) for the proposed micro-inverter takes as analog-to-digital inputs the grid voltage v_{grid}, the grid current i_g, the input voltage V_{IN} and the input current I_{IN}. The MPPT controller is based on the perturb and observe (P&O) MPPT algorithm [16]. This controller determines the amplitude of the reference grid current I_{g_ref} by using I_{IN} and V_{IN} to maximize solar power generation. In the P&O MPPT algorithm used (Figure 6), I_{g_ref} is increased if $\Delta P_{IN} > 0$ and $\Delta V_{IN} > 0$ or if $\Delta P_{IN} < 0$ and $\Delta V_{IN} < 0$. I_{g_ref} is decreased if $\Delta P_{IN} > 0$ and $\Delta V_{IN} < 0$ or if $\Delta P_{IN} < 0$ and $\Delta V_{IN} > 0$. This process is repeated until the maximum power point (MPP) is reached, i.e., $\Delta P_{IN} = 0$.

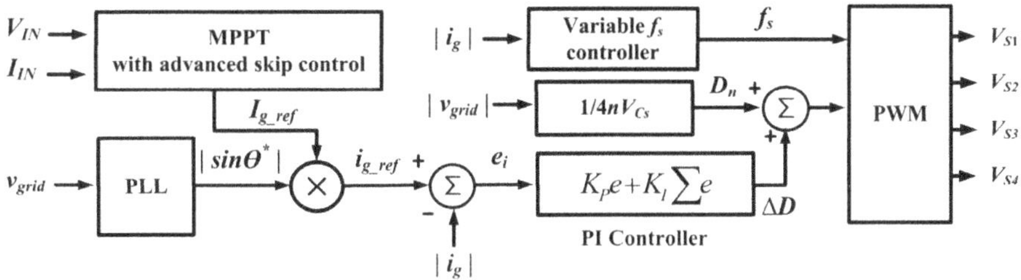

Figure 5. Block diagram of the main controller for the proposed micro-inverter.

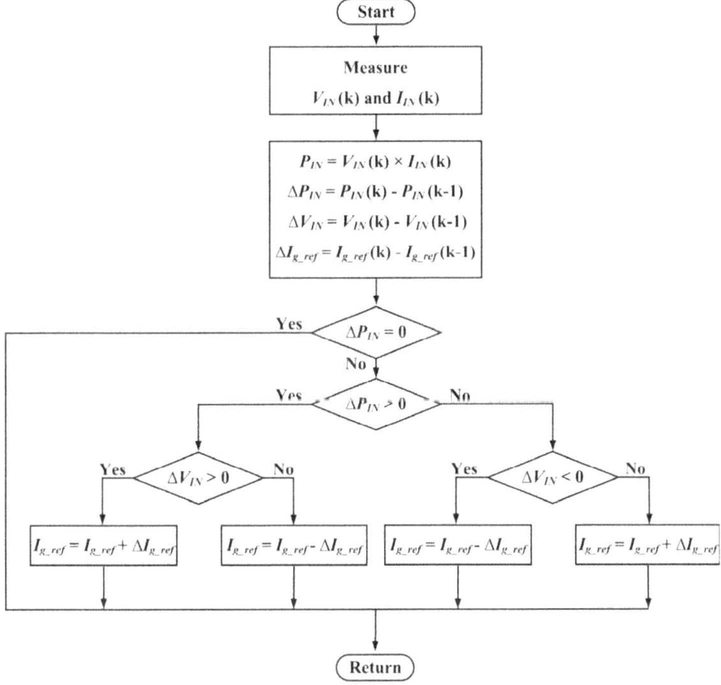

Figure 6. The perturb and observe MPPT algorithm.

The phase-locked loop (PLL) generates the phase information $|\sin \theta^*|$ by using v_{grid}. In the PLL, virtual voltage v_{q1} is derived from v_{grid} for phase detection.

$$v_{q1}(s) = G_{PLL}(s)v_{grid}(s) = V_{grid}\left(-\frac{1}{s+\omega} + \frac{s}{s^2+\omega^2} + \frac{\omega}{s^2+\omega^2}\right), \tag{19}$$

where $G_{PLL}(s)$ is PLL gain and V_{grid} is the amplitude of v_{grid}.

From the inverse Laplace transform of $v_{q1}(s)$,

$$v_{q1}(t) = V_{grid}\left(-e^{\omega t} + \cos \omega t + \sin \omega t\right) \approx V_{grid}(\cos \omega t + \sin \omega t), \tag{20}$$

where ωt is the actual phase of the grid.

Using equation (20), the other virtual voltage v_{q2} is obtained as

$$v_{q2}(t) = v_{q1}(t) - v_{grid}(t) = V_{grid} \sin \omega t. \tag{21}$$

v_{grid} and v_{q2} are transformed into the synchronous reference frame as follows:

$$\begin{bmatrix} v^e_{grid} \\ v^e_{q2} \end{bmatrix} = \begin{bmatrix} \cos \theta^* & \sin \theta^* \\ -\sin \theta^* & \cos \theta^* \end{bmatrix} \begin{bmatrix} v_{grid} \\ v_{q2} \end{bmatrix}, \tag{22}$$

where θ^* is a phase output from the PLL. From Equation (22),

$$v^e_{grid} = V_{grid} \cos(\omega t - \theta^*) \approx V_{grid}, \tag{23}$$

$$v^e_{q2} = V_{grid} \sin(\omega t - \theta^*) \approx V_{grid}(\omega t - \theta^*). \tag{24}$$

The PLL generates θ^* to follow ωt through PI control inside the PLL. The reference current signal i_{g_ref} is the product of I_{g_ref} and $|\sin \theta^*|$:

$$i_{g\_ref} = I_{g\_ref}\left|\sin \theta^*\right|. \tag{25}$$

The proportional-integral (PI) controller determines the duty ratio variation ΔD by using the difference between i_{g_ref} and $|i_g|$ as follows:

$$\Delta D = K_P\left(i_{g\_ref} - \left|i_g\right|\right) + K_I \sum \left(i_{g\_ref} - \left|i_g\right|\right) \tag{26}$$

ΔD compensates for the voltage drop of L_{lk}, so that i_g follows i_{g_ref}. The nominal duty ratio

$$D_n = \frac{\left|v_{grid}\right|}{G_v} = \frac{\left|v_{grid}\right|}{4nV_{Cs}} \tag{27}$$

provides stable system dynamics for nonlinear sinusoidal waves which are difficult to control using only ΔD. The total duty ratio

$$D = D_n + \Delta D = \frac{\left|v_{grid}\right|}{4nV_{Cs}} + K_P(i_{g\_ref} - |i_g|) + K_I \sum (i_{g\_ref} - |i_g|) \tag{28}$$

where D_n is duty ratio generated by the grid voltage and ΔD is a duty ratio variation generated by the grid current. D is given to the pulse-width-modulation (PWM) controller. The PWM controller generates gate signals for switches to track the reference power.

Operating modes (Figure 7) depend on the grid current level when grid voltage is positive. When i_g is low, the proposed micro-inverter operates in discontinuous conduction mode (DCM) because i_g becomes zero before the end of the switching cycle with the period T_s. When i_g is high, continuous conduction mode (CCM) is applied.

If a fixed switching frequency is used for the operating modes, especially the DCM mode, two problems occur: (1) High grid current ripples at low grid currents increase total harmonic distortion (THD); (2) as the output power decreases, the total DCM operating time can increase over the total CCM operating time, and the power conversion efficiency of the micro-inverter can be reduced by high current stress. To solve these problems, this paper proposes two advanced control schemes: Variable-switching-frequency (VSF) control and the advanced burst (AB) control.

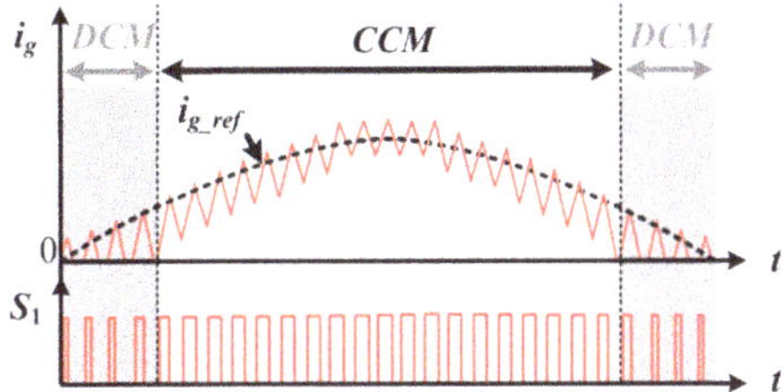

Figure 7. Operating modes depending on the grid current level during the positive grid voltage.

3.1. Variable Switching Frequency Control

During T_s, i_s of T_1 in DCM and CCM modes vary with D (Figure 8). As D decreases, the energy stored in L_{lk} decreases, so time required to demagnetize L_{lk} decreases. Therefore, the micro-inverter is operated in DCM mode. From Equations (9) and (15), the operating condition for DCM is given by

$$0 > \frac{nv_{Lm} - v_{C1}}{Z_r}\sin\omega_r DT_s - \frac{v_{C1}T_s}{2L_{lk}}(1-2D) \tag{29}$$

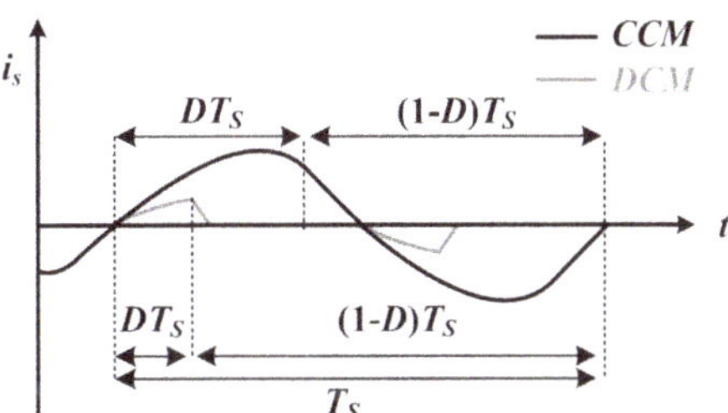

Figure 8. Secondary current i_s of the transformer T_1 depending on the operating mode.

Existing methods to optimize the DCM mode duration have drawbacks. One method is to increase the value of L_{lk}; a large L_{lk} increases the inductive energy and increases the demagnetizing time, but this solution requires a large transformer with a large number of windings. Another solution is to increase the switching frequency f_s; this approach can also increase the power density, but high f_s causes high switching loss. Thus, this paper presents VSF control, which minimizes switching loss without increasing the transformer size. VSF control varies f_s depending on the magnitude $|i_g|$ of the grid current.

Fixed-switching-frequency (FSF) control and VSF controls have distinct attributes (Figure 9). FSF control changes only D depending on v_{grid} (Figure 9a). In contrast, VSF control changes both D and f_s depending on v_{grid} (Figure 9b). When v_{grid} is near zero, the switching loss is very small because i_g is close to zero. Therefore, when VSF control is used, f_s is increased to the maximum switching frequency f_{max} and the time interval between demagnetizings of L_{lk} is reduced (Figure 9b). As v_{grid} increases, f_s is decreased to the minimum switching frequency f_{min} to reduce switching losses. f_s is given by

$$f_s = f_{\max} - (f_{\max} - f_{\max})\frac{v_{grid}}{V_{grid}} \tag{30}$$

where V_{grid} is the peak value of v_{grid}.

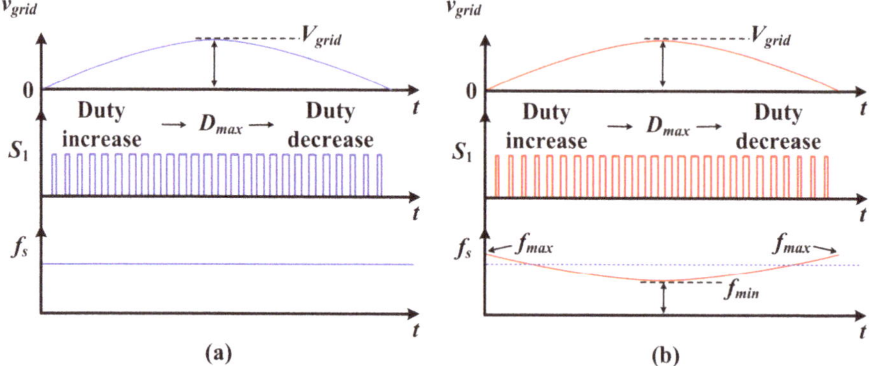

Figure 9. (**a**) Fixed and (**b**) variable switching frequency controls.

3.2. Advanced Burst Control

When solar power generation and load are very small, micro-inverters operate only intermittently to supply the desired power to the grid on an average power basis. This intermittent operation is called "burst control". For the burst control, the micro-inverter supplies i_g to the grid only during the ON state, and stops running during the OFF state. The burst control improves power-conversion efficiency by reducing the ripple of i_g and switching loss when the load is small.

In the conventional burst control scheme, positive and negative grid currents are consecutively supplied to the grid during one ON-state period (Figure 10). Then OFF-state periods follow the ON-state period. During the OFF state, no power is output, so output occurs only during the ON state, and the energy flowing out of C_{IN} is also concentrated. Therefore, the input ripple voltage ΔV_{IN} is increased, the MPPT efficiency is reduced, and additional time is required to charge the input capacitor C_{IN} for the next operation.

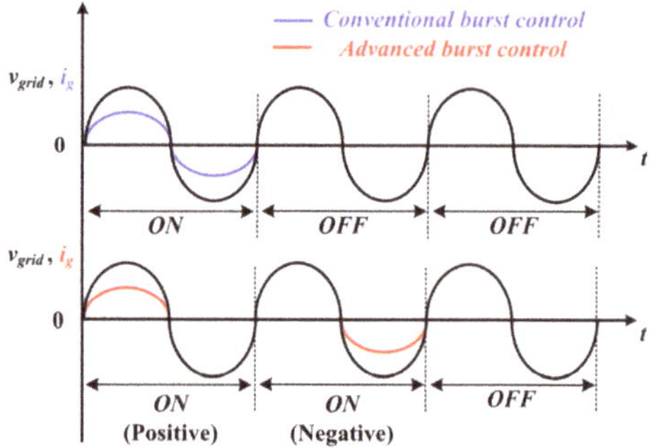

Figure 10. Conventional and advanced burst control schemes.

To further improve the performance of burst control, this paper proposes AB control, which supplies positive grid current during the first ON-state (Figure 10). The negative grid current is supplied during the ON-state that immediately follows the first ON-state period. Then, OFF-state periods follow the ON-state periods. This scheme has the effect of distributing the output current temporally compared with the conventional burst control scheme. Therefore, in the proposed micro-inverter with the advanced burst control scheme, the MPPT efficiency can be improved, and the input capacitance C_{IN} can be reduced due to the reduced ΔV_{IN}.

4. Experimental Results

The proposed grid-connected micro-inverter (Figure 11) was designed to operate at the rated power 320 W, V_{IN} = 25~52 V_{DC}, $I_{IN.max}$ = 12 A_{DC}, and f_s = 60~90 kHz. The grid voltage was 220 V_{rms}, the grid frequency was 60 Hz, and grid current supplied by the proposed micro-inverter is 0~1.45 A_{rms}. The proposed micro-inverter was implemented using the circuit parameters given in Table 2. The microcontroller used was a MN103DF35 (PANASONIC). For the PI controller in the main controller, K_P and K_I were experimentally optimized and set to 9.5 and 200, respectively. The sampling frequency for analog signals is 20 kHz, and the resolution of the analog-to-digital converter is 12 bits. The turns ratio of L_B is 10:10 and that of T_1 is 6:19. The resonant frequency f_r = 35.5 kHz from L_{lk} = 100 µH and C_1 = C_2 = 100 nF. The MOSFET package of S_1–S_4 is PG-TDSON-8 and that of S_5–S_8 is D^{2}PAK. Capacitors C_s, C_1 and C_2 are MPP-film type. The fabricated micro-inverter was compact and slim with 60-mm width, 310-mm length, and 30-mm height.

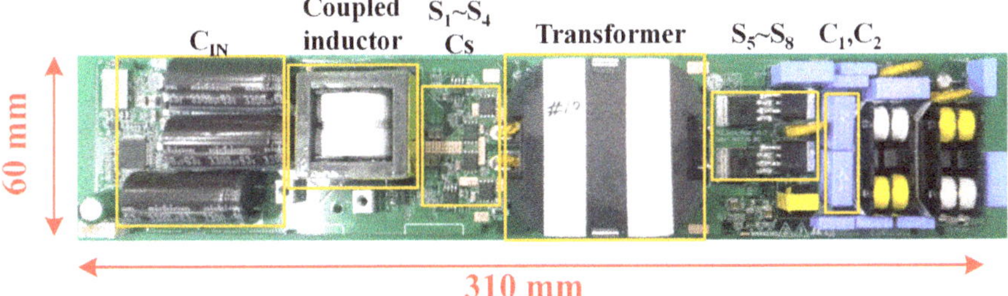

Figure 11. Photograph of the proposed micro-inverter.

Table 2. Hardware specifications and circuit parameters.

Unit Type	Symbol	Value	Note
	P_o	320 W	Output power
	C_{IN}	9900 µF	Input capacitor
	L_1, L_2	190 µH	Self inductance ($k = -0.947$)
	S_1–S_4	BCS035N10NS5	MOSFET (V_{DS} = 100 V, I_D = 100 A)
	C_s	60 µF	Storage capacitor
	L_m	600 µH	Magnetizing inductance
Micro	L_{lk}	100 µH	Leakage inductance
Inverter	S_5–S_8	IPB65R150	MOSFET (V_{DS} = 650 V, I_D = 22.4 A)
	C_1, C_2	100 nF	Doubler capacitors
	f_s	60~90 kHz	Switching frequency
	v_{grid}	220 V_{rms}/60 Hz	Grid voltage
	i_g	~1.45 A_{rms}/60 Hz	Grid current
	V_{IN}	25~52 V_{DC}	Operating voltage range
	$I_{IN.max}$	12 A_{DC}	Max input current
	V_{PV}	40.9 V	Open circuit voltage
PV	V_{MP}	34 V	MPP voltage
module	I_{PV}	10.05 A	Short circuit current
	I_{MP}	9.38 A	MPP current

Instead of an actual PV module, the photovoltaic simulator ETS600X14CPVF TerraSAS from AMETEK was used as an input source. The solar cell *I-V* characteristic curve for the experiment was based on that of the NeON®2 PV module from LG electronics.

Gate-source and drain-source voltages were obtained for S_1 and S_2 at $D \leq 0.5$ (Figure 12a) and at $D > 0.5$ (Figure 12b) at V_{IN} = 34 V and v_{grid} = 220 V_{rms} /60 Hz. The drain-source voltage v$_{DS1}$ of S_1 drops to 0 V before the gate signal v_{S1} is applied, so S_1 turns on with ZVS. S_2 is complementary to S_1

and achieves a ZVS turn-on. The operation of S_3 and S_4 is out of phase with that of S_1 and S_2, so S_3 and S_4 can also achieve the ZVS turn-on.

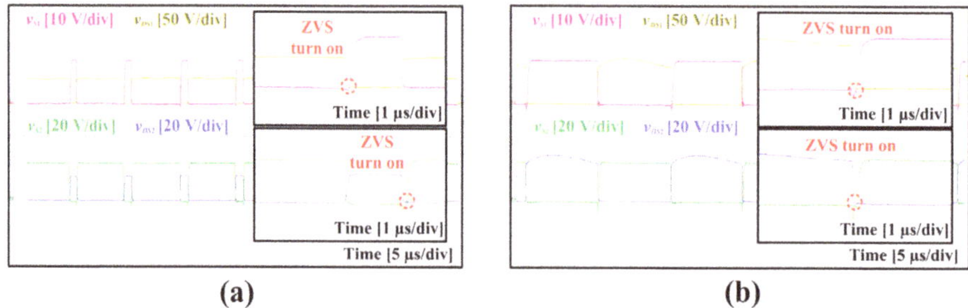

Figure 12. Gate-source and drain-source voltages of S_1 and S_3 for (**a**) $D \leq 0.5$ and (**b**) $D > 0.5$.

Waveforms (Figure 13) were obtained for v_{grid} and i_g at $V_{IN} = 34$ V and $v_{grid} = 220V_{rms}$ / 60 Hz for output power $P_o = 320$ W and 64 W. To maximize efficiency, the proposed micro-inverter operates in normal mode at $P_o \geq 110$ W and in AB control mode at $P_o < 110$ W. The boundary of the output power at which the proposed micro-inverter switches from the normal mode to AB control mode and vice versa is selected to be in a range where the peak value of i_g does not exceed the rated grid current.

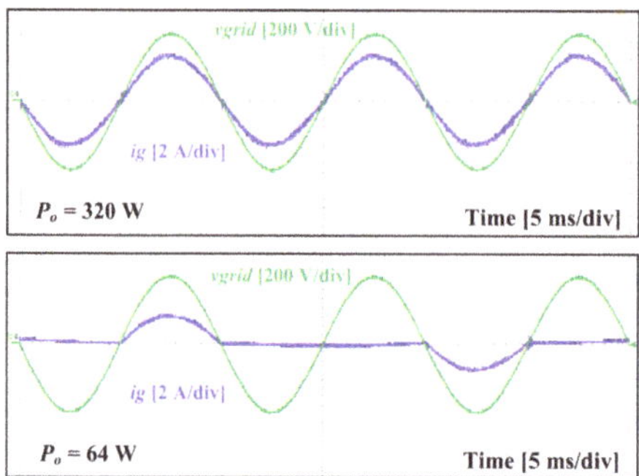

Figure 13. Grid voltage and current waveforms.

Waveforms were obtained for the fixed-frequency (Figure 14a) and the variable-switching-frequency (Figure 14b) controls. Gate signals of S_1 and S_3, i_g and v_{grid} were measured at $V_{IN} = 34$ V, $v_{grid} = 220$ V_{rms} / 60 Hz, and output power $P_o = 320$ W. When fixed-switching frequency control was used, i_g was distorted near zero-crossing, and THD was increased to 5.79%. In contrast, when variable switching frequency control was used, the distortion of i_g was improved near zero-crossing, and THD was reduced to 2.65%, which is below the requirement for distributed power. The switching frequency f_s decreased as i_g increased, so switching loss was also reduced.

ΔV_{IN} is higher when conventional burst control is used (Figure 15a) than when AB control is used (Figure 15b), because AB control reduces the energy supplied by C_{IN} during one ON-state period. At $V_{IN} = 34$ V, $v_{grid} = 220$ V_{rms} / 60 Hz, and $P_o = 32$ W, ΔV_{IN} was 4.2 V when conventional burst control was used, but 2.4 V when AB control was used.

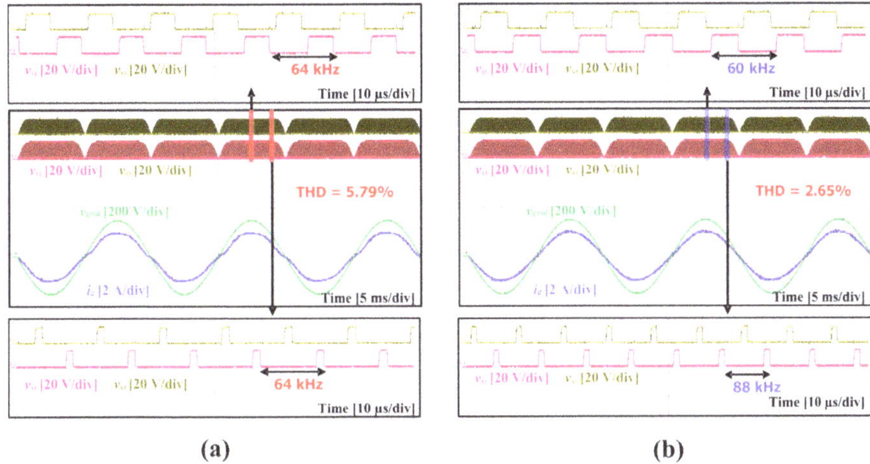

Figure 14. Gate signals of S_1 and S_3, grid voltage and grid current in (**a**) the fixed and (**b**) the variable switching frequency controls.

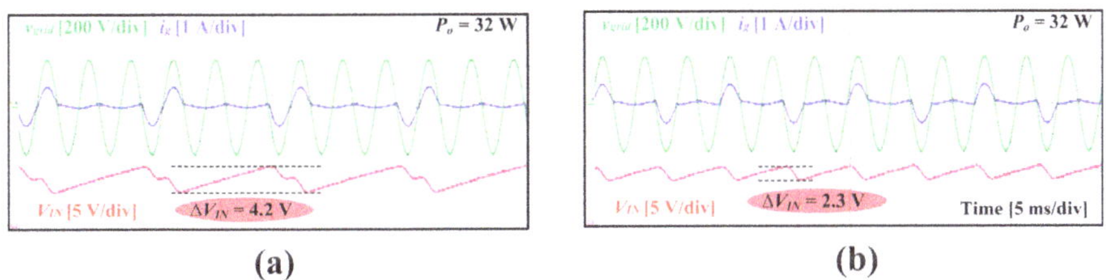

Figure 15. Input ripple voltage in (**a**) the conventional and (**b**) the advanced burst controls.

The MPPT efficiency of the proposed micro-inverter was measured (Figure 16) in the range of irradiance from 50 W/m^2 (P_o = 16 W)–1000 W/m^2 (P_o = 320 W). In the proposed control scheme, for P_o < 110 W (burst mode), the MPPT efficiency was kept >95% because ΔV_{IN} and ΔI_{g_ref} are reduced. However, in the conventional control scheme, the MPPT efficiency was reduced to ~88% because fluctuation of I_{g_ref} increased. During burst mode, the maximum MPPT efficiency was >99% for the proposed control scheme but <97.5% for the conventional control scheme.

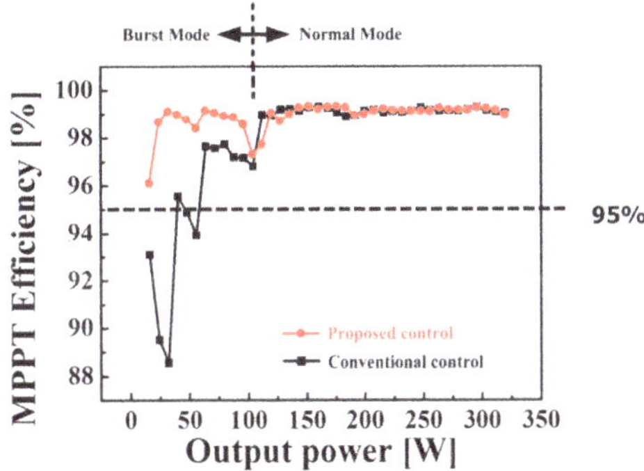

Figure 16. MPPT efficiency depending on control methods.

In a micro-inverter, one of the most important factors is the power conversion efficiency η_e for 50~75% load under actual solar irradiation. Therefore, the California Energy Commission (CEC) weighted efficiency to represent this fact has been widely used to measure the performance of micro-inverters. The power conversion efficiency η_e (Figure 17) was measured for the proposed micro-inverter; the result indicate that the CEC weighted efficiency [17,18] is 95.55%, in which $\eta_e(10\%)$ = 91.71%, $\eta_e(20\%)$ = 94.42%, $\eta_e(30\%)$ = 95.28%, $\eta_e(50\%)$ = 96.06%, $\eta_e(75\%)$ = 95.8%, and $\eta_e(100\%)$ = 95.72%. The maximum η_e is 96.06% for P_o = 160 W.

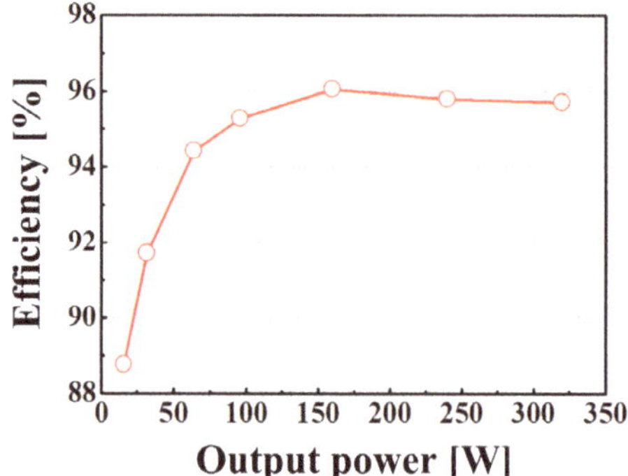

Figure 17. Power conversion efficiency η_e measured at V_{IN} = 34 V and v_{grid} = 220 V_{rms}/60 Hz.

5. Conclusions

A compact single-stage micro-inverter with advanced control schemes for PV systems is described. The proposed micro-inverter achieved a high voltage-conversion ratio and high efficiency by using a new topology that consists of an interleaved boost converter, a full-bridge converter, and a voltage doubler. The leakage inductance of the transformer and the capacitors of the voltage doubler ensure ZVS condition without any additional components. A variable-switching-frequency control scheme is applied to the micro-inverter to decrease THD by reducing the grid ripple current. An advanced burst-control scheme increases MPPT efficiency with smaller input ripple voltage than the conventional burst control causes. A fabricated 320-W prototype micro-inverter was very compact and slim with 60-mm width, 310-mm length, and 30-mm height. It achieved CEC weighted efficiency of 95.55%, MPPT efficiency > 95% over the entire load rage, and THD 2.65% at V_{IN} = 34 V, v_{grid} = 220 V_{rms}/60 Hz, and P_o = 320 W. These results show that the proposed micro-inverter is well suited for PV micro-inverter applications that require low cost, small and slim size, high efficiency, and low noise.

Author Contributions: Y.-G.C. conceived the main idea for the proposed micro-inverter and performed overall analysis and experiment with H.-S.L., B.K. led the project and gave technical advice. S.-C.L. contributed to determining circuit parameters and fabricating a prototype. S.-J.Y. contributed to analyzing the experimental results and writing the manuscript with Y.-G.C.

Acknowledgments: This research was supported by the Ministry of Science and ICT (MSIT), Korea, under the "ICT Consilience Creative Program" (IITP-2018-2011-1-00783) supervised by Institute for Information & communications Technology Promotion (IITP).

Conflicts of Interest: The authors have no conflict of interest.

References

1. Lee, S.-W.; Cho, B.-H. Master–Slave Based Hierarchical Control for a Small Power DC-Distributed Microgrid System with a Storage Device. *Energies* **2016**, *9*, 880. [CrossRef]
2. Han, H.; Luo, C.; Hou, X.; Su, M.; Yuan, W.; Liu, Z.; Guerrero, J.M. A Cost-Effective Decentralized Control for AC-Stacked Photovoltaic Inverters. *Energies* **2018**, *11*, 2262. [CrossRef]

3. Huang, L.; Qiu, D.; Xie, F.; Chen, Y.; Zhang, B. Modeling and Stability Analysis of a Single-Phase Two-Stage Grid-Connected Photovoltaic System. *Energies* **2017**, *10*, 2176. [CrossRef]

4. Jeong, H.-G.; Kim, G.-S.; Lee, K.-B. Second-Order Harmonic Reduction Technique for Photovoltaic Power Conditioning Systems Using a Proportional-Resonant Controller. *Energies* **2013**, *6*, 79–96. [CrossRef]

5. Blaabjerg, F.; Chen, Z.; Kjaer, S.B. Power electronics as efficient interface in dispersed power generation system. *IEEE Trans. Power Electron.* **2004**, *19*, 1184–1194. [CrossRef]

6. Nanakos, A.C.; Christidis, G.C.; Tatakis, E.C. Weighted efficiency optimization of flyback microinverter under improved boundary conduction mode (i-BCM). *IEEE Trans. Power Electron.* **2015**, *30*, 5548–5564. [CrossRef]

7. Rezaei, M.A.; Lee, K.J.; Huang, A.Q. A high-efficiency flyback micro-inverter with a new adaptive snubber for photovoltaic applications. *IEEE Trans. Power Electron.* **2016**, *31*, 318–327. [CrossRef]

8. Chen, Y.; Liao, C. A PV micro-inverter with PV current decoupling strategy. *IEEE Trans. Power Electron.* **2017**, *32*, 6544–6557.

9. Voglitsis, D.; Papanikolaou, N.; Kyritsis, A.C. Incorporation of harmonic injection in an interleaved flyback inverter for the implementation of an active anti-islanding technique. *IEEE Trans. Power Electron.* **2017**, *32*, 8526–8543. [CrossRef]

10. Milad, K.; Ehsan, A.; Hosein, F. Micro-inverter based on single-ended primary-inductance converter topology with an active clamp power decoupling. *IET Power Electron.* **2018**, *11*, 73–81.

11. Jafari, M.; Malekjamshidi, Z.; Li, L.; Zhu, J.G. Performance analysis of full bridge, boost half bridge and half bridge topologies for application in phase shift converters. In Proceedings of the International Conference on Electrical Machines and Systems, Busan, Korea, 26–29 October 2013; pp. 1589–1594.

12. Zhao, Z.; Wu, K.-H.; Lai, J.-S.; Yu, W. Utility grid impact with high penetration PV micro-inverters operating under burst mode using simplified simulation model. In Proceedings of the Energy Conversion Congress and Exposition (ECCE), Phoenix, AZ, USA, 17–22 September 2011; pp. 3928–3932.

13. Du, Y.; Xiao, W.; Hu, Y.; Lu, D.D.-C. Control approach to achieve burst mode operation with DC-link voltage protection in single-phase two-stage PV inverters. In Proceedings of the Energy Conversion Congress and Exposition (ECCE), Pittsburgh, PA, USA, 14–18 September 2014; pp. 47–52.

14. Thiyagarajan, A.; Praveen Kumar, S.G.; Nandini, A. Analysis and comparison of conventional and interleaved DC/DC boost converter. In Proceedings of the Second International Conference on Current Trends In Engineering and Technology—ICCTET 2014, Coimbatore, India, 8 July 2014; pp. 198–205.

15. Wong, P.; Wu, Q.; Xu, P.; Bo, Y.; Lee, F.C. Investigating coupling inductors in the interleaving QSW VRM. In Proceedings of the APEC 2000, Fifteenth Annual IEEE Applied Power Electronics Conference and Exposition (Cat. No.00CH37058), New Orleans, LA, USA, 6–10 February 2000; pp. 973–978.

16. Ahmed, A.S.; Abdullah, B.A.; Abdelaal, W.G.A. Mppt algorithms: Performance and evaluation. In Proceedings of the 11th International Conference on Computer Engineering & Systems (ICCES), Cairo, Egypt, 20–21 December 2016; pp. 461–467.

17. *Building Energy Efficiency Standards*; California Energy Commission: Sacramento, CA, USA, 2019.

18. Zhang, L.; Sun, K.; Hu, H.; Xing, Y. A system-level control strategy of photovoltaic grid-tied generation systems for European efficiency enhancement. *IEEE Trans. Power Electron.* **2014**, *29*, 3445–3453. [CrossRef]

Article

Five-Level T-type Cascade Converter for Rooftop Grid-Connected Photovoltaic Systems

Cristian Verdugo [1,‡], Samir Kouro [2,†,‡], Christian A. Rojas [2,*,†,‡], Marcelo A. Perez [2,†,‡], Thierry Meynard [3,‡] and Mariusz Malinowski [4,‡]

[1] Electrical Engineering Department, Polytechnic University of Catalonia, 08222 Barcelona, Spain; cristian.andres.verdugo@upc.edu

[2] Electronics Engineering Department, Universidad Técnica Federico Santa María, Valparaiso 2390123, Chile; samir.kouro@usm.cl (S.K.); marcelo.perez@usm.cl (M.A.P.)

[3] Institut National Polytechnique de Toulouse, 31071 Toulouse, France; Thierry.Meynard@laplace.univ-tlse.fr

[4] Institute of Control & Industrial Electronics, Warsaw University of Technology, 00-662 Warsaw, Poland; malin@isep.pw.edu.pl

* Correspondence: christian.rojas@usm.cl

† Current address: Av. España 1680, Valparaíso 2390123, Chile.

‡ These authors contributed equally to this work.

Received: 12 April 2019; Accepted: 30 April 2019; Published: 8 May 2019

Abstract: Multilevel converters are widely considered to be the most suitable configurations for renewable energy sources. Their high-power quality, efficiency and performance make them interesting for PV applications. In low-power applications such as rooftop grid-connected PV systems, power converters with high efficiency and reliability are required. For this reason, multilevel converters based on parallel and cascaded configurations have been proposed and commercialized in the industry. Motivated by the features of multilevel converters based on cascaded configurations, this work presents the modulation and control of a rooftop single-phase grid-connected photovoltaic multilevel system. The configuration has a symmetrical cascade connection of two three-level T-type neutral point clamped power legs, which creates a five-level converter with two independent string connections. The proposed topology merges the benefits of multi-string PV and symmetrical cascade multilevel inverters. The switching operation principle, modulation technique and control scheme under an unbalanced power operation among the cell are addressed. Simulation and experimental validation results in a reduced-scale power single-phase converter prototype under variable conditions at different set points for both PV strings are presented. Finally, a comparative numerical analysis between other T-type configurations to highlight the advantages of the studied configuration is included.

Keywords: grid-connected photovoltaic systems; cascade multilevel converters; multistring converters; T-type converters

1. Introduction

Rooftop photovoltaic (PV) energy conversion systems (less than 20 kW), have become a well-established technology in the industry. The most common configurations for single-phase grid-connected PV systems commercially found are the string, multistring and ac-module integrated topologies. Central and string inverters have been widely applied to manage and control PV energy systems [1]. Among the string topologies, the transformerless H5, H6, HERIC, neutral point clamped (NPC) and T-type NPC converters have been successfully commercialized [2]. In fact, multilevel inverters (MLI) are designed to produce a stepped voltage waveform by reducing the Total Harmonic Distortion (THD) and the voltage stress across semiconductor devices. Secondly, reduction of the output filter size and power footprint also permit an important improvement in terms of costs, weight and efficiency [3]. These technical features have led to the massive adoption of MLI over the last thirty years for high-power medium voltage (MV) motor drive applications. In the last years, three-level neutral point clamped (3L-NPC) converters have been used for interfacing PV systems into the grid, where a higher PV incorporation has brought substantial concerns on power efficiency, power quality and grid code compliance [1] as well as power grid services [4].

T-type neutral point clamped inverters (3L-TNPC), also known as neutral point piloted converters (3L-NPP), [5] have gained a wide presence in the industry sector due to several advantages as symmetrical loss distribution, higher overall efficiency, small footprints [6] and low harmonic injection in relation to the conventional 3L-NPC [7]. In fact, many manufactures such as Fuji, On Semiconductor, Mitsubishi and Semikron have commercial T-type legs used in central PV inverters and motor drive applications [8–10]. For the three-level inverter, based on the T-type leg, was presented thirty-five years ago for motor drives, with the bidirectional medium switch being realized with thyristors and improved with GTO-thyristors [11]. After some years, many configurations based on the well-known three-level T-type NPC leg can be found in the literature [12]. In [13] a five-level TNPC (5L-TNPC) was introduced, which corresponds to the parallel connection of two 3L-TNPC legs [14,15]. Furthermore, a variation of this configuration with reduced switches, also known as five-level hybrid T-type NPC (5L-HTNPC), was presented in the recent literature [16]. This topological variation is built with a 3L-TNPC leg and a two-level leg inverter, forming a five-stepped voltage waveform in the AC terminals.

In the literature there are two main possibilities for increasing the number of levels in the power converter field, which is by increasing the internal DC capacitors connected to a single DC source or by connecting several converters in the series at the AC side, in which each converter has an independent DC source. Focusing on the second alternative, cascade MLI can be developed by using symmetrical or asymmetrical voltage levels and by using different type of topologies such as: Full H-Bridge, 3L-TNPC converters or by performing a hybrid configuration [17]. Note that symmetrical cascade configurations have had a more industrial presence as the case of Cascade H-Bridge (CHB) converters [3] due to modulation and control simplicity compared with asymmetrical configurations [18]. In fact, in [19] a symmetrical nine-level T-type converter (9L-TNPC) is presented for motor drive applications, which is based on the cascade connection of two 5L-TNPC converters. The same number of levels can be generated with advanced hybrid topologies as presented in [6,12].

Considering the advantages and features previously presented regarding the 3L-TNPC and symmetrical cascaded configurations, this paper described and validated the 5L-CTNPC topology for rooftop PV applications by using a cascaded connection of two 3L-TNPC legs which was firstly introduced in [20] as a cascade 3L-TNPC converter. Thus, the advantages of symmetrical cascade configurations with multistring inputs are merged. Each 3L-TNPC converter can interface a dedicated DC bus, and consequently two separate maximum power point tracking (MPPT) algorithms are allowed to obtain the maximum power of each PV string. Note that the PV string of each module can be sized to handle half the entire PV string in the conventional 3L-TNPC converter, providing better MPPT efficiency since less modules are combined in a series per string. The main contribution of this paper is the experimental validation of a simplified control scheme to alleviate the power unbalancing mismatch between 3L-TNPC modules and to compensate capacitor voltage variations per each converter, which was presented earlier in [20]. Furthermore, a brief comparison between five-level voltage waveform converters based on the conventional 3L-TNPC is performed as a second contribution in terms of the main electrical features.

The rest of the document is organized as follows. In Section 2, a hardware description of the proposed converter, switching states and implemented modulation is presented. In Section 3, a simple stationary reference-frame voltage-oriented control and a voltage control loop to compensate a possible power unbalance operation are included. Then, in Section 4, simulation results and experimental verification of the proposed multilevel converter and its control system behavior are added. Furthermore, a brief comparison with the 5L-TNPC and 5L-HTNPC is performed to highlight the main advantages of the proposed configuration. Finally, in Section 5, the conclusions of the paper summarize the work done.

2. The 5L-CTNPC Converter Topology

The power topology of the analyzed cascade 5L-CTNPC for a rooftop grid-connected PV system is depicted in Figure 1. The configuration is composed of a series connection of two 3L-TNPC legs, where each of them is built with two conventional IGBTs and one bidirectional switch. This bidirectional switch could be formed either by two conventional IGBTs in common-emitter or by a common-collector and reverse blocking IGBT connection. Actually, a classical IGBT semiconductor structure could be replaced by a reverse blocking MOSFETs for a high-voltage [21] and high-switching frequency operation [22]. Although more than two cells in a series connection are conceptually feasible, for the sake of simplicity, two-cell 3L-TNPC converters have been introduced as a proof-of-concept applied to conventional string rooftop PV applications.

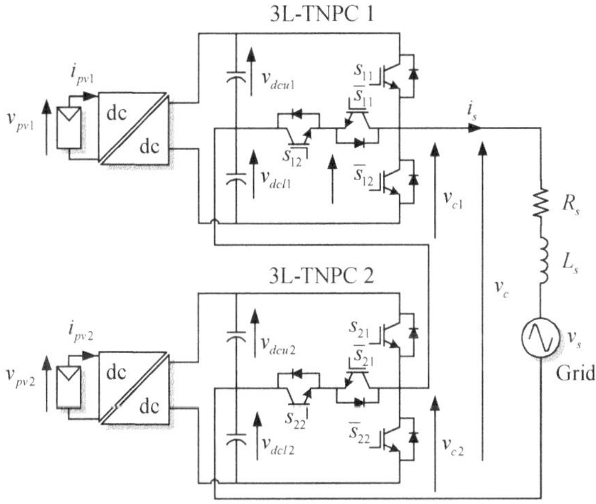

Figure 1. Proposed 5L-CTNPC power topology.

Each 3L-TNPC leg operates as a string inverter connected to a single potential-induced degradation converter which is fed by one PV string. This potential-induced degradation stage can be designed to boost the DC voltage and perform the MPPT. Furthermore, it could be isolated [23] to avoid leakage currents due to the PV aluminium metallic frame grounded [1]. To reduce leakage currents paths and avoid high-frequency transformers there are three successful options well-documented in the literature: Changing the modulation stage to avoid switched common mode [16], by reducing surface conductivity of PV modules to avoid potential-induced degradation (PID) and by including extra switches between the PV array and the inverter, also well-known as transformerless inverters [24]. Although, a potential-induced degradation stage is desired to provide an independent DC voltage control, in this work, the validation of the proposed configuration does not integrate a potential-induced degradation stage, giving place to the worst case scenario under study where the MPPT control is fulfilled directly from each 3L-TNPC power cell, instead of using a high frequency galvanic isolated converter with its appropriate MPPT control. Thus, the overall control loops are more challenging since the voltage fluctuations in the PV panel is directly presented in the DC-side of each 3L-TNPC module, i.e., $v_{dck} = v_{pvk}$, where $k = \{1, 2\}$ is given by the number of cell. Furthermore, in order to extract the maximum power from the PV panels, an integration of an external

MPPT algorithm is required so as to define the appropriate DC voltage reference in each cell. Note that the proposed topology is modeled without affecting the basic control objectives.

2.1. Fundamental Principle of the 5L-CTNPC

The 3L-TNPC provides three-output voltage levels: $v_{dck}/2$, 0 and $-v_{dck}/2$, where k is the cell or module number. These voltage steps are generated by connecting the AC terminals to the positive, neutral and negative pole of the DC-link terminals. Although, the 3L-TNPC configuration gives rise to four switching states, in order to avoid a short-circuit to the DC side there are only three of them possible. The cascade connection of two 3L-TNPC cells permits the generation of five voltage steps, where the zero level is combined into just one at the AC converter output voltage v_c. According to the switching states presented in Table 1, the output voltage in the 5L-CTNPC can be modeled as:

$$v_c = \underbrace{(S_{11} + S_{12} - 1)\frac{v_{dc1}}{2}}_{v_{c1}} + \underbrace{(S_{21} + S_{22} - 1)\frac{v_{dc2}}{2}}_{v_{c2}},\tag{1}$$

where v_c is the addition of the converter voltages of both modules, S_{1k} and S_{2k} are the switching states of the k-th 3L-TNPC unit and $v_{dck}/2$ is the total DC-link voltage of each cell. Furthermore, the dynamic model of the AC current in terms of the output voltage is governed by the next expression:

$$v_c = i_s R_s + L_s \frac{di_s}{dt} + v_s,\tag{2}$$

with v_s as the grid voltage measured at the point of common coupling (PCC), i_s as the grid current, L_s the grid filter inductance and R_s is the filter resistance included for modeling purposes. According to Table 1, the switching states are able to generate nine voltage levels in the output voltage v_c where each state has an associated voltage level in function to the DC-link v_{dc1} and v_{dc2}. Note that in this rooftop PV application both strings will be considered to work with similar DC-link voltages, i.e., $v_{dc1} \approx v_{dc2} = v_{dc}$. By doing this, the output voltage v_c can be reduced just to $\pm v_{dc}$, $\pm v_{dc}/2$ and 0. This assumption leads to five switching states with similar output voltage steps between two consecutive levels [20]. The redundant switching states will be used to balance the voltage in the DC-link capacitors by adjusting the power mismatch between the converter cells. The computed peak amplitude of the converter output voltage $\hat{v}_c$ is equal to $v_{pv1}/2 + v_{pv2}/2$, i.e., each power cell has a DC-link equal to the maximum level of the converter voltage. Therefore, for a proper grid current regulation, each PV string must be designed to satisfy $v_{pv1} \approx v_{pv2} > v_s$. In fact, this aspect is a practical advantage of cascaded configurations, since the overall DC-link voltage of the central configuration is split among power cells, thus reducing the string size.

Table 1. Switching states and output voltage in the AC terminals.

State	S_{11}	S_{12}	S_{21}	S_{22}	v_c
1	1	1	1	1	$v_{dc1}/2 + v_{dc2}/2$
2	1	1	0	1	$v_{dc1}/2$
3	0	1	1	1	$v_{dc2}/2$
4	1	1	0	0	
5	0	0	1	1	0
6	0	1	0	1	
7	0	1	0	0	$-v_{dc1}/2$
8	0	0	0	1	$-v_{dc2}/2$
9	0	0	0	0	$-v_{dc1}/2 - v_{dc2}/2$

2.2. Proposed Hybrid LS-PWM and PS-PWM Modulation Scheme for 5L-CTNPC Converter

The proposed modulation scheme for the 5L-CTNPC is based on two well-known carriers based on the sinusoidal PWM methods. The first is the Sinusoidal Level-Shifted Pulse Width Modulator (LS-PWM), used in 3L-NPC three-phase converters [25] and the single 3L-TNPC legs [20]. This modulation strategy requires two carrier signals in phase, to generate the three voltage levels in the output terminals of each cell. One carrier signal has a positive polarity (0 to 1) and the other has a negative polarity (–1 to 0). Furthermore, the LS-PWM is merged with the Sinusoidal Phase-Shifted PWM (PS-PWM) conventionally used in cascaded H-bridge power converters [26]. In the PS-PWM modulation, a phase shift between the carrier signals of each series connected to a power cell is introduced to increase the number of voltage levels, giving rise to a five-level stepped voltage waveform. The operation principle of this hybrid modulation technique is illustrated in Figure 2, where m_{ck}^* and v_{ck} are the modulation signal and the output voltage in the k-th cell, respectively. Note that each cell uses two carrier signals defined as v_{cr1} and v_{cr2}. Thus, the stacked connection of both cells creates the converter voltage v_c, which is commanded by its reference v_c^*. The combination of both methods is simpler in respect to the space vector modulation (SVM) [27]. Finally, the implementation of this modulation technique is depicted in the block diagram of Figure 3, where simple comparators and two carrier signals are required to implement the proposed technique.

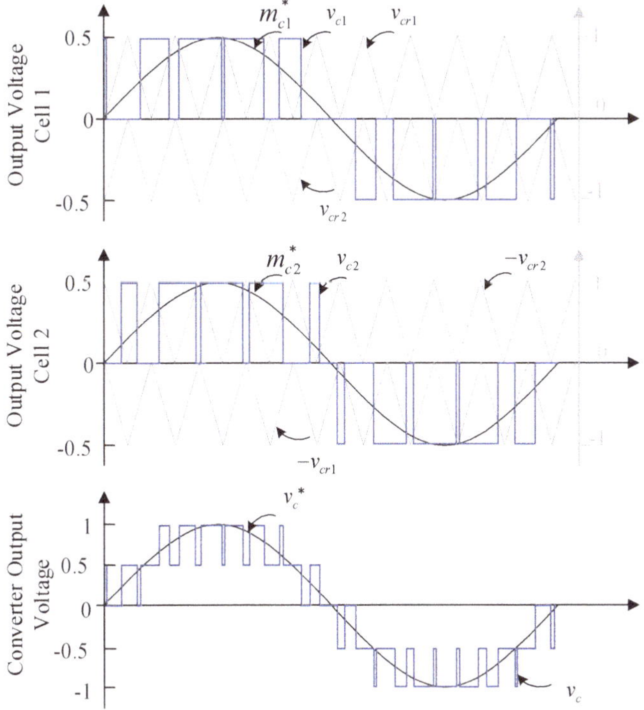

Figure 2. Proposed modulation scheme for 5L-CTNPC based on the hybrid LS-PWM and PS-PWM.

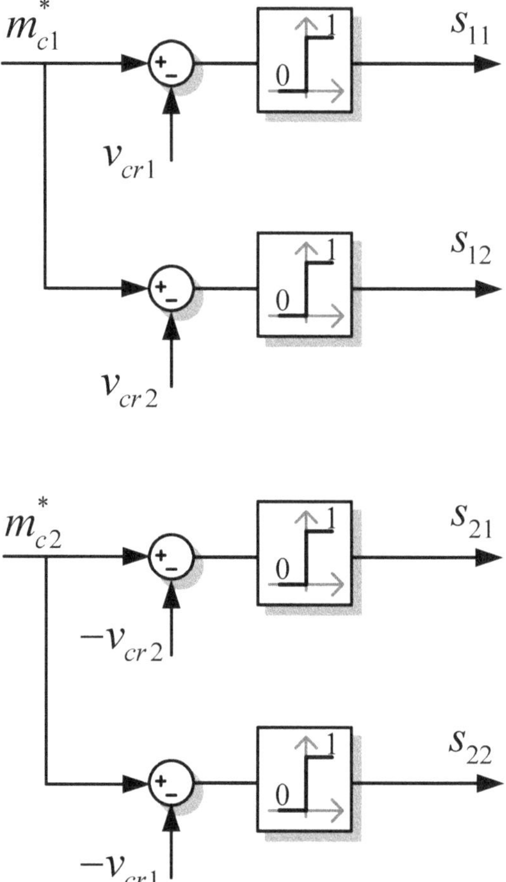

Figure 3. Straightforward implementation of hybrid LS-PWM and PS-PWM modulation for a 5L-CTNPC converter.

3. Overall Control Strategy

In this work three decoupled control stages are programmed to regulate the current injected into the grid, the power generated by each PV string and the power mismatch between cells. The first one is the MPPT, which set the DC-link voltage reference for both cells and is optionally included to extract the maximum power from the PV panels in case of direct connection to the 3L-TNPC modules. This control stage is complemented with the total DC-link control loop based on the energy interchange between the power cells. The second control stage is the single-phase voltage-oriented control loop, which has an embedded stationary current control loop implemented with Proportional Multi-Resonant (PMR) controllers. The last control loop is in charge of attenuating the DC-link voltage differences to compensate power mismatch issues among each cell of the converter. The overall control scheme is presented in Figure 4, where v_{pvk}, i_{pvk}, v_{dck} and s_{ok} are the PV voltage, PV current, DC-link voltage measurement and gating pulses of each k-th module.

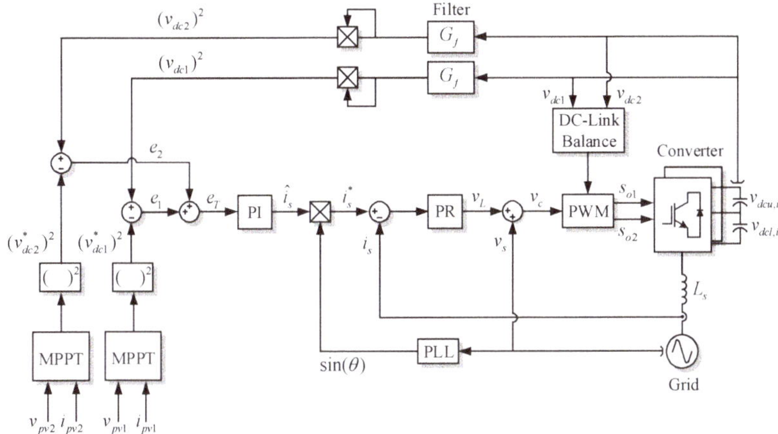

Figure 4. Single-phase voltage-oriented control strategy for the proposed 5L-CTNPC converter.

3.1. MPPT and Outer DC-Link Controller

The well-known Perturb and Observe (P&O) MPPT routine has been implemented for simplicity in this application. As the analyzed power configuration features two separate DC-links or PV string connections, two independent MPPT algorithms are required to obtain its full power operation. The MPPT routines compute the voltage reference for each DC-link v_{dc1}^* and v_{dc2}^*, as illustrated in Figure 4. Then, the DC-link control loop is designed to manage the total energy of the system through the difference between the voltage reference and the voltage measured, i.e., $e_T = e_1 + e_2$. This total energy is governed by using a proportional-integral (PI) controller, which generates the amplitude of the injected grid current $\hat{i}_s$. Note that the DC-link voltage measurements are acquired and processed with a notch filter G_f to eliminate the second harmonic ripple $2\omega_s$ presented in the DC-link capacitors by the rectification of a single-phase grid voltage. In fact, not filtering this harmonic voltage component will generate an undesired third harmonic $3\omega_s$ component in the grid current reference. The MPPT parameters such as voltage step Δv_{pv} and time period T_k are designed according with conventional commercial values. In experimental results, the voltage step $\Delta v_{pv} = 6$ V and the time period $T_k = 2$ s, whereas in simulation results, $\Delta v_{pv} = 6$ V and the time period is ten times smaller than the experimental results. Furthermore, the DC-link compensator has been designed by using a DC-link control bandwidth of 14Hz. Major details about the outer control design can be found in [20].

3.2. PMR Current Control Scheme

The grid current reference is generated by multiplying the amplitude of the injected grid current $\hat{i}_s$ with a unitary sinusoidal signal synchronized to the grid voltage. To avoid voltage measurement noise and low frequency harmonic components, a second order generalized integrator (SOGI) with a synchronous reference frame phase lock loop (SRF-PLL) is implemented to set the synchronous angle. Then, the grid current reference i_s^* is compared with the current measured value i_s, giving rise to a current error which is regulated by using a PMR control scheme. The structure of the implemented controller is included in Figure 5 and expressed as following:

$$C_i(s) = k_p + \sum_{h=1,3,5} \frac{2k_{ih}s}{s^2 + h^2\omega_s^2} \tag{3}$$

where k_p is the proportional gain and k_{ih} is the resonant gain at each selected h-th harmonic. Note that the above resonant controllers have been considered to achieve selective harmonic impedance enhancement at 3st and 5th components. The resonant frequency at ω_s is equal to the grid frequency, hence the compensator $C_i(s)$ has infinite gain at ω_s, providing perfect sinusoidal tracking with zero steady-state error. The PMR compensator in Figure 5 has been designed by a simple pole placement with a crossover frequency of 270 Hz, which corresponds to a rate twenty times faster than the outer control loop. This control scheme is currently adopted for grid-connected PV systems where the grid voltage has important low-frequency harmonics [28].

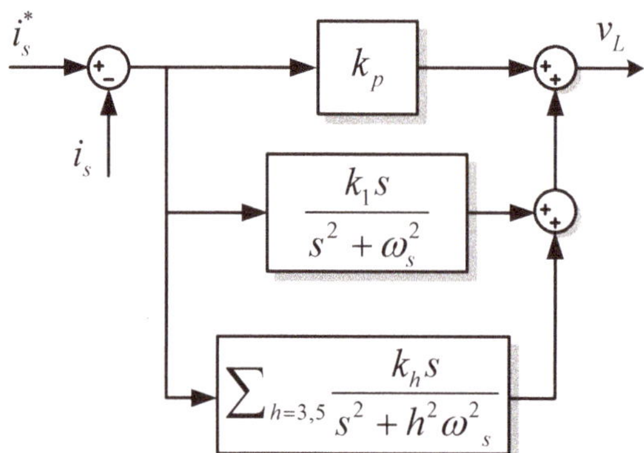

Figure 5. Stationary current control loop implemented with PMR controllers.

The output of the current control loop set the voltage reference across the grid inductor v_L^*. Neglecting the voltage drop in the resistance R_s, the converter voltage is equal to $v_c = v_L + v_s$. Commonly, the obtained inverter voltage reference v_c would be directly connected to the modulation block stage to generate the firing pulses in each semiconductor device. Since the current i_s is the same for both series connected 3L-TNPC converters, the voltage references for each power unit must be modified in advance to allow different power inputs. This important control requirement is performed by including an internal DC-link voltage balance stage, which enables the voltage balancing between capacitors in the DC-link and the power unbalance operation between both cells. In fact, the voltage balancing operation is performed by the DC-link voltage references v_{dc1}^* and v_{dc2}^* naturally delivered by the MPPT algorithm.

3.3. Voltage Balancing Control and Power Balance Scheme

As mentioned, correct power distribution between both cells and a control strategy to avoid voltage unbalancing in the DC-link capacitors is required to provide full operation in the 5L-CTNPC. The compensation block shown in Figure 6 is separated into two parts. The first one is the DC-link balancing control between cells, whose purpose is to regulate the power mismatch by increasing or decreasing the general modulation index amplitude, also referred to as the normalized inverter voltage v_c delivered from the current controller. In this control loop, the DC voltage error is regulated by using a PI controller and then the voltage compensator Δm_c is multiplied by the normalized inverter voltage reference v_c [29]. Therefore, the cell with higher power will increase its modulation amplitude, as the cell with lower power reduces its modulation amplitude. The second part of the control loop is given by the voltage balancing between the internal capacitors in the DC-link, where the voltage error is controlled using another PI controller. The output signal of this compensator Δm_d is added to the modulation index provided by the cell voltage control by moving in the vertical axis the modulation index for capacitor balancing purposes. Both PI controllers have been designed by a pole placement strategy, using a bandwidth of 5 Hz. This dynamic has been imposed to avoid fast disturbances into the modulation signal. Note that the proposed balancing control scheme does not need extra measurements, since they are previously accessible from the outer control stage.

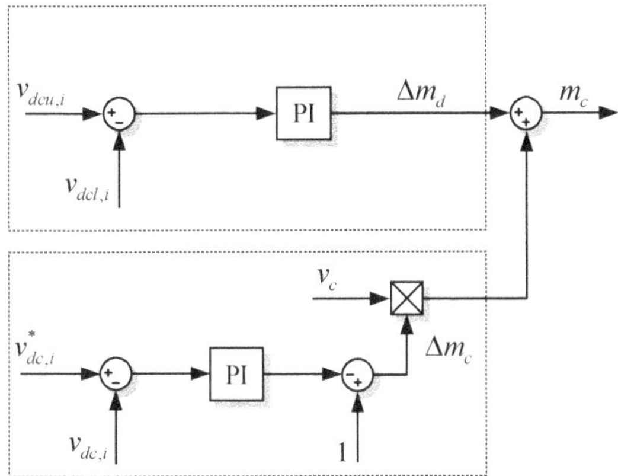

Figure 6. Implemented voltage balancing control strategy per 3L-TNPC cell.

4. Results

Simulation and experimental results of the proposed configuration are presented in this section. The simulation analysis has been performed through MatLab/Simulink for control purposes, while PLECS were used for modelling the modulation stage, power converter, grid voltage and PV strings. The analysis was completed by using the same scenarios of the experimental set-up just to improve the concept verification. The key simulation and experimental parameters are identified in Table 2. It is important to highlight that simulation parameters have been selected according to the reduced power experimental prototype.

Table 2. Simulation and experimental parameters.

Symbol	Parameter	Simulation Value	Experimental Value
Grid Parameters			
$\hat{v}_s$	Peak grid voltage	80 (V)	80 (V)
f_s	Grid frequency	50 (Hz)	50 (Hz)
Converter Parameters			
C_{dc}	DC-link capacitors	1950 (μF)	1950 (μF)
f_{cr}	Carrier frequency	2000 (kHz)	2083 (kHz)
L_s	Grid inductance	5 (mH)	5 (mH)
R_s	Grid resistance	0.01 (Ω)	1 (Ω)
Control Parameters			
T_s	Sample period	10 (μs)	15 (μs)
BW_{vdc}	DC-link control bandwidth	14 (Hz)	14 (Hz)
BW_{is}	Current control bandwidth	270 (Hz)	270 (Hz)
BW_{dv}	Balancing control bandwidth	5 (Hz)	5 (Hz)
MPPT Parameters			
T_k	P&O period	0.24 (s)	2.1 (s)
Δv_{pv}	P&O voltage step	6 (V)	6 (V)
PV String Parameters			
P_{mp}	Maximum power	106 (W)	106 (W)
v_{mp}	Voltage at maximum power	48.4 (V)	48.4 (V)
v_{oc}	Open-circuit voltage	65.0 (V)	65.0 (V)
i_{mp}	Current at maximum power	2.2 (A)	2.2 (A)
i_{sc}	Short-circuit current	2.6 (A)	2.6 (A)

4.1. Simulation Results

The first simulation result presented the injected grid current i_s with reference i_s^* and both capacitor voltages for each power cell under steady-state operation. In Figure 7, it is possible to appreciate the good regulation and synchronization in respect to the grid voltage v_s performed by the grid current control and the synchronization control loop. Furthermore, in Figure 8, the voltage balancing control is demonstrated, where the upper $v_{dcu,k}$ and lower $v_{dcl,k}$ voltage capacitors for each k-th power cell are well balanced.

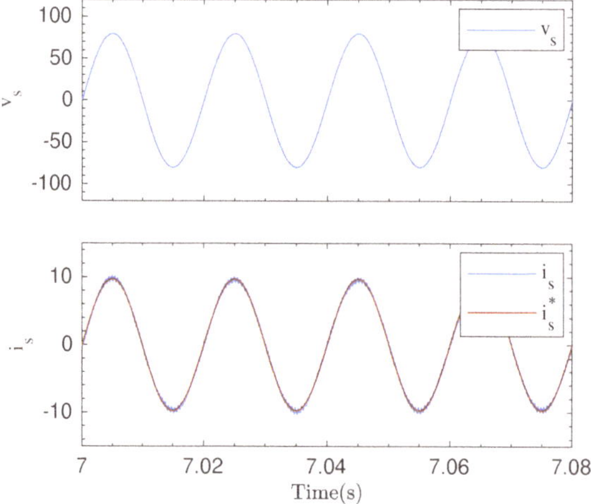

Figure 7. Steady-state operation of the grid current PMR control.

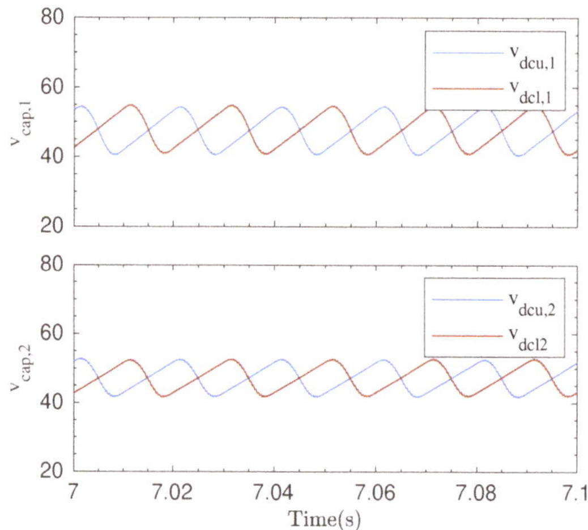

Figure 8. Steady-state operation of capacitor voltages for each power cell.

The second simulation results present a dynamic operation under two different scenarios. An irradiation step from $1\ \mathrm{kW/m^2}$ to $0.8\ \mathrm{kW/m^2}$ was applied to the lower cell, maintaining $1\ \mathrm{kW/m^2}$ of irradiation in the PV string connected to the upper cell. After the irradiation step took place, a temperature step changes was performed from 25 °C to 18 °C to the upper cell, generating an increase in the power. The irradiation and temperature changes were introduced in a simplified PV model provided by PLECS. Figure 9 shows the dynamic operation of the DC-link voltage v_{dck} and the power at DC-side $P_{ck} = v_{dck} \cdot i_{pvk}$ for each k-th cell. It is possible to appreciate how the irradiation step at $t = 3.5$ s only affected to the lower module, producing a voltage perturbation and a power reduction in P_{c2}. Since the reference voltage is provided by the P&O algorithm, the stepped voltage was required to maintain the maximum power operation. In the second scenario, the temperature decreased at $t = 8$ s and the DC voltage as well as the power in the upper module increased. The three-level voltage v_{ck} of each converter cell, and the overall five-level voltage v_c are depicted in Figure 10 under unbalance operation. Additionally, it is possible to appreciate how the power reduction in the lower arm affects the modulation indexes m_{ck}, creating signals with different magnitudes.

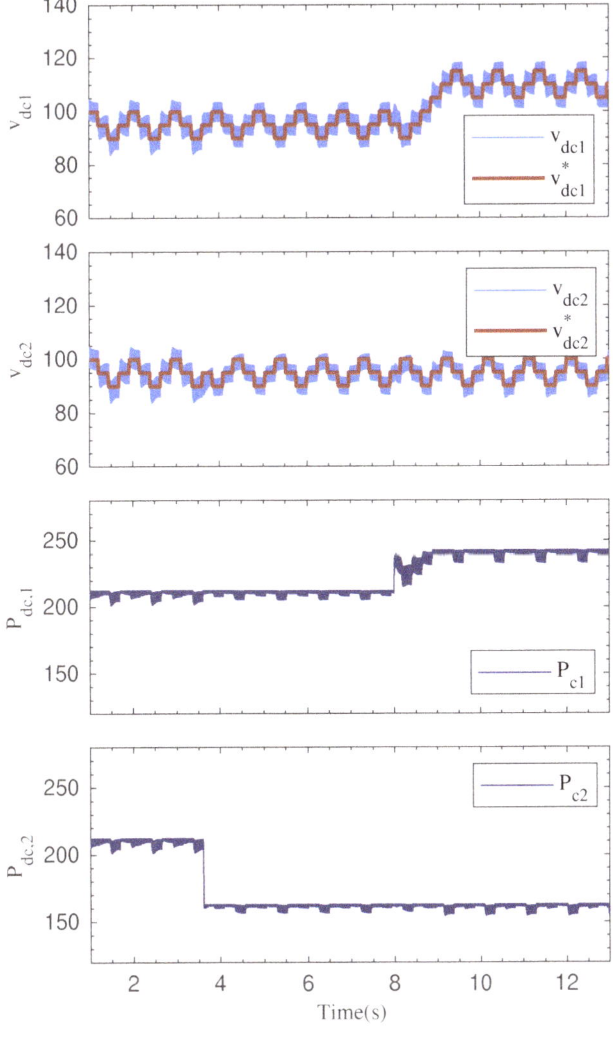

Figure 9. Dynamic operation of DC-link and power on the DC-side under unbalanced power per string.

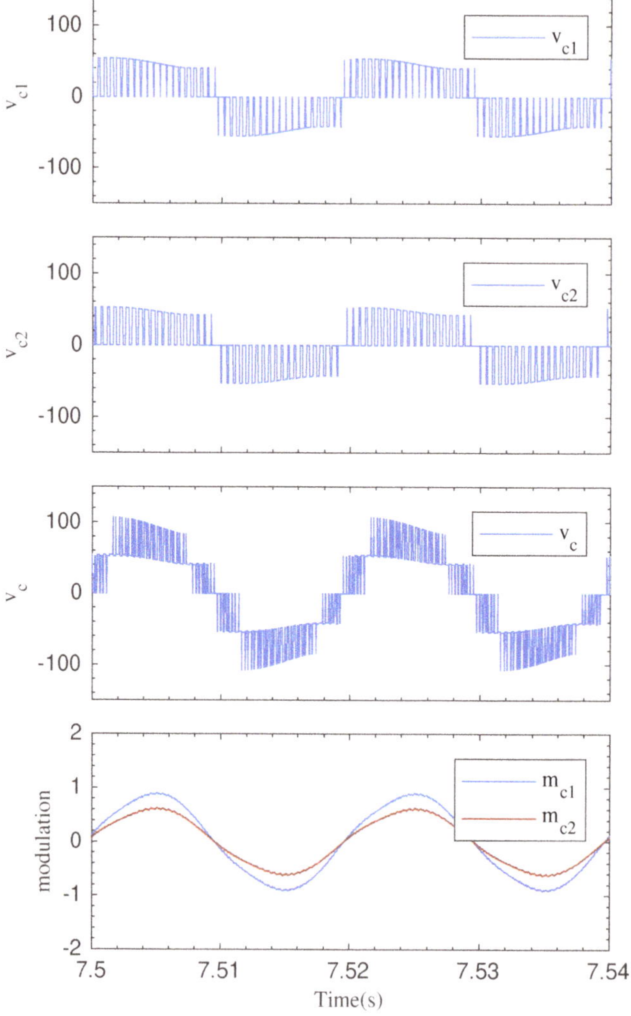

Figure 10. Steady-state converter output voltage performance under unbalanced power per string.

4.2. Experimental Results

The experimental PV system comprises of two 3L-TNPC power cells without isolation and were fed directly by one PV string. Each string was composed by two PV modules emulated with the Agilent E4360 solar array simulator, which enabled a total control of the temperature and irradiation parameters. Each simulator has two output channels connected in series to emulate the PV string generator. The parameters such as maximum power P_m, current at maximum power i_{mp}, short-circuit current i_{sc}, voltage at maximum power v_{mp} and open-circuit voltage v_{oc} are listed in Table 2. The simplified layout of the experimental small-scale setup is depicted in Figure 11. The control algorithm is fully programmed in C code by using a dSPACE 1103 digital control platform running at 15 µs. The modulation stage and dead-time generation is implemented by using a FPGA Spartan3. To experimentally validate the proposed control scheme, three different operation points are evaluated. A steady-state operation, and two dynamic operation under an irradiation and a temperature step.

The first experimental results shown in Figures 12 and 13 are analyzed during steady-state operation. The grid-side variables i.e., grid current and voltage waveforms are presented in Figure 12, where the unitary power factor operation is achieved. Furthermore, Figure 13 captures the output voltage of each cell and the total voltage composed by a five-level waveform. Similar to the simulation results, the steady-state performance of the capacitor voltage balancing in Figure 14 is included.

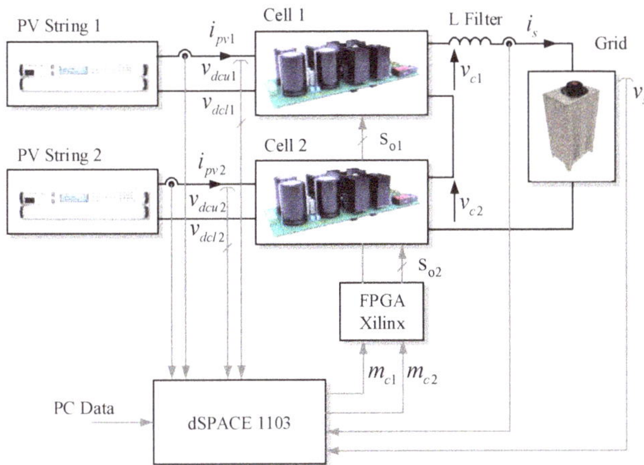

Figure 11. Simplified diagram of implemented experimental setup.

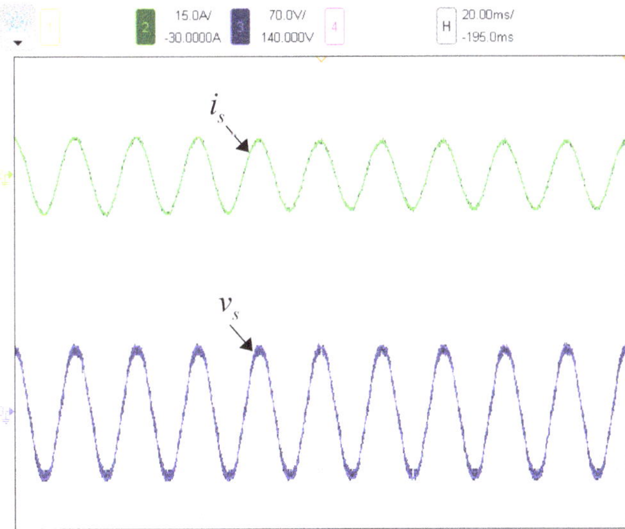

Figure 12. Steady-state experimental results at grid-side variables: Grid current and voltage waveforms.

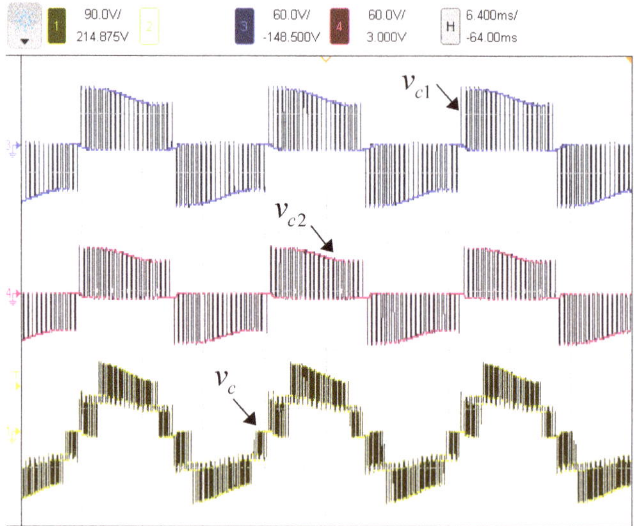

Figure 13. Steady-state experimental results at converter-side variables: Output voltage for each cell and total output voltage.

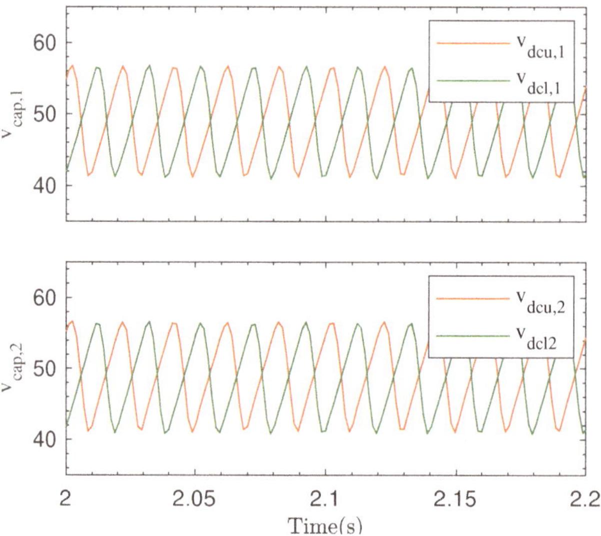

Figure 14. Steady-state experimental results of capacitor voltage across each 3L-TNPC cell.

The second set of experimental results during dynamic operation is shown in Figures 15 and 16. Firstly, an irradiation step from 1 kW/m^2 to 0.8 kW/m^2 was applied to the second PV cell (lower cell operating at reduced power), while the first array irradiation level was retained. After this irradiance step variation, a temperature step change was tested from 25 °C to 18 °C and applied to the first PV cell (upper cell operating at increased power). Under the above conditions, the input voltages generate a three-level waveform signal due to the use of the conventional P&O MPPT method. Note that under both scenarios, the coupling effects from one cell to each other is fully avoided, ensuring a decoupled operation between power cells.

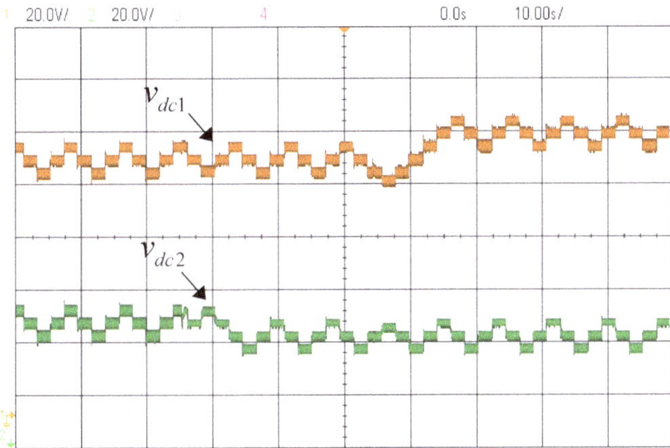

Figure 15. Experimental dynamic operation of DC-link voltage under unbalanced power per string due to solar irradiation and temperature changes.

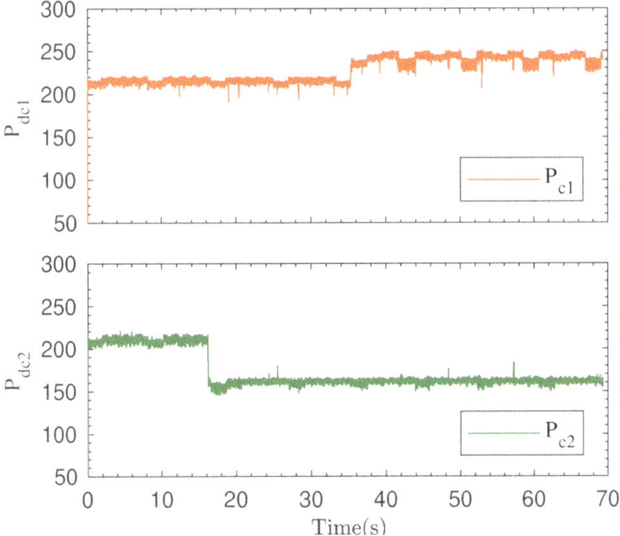

Figure 16. Experimental dynamic operation of DC power under unbalanced power per string due to solar irradiation and temperature changes.

4.3. Brief Comparison with Other Five-Level T-type Converters

A comprehensive comparison between three different five-level T-type converters have been presented in Table 3. The studied topologies are the proposed 5L-CTNPC, the hybrid version 5L-HTNPC and the conventional 5L-TNPC. Each topology presents the same features of the one described in simulation and experimental results. To evaluate the power efficiency of each converter, switching and conduction losses is required in respect to the power operation point for each cell. Semiconductor device losses are included with a thermal model library developed in PLECS, based on the manufacturer datasheet [30]. The resulting efficiency evaluation is depicted in Figure 17, where the obtained efficiency of the proposed configuration 5L-CTNPC is equal to the conventional 5L-TNPC. Due to the fact that there is a reduced number of switches at the second parallel leg in the 5L-HTNPC a slightly higher efficiency was achieved. This analysis is corroborated by counting the number of semiconductor devices used for each evaluated topology, which is summarized in Table 3.

The symmetrical topology configuration and the multiple MPPT possibilities are the main advantages of the studied topology in respect to the rest power converters. Finally, in Figure 18 the current spectrum for each evaluated converter topology is computed. The current THD obtained with the proposed topology is similar to the conventional 5L-TNPC power topology, while the worst value (over 4.9%) was reached in the 5L-HTNPC configuration. In fact, the apparent switching frequency of this topology was equal to the carrier frequency, while in the proposed 5L-CTNPC and 5L-TNPC the apparent switching frequency was twice the switching frequency.

Table 3. Brief comparison between five-level T-type topologies.

Parameter	5L-CTNPC	5L-HTNPC	5L-TNPC
DC-link voltage	v_{dc}	v_{dc}	v_{dc}
IGBT blocking voltage	$4 \times v_{dc}, 4 \times v_{dc}/2$	$4 \times v_{dc}, 2 \times v_{dc}/2$	$4 \times v_{dc}, 4 \times v_{dc}/2$
IGBT switching freq.	8×2 (kHz)	4×2 (kHz), 2×50 (Hz)	8×2 [kHz]
Apparent output voltage Ffreq.	4 (kHz)	2 (kHz)	4 [kHz]
Grid current THD	2.83%	4.91%	2.53%
Switching losses	0.087%	0.078%	0.087%
Cond. losses	1.467%	1.409%	1.469%
Converter efficiency	98.43%	98.51%	98.44%
MPPT efficiency	+++	++	++
Topology configure	Symmetrical	Asymmetrical	Symmetrical
Advantages	2 MPPT High energy yield High power quality	1 voltage control loop Good power quality	1 voltage control loop High power quality
Disadvantages	2 voltage control loops High cond. losses	1 MPPT only DC-voltage offsets	1 MPPT only High cond. losses

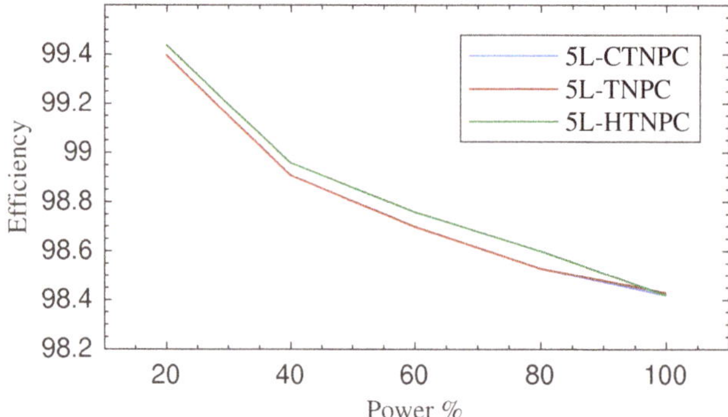

Figure 17. Efficiency comparison between 5L-CTNPC, 5L-HTNPC and 5L-TNPC inverter topologies respect to the power operation.

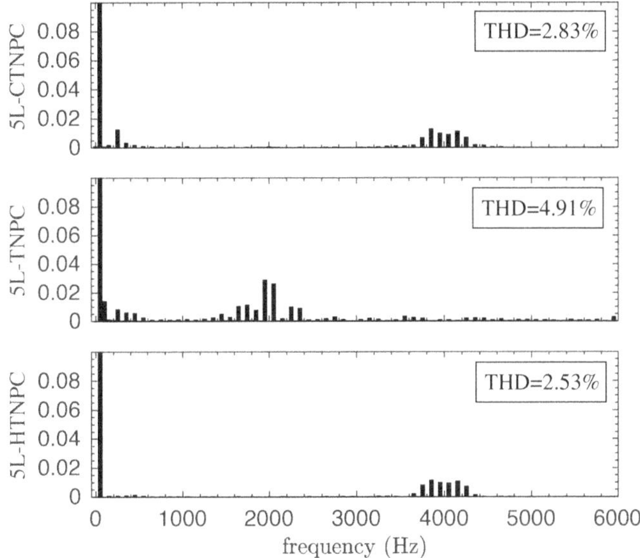

Figure 18. Grid current FFT comparison between 5L-CTNPC, 5L-HTNPC and 5L-TNPC inverter topologies.

5. Conclusions

In this work a grid-tied PV system configuration based on a series connection of three-level T-type inverter cells was described and validated. The proposed power topology merged the benefits of multistring configurations with more than one independent MPPT capability and the benefits of cascade H-bridge converters, generating a five-level output voltage waveform in the AC terminals. The proposed topology, modulation and control scheme were validated experimentally in a reduced scale power prototype with a straightforward implementation. Additionally, the proposed control strategy was evaluated under steady-state and unbalanced power conditions, ensuring a decoupled operation between both power cells. Finally, a brief comparison for key merit figures was included, where the main advantages of the proposed topology among other T configurations were highlighted, giving rise to the possibility of enhancing the power extraction from the PV side due to the multi-string configuration.

Author Contributions: C.V. conceived, designed and performed the experimental evaluations; S.K. provided insight on the power topology and conceptual solution, and was responsible of supervision throughout the work; C.A.R. wrote the manuscript and was part of the development team of the experimental test-bench, M.A.P. was responsible for the guidance during the prototype design stage, T.M. and M.M. provided relevant key theoretical and technical suggestions.

Funding: The authors gratefully acknowledge the financial support provided by CONICYT REDES 170217, AC3E (CONICYT/BASAL/FB0008), SERC Chile (CONICYT/FONDAP/15110019), CONICYT + PAI CONVOCATORIA NACIONAL SUBVENCION A INSTALACION EN LA ACADEMIA CONVOCATORIA 2018 + PAI77180032.

Conflicts of Interest: The authors declare no conflict of interest.

Abbreviations

The following abbreviations are used in this manuscript:

AC	Alternating Current
DC	Direct Current
PV	Photovoltaic
NPC	Neutral Point Clamped
MLI	Multi Level Inverter
THD	Total Harmonic Distortion
MV	Medium-Voltage
3L-NPC	Three-Level Neutral Point Clamped
3L-TNPC	Three-Level T-type Neutral Point Clamped
3L-NPP	Three-Level Neutral Point Piloted
5L-TNPC	Five-Level T-type Neutral Point Clamped
5L-HTNPC	Five-Level Hybrid T-type Neutral Point Clamped
CHB	Cascade H-Bridge
9L-TNPC	Nine-Level T-type Neutral Point Clamped
5L-CTNPC	Five-Level Cascade T-type Neutral Point Clamped
IGBT	Isolated Gate Bipolar Transistor
MOSFET	Metal Oxide Semiconductor Field Effect Transistor
MPPT	Maximum Power Point Tracking
PID	Potential-Induced Degradation
PWM	Pulse Width Modulation
LS-PWM	Level-Shifted Pulse Width Modulation
PS-PWM	Phase-Shifted Pulse Width Modulation
SVM	Space Vector Modulation
PMR	Proportional Multiresonant
P&O	Perturb and Observe
PI	Proportional-Integral
SOGI	Second Order Generalized Integrator
SRF-PLL	Synchronous Reference Frame Phase Lock Loop

Nomenclature

The following variable nomenclature is used along figures and tables of this manuscript:

v_s	Grid voltage
i_s	Grid current
L_s	Filter inductor
R_s	Filter resistor for modelling purposes
v_c	Total converter voltage
v_{ck}	Converter voltage per cell
C_{dc}	Capacitance per cell
S_{k1}, S_{k2}	Switching signals per cell
$v_{dcu,k}, v_{dcl,k}$	Upper and lower capacitor voltages per cell
$v_{dc,k}$	DC-link voltage per cell
v_{dck^*}	DC-link voltage reference per cell
$v_{pv,k}$	PV voltage per cell
$i_{pv,k}$	PV current per cell
m_{ck}	Modulation reference signal per cell
v_{cr1}, v_{cr2}	Carrier signals
v_c^*	Per-unit converter voltage reference
e_k	Energy error per cell
e_T	Total error energy
$\hat{i}_s$	Reference current magnitude
ω_s	Angular grid frequency
θ	Grid angle
f_s	Grid frequency

G_f	Notch filter
Δv_{pv}	MPPT voltage step
T_k	MPPT time step
i_s^*	Current reference
$C_i(s)$	Current controller
k_p	Proportional gain of the PMR controller
k_{ih}	Integral gain of the PMR controller for each h-th frequency component
h	Grid harmonic
v_L	Inductor voltage
v_L^*	Inductor voltage reference
f_{cr}	Carrier frequency
T_s	Sampling period
BW_{vdc}	DC-link control bandwidth
BW_{is}	Current control bandwidth
BW_{dv}	Balancing control bandwidth
P_{mp}	Maximum power
v_{mp}	Voltage at maximum power
v_{oc}	Open-circuit voltage
i_{mp}	Current at maximum power
i_{sc}	Short-circuit current
Δm_d	Voltage drift modulation component
Δm_c	Cell voltage control
P_{ck}	Power per cell

References

1. Kouro, S.; Leon, J.I.; Vinnikov, D.; Franquelo, L.G. Grid-Connected Photovoltaic Systems: An Overview of Recent Research and Emerging PV Converter Technology. *IEEE Ind. Electron. Mag.* **2015**, *9*, 47–61. [CrossRef]
2. Jedtberg, H.; Pigazo, A.; Liserre, M.; Buticchi, G. Analysis of the Robustness of Transformerless PV Inverter Topologies to the Choice of Power Devices. *IEEE Trans. Power Electron.* **2017**, *32*, 5248–5257. [CrossRef]
3. Kouro, S.; Malinowski, M.; Gopakumar, K.; Pou, J.; Franquelo, L.G.; Wu, B.; Rodriguez, J.; Perez, M.A.; Leon, J.I. Recent Advances and Industrial Applications of Multilevel Converters. *IEEE Trans. Ind. Electron.* **2010**, *57*, 2553–2580. [CrossRef]
4. Muñoz-Cruzado-Alba, J.; Rojas, C.A.; Kouro, S.; Díez, E.G. Power production losses study by frequency regulation in weak-grid-connected utility-scale photovoltaic plants. *Energies* **2016**, *9*, 317. [CrossRef]
5. Wang, Y.; Shi, W.W.; Xie, N.; Wang, C.M. Diode-Free T-Type Three-Level Neutral-Point-Clamped Inverter for Low-Voltage Renewable Energy System. *IEEE Trans. Ind. Electron.* **2014**, *61*, 6168–6174. [CrossRef]
6. Shi, Y.; Wang, L.; Xie, R.; Shi, Y.; Li, H. A 60-kW 3-kW/kg Five-Level T-Type SiC PV Inverter With 99.2% Peak Efficiency. *IEEE Trans. Ind. Electron.* **2017**, *64*, 9144–9154. [CrossRef]
7. Schweizer, M.; Kolar, J.W. Design and Implementation of a Highly Efficient Three-Level T-Type Converter for Low-Voltage Applications. *IEEE Trans. Power Electron.* **2013**, *28*, 899–907. [CrossRef]
8. Semikron: 3L NPC and TNPC Topology, 2015. Available online: https://www.semikron.com/zh/service-support/downloads/detail/semikron-application-note-3l-npc-tnpc-topology-en-2015-10-12-rev-05.html (accessed on 1 March 2019).
9. Infineon: 3-Level configurations, 2016. Available online: https://www.infineon.com/dgdl/Infineon-3_Level_Inverter-PB-v08_00-EN.pdf?fileId=db3a30432a14dd54012a31d664d90273 (accessed on 1 March 2019).
10. OnSemiconductor: Q0PACK Modules, 2018. Available online: https://www.onsemi.com/pub/Collateral/NXH80T120L2Q0S2G-D.PDF (accessed on 1 March 2018).
11. Joetten, R.; Gekeler, M.; EIBEL, J. AC drive with three level voltage source inverter and high dynamic performance micropocessor control. In Proceedings of the First European Conference on Power Electronics and Applications, Brussels, Belgium, September 1985; p. 6.
12. Salem, A.; Abido, M.A. T-Type Multilevel Converter Topologies: A Comprehensive Review. *Arab. J. Sci. Eng.* **2018**, *44*, 1713–1735. [CrossRef]

13. Xue, Y.; Manjrekar, M. A new class of single-phase multilevel inverters. In Proceedings of the 2nd International Symposium on Power Electronics for Distributed Generation Systems, Hefei, China, 16–18 June 2010; pp. 565–571.

14. Almeida Cacau, R.; Torrico-Bascope, R.P.; Neto, J.A.F.; Torrico-Bascope, G.V. Five-Level T-Type Inverter Based on Multistate Switching Cell. *IEEE Trans. Ind. Appl.* **2014**, *50*, 3857–3866. [CrossRef]

15. Aly, M.; Ahmed, E.M.; Shoyama, M. Modulation Method for Improving Reliability of Multilevel T-type Inverter in PV Systems. *IEEE J. Emerg. Sel. Top. Power Electron.* **2019**, *1*. [CrossRef]

16. Valderrama, G.E.; Guzman, G.V.; Pool-Mazun, E.I.; Martinez-Rodriguez, P.R.; Lopez-Sanchez, M.J.; Zuniga, J.M.S. A Single-Phase Asymmetrical T-Type Five-Level Transformerless PV Inverter. *IEEE J. Emerg. Sel. Top. Power Electron.* **2018**, *6*, 140–150. [CrossRef]

17. Limones-Pozos, C.A.; Martinez-Rodriguez, P.R.; Sosa, J.M.; Vazquez, G.; Izaguirre-Vera, A. Design and analysis of a single-phase transformerless multilevel 7L-TT-HB cascade inverter for renewable energy applications. In Proceedings of the 2018 IEEE International Autumn Meeting on Power, Electronics and Computing (ROPEC), Ixtapa, Mexico, 14–16 November 2018; pp. 1–6. [CrossRef]

18. Chang, W.-N.; Liao, C.-H. Design and Implementation of a STATCOM Based on a Multilevel FHB Converter with Delta-Connected Configuration for Unbalanced Load Compensation. *Energies* **2017**, *10*, 921. [CrossRef]

19. Yoo, A. Multi-level Medium-Voltage Inverter. 2015/EP2822164A2, 2015. Available online: https://patents.google.com/patent/EP2822164A2/en (accessed on 1 May 2018).

20. Verdugo, C.; Kouro, S.; Perez, M.A.; Malinowski, M.; Meynard, T. Series-connected T-type Inverters for single-phase grid-connected Photovoltaic Energy System. In Proceedings of the 39th Annual Conference of the IEEE Industrial Electronics Society (IECON 2013), Vienna, Austria, 10–13 November 2013; pp. 7021–7027. [CrossRef]

21. Mori, S.; Aketa, M.; Sakaguchi, T.; Asahara, H.; Nakamura, T.; Kimoto, T. Demonstration of 3 kV 4H-SiC reverse blocking MOSFET. In Proceedings of the 2016 28th International Symposium on Power Semiconductor Devices and ICs (ISPSD), Prague, Czech Republic, 12–16 June 2016; pp. 271–274. [CrossRef]

22. Gurpinar, E.; Castellazzi, A. Single-Phase T-Type Inverter Performance Benchmark Using Si IGBTs, SiC MOSFETs, and GaN HEMTs. *IEEE Trans. Power Electron.* **2016**, *31*, 7148–7160. [CrossRef]

23. Chub, A.; Vinnikov, D.; Blaabjerg, F.; Peng, F. Z. A Review of Galvanically Isolated Impedance-Source DC-DC Converters. *IEEE Trans. Power Electron.* **2016**, *31*, 2808–2828. [CrossRef]

24. Dargahi, V.; Abarzadeh, M.; Corzine, K.A.; Enslin, J.H.; Sadigh, A.K.; Rodriguez, J.; Blaabjerg, F.; Maqsood, A. Fundamental Circuit Topology of Duo-Active-Neutral-Point-Clamped, Duo-Neutral-Point-Clamped, and Duo-Neutral-Point-Piloted Multilevel Converters. *IEEE J. Emerg. Sel. Top. Power Electron.* **2018**, *1*. [CrossRef]

25. Serban E.; Pondiche, C.; Ordonez, M. Modulation Effects on Power-Loss and Leakage Current in Three-Phase Solar Inverters. *IEEE Trans. Energy Convers.* **2019**, *34*, 339–350. [CrossRef]

26. Leon, J.I.; Kouro, S.; Franquelo, L.G.; Rodriguez, J.; Wu, B. The Essential Role and the Continuous Evolution of Modulation Techniques for Voltage-Source Inverters in the Past, Present, and Future Power Electronics. *IEEE Trans. Ind. Electron.* **2016**, *63*, 2688–2701. [CrossRef]

27. Lee, T.; Bu, H.; Cho, Y. Hybrid PWM Strategy for Power Efficiency Improvement of 5-Level TNPC Inverter and Current Distortion Compensation Method. *Electronics* **2019**, *8*, 76. [CrossRef]

28. Teodorescu, R.; Liserre, M.; Rodriguez, P. *Grid Converters for Photovoltaic and Wind Power Systems*; John Wiley & Sons: Hoboken, NJ, USA, 2011.

29. Kouro, S.; Wu, B.; Moya, A.; Villanueva, E.; Correa, P.; Rodriguez, J. Control of a cascaded H-bridge multilevel converter for grid connection of photovoltaic systems. In Proceedings of the 2009 35th Annual Conference of IEEE Industrial Electronics, Porto, Portugal, 3–5 November 2009, pp. 3976–3982. [CrossRef]

30. IGBT Model IXXK100N60C3H1 Datasheet. 2015. Available online: https://hr.mouser.com/ProductDetail/IXYS/IXXK100N60C3H1/?qs=DTxAbJQb9nVqmCmAy76OGQ%3D%3D (accessed on 1 May 2018).

Article

A Novel Three-Phase Six-Switch PFC Rectifier with Zero-Voltage-Switching and Zero-Current-Switching Features

Chun-Wei Lin [1,*], Chang-Yi Peng [2] and Huang-Jen Chiu [1]

[1] Department of Electronic and Computer Engineering,
    National Taiwan University of Science and Technology, Taipei 10607, Taiwan; hjchiu@mail.ntust.edu.tw
[2] Department of Electrical Engineering, Chung-Yuan Christian University, Taoyuan 32023, Taiwan;
    Kingpeng50@yahoo.com.tw
* Correspondence: d10402204@mail.ntust.edu.tw; Tel.: +886-937-880-025

Received: 18 March 2019; Accepted: 19 March 2019; Published: 22 March 2019

Abstract: A novel three-phase power-factor-correction (PFC) rectifier with zero-voltage-switching (ZVS) in six main switches and zero-current-switching (ZCS) in the auxiliary switch is proposed, analyzed, and experimentally verified. The main feature of the proposed auxiliary circuit is used to reduce the switching loss when the six main switches are turned on and the one auxiliary switch is turned off. In this paper, a detailed operating analysis of the proposed circuit is given. Modeling and analysis are verified by experimental results based on a three-phase 7 kW rectifier. The soft-switched PFC rectifier shows an improvement in efficiency of 2.25% compared to its hard-switched counterpart at 220 V under full load.

Keywords: three-phase rectifier; PFC; switch-mode rectifier; ZVS; ZCS

---

1. Introduction

Power electronic converters play a critical role in the energy industry due to their ability to optimally control and condition the power they deliver to a load. In addition, they are required to control and condition the power they draw from energy sources to support their optimal operation. This is achieved by compliance to EMI and harmonic standards such as EN6100-3-2 and efficiency standards such as 80Plus [1]. Soft-switching technologies are a primary enabler for improving efficiency by minimizing switching losses and reducing EMI and harmonics by "soft" ending the edges of the switching transitions [2–12]. References [8] and [9] report soft-switching techniques that includes zero-voltage-switching (ZVS) and zero-current-switching (ZCS). Three-phase rectifiers with active power-factor-correction (PFC) control achieve an improved power factor and lower harmonic content [10–13]. Active PFC rectifiers using a boost (current source) front end achieve better input current wave-shaping and lower harmonic distortion compared to their buck-derived counterparts [14]. Three single-phase PFC rectifiers are used in [15] to synthesize a three-phase PFC rectifier. Reference [16] reports the use of space vector modulation (SVM) to achieve a high power factor in a three-phase six-switch rectifier. Soft-switching techniques employed in three-phase rectifiers are reported in [17–19] to improve efficiency and EMI performance. Soft-switching using a passive lossless snubber is presented in [17]. Although this approach can improve the efficiency, the circuit suffers from higher component stress. In [18], an active snubber is used to achieve soft-switching at the expense of higher control complexity and switching stress in the auxiliary switch. In [19–22], the zero-voltage-transition and control technique was applied in a three-phase PFC rectifier. Although the main switches can achieve ZVS at turn-on, the auxiliary switch was hard-switched operated at turn-off.

A conventional three-phase six-switch PFC rectifier is shown in Figure 1. A novel soft-switched three-phase active rectifier using an active auxiliary circuit is proposed in this paper. The principal performance improvement is the achievement of ZVS at turn-on for the six rectifier switches and ZCS at turn-off for the one auxiliary switch. A detailed description of the operation of the proposed soft-switched rectifier is presented in Section 2. Validation of the design through simulation and experimental results are shown in Section 3 followed by concluding remarks in Section 4.

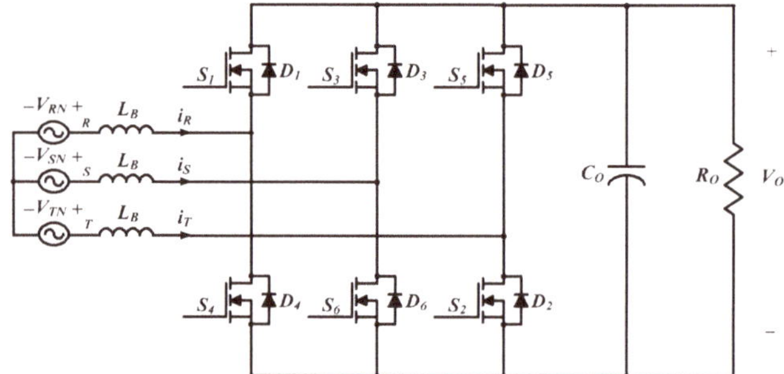

Figure 1. A conventional three-phase six-switch power-factor-correction (PFC) rectifier.

2. Proposed Three-Phase Six-Switch Soft-switching PFC Rectifier

The proposed three-phase six-switch soft-switching PFC rectifier is shown in Figure 2. The circuit inside the dotted box is a soft-switching assist circuit to achieve ZVS in the main switches and ZCS in the auxiliary switch. The soft-switching assist circuit consists of the auxiliary switch S_A, resonant inductor L_R, transformer T_r, barrier diode D_{R1}, clamp circuit R_C–D_C–C_C, and resonant capacitor (the capacitance employs the parasitic capacitance of main switch).

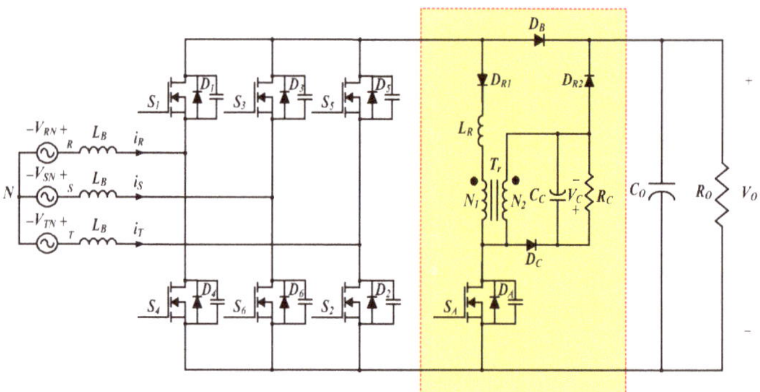

Figure 2. The circuit of the proposed soft-switching PFC rectifier.

Three phase line voltages V_{RN}, V_{SN}, V_{TN} for a balanced three-phase system are shown in Figure 3. The 60° symmetry in the three-phase voltages is evident from Figure 3. The operation of the three-phase PFC using the 60° symmetry is described in detail in [11].

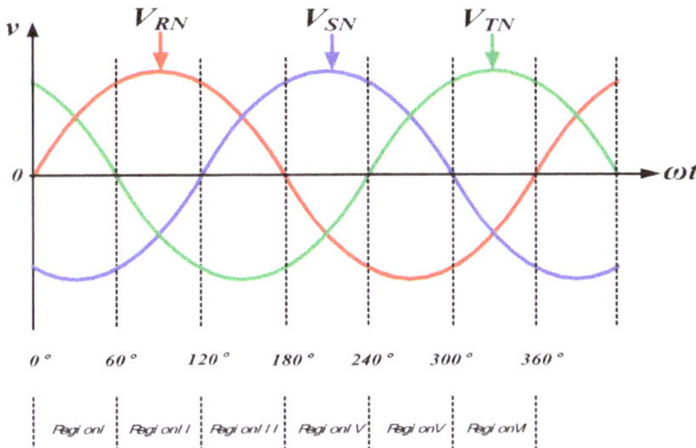

Figure 3. The line cycle in three-phase balance power system.

In order to simplify the analysis, Interval 1 (0°–60°) can be selected for the analysis of the switching cycles as the operation over the rectifier is identical in the other 60° segments. The following assumptions are made to support the operating analysis:

(1) Input inductance L_B is large enough to allow the input current to be considered as a current source over a switching period;
(2) Input capacitance C_L is large enough to be equivalent to the ideal voltage source V_O; and
(3) The output capacitance of the clamp circuit C_C is large enough to allow its voltage V_C to be considered a voltage source over a switching period.

Under the assumptions listed above, the simplified circuit diagram is shown in Figure 4 and the voltage polarity and current direction for each main component are defined.

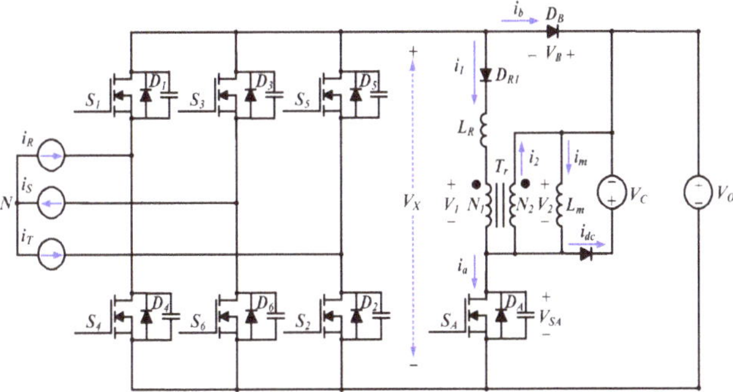

Figure 4. The simplified circuit of the proposed soft-switched rectifier.

A detailed description of circuit operation is provided in this section. The key waveforms of the circuit for Interval 1 are shown in Figure 5, and equivalent circuits for each operating mode are shown in Figure 6. There are 12 operating modes to be analyzed over a switching cycle.

2.1. Mode 0: ($t \leqq T_0$)

This mode is based on the analysis of the switching cycle in Interval 1 ($V_{RN} > 0$, $V_{TN} > 0$, and $V_{SN} < 0$). Before T_0, as in Figure 6a, the diode D_1, D_6, and D_5 are in the state of conduction. The main switches S_1 to S_6 and auxiliary switch S_A are turned off. The currents i_R and i_T flow through diode D_B to the load and return to the AC source as the current i_S. Under this condition, the voltage across the active rectifier bridge is $V_X = V_O$.

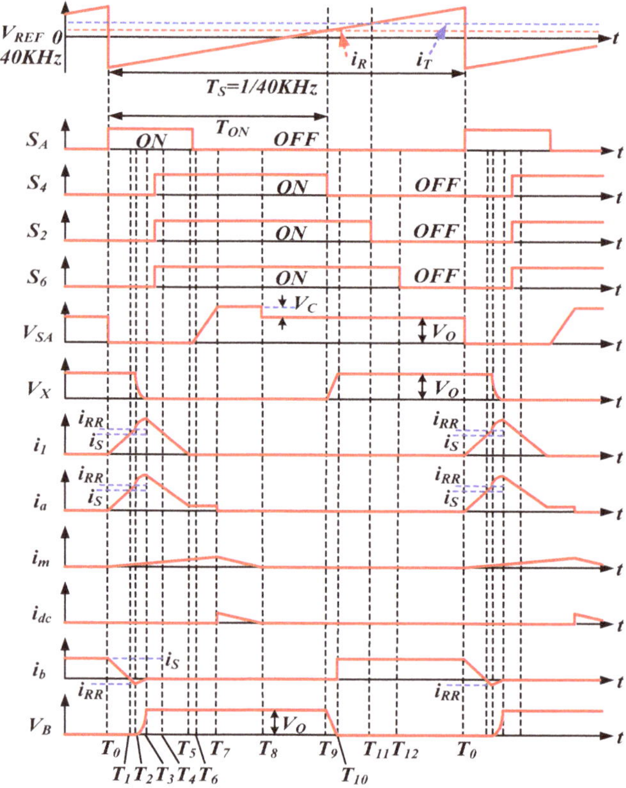

Figure 5. The key waveforms of the proposed soft-switched rectifier.

2.2. Mode 1 ($T_0 < t \leqq T_1$)

At T_0, the auxiliary switch S_A is turned on to go into Mode 1. The current i_1 of resonant inductor L_R starts to increase, and current i_1 flows through the primary winding N_1 of the transformer T_r. The induced current i_2 and excitation current i_m outflow through secondary coil N_2, as shown in Figure 6b. The voltage on the seconding winding N_2 is the output voltage V_O. The voltage V_1 and V_2 across the windings of transformer T_r are obtained as follows:

$$V_2 = V_O \tag{1}$$

$$V_1 = \frac{N_1}{N_2}V_2 = nV_O \tag{2}$$

The current in the resonant inductor i_1 increases linearly with the slope given by

$$\frac{di_1}{dt} = \frac{V_O - V_1}{L_R} = \frac{V_O - nV_O}{L_R} = (1-n)\frac{V_O}{L_R} \tag{3}$$

Similarly to i_1, the excitation current i_m also displays a linear increase, and the slope is

$$\frac{di_m}{dt} = \frac{V_O}{L_m} \tag{4}$$

When the current i_1 ascends to i_S current, this mode ends. The time interval is given as below:

$$t_{01} = \frac{i_S}{V_o(1-n)L_R}$$

(5)

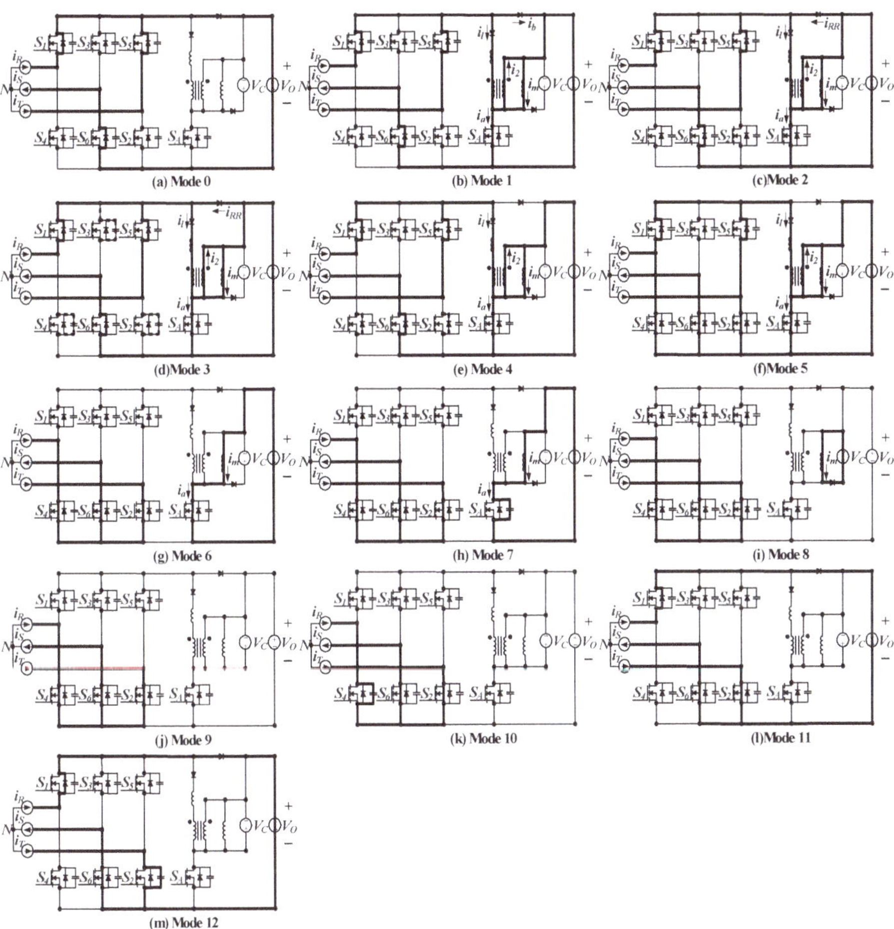

Figure 6. Operation modes of the proposed soft-switched rectifier.

2.3. Mode 2 ($T_1 < t \leqq T_2$)

At $t = T_1$, the DC link diode current i_b reaches zero. The reverse recovery current of diode D_B flows through diode D_B in a negative direction. The resonant inductor current keeps increasing, as shown in Figure 6c.

2.4. Mode 3 ($T_2 < t \leqq T_3$)

At $t = T_2$, the parasitic capacitance of diode D_b along with the resonant capacitor C_R, which includes the parasitic capacitor of the main switches and the resonant inductor L_R, start resonating, as shown in Figure 6d. When the equivalent voltage V_X of the main switch is decreased to zero at $t = T_3$, the mode ends. The equivalent voltage V_X and resonant current i_1 are shown in Equations (6) and (7).

$$V_X = V_O - (1 - n)V_O(1 - \cos(\omega_R t)) \tag{6}$$

$$i_1 = i_s + i_{RR} + \frac{(1 - n)V_O}{Z_C} \sin(\omega_R t) \tag{7}$$

$$C_R = C_2 + C_3 + C_4 \tag{8}$$

$$\omega_R = \frac{1}{\sqrt{L_R(C_R + C_b)}} \tag{9}$$

$$Z_C = \sqrt{\frac{L_R}{C_R + C_b}} \tag{10}$$

2.5. Mode 4: ($T_3 < t \leqq T_4$)

When $t > T_3$, the bridge rectifier voltage V_X is decreased to zero and the auxiliary switch S_A continues to conduct. The corresponding equivalent circuit is shown in Figure 6e. The body diodes D_4, D_3, and D_2 of the main switches S_4, S_3, and S_2 are conducting. Turning on the main switches S_4, S_6, and S_2 when the bridge voltage reaches zero achieves ZVS turn-on. The detection circuitry to turn on the main switches at zero voltage also enables the minimization of duty-cycle loss and, thus, loss of efficiency. After the main switches turn on at ZVS, the resonant inductor current i_1 decreases linearly with the slope given by Equation (11). When the current of the main switches S_4 and S_2 reaches zero at $t = T_4$, the mode ends.

$$\frac{di_1}{dt} = -\frac{nV_O}{L_R} \tag{11}$$

2.6. Mode 5: ($T_4 < t \leqq T_5$)

As shown in Figure 6f, when $t > T_4$, then S_4, S_6, and S_2 keep conducting. The current i_l is continuously decreased to zero until $t = T_5$.

2.7. Mode 6: ($T_5 < t \leqq T_6$)

When $t > T_5$, the input currents i_R and i_T flow through the main switches S_4 and S_2, as shown in Figure 6g. When the current i_a flows through the auxiliary switch S_A, it consists mostly of the magnetizing current, i_m, of the transformer. If the magnetizing inductance L_m is designed to be relatively large, the current i_a of the auxiliary switch S_A is extremely close to zero. When $T_5 < t \leqq T_6$, the auxiliary switch is set to be turned off so that it can effectively achieve the purpose of ZCS.

2.8. Mode 7: ($T_6 < t \leqq T_7$)

When $t = T_6$, the auxiliary switch S_A is turned off, as shown in Figure 6h. Subsequently, the magnetizing current i_m of the transformer charges the parasitic capacitance C_{oss1} of the auxiliary switch S_A so that the auxiliary switch voltage will increase continuously.

2.9. Mode 8: ($T_7 \leqq t \leqq T_8$)

At $t = T_7$, the auxiliary switch voltage V_{SA} increases to $V_O + V_C$ and the clamp diode D_C is conducting. The magnetizing current i_m discharges through the clamp circuit D_C–V_C, as shown in Figure 6i. The slope of the excitation current in this mode is given by

$$\frac{di_m}{dt} = -\frac{V_C}{L_m} \tag{12}$$

2.10. Mode 10: ($T_8 \leq t \leq T_9$)

At $t = T_8$, the magnetizing current i_m is decreased to zero which resets the transformer, as shown in Figure 6j.

2.11. Mode 10: ($T_9 < t \leq T_{10}$)

At $t = T_9$, the main switch S_4 is turned off. The input current i_R charges the parasitic capacitance of the main switch S_4 and the equivalent voltage V_X of the main switch is increased, as shown in Figure 6k.

2.12. Mode 11: ($T_{10} \leq t \leq T_{11}$)

At $t = T_{10}$, the equivalent voltage V_X of the main switch is increased to V_O and the diode D_B is conducting as shown in Figure 6l. Subsequently, the antiparallel diode D_1 of the main switch S_1 is conducting. The input current i_R flows through diodes D_1 and D_B and flows back from i_S through the load.

2.13. Mode 12: ($T_{11} \leq t \leq T_{12}$)

At $t = T_{11}$, the main switch S_2 is turned off. The input current i_T starts to charge the parasitic capacitance of the main switch S_2, as shown in Figure 6m.

3. Experimental Verifications

Based on the design described, a prototype was built. When the line voltage in the three-phase input was 220 V, namely, the phase voltage was 127 V, the switching frequency 40 kHz, and the output load 7 kW, then the measured waveforms for three-phase V_{RN}, V_{SN}, V_{TN}, line voltage and the line current were at low line and full load. The current waveform, as in Figure 7, is practically sinusoidal with low THD and a high power factor (the A-THD is shown in Figure 8 and power factor is shown in Figure 9).

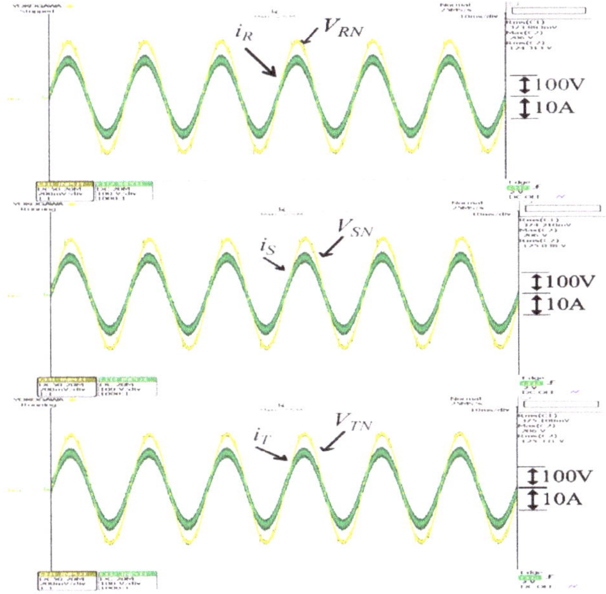

Figure 7. Measured waveforms of input voltage and input current at full load.

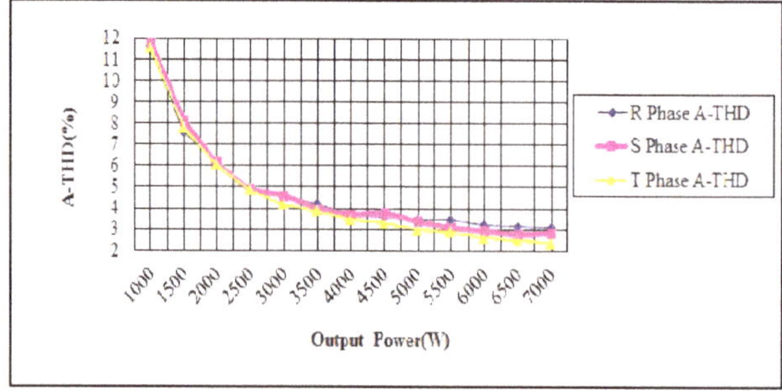

Figure 8. A-THD measured results.

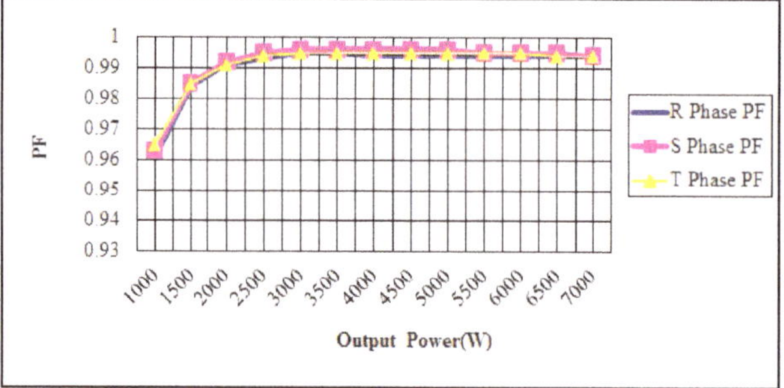

Figure 9. Power factor measured results.

Figure 10 shows the simulation waveforms using Isspice at half load and full load. It can be seen that the main switches S_4 and S_2 on the bottom sides turned on when their V_X was down to zero by the resonant circuit, after the auxiliary switch S_A was turned on.

Measured waveforms of the gate drive signals and voltage across the main switch are shown in Figure 11. The captured waveforms indicate the need to turn on the auxiliary switch S_A before the turning on the main switches S_4 and S_2 to achieve ZVS.

As shown in Figure 12, when the current i_1 of resonant inductor L_R decreases to zero, the auxiliary switch S_A can be turned off under ZCS conditions.

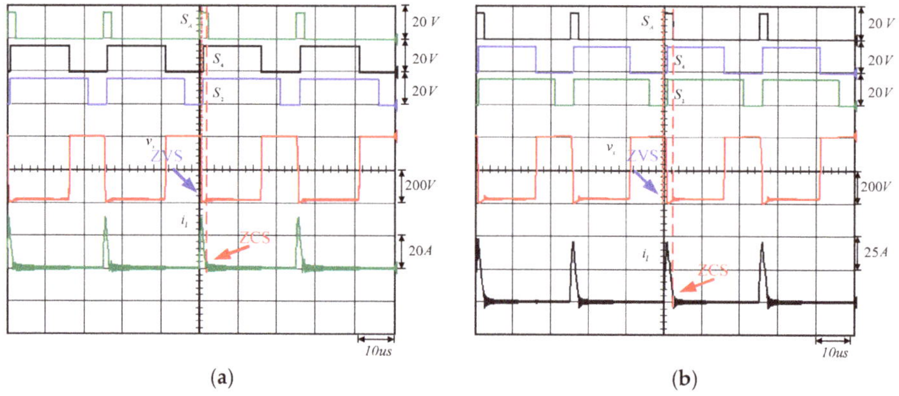

(a)   (b)

Figure 10. Simulation of key waveforms of the main switch voltage V_X and current i_l at (**a**) half load and (**b**) full load.

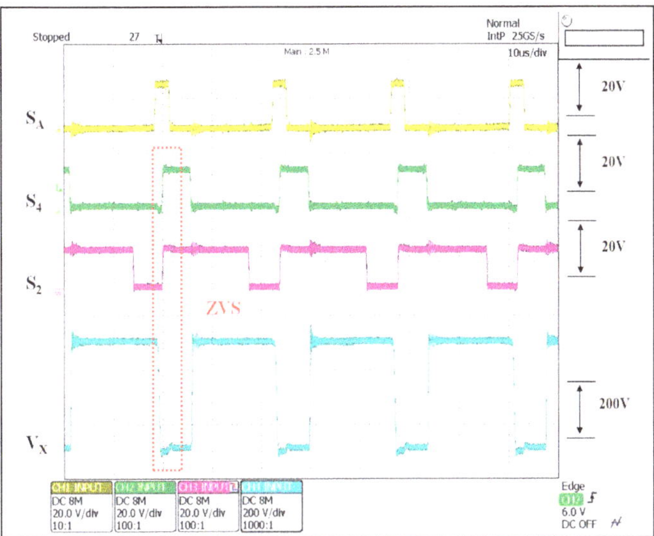

Figure 11. Measured waveforms of drive signal and voltage for the main switch of the soft-switched rectifier.

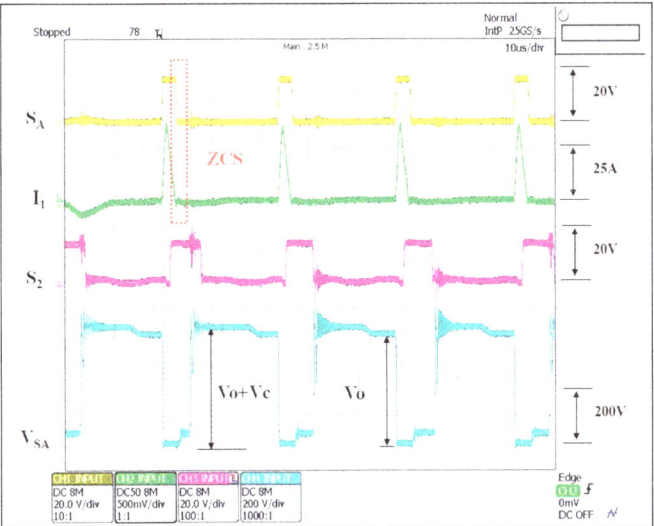

Figure 12. Measured waveforms of drive signal and resonant current for the main switch of the soft-switched rectifier.

A comparison between the hard-switched and proposed soft-switched rectifier was performed at a load of 7 kW over an input voltage range of 190–250 V. The hard-switched rectifier was tested by simply disabling the soft-switch assist circuitry. Figure 13 shows the efficiency improvement from the soft-switched rectifier. The largest efficiency difference between hard-switch and soft-switch rectifier was 2.55% when the input line voltage was 220 V, and Figure 14 shows the soft-switched rectifier efficiency from 1–7 kW.

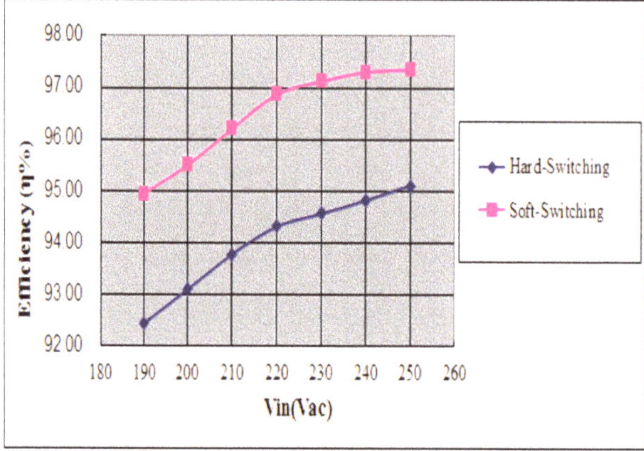

Figure 13. The efficiency comparison between hard-switching and soft-switching rectifier at 7 kW load.

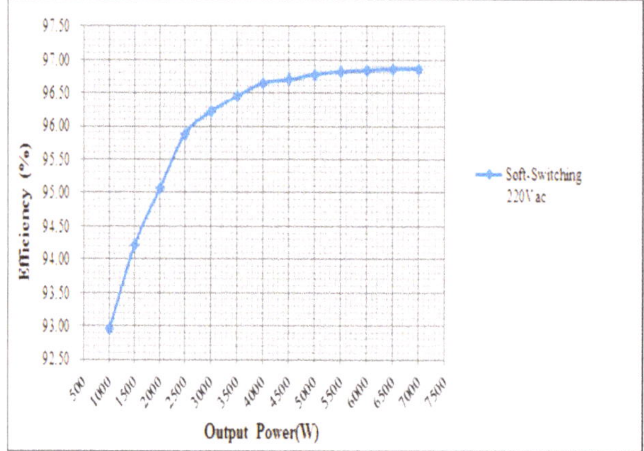

Figure 14. Measured efficiency from 1 to 7 kW with the soft-switched rectifier.

4. Conclusions

A novel three-phase rectifier with zero-voltage-switching and zero-current-switching features was proposed. The design has been validated with simulation and experimental data captured on a 7 kW three-phase rectifier prototype. Efficiency improvement between hard-switch and soft-switch rectifiers peaks at 2.55% when the input line voltage is 220 V at full load. Mathematical equations to explain circuit operation have been derived and analyzed under a sequence of operating modes. The experimental results have confirmed the proposed design of the soft-switched rectifier in achieving high efficiency, high power factor, and low THD.

Author Contributions: C.-W.L. and C.-Y.P. designed, debugged the system, built some part of hardware and performed the experiment. C.-W.L. also mainly responsible for preparing the paper. H.-J.C., supervised the design, analysis, experiment, and editing the paper.

Funding: The authors would like to acknowledge the financial support of the Ministry of Science and Technology of Taiwan through grant number NSC 103-2221-E-011 -064 -MY3.

References

1. Ecova Plug Load Solutions Website. 80 PLUS Certified Power Supplies and Manufacturers. Available online: http://www.plugloadsolutions.com/80PlusPowerSupplies.aspx (accessed on 1 March 2019).
2. Kolar, J.W.; Friedli, T. The essence of three-phase PFC rectifier systems—Part I. *IEEE Trans. Power Electron.* **2013**, *28*, 176–198. [CrossRef]
3. Friedli, T.; Hartmann, M.; Kolar, J.W. The essence of three-phase PFC rectifier systems—Part II. *IEEE Trans. Power Electron.* **2014**, *29*, 543–560. [CrossRef]
4. Chang, C.H.; Cheng, C.A.; Chang, E.C.; Cheng, H.L.; Yang, B.E. An integrated high-power-factor converter with ZVS transition. *IEEE Trans. Power Electron.* **2016**, *31*, 2362–2371. [CrossRef]
5. Martins, M.L.S.; Hey, H.L. Self-commutated auxiliary circuit ZVT PWM converters. *IEEE Trans. Power Electron.* **2004**, *19*, 1435–1445. [CrossRef]
6. Bodur, H.; Bakan, A.F. A new ZVT-ZCT-PWM dc/dc converter. *IEEE Trans. Power Electron.* **2004**, *19*, 676–684. [CrossRef]
7. Ivanovic, B.; Stojiljkovic, Z. A novel active soft Switching snubber designed for boost converter. *IEEE Trans. Power Electron.* **2004**, *19*, 658–665. [CrossRef]
8. Moran, L.; Werlinger, P.; Dixon, J.; Wallace, R. A Series active power filter which compensates current harmonics and voltage unbalance simultaneously. In Proceedings of the IEEE PESC Conference, Atlanta, GA, USA, 18–22 June 1995; pp. 222–227.
9. Chang, Y. Boost Converter with Zero Voltage Main Switch and Zero Current Auxiliary Switches. U.S. Patent 6,498,463 B2, 24 December 2002.
10. Tsai, H.; Hsia, T.; Chen, D. A novel soft-switching bridgeless power factor correction circuit. In Proceedings of the European Power Electronics and Applications Conference, Aalborg, Denmark, 2–5 September 2007.
11. Jang, Y.; Jovanovic, M.M.; Fang, K.H.; Chang, Y.M. High-Power-Factor Soft-Switched Boost Converter. *IEEE Trans. Power Electron.* **2006**, *21*, 98–104. [CrossRef]
12. Tsai, H.Y.; Hsia, T.H.; Chen, D. A Family of Zero-Voltage Transition Bridgeless Power Factor CorrectionCircuits with a Zero-Current-Switching Auxiliary Switch. *IEEE Trans. Ind. Electron.* **2011**, *58*, 1848–1855. [CrossRef]
13. Wei, H.; Bataresh, I. Comparison of basic converter topologies for power factor correction. In Proceedings of the Southeastcon Conference, Orlando, FL, USA, 24–26 April 1998; pp. 348–353.
14. Mohan, N.; Undeland, T.M.; Robbins, W.P. *Power Electronics Converters, Applications and Design*, 2nd ed.; Wiley: New York, NY, USA, 1995.
15. Jiang, Y.; Mao, H.; Lee, F.C.; Borojevic, D. Simple high Performance three-phase boost rectifiers. In Proceedings of the IEEE Power Electronics Specialists Conference, Taipei, Taiwan, 20–25 June 1994; pp. 1158–1163.
16. Li, R.; Ma, K.; Xu, D. A novel 40 KW ZVS-SVM controlled three-phase boost PFC converter. In Proceedings of the IEEE Applied Power Electronics Conference and Exposition, Washington, DC, USA, 15–19 February 2009; pp. 376–382.
17. Hengchun, M.; Lee, C.Y.; Boroyevich, D.; Hiti, S. Review of high-performance three-phase power-factor correction circuits. *IEEE Trans. Ind. Electron.* **1997**, *44*, 437–446. [CrossRef]
18. Li, Q.; Zhou, X.; Lee, F.C. A novel ZVT three-phase rectifier/inverter with reduced auxiliary switch stresses and losses. In Proceedings of the IEEE PESC PESC'96, Baveno, Italy, 23–27 June 1996; pp. 153–158.
19. Vlatkovic, V.; Borojevic, D.; Lee, F.C.; Cuadros, C.; Gataric, S. A new zero-voltage transition, three-phase PWM rectifier/inverter circuit. In Proceedings of the IEEE PESC'93, Seattle, WA, USA, 20–24 June 1993; pp. 868–873.
20. Kennel, R.; Schröder, D. Predictive control strategy for converters. In Proceedings of the 3rd IFAC Symposium on Control in Power Electronics and Electrical Drives, Lausanne, Switzerland, 12–14 September 1983; pp. 415–422.

21. Mercorelli, P.; Kubasiak, N.; Liu, S. Multilevel bridge governor by using model predictive control in wavelet packets for tracking trajectories. In Proceedings of the IEEE International Conference on Robotics and Automation, New Orleans, LA, USA, 26 April–1 May 2004; pp. 4079–4084.
22. Mercorelli, P.; Kubasiak, N.; Liu, S. Model predictive control of an electromagnetic actuator fed by multilevel PWM inverter. In Proceedings of the IEEE International Symposium on Industrial Electronics, Ajaccio, France, 4–7 May 2004; pp. 531–535.

Article

Energy Storage Sizing Strategy for Grid-Tied PV Plants under Power Clipping Limitations

Nicolás Müller [1,2,*,†], Samir Kouro [2,†], Pericle Zanchetta [1,†], Patrick Wheeler [1,†], Gustavo Bittner [2,†] and Francesco Girardi [3,†]

1   Department of Electrical and Electronic Engineering, University of Nottingham, Nottingham NG7 2RD, UK; pericle.zanchetta@nottingham.ac.uk (P.Z.); pat.wheeler@nottingham.ac.uk (P.W.)
2   Electronics Engineering Department, Universidad Técnica Federico Santa María, Valparaiso 2390123, Chile; samir.kouro@usm.cl (S.K.); gbittnerhofmann@gmail.com (G.B.)
3   Bluefield Services, Bristol BS1 6DZ, UK; fgirardi@bluefieldservices.com
*   Correspondence: nicolas.mullerpollmann@nottingham.ac.uk
†   These authors contributed equally to this work.

Received: 4 April 2019; Accepted: 9 May 2019; Published: 13 May 2019

Abstract: This paper presents an analyses of an Energy Storage System (ESS) for grid-tied photovoltaic (PV) systems, in order to harness the energy usually lost due to PV array oversizing. A real case of annual PV power generation analysis is presented to illustrate the existing problem and future solutions. Three PV modeling techniques have been applied to estimate non-measured non-harnessed PV power to provide an ESS energy and power sizing strategy. Moreover, a control strategy to store or release power from the DC-link, without modifying the Maximum Power Point Tracking (MPPT) strategy, is presented. The results show an estimation of the annual power loss caused by oversizing the PV array. The ESS sizing strategy gives insight into not only the energy requirements, but also the power requirements of the system. Simulation results show that the proposed ESS control strategy is capable of harnessing the extra power without modifying the existing power converter of the PV plant nor its control strategy.

Keywords: power clipping; ESS sizing; grid-tied PV plant

---

1. Introduction

Perpetration of renewable energy in electric markets has reached an impressive 26.5%, being wind, bio and solar power at the forefront of modern renewables development and integration to electric retail [1]. A crucial parameter when designing renewable energy plants is its load factor, also known as capacity factor or plant (load) factor, which corresponds to the ratio between the generated and rated energy of the plant during a certain amount of time. In UK, PV plants present annual load factors close to 10% [2], which are calculated considering the rated power of the converter. A common practice is to increase annual plant factor by oversizing the power rating of the PV array, with respect to the converter [3]. The ratio between PV array rated power and the inter AC rated output power is known as Inverter Loading Ratio (ILR) [4]; in places with high irradiation variability such as UK, PV array power ILR oversizing can reach as much as 40%, whereas, in places with lower irradiation variability, such as central Chile, oversizing is closer to 15%. Moreover, the continuous drop on PV module prices have encouraged the increase of ILR in PV plants [3]; some authors have even proposed ILR oversizing up to 80% [4].

When an oversized PV array reaches the power rating of the converter, the converter loses the ability to increase its current and therefore is unable to reduce the DC-link voltage and loses the ability to track the Maximum Power Point (MPP). This behavior is called clipping, and it forces the system to waste available PV power. Clipped power is the name assigned to this wasted power. Figure 1 presents a grid-tied central inverter PV plant with PV array oversizing, where the available PV power is truncated at the rating of the inverter (clipped), limiting the exported PV power. Both power curves were normalized to the inverter rating.

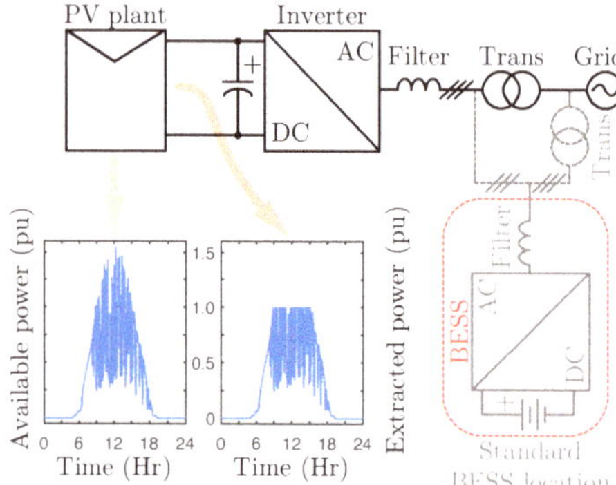

Figure 1. Central inverter grid-tied PV plant with additional Battery Energy Storage System (BESS).

Generation–demand matching paradigm, where intermittency and high variability of renewable resources play a major role, can be achieved by relying on other systems connected to the electrical network and/or by the addition of an Energy Storage System (ESS). The first solution is not suitable for harnessing clipped power, since it requires the clipped power to be transferred to the electric network through the inverter, which is already operating at its rated power. The latter solution presents a more promising alternative to enhance existing PV plants, enabling them to harness clipped power. This solution has been widely researched as an alternative not only to deal with generation–demand mismatch, but also as means for renewables to provide complementary services, such as load shifting [5], global maximum power point tracking [6] and peak-shaving [7]. Note that the standard location for ESSs is, as shown in Figure 1, beside the transformer (before [8] or after [9]).

Sizing the ESS is a fundamental part of designing a tailored solution to handle clipped power. ESS sizing strategies for PV applications have been previously proposed in the literature: in [10,11], sizing strategies to comply maximum power ramp rate regulation were proposed; in [12], a sizing strategy to provide support for household PV applications; in [5], a sizing strategy to balance the peak and off-peak electricity consumption; and, in [13], a sizing strategy for smoothing power output and storing clipped power at PV plant level. This latter sizing strategy consists on averaging the PV power beyond a certain power limit (clipping level), hence hiding power dynamics to the sizing process. Additionally, the analysis is based on a single sunny day. It must be noted that the power limitation is imposed by contract with the grid operator. In addition, this sizing strategy aims at providing a concentrated solution for a full PV plant, where energy storage is connected at the point of common coupling and inverter ratings are not a limitation for the power exceeding PV plant.

A much wider variety of ESS sizing can be found in the literature related to wind power applications. In [14], a sizing strategy to maximize service-hours per BESS unit cost is presented. The strategy forecasts power generation based in long term historic data and statistical noise; this prediction is then low pass filtered, allowing to obtain an ESS power reference curve, which is later processed by a cost function obtaining the ESS energy rating. Nevertheless, low pass filtering generates phase delay depending on frequency, consequently reshaping the power curve and leading to over or under sizing of the ESS. A sizing strategy to minimize penalties caused by not complying day-ahead power bidding is presented in [15]. The strategy generates 25 initial ESS power references by subtracting bid power from 43-hour-power generation forecasts. The initial references are then presented in a histogram, together with a compliance level, which can be used to generate the ESS sizing. The power generation forecast and bidding strategy are not described in the paper. Moreover, this method considers a 43-hour horizon, which is not ideal for PV systems' daily cycles. A hybrid ESS sizing strategy to comply with maximum power ramp rate regulation is presented in [16]. For this purpose, wind forecast and uncertain load behavior are subtracted, generating a power reference which is later transformed into frequency-domain by Discrete Fourier Transform (DFT). The result is later separated into low, medium and high frequencies, corresponding to the desired power output, the power reference for BESS and the power reference for Supercapacitors, respectively. However, DFT strategy decomposes the full signal into periodic sinusoids losing information regarding the time location of frequencies, therefore leading to a wrong sizing of the ESS. Another strategy to size a hybrid ESS while complying with maximum power ramp rate is proposed in [9]. Here, several historical datasets are filtered by wavelet discrete transform, generating a maximum power ramp rate compliant power curve. ESS power reference curve is obtained by averaging the differences between all original curves and their filtered version. The result is later filtered selecting high and low frequencies as supercapacitors and BESS power references, respectively. The strategy relies on averaging the results, hence masking some dynamic behaviors.

This document presents an analysis of the annual power generated by a PV plant. An analytical model was applied to estimate annual clipping losses. An ESS sizing strategy, based in historic data, was proposed; this strategy considers efficiency of the technology (energy storage technology and power electronics) and provides ESS energy and power sizing, required to recover a certain percentage of the annual clipped power. Additionally, configuration and control strategies were proposed to retrofit an existing PV plant, in order to handle clipped power without modifying the existing MPPT strategy. To validate, at power converter level, the technical feasibility of performing the clipping energy storage service, real PV system and power converters models including control strategies were simulated. The simulations shows specifically that existing central inverter based PV plants can be retrofitted to perform this service (without modifications to the central inverter topology and control). Moreover, the study provides an insight into the daily and seasonal behavior of PV power generation, hence suggesting the advantage of additional usage of ESS, as ancillary services, during idle hours.

To the best knowledge of the authors, the estimation of clipped power, the ESS sizing strategy, the proposed ESS configuration enabling fully usage of a Battery ESS (BESS) and the proposal of a control strategy to harness such power, are novel.

The document is arranged as follows: Section 2 presents a brief description of problem. Section 3 describes the PV model applied to estimate available MPP and a comparison between predicted power and empiric power measurements. Section 4 section presents the ESS sizing strategy. The selection of an Energy Storing Technologies (ESTs), capable of handling clipped power, is presented in Section 5. The control strategy, configuration and simulation of the ESS connected to the PV plant is presented in Section 6.

2. Problem Description

To emphasize the consequences of oversizing the PV array and highlight the effects of clipping, data from a PV plant located in UK with 39% PV array oversizing is presented in Figure 2. This PV plant has an empiric annual plant factor of 15.43%; if oversizing was neglected a yearly plant factor of 11.11% would have been obtained instead. A year of PV power generation from a central inverter grid-tied PV plant is shown in Figure 2a, where the installed capacity of the PV array is 2 MW, while the DC rating of inverter is 1.54 MW. Power measurements were taken at the DC side of the inverter once per minute during a full year, from 1 October 2016 to 30 September 2017. These measurements have been normalized with respect to the inverter rating. The surface presented in Figure 2b (lateral view of Figure 2a) is equivalent to overlapping all daily DC power generation curves, showing the power limitation, hereafter power clipping, caused by the power-oversized PV array reaching the power rating of the inverter. Seasonal behavior is shown in Figure 2c,d, where yearlong dawn, dusk and daily maximum power generation are presented. These characteristics were exploited to propose alternatives to obtain further usage and revenue from the ESS. Autumnal (Autumn) Equinox, Summer Solstice,

Vernal (Spring) Equinox and Winter Solstice are identified in Figure 2c,d with the abbreviations AE, SS, VE and WS, respectively.

Irradiance, temperature and DC power measurements are the only data available for the PV plant located in UK. Since the power limitation applied to the system is imposed by the converter rating, power losses caused by clipping are neither estimated nor accounted for. Estimating those losses would result in a mandatory effort to assess clipping effects.

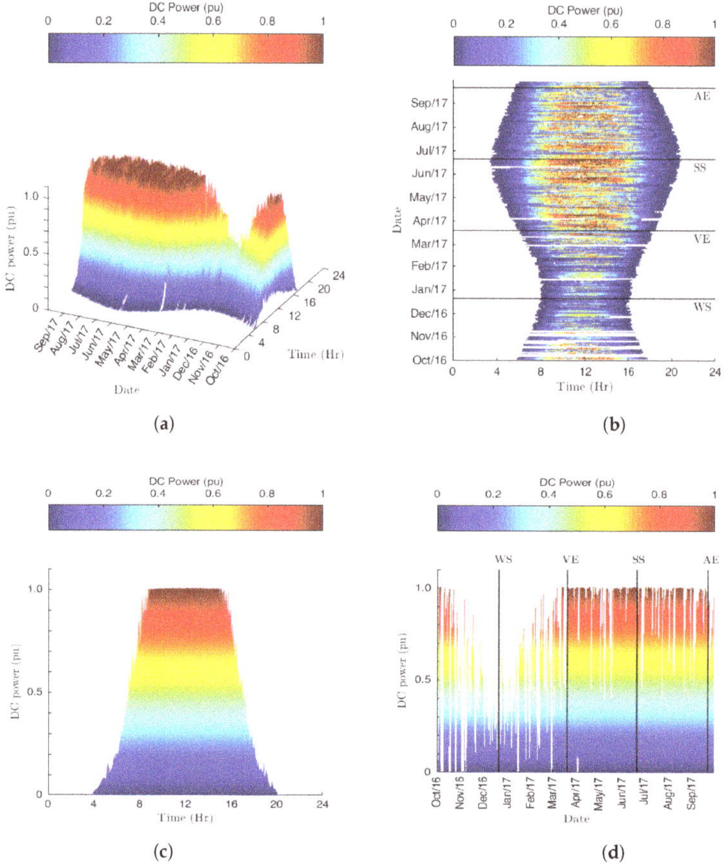

Figure 2. PV plant power generation at the DC-side of a central inverter configuration, annual measurements taken every minute (1 October 2016 to 30 September 2017): (**a**) per day and per minute power generation; (**b**) daily power generation and dawn to dusk daily daylight (top view of Figure 2a); (**c**) overlapping of daily power generation (lateral view of Figure 2a); and (**d**) daily maximum generated power (lateral view of Figure 2a).

Proposed Sizing Strategy

The following steps correspond to the ESS sizing strategy applied to enable harnessing the PV plant clipped power.

1. PV modeling selection: Select a model to estimate clipped power. For this purpose, an analytical model depending on irradiance, module temperature and PV module parameters was chosen. The model predicts the MPP, which is later limited (clipped) at the rating of the converter (clipping

level) and compared to the DC power measurements from the PV plant. Four error metrics and additional features were considered to show the accuracy of the model.

2. Clipped power estimation: Subtract the measured DC power from the selected model predicted DC power (not clipped) to estimate the clipped energy resulting from clipping the power curve. These calculations generate annual clipped power and energy curves with one minute spanning.

3. ESS sizing: Analyze annual clipped power and energy curves. For this purpose, a statistical analysis of daily clipped power and power-limited daily clipped energy is presented. Power and energy sizing are performed by setting a recovery ratio of the annual clipped energy; from here, ESS energy and power ratings are obtained.

3. PV Plant Model

There are several modeling techniques that can be used to estimate maximum power generation from a PV plant, such as single diode circuital model [17], double diode circuital model [18], artificial neural networks [19] and analytic model [20]. An analytical modeling method was considered to estimate power loss due to clipping in a grid-tied PV plant.

To validate the model, one year of data (Figure 2) were compared to the power predicted by the model. Clipped DC power in kW, irradiance in kW/m^2, module temperature in $^\circ$K measurements and PV module data sheet parameters were available. Uniform irradiance and temperature conditions among PV modules were considered to estimate the output power. A description of the analytical modeling method and a quantitative comparison, with the empiric data, is given below.

3.1. Analytical Model

This mathematical model predicts the MPP (P_{mpp} in W) as a function of the irradiance (G in W/m^2) and the module temperature (T in $^\circ$K).

$$P_{mpp} = \left(\left[\frac{k_p}{100} \cdot \Delta T + 1 \right] \cdot G \cdot A \cdot \eta \right) \cdot \eta_{mppt} \cdot N_{ms} \cdot N_{sp} \tag{1}$$

where k_p, ΔT, A, η, η_{mppt}, N_{ms} and N_{sp} are, respectively, the temperature coefficient of P_{mpp} in %/$^\circ$K [21] (or maximum power correction factor for temperature [20]), module temperature difference between the module temperature T and the STC module temperature T_{stc} in $^\circ$K ($\Delta T = T - T_{stc}$), area covered by PV cells in m^2, STC module efficiency, MPPT efficiency [22], number of modules in series in each string and number of strings in parallel. A similar alternative is presented in [20] where $A \cdot \eta$ is replaced by $P_{mpp\ stc}/G_{stc}$.

3.2. Model Validation

The analytical model used to predict the clipped power was chosen, since it presents a low modeling error below clipping, presents a low model error compared data sheet stated STC, presents a low computational cost, does not require an optimization stage (neither parameter identification nor training) and is conceptually simple. The model prediction was clipped at the DC power rating of the converter to match the maximum DC power level (clipping level), and then the error between the clipped prediction and the measured DC power was calculated. The technical details applied to model the PV plant are presented in Table 1. PV modules correspond to the Jinko model JKM260-PP.

Figure 3 shows the irradiance, module temperature, DC power model predictions (clipped) and DC power measurements on four different days (18 March 2017, 3 June 2017, 10 August 2017 and 20 September 2017). The top plots in Figure 3a–d show the daily irradiance and temperature measurements, while the lower plots show the DC power (clipped) and measured DC power. The analytical model display an adequate tracking of the measured DC power; to present a complete analysis, some error metrics and other characteristics of the models were considered.

Table 1. PV plant and PV modules parameters.

Symbol	Parameter	Value
	PV plant	
$P_{\text{pv mpp stc}}$	PV array rated power at STC	2 MW
P_{clip}	Clipping level (inverter rated power)	1.54 MW
N_{ms}	Number of modules in series	20
N_{sp}	Number of strings in parallel	386
A_{pv}	Area of the PV plant	4 Ha
	PV module at STC	
$P_{\text{mpp stc}}$	Maximum power point	260 W
$V_{\text{mpp stc}}$	MPP voltage	31.1 V
$i_{\text{mpp stc}}$	MPP current	8.37 A
$V_{\text{oc stc}}$	Open circuit voltage	38.1 V
$i_{\text{sc stc}}$	Short circuit current	8.98 A
G_{stc}	Irradiance	1000 W/m^2
T_{stc}	Temperature	298.15 °K
k_{i}	Temperature coefficient of i_{sc}	0.06%/°K
k_{p}	Temperature coefficient of P_{mpp}	−0.40%/°K
k_{v}	Temperature coefficient of v_{oc}	−0.30%/°K
A	Area	1.6368 m^2
η	STC efficiency	15.58%
η_{mppt}	MPPT efficiency	98%

The error metrics applied to validate the PV plant model were Normalized Root Mean Squared Error ($NRMSE$), Normalized Mean Absolute Error ($NMAE$), Pearson linear correlation factor ($Pearson$) and Normalized Root Mean Squared Error Fitness ($NRMSEF$). For the first two error metrics, $NRMSE$ and $NMAE$, the optimal value is 0%, while, in the second pair of error metrics, $Pearson$ and $NRMSEF$, the optimum is 100%. Equations (2)–(5) correspond to the mathematical description of the metrics applied to analyze the error between the DC power measurement (X_i) and the clipped DC output power predicted by each model ($\tilde{X}_i$). Variables μ_X, $\mu_{\tilde{X}}$, σ_X and $\sigma_{\tilde{X}}$ in Equation (4) correspond, respectively, to the mean of X_i and $\tilde{X}_i$, and the standard deviation of X_i and $\tilde{X}_i$. N corresponds to the number of samples. Normalized metrics were measured respect to the clipping power level (P_{clip}).

$$NRMSE = \frac{\sqrt{\frac{1}{N} \cdot \sum_{i=1}^{N}(X_i - \tilde{X}_i)^2}}{P_{\text{clip}}} \cdot 100 \tag{2}$$

$$NMAE = \frac{\frac{1}{N} \cdot \sum_{i=1}^{N}|X_i - \tilde{X}_i|}{P_{\text{clip}}} \cdot 100 \tag{3}$$

$$Pearson = \frac{\sum_{i=1}^{N}\left[\left(\frac{X_i - \mu_X}{\sigma_X}\right) \cdot \left(\frac{\tilde{X}_i - \mu_{\tilde{X}}}{\sigma_{\tilde{X}}}\right)\right]}{N-1} \cdot 100 \tag{4}$$

$$NRMSEF = \left(1 - \frac{||X - \tilde{X}||}{||X - \mu_X||}\right) \cdot 100 \tag{5}$$

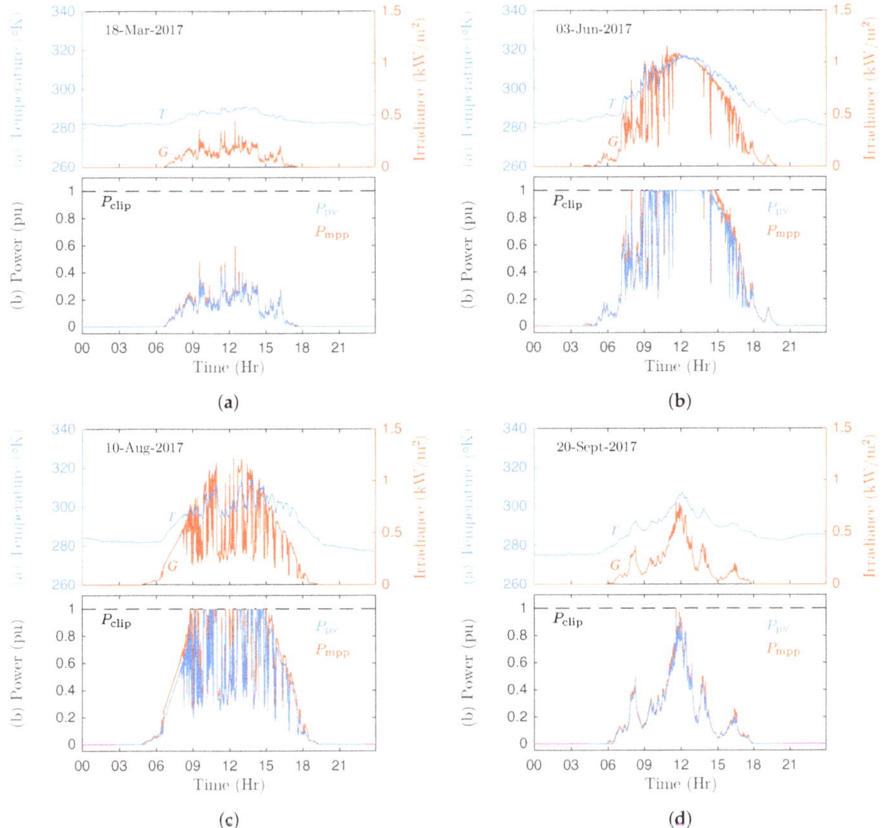

Figure 3. Daily analysis of measured DC power versus models, top plot module temperature and irradiance daily measurements, bottom plot clipping level (P_{clip}), measured DC power (P_{pv}) and analytical model estimated power (P_{mpp}): (**a**) 18 March 2017; (**b**) 3 June 2017; (**c**) 10 August 2017; and (**d**) 20 September 2017.

Table 2 summarizes the error metrics applied to the model. The analytical model presents a low calculation time, a simple approach and small errors metrics. Therefore, this method was chosen to predict PV clipped power. It is also useful for predicting the clipped power in real time.

Aging and soiling effects were not considered in the previously described analytical model, nonetheless the model enables a straight forward update by fitting the STC efficiency term (η) in Equation (1).

Table 2. Modeling error metrics.

Parameters	Analytical Model
NRMSE	4.64%
NMAE	1.85%
Pearson	98.91%
NRMSEF	81.95%
Percentage Error at STC	−1.97%
Execution time	1.98 µs
Optimization stage	No
Conceptual complexity	Low

4. ESS Sizing

ESS are composed by a bidirectional power converter and an EST, such as batteries, super capacitors, flywheels, CAES, HPS, among others [23]. Energy restrictions are imposed by EST, while charging and discharging power ratings depend on both EST and the bidirectional converter. Some ESTs, such as batteries, usually present different charge vs discharge power ratings. In this application, ESS charge power rating is limited by the clipped power.

PV power generation is strongly dependant on solar irradiance and the module temperature. The PV module temperature behaves as a low pass filter of the incident irradiance, with an equivalent time constant of a few minutes depending on the wind speed [24]. In addition, according to Vernica et al. [25], the PV plant output power can be modeled as a low-pass filtered version of the solar irradiance, where the cut-off frequency of the equivalent filter is determined by the area of the PV plant. Specifically, for the PV plant being analyzed in this work, the cut-off frequency of the equivalent filter is 0.01 Hz. Additionally, most grid code requirements related to power fluctuation (maximum power Ramp Rate) regulate power fluctuations per minute [26]. Moreover, standard PV plant available data range in sampling times between 1 and 30 min. Therefore, a data sampling rate of 1 min was selected to calculate the clipped energy and the ESS sizing strategy presented in this work.

Clipped power (P_{clipped}) is calculated according to Equation (6), where P_{mpp} and P_{pv} are, respectively, the predicted PV power generation from Equation (1) and measured PV power, both in kW. The clipped energy in kWh is estimated according to Equation (7), where dt corresponds to the sampling time in min (1 min).

$$P_{\text{clipped}} = \begin{cases} P_{\text{mpp}} - P_{\text{pv}} \quad, & P_{\text{pv}} > P_{\text{clip}} \\ \\ 0 \quad\quad\quad\quad, & \text{otherwise} \end{cases} \tag{6}$$

$$E_{\text{clipped}} = P_{\text{clipped}} \cdot \frac{dt}{60} \tag{7}$$

ESS for clipping in PV plants are designed for a daily use cycle (dawn to dusk), which means there is no need to store energy for more than one day, and therefore the stored energy is completely depleted before a new day cycle.

The following efficiencies were considered in the sizing of the ESS: a single stage DC-DC converter with 97% efficiency [27] (94.09% round-trip efficiency), new commercially available batteries present a round-trip efficiency of 95% [28,29] and DC/AC inverter has a nominal efficiency of 97%. Therefore, the energy passing through the ESS (PV to ESS and ESS to Grid) would experience an efficiency of 86.70%.

Figure 4a shows a histogram of the maximum recoverable daily-energy-loss ($\hat{E}_{\text{del}}$) due to clipping per amount of days of occurrence during a year, and two overlapped curves showing the accumulated annual energy loss (due to clipping) and the total recoverable annual energy (including efficiency of the PV-ESS-grid system of 86.7%). Both curves are function of the maximum recoverable daily-energy-loss and were normalized respect to the annual energy loss (33 MWh).

The power rating analysis is shown in Figure 4b, for the case where four ESS power rating design criteria are depicted as a function of the recoverable-daily-energy-loss. All criteria include a 97% efficiency of the DC/DC power converter. The criteria C_1 to C_4 correspond respectively to maximum recoverable-daily-energy-loss in MWh ($\hat{E}_{\text{del}}$), average recoverable daily-energy-loss in MWh ($\overline{E}_{\text{del}}$), mean plus standard deviation of recoverable daily-energy-loss in MWh ($\overline{E}_{\text{del}} + \sigma_{\text{Edel}}$) and mean plus two times the standard deviation of recoverable daily-energy-loss in MWh ($\overline{E}_{\text{del}} + 2 \cdot \sigma_{\text{Edel}}$).

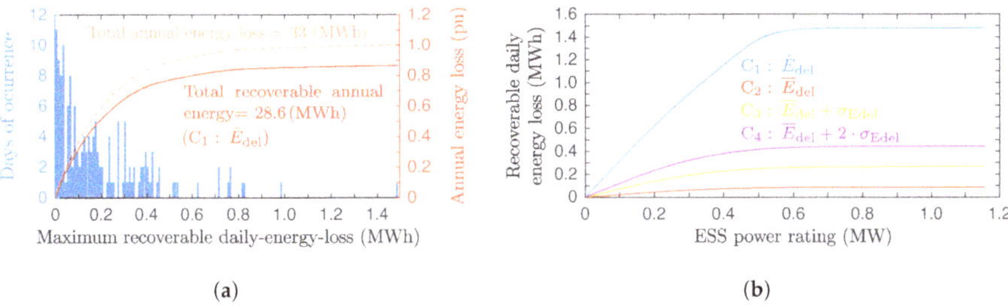

(a)   (b)

Figure 4. PV plant ESS energy and power sizing analysis: (**a**) daily-energy-loss histogram, accumulated annual energy loss and total recoverable annual energy (considering the efficiency of the PV-ESS-grid system); and (**b**) power-limited recoverable daily energy considering the efficiency of the DC/DC power converter.

As an example, to recover 80% of the annual energy lost due to clipping, an ESS of 600 kWh is required to store the maximum daily-energy-loss of (Figure 4a), which considering criteria C_1 (from Figure 4b) leads to a power rating of 200 kW.

5. Energy Storage Technology Selection

ESS are formed by an EST and its power converter. This section focuses on selecting a suitable EST alternative to store clipped energy. According to the type of energy conversion and the nature of the stored energy, ESTs may be classified as electric, chemical, mechanical and thermal [30], as shown in Figure 5. The main characteristics of ESTs, relevant for clipping, are summarized in Table 3 [31–33]. For a detailws description of each ESTs from Figure 5, please refer to [33,34].

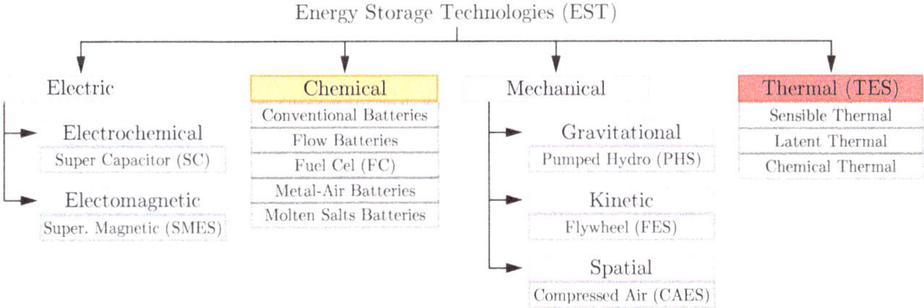

Figure 5. Energy storage technologies classification.

Based on round-trip efficiency and maturity level, the best EST alternatives for handling clipped power are SC, LiIon batteries and FES. Nonetheless, energy cost of FES doubles the energy cost of SC and LiIon batteries. Moreover, installation costs are not included in the table, although they depend on the weight of the equipment that needs to be transported. In addition, considering energy density, we have have concluded that LiIon batteries are an adequate EST technology to be applied to handle clipped power. The following section shows a simulation of a LiIon BESS applied to a PV system, for validation of the sizing methodology.

Table 3. Energy storage technologies specifications [31–33].

Energy Storage Technology	Power Cost (USD/kW)	Energy Cost (USD/kWh)	Round-Trip Efficiency (%)	Lifetime (Years)	Energy Density (Wh/kg)	Maturity
SC	100–300	300–2000	84–98	5–30	0.05–15	Developing
SMES	200–350	1000–10^4	85–98	15–30	0.5–5	Demo
LiIon [1]	1200–4000	400–2500	75–97	5–15	120–230	Commercial
PbA [1]	175–600	150–400	63–90	5–15	30–50	Mature
NiCd [1]	500–1500	600–2400	60–75	10–20	15–55	Mature
VRB [2]	600–3700	150–1000	60–90	5–20	25–35	Developing
ZBB [2]	700–2500	100–1000	60–85	5–10	65–75	Developing
Hydrogen [3]	400–2000	1–15	20–66	5–15	600–1200	Developing
Metal-Air	100–250	10–160	50–65	>1	1000–1300	Demo
NaS [4]	1000–4000	300–500	75–90	10–15	150–240	Commercial
ZEBRA [3]	150–300	230–345	90	5–15	86–140	Commercial
PHS	500–2000	5–100	65–87	30–60	0.5–1.5	Mature
FES	100–350	1000–5000	85–95	15–20	5–80	Commercial
CAES	400–1800	2–400	41–90	20–60	30–300	Developed
Low Temp. [5]	200–300	20–50	30–50	10–40	100–200	Developing
High Temp. [5]	200–300	30–60	80	5–15	80–250	Demo

[1] Conventional battery; [2] flow battery; [3] fuel cell; [4] molten salts batteries; [5] TES.

6. Simulation Results

This section presents the control strategy, configuration and simulation of the ESS connected to the PV plant. BESS was selected as EST to be emulated during the simulation, though the previously described sizing strategy is valid for any EST. BESS was selected due to its modularity, which allows retrofitting each PV inverter at their DC side, and because BESS combine both energy and power density required for this application.

6.1. Configuration

The original PV system consists on a central inverter grid-tied PV plant, connected to the grid through a standard two level voltage source inverter (2LVSI). The addition of the ESS, formed by a battery pack and a single-stage DC/DC converter as in [35], merged to the DC-link of the PV system, enables clipped energy to be stored and released according to the system requirements and limitations. Figure 6 shows a simplified version of the full configuration; colored arrows illustrate the possible power paths. The grey arrow corresponds to a continuous power flow (from PV to the grid), while the green (PV to ESS) and purple (ESS to the grid) arrows correspond to excluding and non necessarily continuous power flows.

This configuration has a single stage DC/DC converter. However, as shown in Figure 6, the system can be implemented with a seconds DC/DC stage (marked as Optional). The difference between both alternatives lies in the percentage utilization of the capacity of the battery pack. A single DC/DC converter performing a standard battery charging strategy applies constant current (CC) charging mode until reaching a certain State of Charge (SoC), usually around 85%. The addition of a second DC/DC stage allows the control of the output voltage of the converter, hence enabling to transition into constant voltage (CV) charging mode, and therefore allowing full charging of the battery. Storing clipped power requires withdrawing power from the DC-link, since the MPP tracking voltage is imposed by the inverter, thus clipped-storable power must be controlled through current.

The simulation results are focused on the control strategy that enables performing power extraction and injection to the DC-link, without modifying the existing control strategy of the inverter. This strategy depends on the SoC of the battery pack, which was estimated through standard Coulomb counting method [36]. This estimation is independent of the battery model applied since it relies in the output current of the battery pack, enabling the utilization of an ideal circuital model [37], to emulate the behavior of the EST.

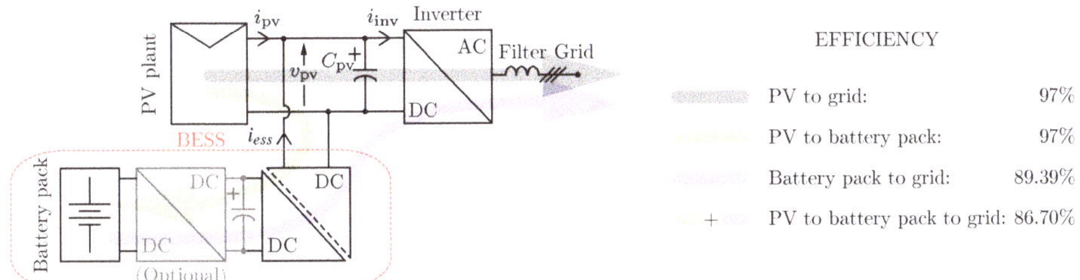

Figure 6. Central inverter grid-tied PV configuration with additional single or double DC/DC stage ESS showing power flow paths and corresponding data sheet efficiency.

6.2. Control Schemes

The central inverter was considered to be controlled through a standard voltage oriented control strategy with MPPT reference, as depicted in the control scheme in Figure 7 [38]. Since the objective is to retrofit an existing PV system, the addition of the ESS is performed at the DC-side of the inverter and does not consider modifying the PV system or its control strategy. It must be noted that the addition of an ESS at the point of common coupling would not allow harnessing the power loss caused by power clipping, since the inverter power limitation would remain an issue, which is not the situation in other similar cases that can benefit from the same ESS sizing strategy, such as power curtailment, where the power limit is imposed by the control strategy and not by the rating of the converter.

Retrofitting forces the ESS control strategy to extract or inject power to the DC link without modifying the existing MPPT strategy of the PV inverter. For this purpose, a tailored ESS control strategy, shown in the flow chart in Figure 8, was designed. P_{pv}, P_{inv}, P_{ess}, P_{mpp}, P_{clip} and P_{pre} correspond, respectively, to PV plant output power ($P_{pv} = i_{pv} \cdot v_{pv}$), inverter input power ($P_{inv} = i_{inv} \cdot v_{pv}$), ESS power ($P_{ess} = i_{ess} \cdot v_{pv}$), estimated MPP according to Equation (1), clipping level, and pre-clipping level. This latter parameter is used as margin to perform power injection or subtraction to/from the DC link ($P_{pre} - P_{mpp}$). $\check{P}_{ess}$ (< 0) and $\hat{P}_{ess}$ (> 0) correspond to the minimum and maximum ESS power. Cases I–V match those in Figure 9. This control strategy can be adapted to handle any EST. For this purpose, estimation of the ESS SoC value (SoC) in the control scheme of Figure 8 must be replaced by an estimation of the available energy in the applied EST.

The ESS output current reference (i_{ess}^*) is provided to a standard PI controller, generating the modulation index and switching pattern according to the power flow direction (storing or releasing energy). The SoC estimation was performed through standard Coulomb counting method [39].

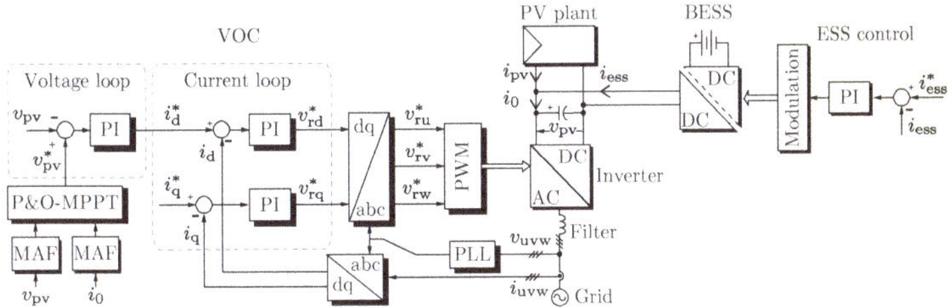

Figure 7. Configuration control schemes: (**left**) Perturb and Observe MPPT and Voltage Oriented Control (VOC) scheme applied to the central inverter; and (**right**) CC charging mode controller scheme applied to the ESS.

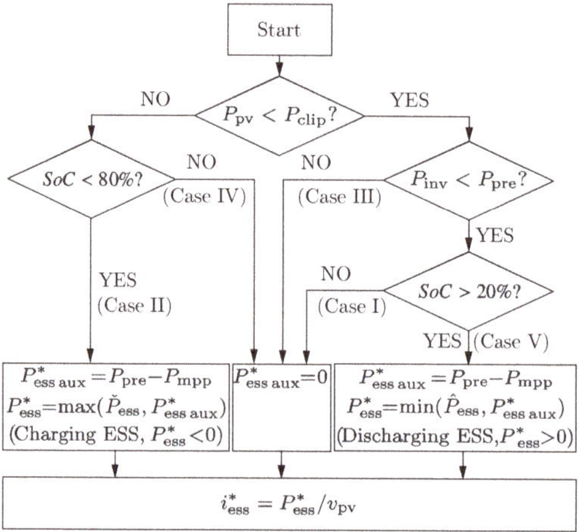

Figure 8. ESS control strategy to enable storing and releasing of PV clipped power.

6.3. Simulation

A simulation of the full system, formed by a 2LVSI operating as central inverter grid-tied PV system, and ESS formed by a Battery ESS (BESS) and an Isolated Bidirectional Boost Converter [35], is shown in Figure 9. The control strategies previously described in Section 6.2 were applied. It must be noted that the ESS was undersized to allow its control system to be tested in all possible scenarios and display those results in a single figure; this was performed considering the dynamics of the systems and an ESS operating range of 20–80% of the SoC. Several irradiance conditions were tested in the simulations, all with PV cell temperature of 298.15 °K.

The PV plant parameters correspond to those presented in Table 1, while Table 4 presents the parameters of the converters and ESS. To calculate per-unit values in Figure 9, the following base values were considered: $G_{base} = 1000$ W/m^2, $P_{base} = 1540$ kW, $i_{base} = 3.767$ kA and $v_{base} = 800$ V.

Table 4. Simulation parameters.

Symbol	Parameter	Value
	Central inverter grid-tied PV system	
C_{pv}	DC-link capacitance	4.4 mF
v_{pv}	DC-link voltage	510–800 V
v_{uvw}	Grid voltage	320 V$_{ll\ rms}$
i_{uvw}	Grid current	2664 A$_{rms}$
f	Grid frequency	50 Hz
f_{sw}	Switching frequency	5 kHz
L_g	Line filter	0.35 mH
	EST and DC/DC converter	
C_{ess}	ESS equivalent capacitance	29 F
v_{ess}	ESS voltage range	120–200 V
L	Inductance	0.1 mH
R	Inductance resistance	10 mΩ
n_t	Turns ratio	1:1
$f_{dc\ sw}$	Switching frequency DC/DC converter	50 kHz

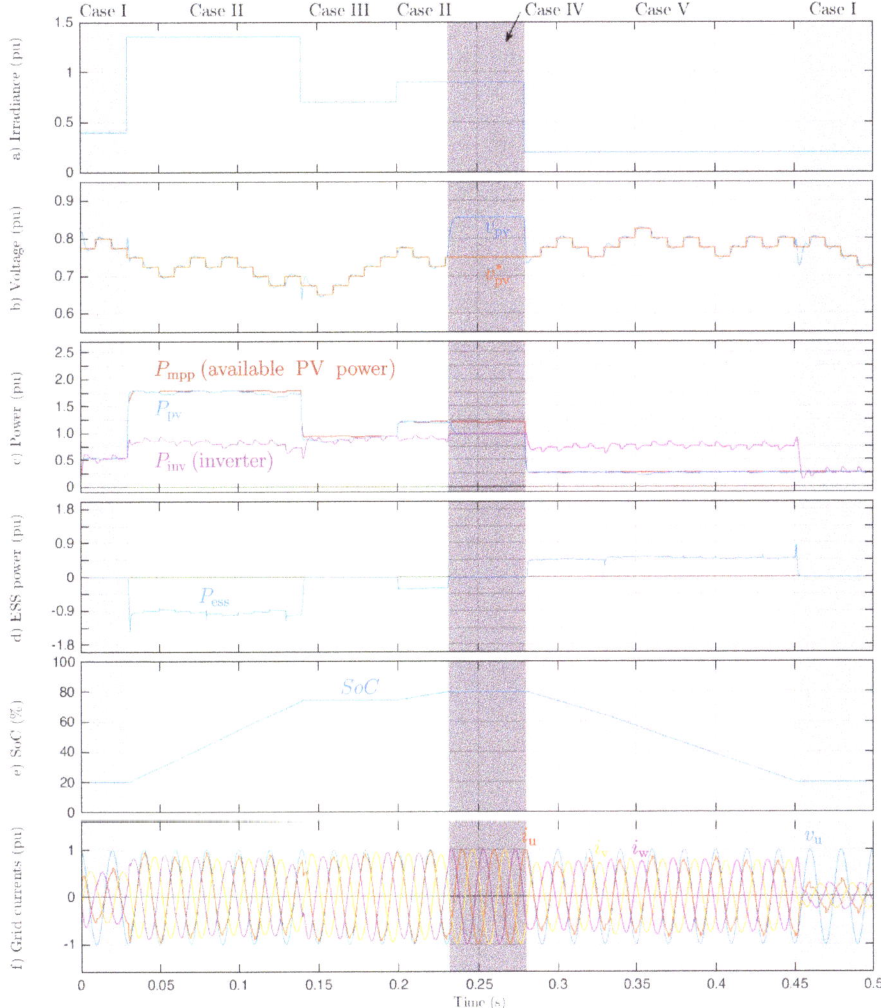

Figure 9. PV plant with ESS connected at the DC-link simulation results: (**a**) irradiance in kW/m^2; (**b**) DC-link voltage and reference; (**c**) generated PV power (P_{pv}), available PV power without considering clipping limitation (P_{mpp} (available PV power)) and inverter power (P_{inv} (inverter)); (**d**) ESS power (P_{ess}); (**e**) ESS SoC (*SoC*); and (**f**) grid currents (i_u, i_v and i_w) and phase u voltage (v_u).

From 0 to 0.03 s (Case I), the system is operating at a power rating below P_{clip} (1540 kW) and the ESS is at a SoC level of 20%, hence ESS power reference is zero. From 0.03 to 0.14 s (Case II), a step in solar irradiance causes the PV power (P_{pv}) to reach P_{clip}, since *SoC* < 80%, P_{ess}^* power reference is set to $P_{ess}^* = P_{pre} - P_{mpp}$ (and $i_{ess}^* = P_{ess}^*/v_{pv}$), which reduces power flowing through the inverter (P_{inv}) enabling MPP tracking. From 0.14 to 0.20 s (Case III), a step down in solar irradiance causes the PV power (P_{pv}) to be lower than P_{clip}, but, since $P_{pre} < P_{pv} < P_{clip}$, the ESS power reference (P_{ess}^*) is set to zero. From 0.2 to 0.25 s (Case II), a second step up in irradiance generates for the PV power to surpass P_{clip}, the system behaves exactly as from 0.03 to 0.14 s. Once the ESS reaches 80% of the SoC, the ESS stops drawing power and the inverter losses momentarily the capability to track the MPP (Case IV from 0.25 to 0.28 s). Note that the ESS was undersized, to show the behavior of the system when reaching its maximum and minimum permitted SoC. From 0.28 to 42 s (Case V), a step down in solar irradiance causes the PV power to be below pre clipping level ($P_{pv} < P_{pre} < P_{clip}$), and the ESS releases power towards the DC-link ($P^*_{ess} = P_{pre} - P_{mpp}$) and through the inverter into the grid. From 0.42 to 0.5 s, the ESS reaches 20% of the SoC, hence power flow from ESS towards the DC-link is stopped and the system goes back to operating in Case I; in this part, grid currents show high harmonic content due the comparison of power flowing to the grid and power required to generate voltage steps to perform MPPT. In a real case, the MPPT period is longer, hence harmonic content would be lower.

Note that the instantaneous power balance is given by Equation (8), where the corresponding terms are given by Equations (9)–(11). This can also be verified in power curves shown in Figure 9c,d. Variables v_{gd}, v_{gq}, i_d and i_q in Equation (9) correspond to the grid voltage and current in rotational coordinates.

$$P_{inv}(t) = P_{pv}(t) + P_{ess}(t) \tag{8}$$

$$P_{inv}(t) = \frac{3}{2} \cdot \left(v_{gd}(t) \cdot i_d(t) + v_{gq}(t) \cdot i_{q(t)} \right) \tag{9}$$

$$P_{pv}(t) = i_{pv}(t) \cdot v_{pv}(t) \tag{10}$$

$$P_{ess}(t) = i_{ess}(t) \cdot v_{pv}(t) \tag{11}$$

7. Conclusions

The work described in this paper focused on the effects clipping, caused by oversizing a PV array with respect to the power rating of its converter. An insight into the benefits and drawbacks in terms of annual plant factor and energy losses of this commercial practice is presented. An analytical model was used to predict the annual power loss due to clipping; the validation of the model was performed by comparing the predicted power, limited at the clipping level, with real data over a time horizon of one year sampled every minute. An ESS sizing strategy based on recovering annual clipping losses is proposed. The strategy uses a power and energy approach, which considers statistical data to select the best fitting ESS ratings. For the analyzed PV plant, the sizing method determined that a ESS of 600 kWh per central inverter enables the retention of 80% of energy that would be lost due to clipping. From a power perspective, each central inverter, rated at 1.4 MW, requires being retrofitted with a 200 kW DC-DC converter for the ESS to enable the aforementioned clipped energy storage. In relation to the EST, LiIon based BESS was selected to perform a validation of the ESS and control system due to comparison of several criteria (LiIon have 10 times higher energy density than FC, about half the cost of FES among other values). Simulation results at power electronics level of the ESS, the central inverter, the grid connection, and their control show that the selected ESS can be retrofitted to existing central inverters, and provide clipping energy storage while still performing properly (perform MPPT, store/deliver energy, retain a controlled DC link and inject energy to the grid with high power quality).

The proposed methodology and analysis can be applied to determine, depending on their plant measurements and parameters, the size of a ESS to retrofit their PV plant for clipping energy storage. This information is necessary to perform an economic assessment, and how it can impact the levelized cost of energy (LCOE), and assist in the decision-making process. In addition, the sizing methodology can be adapted to perform new analysis on other ESS applications such as: load shifting, power curtailment, frequency and voltage regulation, base load generation, capacity firming, etc., providing further value for the ESS to the retrofitting of a PV plant. Furthermore, provided the proper model and data, it can also be extended to other renewable energy sources, such as wind energy and ocean energy.

Author Contributions: N.M. conceived, designed and applied the sizing methodology, and wrote the original draft; S.K. provided insight, resources, and supervision and reviewed the manuscript; P.Z. provided insight, resources, supervision and analysis; P.W. provided relevant key theoretical and technical suggestions and reviewed the manuscript; G.B contributed with mathematical modeling tools and insight; and F.G. provided experimental data, problem conceptualization and practical insight.

Funding: The authors gratefully acknowledge the financial support provided by FONDECYT 1191532, AC3E (CONICYT/BASAL/FB0008), SERC Chile (CONICYT/FONDAP/15110019), CONICYT-PCHA/Doctorado Nacional/2014-21141092, and PIIC of UTFSM.

Conflicts of Interest: The authors declare no conflict of interest.

Abbreviations

The following abbreviations are used in this manuscript:

AC	Alternating Current
ANN	Artificial Neural Network
BESS	Battery Energy Storage System
CAES	Compressed Air Energy Storage
CC	Constant Current
CV	Constant Voltage
DC	Direct Current
DFT	Discrete Fourier Transform
ESS	Energy Storage System
EST	Energy Storage Technology
FES	Flywheel Energy Storage
ILR	Inverter Loading Ratio
LiIon	Lithium Ion Battery
MAF	Moving Average Filter
MPP	Maximum Power Point
MPPT	Maximum Power Point Tracking
NaS	Sodium-Sulphur Battery
NiCd	Nickel Cadmium Battery
NOCT	Normal Operating Cell Temperature
PbA	Lead Acid Battery
PHS	Pumped Hydro Energy Storage
P&O	Perturb and Observe
PV	Photovoltaic
SC	Super Capacitor
SMES	Super Conducting Magnetic Energy Storage
SoC	State of Charge
STC	Standard Test Condition
TES	Thermal Energy Storage
UK	United Kingdom
VOC	Voltage Oriented Control
VRB	Vanadium Redox Battery
ZBB	Zinc Bromine Battery
2LVSI	Two Level Voltage Source Inverter

References

1. Hales, D. REN21. Renewables 2018-Global Status Report. Available online: http://www.ren21.net/gsr-2018/ (accessed on 2 May 2019).
2. BEIS. *Digest of United Kingdom Energy Statistics 2018*; Department for Business, Energy & Industrial Strategy: London, UK, 2018; p. 260,
3. Serdio Fernández, F.; Muñoz-García, M.A.; Saminger-Platz, S. Detecting clipping in photovoltaic solar plants using fuzzy systems on the feature space. *Sol. Energy* **2016**, *132*, 345–356. [CrossRef]

4. Good, J.; Johnson, J.X. Impact of inverter loading ratio on solar photovoltaic system performance. *Appl. Energy* **2016**, *177*, 475–486. [CrossRef]

5. Ke, B.R.; Ku, T.T.; Ke, Y.L.; Chung, C.Y.; Chen, H.Z. Sizing the Battery Energy Storage System on a University CampusWith Prediction of Load and Photovoltaic Generation. *IEEE Trans. Ind. Appl.* **2016**, *52*, 1136–1147. [CrossRef]

6. Muller, N.; Kouro, S.; Renaudineau, H.; Wheeler, P. Energy storage system for global maximum power point tracking on central inverter PV plants. In Proceedings of the 2016 IEEE 2nd Annual Southern Power Electronics Conference (SPEC 2016), Auckland, New Zealand, 5–8 December 2016.

7. Dong, J.; Gao, F.; Guan, X.; Zhai, Q.; Wu, J. Storage Sizing With Peak-Shaving Policy for Wind Farm Based on Cyclic Markov Chain Model. *IEEE Trans. Sustain. Energy* **2017**, *8*, 978–989. [CrossRef]

8. Luo, F.; Meng, K.; Dong, Z.Y.; Zheng, Y.; Chen, Y.; Wong, K.P. Coordinated operational planning for wind farm with battery energy storage system. *IEEE Trans. Sustain. Energy* **2015**, *6*, 253–262. [CrossRef]

9. Jiang, Q.; Hong, H. Wavelet-Based Capacity Configuration and Coordinated Control of Hybrid Energy Storage System for Smoothing Out Wind Power Fluctuations. *IEEE Trans. Power Syst.* **2013**, *28*, 1363–1372. [CrossRef]

10. Zhao, Q.; Wu, K.; Khambadkone, A.M. Optimal sizing of energy storage for PV power ramp rate regulation. In Proceedings of the 2016 IEEE Energy Conversion Congress and Exposition (ECCE), Milwaukee, WI, USA, 18–22 September 2016; pp. 1–6. [CrossRef]

11. Galtieri, J.; Krein, P. Solar Variability Reduction Using Off-Maximum Power Tracking and Battery Storage. In Proceedings of the 2017 IEEE 44th Photovoltaic Specialist Conference (PVSC), Washington, DC, USA, 25–30 June 2017; pp. 1–4. [CrossRef]

12. Aichhorn, A.; Greenleaf, M.; Li, H.; Zheng, J. A cost effective battery sizing strategy based on a detailed battery lifetime model and an economic energy management strategy. In Proceedings of the 2012 IEEE Power and Energy Society General Meeting, San Diego, CA, USA, 22–26 July 2012; pp. 1–8. [CrossRef]

13. Rallabandi, V.; Akeyo, O.M.; Jewell, N.; Ionel, D.M. Incorporating battery energy storage systems into multi-MW grid connected PV systems. *IEEE Trans. Ind. Appl.* **2019**, *55*, 638–647. [CrossRef]

14. Li, Q.; Choi, S.S.; Yuan, Y.; Yao, D.L. On the determination of battery energy storage capacity and short-term power dispatch of a wind farm. *IEEE Trans. Sustain. Energy* **2011**, *2*, 148–158. [CrossRef]

15. Pinson, P.; Papaefthymiou, G.; Klöckl, B.; Verboomen, J. Dynamic sizing of energy storage for hedging wind power forecast uncertainty. In Proceedings of the 2009 IEEE Power & Energy Society General Meeting, Calgary, AB, Canada, 26–30 July 2009; pp. 1–8. [CrossRef]

16. Makarov, Y.V.; Du, P.; Kintner-Meyer, M.C.W.; Jin, C.; Illian, H.F. Sizing energy storage to accommodate high penetration of variable energy resources. *IEEE Trans. Sustain. Energy* **2012**, *3*, 34–40. [CrossRef]

17. Ding, K.; Bian, X.; Liu, H.; Peng, T. A MATLAB-simulink-based PV module model and its application under conditions of nonuniform irradiance. *IEEE Trans. Energy Convers.* **2012**, *27*, 864–872. [CrossRef]

18. Masmoudi, F.; Ben Salem, F.; Derbel, N. Single and double diode models for conventional mono-crystalline solar cell with extraction of internal parameters. In Proceedings of the 13th International Multi-Conference on Systems, Signals & Devices (SSD 2016), Leipzig, Germany, 21–24 March 2016, pp. 720–728. [CrossRef]

19. Lopez-Guede, J.M.; Ramos-Hernanz, J.A.; Zulueta, E.; Fernadez-Gamiz, U.; Oterino, F. Systematic modeling of photovoltaic modules based on artificial neural networks. *Int. J. Hydrogen Energy* **2016**, *41*, 12672–12687. [CrossRef]

20. Mellit, A.; Saglam, S.; Kalogirou, S. Artificial neural network-based model for estimating the produced power of a photovoltaic module. *Renew. Energy* **2013**, *60*, 71–78. [CrossRef]

21. JinkoSolar. PV Module JKM260PP Datasheet. Available online: https://www.jinkosolar.com/ (accessed on 2 May 2019).

22. Pakkiraiah, B.; Sukumar, G.D. Research Survey on Various MPPT Performance Issues to Improve the Solar PV System Efficiency. *J. Sol. Energy* **2016**, *2016*, 1–20. [CrossRef]

23. Díaz-González, F.; Sumper, A.; Gomis-Bellmunt, O.; Villafáfila-Robles, R. A review of energy storage technologies for wind power applications. *Renew. Sustain. Energy Rev.* **2012**, *16*, 2154–2171. [CrossRef]

24. Armstrong, S.; Hurley, W.G. A thermal model for photovoltaic panels under varying atmospheric conditions. *Appl. Therm. Eng.* **2010**, *30*, 1488–1495. [CrossRef]

25. Vernica, I.; Wang, H.; Blaabjerg, F. Impact of Long-Term Mission Profile Sampling Rate on the Reliability Evaluation of Power Electronics in Photovoltaic Applications. In Proceedings of the 2018 IEEE Energy Conversion Congress and Exposition (ECCE 2018), Portland, OR, USA, 23–27 September 2018; pp. 4078–4085. [CrossRef]

26. Martins, J.; Spataru, S.; Sera, D.; Stroe, D.I.; Lashab, A. Comparative Study of Ramp-Rate Control Algorithms for PV with Energy Storage Systems. *Energies* **2019**, *12*, 1342. [CrossRef]

27. Tame Power. Smart DC / DC Converter Reversible COMET productline—Liquid Cooled Version. Available online: https://www.tame-power.com/sites/tame-power.com (accessed on 2 May 2019).

28. LG Chem. Resu Data Sheet. Available online: https://www.europe-solarstore.com/storage-and-system-solutions/solar-storage-batteries/lg-chem/lg-chem-resu-10h-400v-lithium-ion-storage-battery.html (accessed on 2 May 2019).

29. BYD. New Energy. Available online: http://www.byd.com/en/NewEnergy.html (accessed on 2 May 2019).

30. Luo, X.; Wang, J.; Dooner, M.; Clarke, J. Overview of current development in electrical energy storage technologies and the application potential in power system operation. *Appl. Energy* **2015**, *137*, 511–536. [CrossRef]

31. Khodadoost Arani, A.A.; B. Gharehpetian, G.; Abedi, M. Review on Energy Storage Systems Control Methods in Microgrids. *Int. J. Electr. Power Energy Syst.* **2019**, *107*, 745–757. [CrossRef]

32. Wong, L.A.; Ramachandaramurthy, V.K.; Taylor, P.; Ekanayake, J.B.; Walker, S.L.; Padmanaban, S. Review on the optimal placement, sizing and control of an energy storage system in the distribution network. *J. Energy Storage* **2019**, *21*, 489–504. [CrossRef]

33. Nadeem, F.; Hussain, S.M.; Tiwari, P.K.; Goswami, A.K.; Ustun, T.S. Comparative review of energy storage systems, their roles, and impacts on future power systems. *IEEE Access* **2019**, *7*, 4555–4585. [CrossRef]

34. Chen, H.; Cong, T.N.; Yang, W.; Tan, C.; Li, Y.; Ding, Y. Progress in electrical energy storage system: A critical review. *Prog. Nat. Sci.* **2009**, *19*, 291–312. [CrossRef]

35. Muller, N.; Renaudineau, H.; Flores-Bahamonde, F.; Kouro, S.; Wheeler, P. Ultracapacitor storage enabled global MPPT for photovoltaic central inverters. In Proceedings of the 2017 IEEE 26th International Symposium on Industrial Electronics (ISIE), Edinburgh, UK, 19–21 June 2017; pp. 1046–1051. [CrossRef]

36. Nayak, A.; Kasturi, K.; Nayak, M.R. Cycle-charging dispatch strategy based performance analysis for standalone PV system with DG & BESS. In Proceedings of the 2018 Technologies for Smart-City Energy Security and Power (ICSESP), Bhubaneswar, India, 28–30 March 2018; pp. 1–5. [CrossRef]

37. Sparacino, A.R.; Reed, G.F.; Kerestes, R.J.; Grainger, B.M.; Smith, Z.T. Survey of battery energy storage systems and modeling techniques. In Proceedings of the 2012 IEEE Power and Energy Society General Meeting, San Diego, CA, USA, 22–26 July 2012; pp. 1–8. [CrossRef]

38. Muller, N.; Kouro, S.; Zanchetta, P.; Wheeler, P. Bidirectional partial power converter interface for energy storage systems to provide peak shaving in grid-tied PV plants. In Proceedings of the 2018 IEEE International Conference on Industrial Technology (ICIT), Lyon, France, 20–22 February 2018.

39. Chang, W.Y. The State of Charge Estimating Methods for Battery: A Review. *ISRN Appl. Math.* **2013**, *2013*, 1–7. [CrossRef]

Article

A Research on Cascaded H-Bridge Module Level Photovoltaic Inverter Based on a Switching Modulation Strategy

Wang Mao [1], Xing Zhang [1,*], Yuhua Hu [1], Tao Zhao [1], Fusheng Wang [1], Fei Li [1] and Renxian Cao [2]

[1] School of Electrical Engineering and Automation, Hefei University of Technology, Hefei 230009, China; hfut_maow@163.com (W.M.); hfut_hyh@163.com (Y.H.); zt_kyyx@163.com (T.Z.); wangfusheng@hfut.edu.cn (F.W.); lifei@hfut.edu.cn (F.L.)

[2] Sungrow Power Supply Co., Ltd., Hefei 230088, China; crx@sungrow.cn

* Correspondence: honglf@ustc.edu.cn; Tel.: +86-136-0560-1932

Received: 28 April 2019; Accepted: 13 May 2019; Published: 15 May 2019

Abstract: The stable operating region of a photovoltaic (PV) cascaded H-bridge (CHB) grid-tied module level inverter is extended by adopting the hybrid modulation strategy. However, the traditional single hybrid modulation method is unable to regulate the DC-side voltage of each module precisely, which may aggravate the fluctuation of modules' DC-side voltages or even cause the deviation of modules' DC-side voltages under some fault conditions and, thus, degrade the energy harvesting of PV panels. To tackle this problem, a switching modulation strategy for PV CHB inverter is proposed in this paper. When the CHB inverter is operating in normal mode, the hybrid modulation strategy containing the zero state is adopted to suppress DC-side voltage fluctuation, thereby, improving the output power of PV modules. When the CHB inverter is operating in fault mode owing to failing solar panels, the hybrid modulation strategy without the zero state is utilized to make the DC-side voltages reach the references and, thus, maintain a higher energy yield under fault conditions. Experimental results achieved by a laboratory prototype of a single-phase eleven-level CHB inverter demonstrate both the feasibility and effectiveness of the proposed method.

Keywords: cascaded H-bridge; photovoltaic inverter; module level; switching modulation strategy; energy yield

---

1. Introduction

With the increasing demand for conversion efficiency during recent years, multilevel converter topologies have become more and more applied to the grid-connected PV generation system [1–4]. Among all kinds of multilevel inverters, the cascaded H-bridge (CHB) has many advantages, such as modularization, simplicity, and high reliability, which makes it become the most attractive topology [5–9]. In addition, the CHB topology makes it possible to, respectively, regulate each DC-side voltage and, thus, realizing the maximum power point tracking (MPPT) of each PV module. Therefore, the CHB inverter is considered to be one of the most suitable candidates for next generation PV inverters [10,11].

The energy loss caused by non-uniform solar radiation, degradation of PV modules and different types of partial shading is greatly decreased by module level MPPT. However, due to the unequal temperature and irradiance of PV modules, the output power of H-bridge unit varies greatly in the case of severe mismatching. This may lead to over-modulation of H-bridge units with high output power, thus, resulting in system instability and injection current distortion [12].

For the sake of expanding the stable operation range of the CHB inverter, some methods have been proposed. The power balance control strategy based on active component modification of duty cycles, and its effective power balance area are proposed in [13–15]. Although the system can operate stably by utilizing this method under slight mismatch conditions, there are still instability problems under severe mismatch conditions. In [16,17], a control strategy of reactive power compensation is proposed, which utilizes the power factor as the degree of freedom to stabilize the operation of the system. However, an increase in reactive power injection will lead to a decrease of the system power factor, which may be undesirable from the perspective of electric dispatchers. A new, improved MPPT method is presented in [18], which changes the working point of the over-modulated H-bridge unit to ensure that all H-bridge units work in the stable operation region to improve the stability margins of the system. However, this will result in lower output power and make the system less-efficient. This is contrary to the original intention that all PV modules operate at a maximum power point to realize higher efficiency in energy harvesting.

Hopefully, In [19–21], the hybrid modulation strategy (HMS) is proposed, which utilizes a mixture of low-frequency PWM and high-frequency PWM methods to regulate the DC-side voltage and control the AC current separately. The HMS has been proved to maintain the stabilization of power rectifiers effectively even under critical operating conditions, because it provides higher DC-side utilization (the maximum modulation index of square wave can reach $4/\pi$) compared with conventional sinusoidal pulse width modulation (SPWM). In [22,23], a control method of a grid-connected PV CHB inverter is proposed, which is based on the hybrid modulation strategy containing the zero state (HMSCZS). With HMSCZS, the system can operate stably under heavy mismatching conditions. Once the CHB inverter is operating in the fault mode, owing to failing solar panels, the HMSCZS fails to regulate the DC-side voltages to the references and may lead to low output power of the inverter. In [24], the hybrid modulation strategy without the zero state (HMSWZS) is proposed to maximize the range of stable operation of system. However, the HMSWZS cannot control the DC-side voltage of each module accurately. This may aggravate the fluctuation of modules' DC-side voltages, thus resulting in the decrease of the generated energy of PV modules.

Therefore, a switching hybrid modulation strategy (SHMS) is proposed in this paper. The HMSCZS is selected to suppress DC-side voltages fluctuation when the CHB inverter is operating in normal mode. Once the CHB inverter is operating in fault mode, owing to failing solar panels, the HMSWZS is utilized to control the DC-side voltages to track the references, thus maintaining a higher energy yield under fault condition. With this method, the average output power of the PV modules will be improved both in the normal and fault modes.

The paper is arranged as follows: In Section 2, the system configuration and control method are introduced. In Section 3, the existing problem of the HMSCZS and the HMSWZS is shown. In Section 4, the switching hybrid modulation strategy is put forward. In Sections 5 and 6, the performance of the proposed strategy is verified by simulation and experimental results. Finally, in Section 7, a conclusion is drawn to summarize the paper.

2. System Configuration

The structure of the single-phase PV CHB module level grid-tied inverter is shown in Figure 1. The CHB module level inverter consists of m H-bridge units. Each H-bridge unit is connected to a PV module. V_{PVi} and I_{PVi} ($I = 1, 2, \ldots, m$) stand for the output voltage and current of PV module of ith H-bridge. I_{Ci} represents the current flowing through the capacitor on the DC-side of ith H-bridge. V_G stands for grid voltage. I_S stands for grid current. V_{Hi} ($I = 1, 2, \ldots, m$) is the ith H-bridge output voltage and V_{HT} is the total output voltage of H-bridges. Figure 2 shows the configuration of the control diagram of CHB module level inverter. The voltage and current double closed-loop control are utilized to obtain the goal of the controlling. In order to decrease third harmonic component in the grid current, a digital 100 Hz notch filter is utilized to eliminate the second order harmonic in the total DC-side voltage $V_{PV1} + V_{PV2} \bullet\bullet\bullet + V_{PVm}$. The external voltage loop is in charge of controlling the filtered total DC-side voltages to the sum of the references $V_{PV1}{}^* + V_{PV2}{}^* \bullet\bullet\bullet + V_{PVm}{}^*$ by a conventional PI controller. The internal current loop is responsible for controlling the grid current to a sinusoidal shape in phase with the grid voltage. In the paper, a direct-quadrature rotating frame control method for the single-phase inverter is used to achieve this goal because it can achieve zero steady-state error by utilizing a traditional PI regulator [25]. The output of the current loop regulator, V_r, serves as the modulation wave. In order to compensate for the harmonic component of the CHB inverter caused by the distorted grid voltage, the third, fifth, and seventh harmonic compensation algorithm is utilized [26,27].

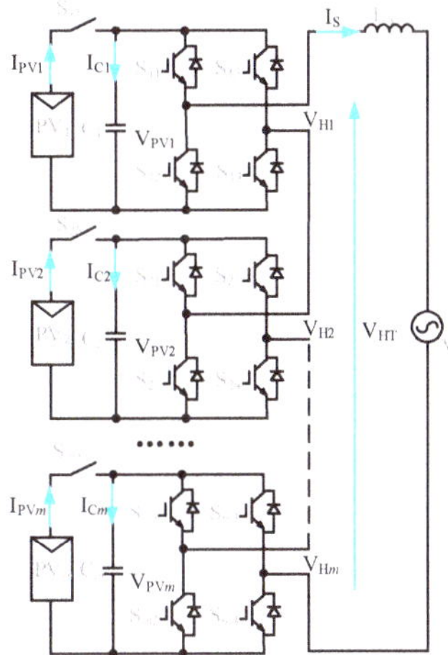

Figure 1. Structure of single-phase PV CHB module level inverter.

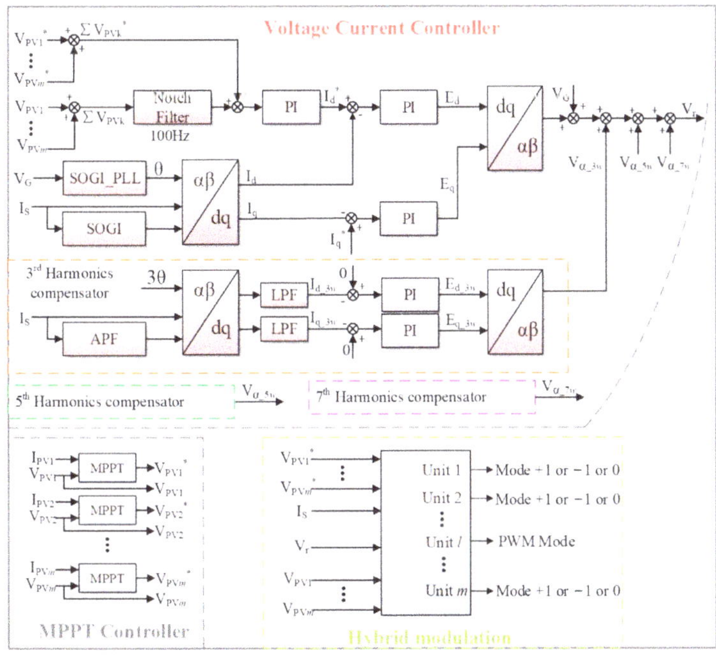

Figure 2. The control diagram of single-phase PV CHB module level grid-tied inverter.

The balancing algorithm allocates suitable switching modes to each unit during every switching period based on the instantaneous value of modulation wave. For HMSCZS, only one unit operates in PWM state, the $K - 1(K = 1, 2, \ldots, m)$ units operate in the "+1" or "−1" state and other units work in the "0" state. However, for HMSWZS, only one unit operates in the PWM state, while the other units work in the "+1" or "−1" mode. Probable switching states of the ith unit, S_i, and their corresponding output voltages are given in Table 1.

Table 1. Switch states of the ith unit.

Mode (S_i)	Unit Output Voltage
+1	$+V_{PVi}$
−1	$-V_{PVi}$
0	0
PWM	PWM Reference

3. Review of The Existing Hybrid Modulation Strategy

3.1. The Hybrid Modulation Strategy Containing the Zero State (HMSCZS)

The HMSCZS proposed in [19,23] maximizes the steady operation range of the system, because it provides higher DC-side utilization by adopting a square wave modulation. The major procedures of HMSCZS can be summarized as follows:

(1) Calculating the voltage error ΔV_{PVi} ($I = 1, 2, \ldots, m$) at the DC-side of each H-bridge unit:

$$\Delta V_{PVi} = V_{PVi} - V_{PVi}^*$$ (1)

(2) Sorting the voltage errors in ascending order ($[\Delta V_{PV1}, \Delta V_{PV2}, \ldots, \Delta V_{PVm}]$) at a fixed frequency, f_{sort}.
(3) Mapping for the filtered DC-side voltages derived from the sorted vector of voltage errors ($[V_1, V_2, \ldots, V_m]$).
(4) Identifying the voltage area l of V_r based on Equation (2):

$$\sum_{i=1}^{l-1} V_i < |V_r| < \sum_{i=1}^{l} V_i$$ (2)

(5) Updating the H-bridge units' operating state. The $l - 1$ ($l = 1, 2, \ldots, m$) units with the higher DC-side voltage are selected to be discharged in state "+1" or "−1" (according to the direction of grid-connected current), the lth unit works in the PWM state, and the rest operate in state "0".

As shown in Figure 3, the switching modes for an eleven-level CHB inverter with the HMSCZS method is presented. For instance, when $\sum_{i=1}^{3} V_i < V_r < \sum_{i=1}^{4} V_i$ and $I_S > 0$, the switching modes of the five H-bridge modules are, respectively, "0", "+PWM", "+1", "+1", and "+1". According to the rules of HMSCZS, the switching mode of each H-bridge unit can only be "+1", "0", or "+PWM" when V_r is positive. Similarly, it can only be "−1", "0", or "−PWM" when V_r is negative. As shown in Figure 4, once the CHB inverter is operating in fault mode owing to failing PV modules, the fault H-bridge unit is always in the discharge state, whether the switching mode is "+1", "+PWM", "−1" or "−PWM". The result is that the DC-side voltages of all H-bridge units diverge from the reference value and the generation of CHB inverter is low.

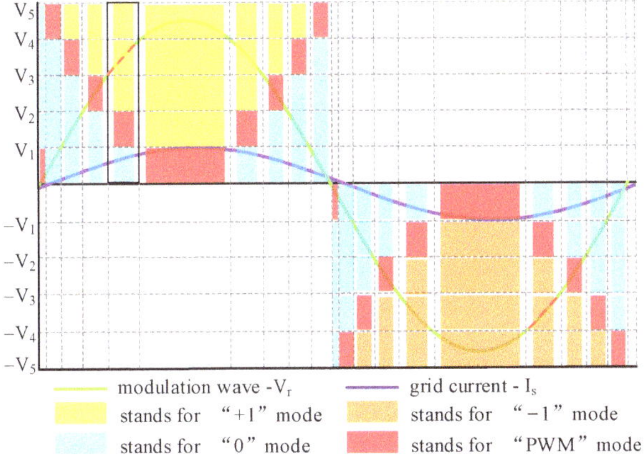

Figure 3. The switching modes for an eleven-level CHB inverter with the HMSCZS method.

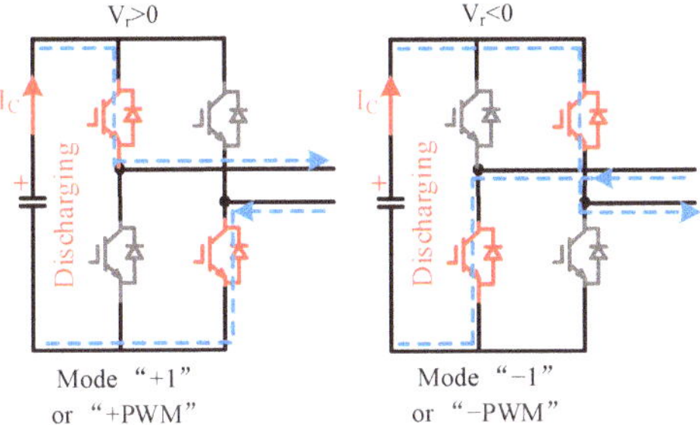

Figure 4. The switching modes of the fault H-bridge unit with the HMSCZS method.

3.2. The Hybrid Modulation Strategy without the Zero State (HMSWZS)

To overcome the shortcoming of the HMSCZS, the HMSWZS is presented in [21,24]. Different from the HMSCZS, in the HMSWZS the units with the lower voltage errors are selected to be charged in state "+1" or "−1" (according to the direction of grid-connected current) and no unit operates in the "0" mode. Figure 5 shows the switching modes for an eleven-level CHB inverter with the HMSWZS method. As depicted in the Figure 5, when $\sum_{i=1}^{1} V_i < V_r < \sum_{i=1}^{2} V_i$ and $I_S > 0$, the switching modes of the five H-bridge modules are, respectively, "−1", "−PWM", "+1", "+1", and "+1". According to the rules of HMSWZS, as shown in Figure 6, the switching mode of the fault H-bridge unit can be "+1", "+PWM", "−1", or "−PWM" regardless of the polarity of V_r. Therefore, compared with the HMSCZS, the HMSWZS is able to maintain the DC-side voltages balance and a higher energy yield under fault condition.

However, as is illustrated in [23], the HMSWZS may aggravate the DC-side voltages fluctuation of H-bridge units and, thus, lead to losses in energy harvesting of PV modules. As shown in Equation (3), ΔU_{ci} (i = 1, 2, ..., m), the fluctuation of the DC-side voltage of the *i*th H-bridge unit during a sorting cycle is composed of two parts: ΔU_{ci1} and ΔU_{ci2}. Based on the superposition theorem of linear circuits, the fluctuation of the DC-side voltage could be regarded as the sum of fluctuations produced by two separate parts:

$$\begin{aligned} \Delta U_{ci} &= \frac{1}{C_i} \int_0^{\frac{1}{f_{sort}}} (I_{PVi} - S_i I_s) dt \\ &= \frac{I_{PVi}}{C_i f_{sort}} - \frac{1}{C_i} \int_0^{\frac{1}{f_{sort}}} S_i I_s dt \\ &= \Delta U_{ci1} + \Delta U_{ci2} \end{aligned} \tag{3}$$

where:

$$\Delta U_{ci1} = \frac{I_{PVi}}{C_i f_{sort}} \tag{4}$$

$$\Delta U_{ci2} = -\frac{1}{C_i} \int_0^{\frac{1}{f_{sort}}} S_i I_s dt \tag{5}$$

If the HMSWZS is utilized, ΔU_{ci} has two possible values: ΔU_{1min} and ΔU_{1max} (if the polarity of S_i and I_s is the same, $\Delta U_{ci} = \Delta U_{1min}$, otherwise, $\Delta U_{ci} = \Delta U_{1max}$), where:

$$\begin{aligned} \Delta U_{1min} &= \Delta U_{ci1} - |\Delta U_{ci2}| \\ \Delta U_{1max} &= \Delta U_{ci1} + |\Delta U_{ci2}| \end{aligned} \tag{6}$$

If the HMSCZS is employed, ΔU_{ci} also has two possible values: ΔU_{2min} and ΔU_{2max} (if the value of S_i is not equal to zero, $\Delta U_{ci} = \Delta U_{2min}$, otherwise, $\Delta U_{ci} = \Delta U_{2max}$), where:

$$\begin{aligned} \Delta U_{2min} &= \Delta U_{ci1} - |\Delta U_{ci2}| \\ \Delta U_{2max} &= \Delta U_{ci1} \end{aligned} \tag{7}$$

It is obvious that ΔU_{1min} is equal to ΔU_{2min} and ΔU_{1max} is greater than ΔU_{2max}. Compared with the HMSCZS, the HMSWZS may lead to larger fluctuation of DC-side voltages of H-bridge units and thus more energy will be lost.

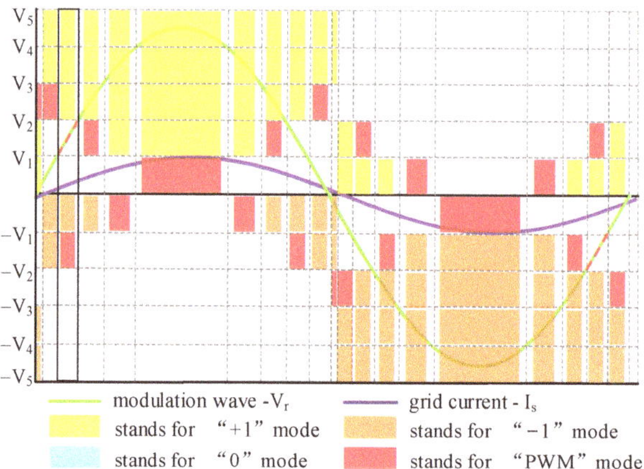

Figure 5. The switching modes for an eleven-level CHB inverter with the HMSWZS method.

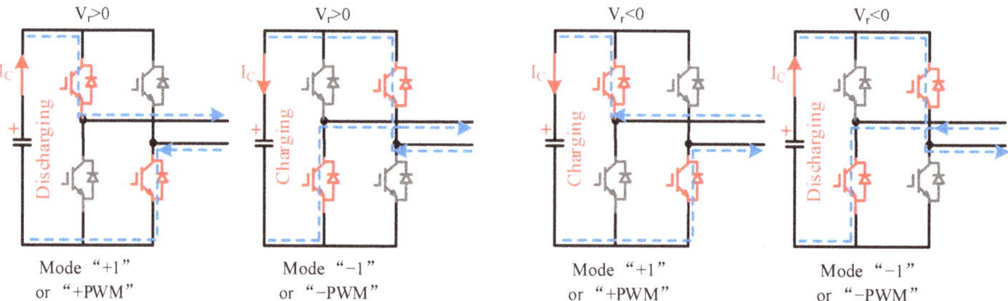

Figure 6. The switching modes of the fault H-bridge unit with the HMSWZS method.

4. The Switching Hybrid Modulation Strategy

A switching hybrid modulation strategy (SHMS) based on the HMSCZS and HMSWZS is proposed to maximize the output power of PV panels. When the CHB inverter is operating in the normal mode, the HMSCZS is adopted to suppress DC-side voltages fluctuation and, thus, realizing higher efficiency in energy harvesting. When the CHB inverter is operating in the fault mode owing to failing solar panels, the HMSWZS is utilized to control the DC-side voltages to reach the references, thus, maintaining a higher energy yield under the fault condition. Figure 7 shows the major procedures of the SHMS, which are basically the same as HMSCZS.

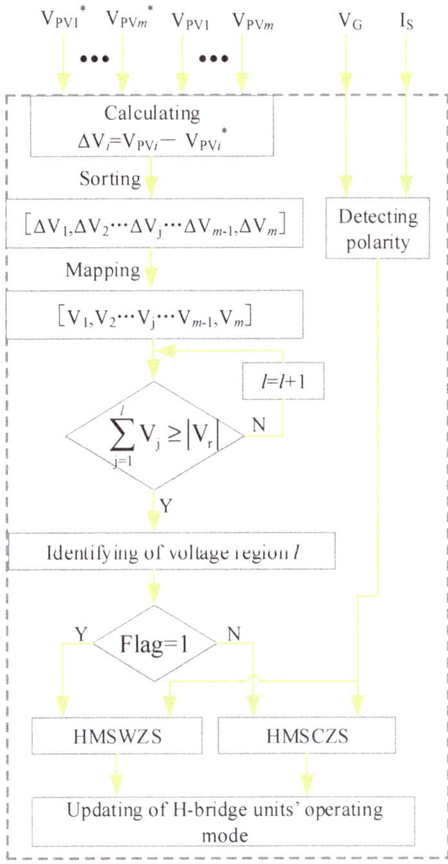

Figure 7. Flowchart of the proposed switching hybrid modulation strategy.

In consideration of the direction of I_S and V_r, and assuming that the modulation wave V_r is in area l, the SHMS method allocates suitable switching modes to each H-bridge unit based on the following rules:

4.1. Normal Mode

(1)   When ($V_r > 0$ and $I_S > 0$), ($l - 1$) units with higher voltage errors are discharging in the "+1" state, ($m - l$) units with lower voltage errors operate in the "0" state, and the lth unit works in PWM state.

(2)   When ($V_r > 0$ and $I_S \leq 0$), ($l - 1$) units with lower voltage errors are charging in the "+1" state, ($m - l$) units with higher voltage errors operate in the "0" state, and the lth unit works in the PWM state.

(3)   When ($V_r \leq 0$ and $I_S > 0$), ($l - 1$) units with lower voltage errors are charging in the "−1" state, ($m - l$) units with higher voltage errors operate in the "0" state, and the lth unit works in the PWM state.

(4)   When ($V_r \leq 0$ and $I_S \leq 0$), ($l - 1$) units with higher voltage errors are discharging in the "−1" state, ($m - l$) units with lower voltage errors operate in the "0" state, and the lth unit works in the PWM state.

4.2. Fault Mode

(1)   When ($V_r > 0$ and $I_S > 0$), and the value of ($m - l$) is even, ($m - l$)/2 units with lower voltage errors are charging in the "−1" state, the (($m - l + 2$)/2)th unit works in the PWM state, and the remaining units operate in the "+1" state.

(2)   When ($V_r > 0$ and $I_S > 0$), and the value of ($m - l$) is uneven, ($m - l - 1$)/2 units with lower voltage errors are charging in the "−1" state, the (($m - l + 1$)/2)th unit works in the PWM state, and the remaining units operate in the "+1" state.

(3)   When ($V_r > 0$ and $I_S \leq 0$), and the value of ($m - l$) is even, ($m + l - 2$)/2 units with lower voltage errors are charging in the "+1" state, the (($m + l$)/2)th unit works in the PWM state, and the remaining units operate in the "−1" state.

(4)   When ($V_r > 0$ and $I_S \leq 0$), and the value of ($m - l$) is uneven, ($m + l - 1$)/2 units with lower voltage errors are charging in the "+1" state, the (($m + l + 1$)/2)th unit works in the PWM state, and the remaining units operate in the "−1" state.

(5)   When ($V_r \leq 0$ and $I_S > 0$), and the value of ($m - l$) is even, ($m + l - 2$)/2 units with lower voltage errors are charging in the "−1" state, the (($m + l$)/2)th unit works in the PWM state, and the remaining units operate in the "+1" state.

(6)   When ($V_r \leq 0$ and $I_S > 0$), and the value of ($m - l$) is uneven, ($m + l - 1$)/2 units with lower voltage errors are charging in the "−1" state, the (($m + l + 1$)/2)th unit works in the PWM state, and the remaining units operate in the "+1" state.

(7)   When ($V_r \leq 0$ and $I_S \leq 0$), and the value of ($m - l$) is even, ($m - l$)/2 units with lower voltage errors are charging in the "+1" state, the (($m - l + 2$)/2)th unit works in the PWM state, and the remaining units operate in the the "−1" state.

(8)   When ($V_r \leq 0$ and $I_S \leq 0$), and the value of ($m - l$) is uneven, ($m - l - 1$)/2 units with lower voltage errors are charging in the "+1" state, the (($m - l + 1$)/2)th unit works in the PWM state, and the remaining units operate in the "−1" state.

5. Simulation Verification

To verify the performance of the proposed method, an CHB inverter which consists of five modules is simulated in MATLAB/Simulink. The PV panel is a JAP6-60-255/4BB and the specifications and equivalent circuit parameters are given in Table 2. Table 3 shows the inverter and grid parameters.

Table 2. Parameters of the PV panel.

Symbol	Parameter	Value
P_m	Max power	255 W
V_{oc}	Open circuit voltage	37.61 V
V_{mp}	Max power voltage	30.59 V
I_{sc}	Short circuit current	8.90 A
I_{mp}	Max power current	8.34 A

Table 3. Parameters of power grid and inverter.

Symbol	Parameter	Value
C_i	DC-side capacitor	14.1 mF
L_s	Filter inductance	1.8 mH
V_m	Peak value of grid voltage	130 V
f_{grid}	Frequency of grid voltage	50 Hz
f_c	PWM frequency	2500 Hz
f_{sort}	Frequency of mode change	500 Hz

5.1. Normal Mode

In order to investigate the performance of both HMSWZS and SHMS in terms of DC-side voltage fluctuation and average output power of the PV module under identical conditions, the DC-side voltage (V_{PV1}) and output power (P_1) of the PV module in the first H-bridge are shown in Figure 8. The simulation starts with the operation of the system under symmetrical condition (E = 1000 W/m^2 and T = 25 °C). In the simulation, since the focus of this paper is to study the influence of DC-side voltage fluctuation on the energy acquisition of PV module, the reference of the DC-side voltage is set directly at the maximum power point of PV module instead of adopting the MPPT algorithm. As shown in Figure 8, the maximum fluctuation of V_{PV1} is 4.95 V when HMSWZS is utilized, while that of SHMS is only 3.40 V, which is reduced by up to 31.30%. With the HMSWZS, the output power of the first PV module ranges from 220–255.1 W, and the average is about 251.9 W. Since the SHMS is utilized to reduce the DC-side voltage fluctuation, the output power ranges from 245–255.1 W and the average is about 253.4 W. Figure 9 shows the total output power (P_T) of the CHB inverter with both methods. As shown in Figure 9, with HMSWZS, the total output power of the CHB inverter ranges from 1243–1275 W and the average is about 1262 W. When SHMS is utilized, the total output power of the CHB inverter ranges from 1256–1277 W and the average is about 1269.2 W. Compared with the HMSWZS, the efficiency of the CHB inverter can be improved about 0.56% by adopting the SHMS. Therefore, under the normal mode, the SHMS has a superior capacity of suppressing the DC-side voltage fluctuation compared with the HMSWZS, thereby improving the energy collection of the PV module.

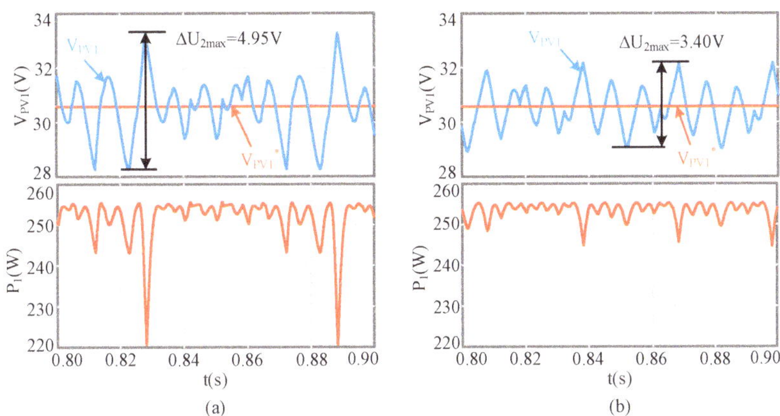

Figure 8. Simulation results under normal mode: the DC-side voltage (V_{PV1}), and output power (P_1) of the PV module in the first H-bridge with: (**a**) HMSWZS and (**b**) SHMS.

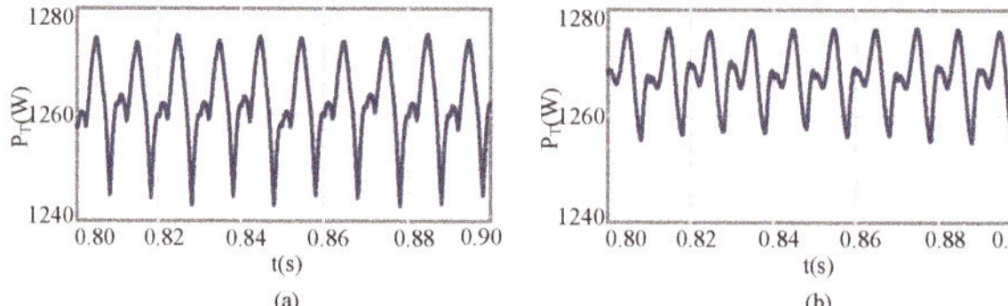

Figure 9. Simulation results of the total output power of the CHB inverter under the normal mode with: (**a**) HMSWZS and (**b**) SHMS.

5.2. Fault Mode

The following simulation evaluates the performance of HMSCZS and SHMS under the fault mode where the second PV module fails. The simulation starts with the operation of the system under symmetrical condition (E = 1000 W/m^2 and T = 25 °C). At t = 1.5 s, the second PV module is removed owing to the fault and, thus, the CHB inverter is operating in the fault condition. Figure 10 depicts the total modulation voltage (V_r) and DC-side capacitor current of the second H-bridge (I_{C2}) with both methods. As shown in Figure 10a I_{C2} is always negative regardless of the polarity of V_r and, thus, V_{PV2} gradually diverges from the reference value. As a result, all the DC-side voltages diverge from the references in steady state, which is visible in Figure 11a. For comparison, the SHMS is also subjected to the same test under the identical condition. As presented in Figure 10b, I_{C2} can be positive and negative whether V_r is positive or negative and, thus, the DC-side capacitor of the second H-bridge could realize the equalization of the charge-discharge. Therefore, the CHB inverter is able to operate properly under fault condition, which is obvious in Figure 11b. Figure 12 shows the output power of the PV module in all H-bridges with both methods. As shown in Figure 12, due to the removal of the second PV module, the output power of the second H-bridge is, therefore, zero. The output power of the PV module in other H-bridges ranges from 243.2–255.1 W by using SHMS and the average is about 253.2 W. However, due to the deviation of the DC-side voltage, the output power of the PV module in other H-bridges ranges from 221.4–255.1 W by utilizing HMSCZS and the average is only about 245.3 W. The total output power (P_T) of the CHB inverter with both methods are shown in Figure 13. As could be seen from Figure 13, with HMSCZS, the total output power of the CHB inverter ranges from 935.9–1016 W and the average is about 982.2 W. When SHMS is utilized, the total output power of the CHB inverter ranges from 1000–1020 W and the average is about 1014 W. Compared with the HMSCZS, the efficiency of the CHB inverter can be improved about 3.12% by adopting the SHMS. Therefore, under the fault mode, the SHMS is still able to make the DC-side voltages reach the references, thus maintaining a higher energy yield.

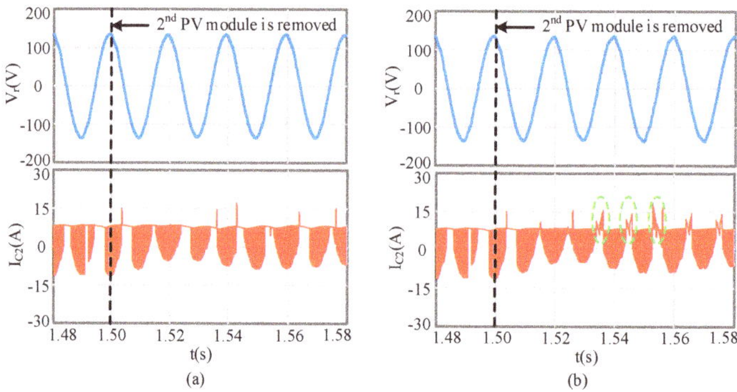

Figure 10. Simulation results under fault mode: the total modulation voltage (V_r), and DC-side capacitor current of the second H-bridge (I_{C2}) with: (**a**) HMSCZS and (**b**) SHMS.

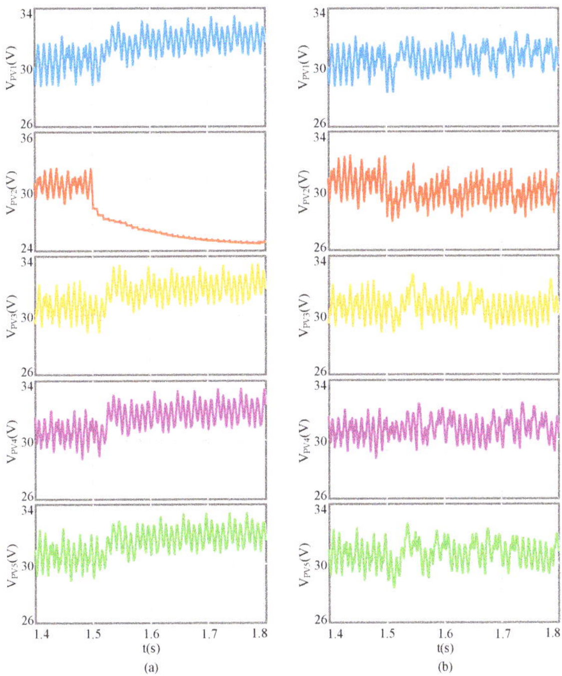

Figure 11. Simulation results under fault mode: DC-side voltages of all H-bridges with: (**a**) HMSCZS, and (**b**) SHMS.

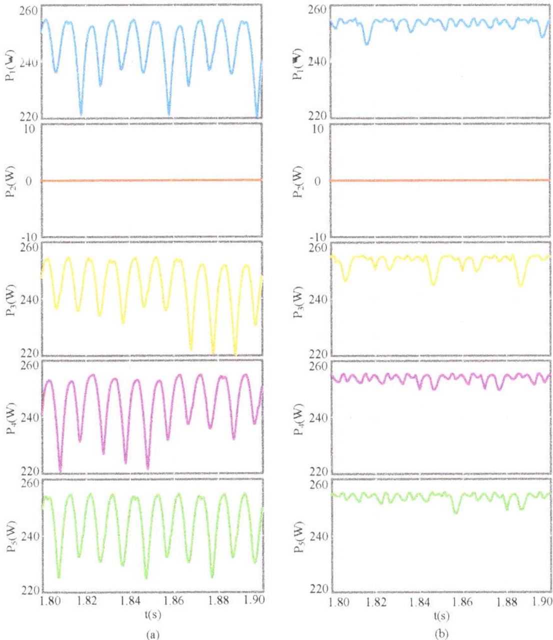

Figure 12. Simulation results under fault mode: the output power of PV module in all H-bridges with: (**a**) HMSCZS and (**b**) SHMS.

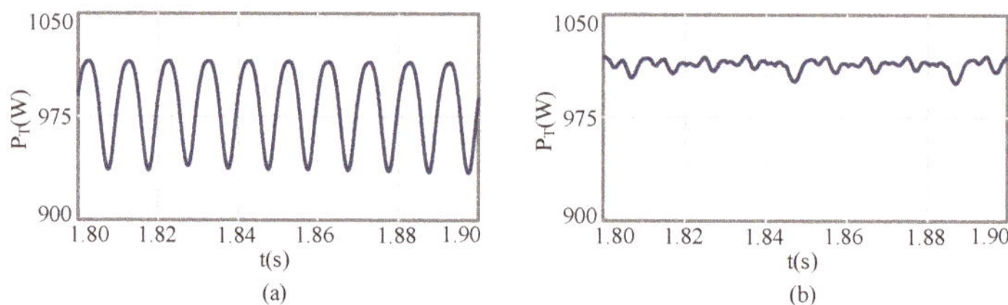

Figure 13. Simulation results of the total output power of the CHB inverter under the fault mode with: (**a**) HMSCZS and (**b**) SHMS.

6. Experimental Results

As shown in Figure 14, an experimental prototype with five H-bridge units has been established to validate the effectiveness of the proposed strategy. The prototype includes a central controller and five local controllers. The BECKHOFF industrial PC (CX2040) is utilized as the central controller and TMS320F28335 processors are used as local controllers which communicate with each other through an EtherCAT real-time communication network. The EtherCAT protocol features not only enhance the effective synchronization between master and slave clocks, but also improves the synchronization between the H-bridge units. Since the testing equipment of the university laboratory is limited, only the first H-bridge unit is connected to a Chroma62020H-150S PV simulator, and the DC side of the four other H-bridge units are connected to a 36 V switching power supply through a 0.5 Ω resistance, respectively, to simulate PV modules. Given that there are only five H-bridge units in the experiment, the output of the inverter is connected to the 130 V AC grid regulated by a voltage regulator. The other parameters of the experimental system are the same as the simulation parameters.

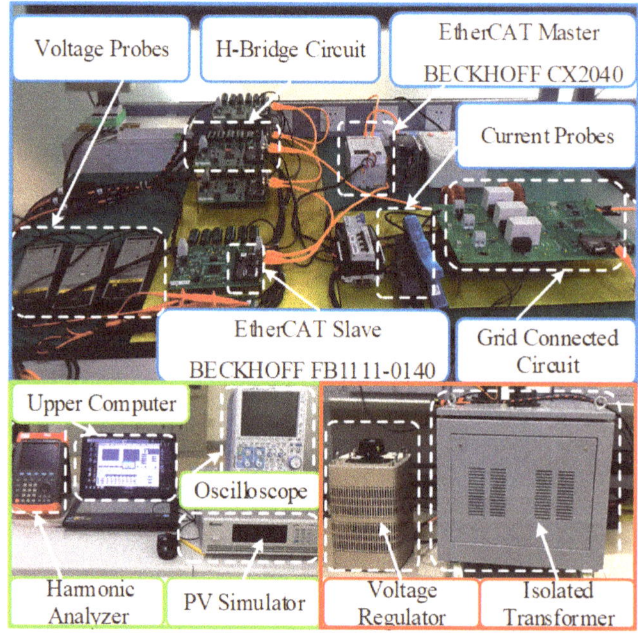

Figure 14. Experimental platform of the single-phase PV CHB module level inverter.

6.1. Normal Mode

The first experiment has been carried out under the initial conditions of PV simulator with irradiance of 1000 W/m^2 and a temperature of 25 °C. In this case, the input power of the first H-bridge unit is approximately 255 W. The reference of the other four H-bridge units' DC-side voltages is 32.02 V, and therefore the output power of the other four H-bridges is about 255 W in theory. As depicted in the Figure 15, both SHMS and HMSWZS enable the CHB inverter to operate at unity power factor with good grid current quality. It can also be seen that SHMS contains four output modes: "+1", "−1", PWM, and "0" mode, which is never adopted in HMSWZS.

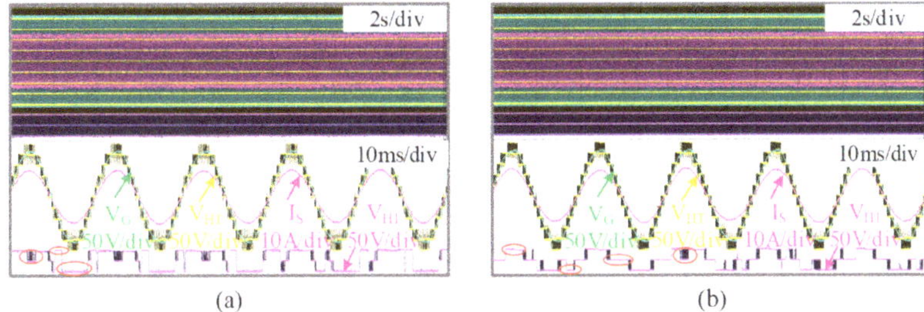

(a)   (b)

Figure 15. Experimental results under normal mode: grid voltage (V_G), grid current (I_S) and the total output voltage of CHB inverter (V_{HT}); the first H-bridge output voltage (V_{H1}) with: (**a**) HMSWZS and (**b**) SHMS.

For the sake of evaluating the characteristic of both two modulation methods in terms of DC-side voltage fluctuation and average output power of the PV module, the following experiments have been performed under the same conditions as the first experiment. As shown in Figure 16, the maximum fluctuation of V_{PV1} by adopting HMSWZS is 5.3 V, but the value is only 4.1V for SHMS, which is reduced by about 22.64%. In order to compare the output power of the two methods under normal mode, the total output power of the CHB inverter is recorded by the upper computer, respectively. As can be seen from Figure 17, with HMSWZS, the total output power of the CHB inverter ranges from 1218–1244.2 W and the average is about 1233.7 W. When SHMS is utilized, the total output power of the CHB inverter ranges from 1230–1249.3 W and the average is about 1239.8 W. Compared with the HMSWZS, the efficiency of the CHB inverter can be improved about 0.48% by adopting the SHMS. Furthermore, the efficiency of the CHB inverter with both methods is presented in Figure 18. As can be seen from Figure 18, the SHMS is capable of improving the efficiency of the CHB inverter compared with the HMSWZS. Therefore, under normal mode, the SHMS is able to effectively suppress the DC-side voltage fluctuation compared with the HMSWZS, thereby improving the energy acquisition of the PV module.

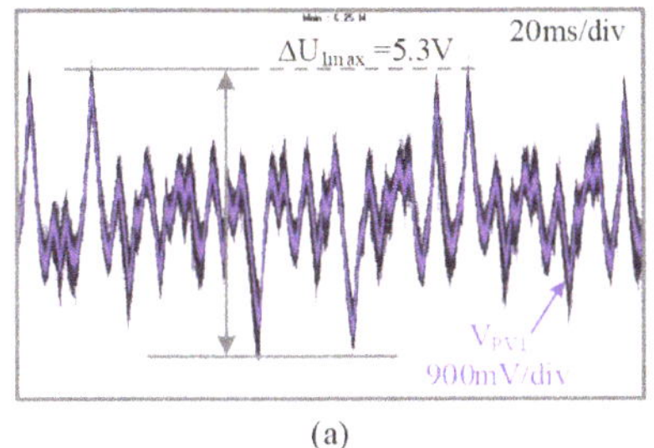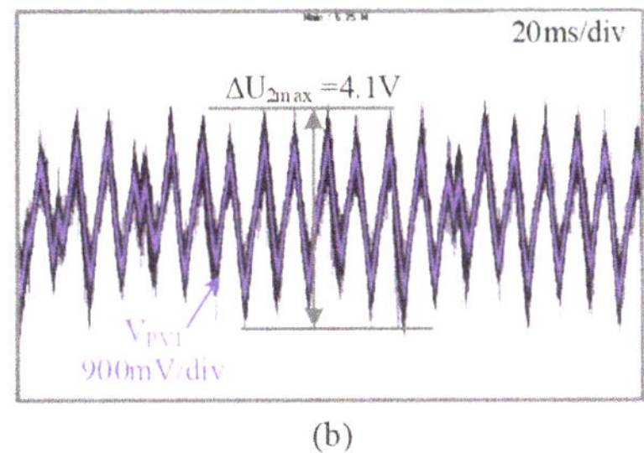

(a)   (b)

Figure 16. Experimental results under normal mode: the DC-side voltage (V_{PV1}) of the first H-bridge with: (**a**) HMSWZS, and (**b**) SHMS.

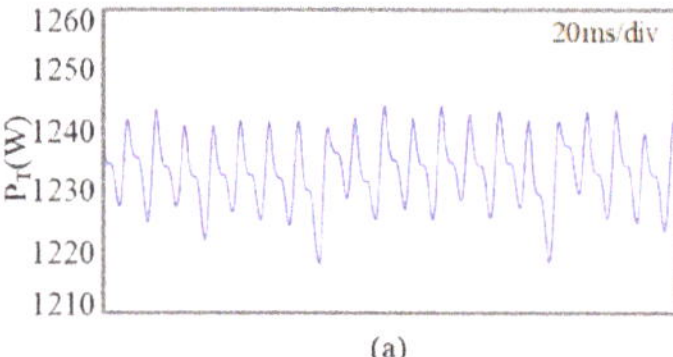
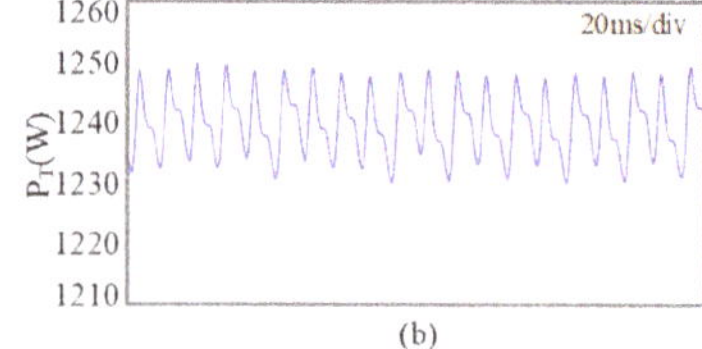

(a)   (b)

Figure 17. Experimental results of the total output power of the CHB inverter under normal mode with: (**a**) HMSWZS and (**b**) SHMS.

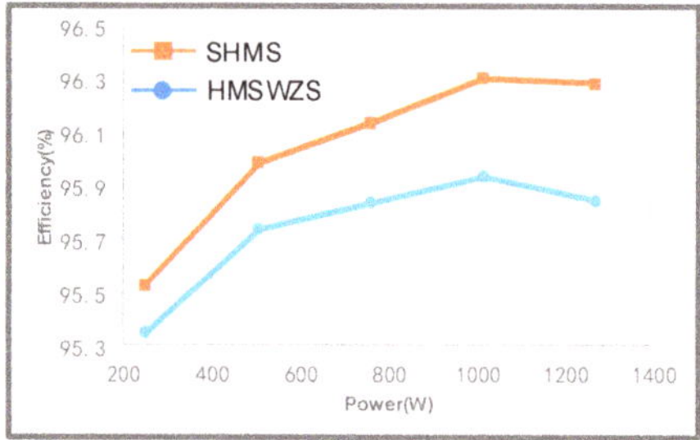

Figure 18. Experimental results of the efficiency of the CHB inverter under normal mode with different methods.

6.2. Fault Mode

The last experiment is conducted to evaluate the performance of SHMS and HMSCZS in the fault mode where the PV module fails. Initially, the CHB inverter and PV modules are operating in normal mode as indicated in the first experiment. Then the second PV module is removed. As can be seen from Figure 19a,b, both the HMSCZS and SHMS can keep the CHB inverter stable and operating at a unity power factor without interruption. As is presented in the Figure 19a, by using the HMSCZS, the fault H-bridge unit can operate only in "0" or "+PWM" mode when V_G is positive and "0" or "−PWM" mode when V_G is negative. In such condition, the fault H-bridge unit is always in discharge status and unable to realize the equalization of the charge-discharge. Consequently, as presented in Figure 19c, the DC-side voltages of all H-bridge units diverge from the references in the steady state, which may result in low generated power of the CHB inverter. However, as presented in the Figure 19b, by utilizing the SHMS, the fault H-bridge unit can operate in "+1", "+PWM", "−1" or "−PWM" mode whether the grid voltage (V_G) is positive or negative. Therefore, as depicted in the Figure 19d, the SHMS is capable of realizing the equalization of the charge-discharge of the fault H-bridge unit and maintaining the DC-side voltage balance of the other H-bridge units. In order to compare the output power of the two methods under fault mode, the total output power of the CHB inverter is recorded by the upper computer respectively. As can be seen from Figure 20, with HMSCZS, the total output power of the CHB inverter ranges from 919.2–986.4 W and the average is about 948.8 W. When SHMS is utilized, the total output power of the CHB inverter ranges from 972.3–988.6 W and the average is about 978.2 W. Compared with the HMSCZS, the efficiency of the CHB inverter can be improved about 2.89% by adopting the SHMS. Furthermore, the efficiency of the CHB inverter with both methods is presented in Figure 21. As can be seen from Figure 21, the SHMS is capable of improving the efficiency of the CHB inverter compared with the HMSCZS. Therefore, under the fault mode, the SHMS is able to make the DC-side voltages reach the references, thus maintaining a higher energy yield.

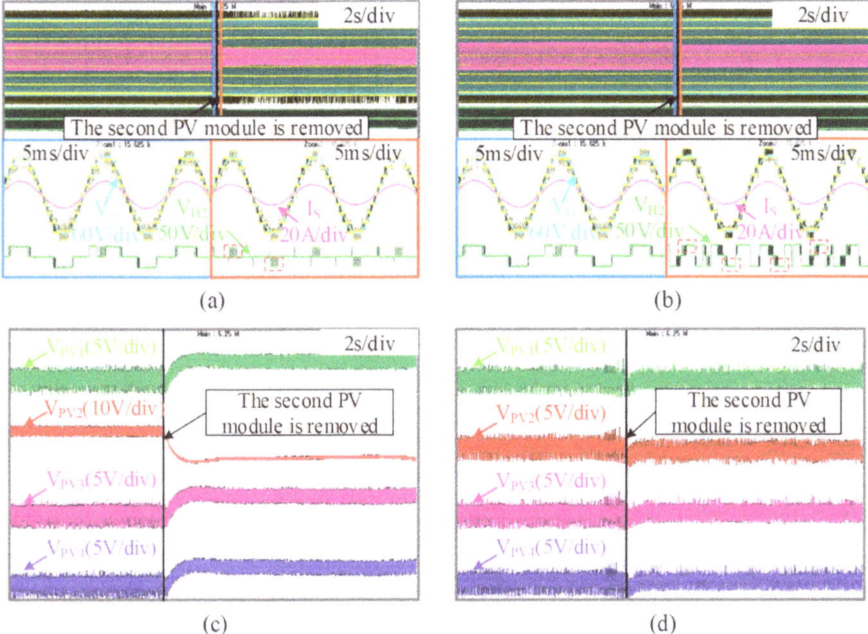

Figure 19. Experimental results under fault mode: grid voltage (V_G), grid current (I_S), and the second H-bridge output voltage (V_{H2}) with: (**a**) HMSCZS and (**b**) SHMS; DC-side voltages of all H-bridge units with: (**c**) HMSCZS and (**d**) SHMS.

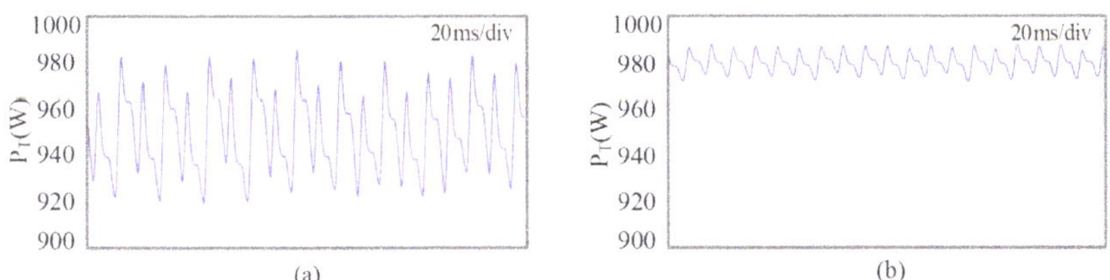

Figure 20. Experimental results of the total output power of the CHB inverter under fault mode with: (**a**) HMSCZS and (**b**) SHMS.

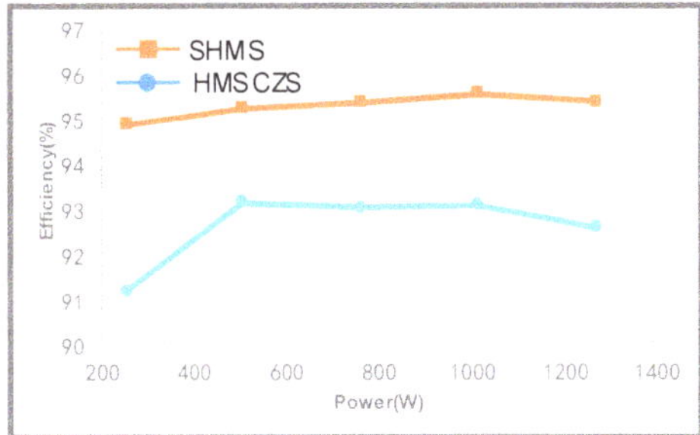

Figure 21. Experimental results of the efficiency of the CHB inverter under the fault mode with different methods.

7. Conclusions

This paper proposes a switching hybrid modulation strategy, which can effectively reduce the fluctuation of DC-side voltage and maximize the output power of a PV CHB grid-connected inverter. When the CHB inverter is operating in the normal mode, the hybrid modulation strategy containing the zero state is adopted to suppress DC-side voltage fluctuation, thereby improving the output power of PV modules. Once the CHB inverter is operating in the fault mode, owing to failing solar panels, the hybrid modulation strategy without the zero state is utilized to make the DC-side voltages reach the references, thus, maintaining a higher energy yield under the fault condition. A set of experimental results demonstrate that, with this method, the output energy of the PV modules is improved both in the normal mode and fault mode.

Author Contributions: Conceptualization: X.Z. and W.M.; data curation: Y.H.; formal analysis: W.M. and T.Z.; funding acquisition: F.W. and F.L.; methodology: X.Z.; resources: X.Z. and R.C.; software: Y.H.; validation: W.M.; writing—original draft: W.M.; writing—review and editing: W.M.

Funding: This work was funded by the National Key Research and Development Program of China (No. 2018YFB1500703) and the National Natural Science Foundation of China (No. 51677049).

References

1. Schweizer, M.; Friedli, T.; Kolar, J.W. Comparative evaluation of advanced three-phase three-level inverter/converter topologies against two-level systems. *IEEE Trans. Ind. Electron.* **2012**, *60*, 5515–5527. [CrossRef]

2. Villanueva, E.; Correa, P.; Rodríguez, J.; Pacas, M. Control of a single-phase cascaded H-bridge multilevel inverter for grid-connected photovoltaic systems. *IEEE Trans. Ind. Electron.* **2009**, *56*, 4399–4406. [CrossRef]

3. Xiao, B.; Hang, L.; Mei, J.; Riley, C.; Tolbert, L.M.; Ozpineci, B. Modular cascaded H-bridge multilevel PV inverter with distributed MPPT for grid-connected applications. In Proceedings of the IEEE International Conference on Robotics and Automation, New Orleans, LA, USA, 26 April–1 May 2004; pp. 1722–1731.

4. Mercorelli, P.; Kubasiak, N.; Liu, S. Multilevel bridge governor by using model predictive control in wavelet packets for tracking trajectories. In Proceedings of the IEEE International Conference on Robotics and Automation (ICRA), New Orleans, LA, USA, 26 April–1 May 2004; 2004; Volume 4, pp. 4079–4084.

5. Townsend, C.D.; Yu, Y.; Konstantinou, G.; Agelidis, V.G. Cascaded H-bridge multilevel PV topology for alleviation of per-phase power imbalances and reduction of second harmonic voltage ripple. *IEEE Trans. Power Electron.* **2016**, *31*, 5574–5586. [CrossRef]

6. Rivera, S.; Kouro, S.; Wu, B.; Leon, J.I.; Rodríguez, J.; Franquelo, L.G. Cascaded H-bridge multilevel converter multistring topology for large scale photovoltaic systems. In Proceedings of the IEEE International Symposium on Industrial Electronics, Gdansk, Poland, 27–30 June 2011; pp. 1837–1844.

7. Sun, D.; Ge, B.; Yan, X.; Bi, D.; Zhang, H.; Liu, Y.; Peng, F.Z. Modeling, impedance design, and efficiency analysis of quasi-Z source module in cascaded multilevel photovoltaic power system. *IEEE Trans. Ind. Electron.* **2014**, *61*, 6108–6117. [CrossRef]

8. Mercorelli, P.; Kubasiak, N.; Liu, S. Model predictive control of an electromagnetic actuator fed by multilevel PWM inverter. In Proceedings of the 2004 IEEE International Symposium on Industrial Electronics, Ajaccio, France, 4–7 May 2004; Volume 1, pp. 531–535.

9. Mercorelli, P. A Multilevel inverter bridge control structure with energy storage using model predictive control for flat systems. *J. Eng.* **2013**, *2013*, 750190. [CrossRef]

10. Yu, Y.; Konstantinou, G.; Hredzak, B.; Agelidis, V.G. Power balance of cascaded H-bridge multilevel converters for large-scale photovoltaic integration. *IEEE Trans. Power Electron.* **2016**, *31*, 292–303. [CrossRef]

11. Vazquez, S.; Leon, J.I.; Carrasco, J.M.; Franquelo, L.G.; Galvan, E.; Reyes, M.; Dominguez, E. Analysis of the power balance in the cells of a multilevel cascaded H-bridge converter. *IEEE Trans. Ind. Electron.* **2010**, *57*, 2287–2296. [CrossRef]

12. Rezaei, M.A.; Farhangi, S.; Iman-Eini, H. Extending the operating range of cascaded H-bridge based multilevel rectifier under unbalanced load conditions. In Proceedings of the 2010 IEEE International Conference on Power and Energy, Kuala Lumpur, Malaysia, 29 November–1 December 2010; pp. 780–785.

13. Wang, S.; Zhao, J.; Yao, X.; Sun, Y. Power balanced controlling of cascaded inverter for grid-connected photovoltaic systems under unequal irradiance conditions. *Trans. Chin. Electrotech. Soc.* **2012**, *28*, 251–261.

14. Kouro, S.; Wu, B.; Moya, Á.; Villanueva, E.; Correa, P.; Rodríguez, J. Control of a cascaded H-bridge multilevel converter for grid connection of photovoltaic systems. In Proceedings of the 2009 35th Annual Conference of IEEE Industrial Electronics, Porto, Portugal, 3–5 November 2009; pp. 3976–3982.

15. Rivera, S.; Wu, B.; Lizana, R.; Kouro, S.; Perez, M.; Rodriguez, J. Modular multilevel converter for large-scale multistring photovoltaic energy conversion system. In Proceedings of the 2013 IEEE Energy Conversion Congress and Exposition, Denver, CO, USA, 15–19 September 2013; pp. 1941–1946.

16. Rezaei, M.A.; Iman-Eini, H.; Farhangi, S. Grid-connected photovoltaic system based on a cascaded h-bridge inverter. *J. Power Electron.* **2012**, *12*, 578–586. [CrossRef]

17. Liu, L.; Li, H.; Xue, Y.; Liu, W. Reactive power compensation and optimization strategy for grid-interactive cascaded photovoltaic systems. *IEEE Trans. Power Electron.* **2015**, *30*, 188–202. [CrossRef]

18. Eskandari, A.; Javadian, V.; Iman-Eini, H.; Yadollahi, M. Stable operation of grid connected cascaded H-bridge inverter under unbalanced insolation conditions. In Proceedings of the 2013 3rd International Conference on Electric Power and Energy Conversion Systems, Istanbul, Turkey, 2–4 October 2013; pp. 1–6.

19. Iman-Eini, H.; Schanen, J.L.; Farhangi, S.; Roudet, J. A modular strategy for control and voltage balancing of cascaded H-bridge rectifiers. *IEEE Trans. Power Electron.* **2008**, *23*, 2428–2442. [CrossRef]

20. Keshavarzian, A.; Iman-Eini, H. A redundancy-based scheme for balancing DC-link voltages in cascaded H-bridge rectifiers. *IET Power Electron.* **2013**, *6*, 235–243. [CrossRef]

21. Moosavi, M.; Farivar, G.; Iman-Eini, H.; Shekarabi, S.M. A voltage balancing strategy with extended operating region for cascaded H-bridge converters. *IEEE Trans Power Electron.* **2014**, *29*, 5044–5053. [CrossRef]

22. Coppola, M.; Di Napoli, F.; Guerriero, P.; Iannuzzi, D.; Daliento, S.; Del Pizzo, A. An FPGA-based advanced control strategy of a grid-tied PV CHB inverter. *IEEE Trans. Power Electron.* **2016**, *31*, 806–816. [CrossRef]

23. Zhao, T.; Zhang, X.; Mao, W.; Wang, F.; Xu, J.; Gu, Y. A modified hybrid modulation strategy for suppressing DC voltage fluctuation of cascaded H-bridge photovoltaic inverter. *IEEE Trans. Ind. Electron.* **2018**, *65*, 3932–3941. [CrossRef]

24. Miranbeigi, M.; Iman-Eini, H. Hybrid modulation technique for grid-connected cascaded photovoltaic systems. *IEEE Trans. Ind. Electron.* **2016**, *63*, 7843–7853. [CrossRef]

25. Zmood, D.N.; Holmes, D.G. Stationary frame current regulation of PWM inverters with zero steady-state error. *IEEE Trans. Power Electron.* **2003**, *18*, 814–822. [CrossRef]

26. Kim, S.Y.; Park, S.Y. Compensation of dead-time effects based on adaptive harmonic filtering in the vector-controlled AC motor drives. *IEEE Trans. Ind. Electron.* **2007**, *54*, 1768–1777. [CrossRef]

27. Zhu, D.; Zou, X.; Deng, L.; Huang, Q.; Zhou, S.; Kang, Y. Inductance-emulating control for DFIG-based wind turbine to ride-through grid faults. *IEEE Trans. Power Electron.* **2017**, *32*, 8514–8525. [CrossRef]

Article

A Novel Photovoltaic Virtual Synchronous Generator Control Technology Without Energy Storage Systems

Guangqing Bao [1], Hongtao Tan [2,*], Kun Ding [3], Ming Ma [3] and Ningbo Wang [3]

[1] Information Engineering and College of Electrical, Lanzhou University of Technology, Lanzhou 730050, Gansu Province, China; gqbao@lut.com
[2] State Key Laboratory of Power Transmission Equipment & System Security and Technology, ChongQing University, Chongqing 400044, China
[3] Wind Power Technology Center, State Grid Gansu Electric Power Corporation, Lanzhou 730070, Gansu Province, China; dingk02@foxmail.com (K.D.); m15214063389@163.com (M.M.); wangnb@gs.sgcc.com.cn (N.W.)
* Correspondence: 13679261064@163.com; Tel.: +86-136-7926-1064

Received: 20 February 2019; Accepted: 3 June 2019; Published: 12 June 2019

Abstract: Photovoltaic virtual synchronous generator (PV-VSG) technology, by way of simulating the external characteristics of a synchronous generator (SG), gives the PV energy integrated into the power grid through the power electronic equipment the characteristics of inertial response and active frequency response (FR)—this attracts much attention. Due to the high volatility and low adjustability of PV energy output, it does not have the characteristics of a prime mover (PM), so it must be equipped with energy storage systems (ESSs) in the DC or AC side to realize the PV-VSG technology. However, excessive reliance on ESSs will inevitably affect the application of VSG technology in practical PV power plants (PV-PPs). In view of this, this paper proposes the PV power reserve control type VSG (PV-PRC-VSG) control strategy. By reducing the active power output of part of the PV-PPs, the internal PV-PPs can maintain a part of the active power up/down-regulation ability in real time, instead of relying on external ESSs. By adjusting the active reserve power of this part, the output of the PV-PPs can be controlled within a certain range, and the PV-PPs can better simulate the PM characteristics and realize the FR of the grid by combining the VSG technology. At the same time, the factors affecting the reserve ratio are analyzed, and the position of the voltage operating point in PRC mode is deduced. Finally, the simulation results show that the proposed control strategy is effective and correct.

Keywords: photovoltaic (PV); virtual synchronous generator (VSG); frequency response (FR); power reserve control (PRC); active power up-regulation

---

1. Introduction

China's new-generation energy revolution advocates the development of non-fossil energy, and PV energy has become an important part of non-fossil energy due to its inexhaustible advantages. It is estimated that in 2035, photovoltaic virtual (PV) energy will account for 26% of total installed capacity and 14% of electricity generation [1].

Power system frequency is an important standard to reflect the power surplus and deficiency of a power system. Frequency response control mainly reflects the power support function of source side to grid side. The influence of large-scale PV power plants (PV-PPs) on power system frequency is mainly reflected in two aspects [2–4]:

- Reduce the equivalent moment of inertia of the power system. The PV cell is a static original, which does not have any rotating standby. After being connected to the power grid, the original equivalent inertia will be less;

- The primary frequency response capability of the system is weakened. Under the action of maximum power point tracking (MPPT), the PV output is uncontrollable and cannot provide power support for the system.

It is obvious that large-scale PV-PPs will weaken this support after they are connected to the power grid [5–7]. The proposal and promotion of virtual synchronous generator (VSG) control technology can effectively solve the above problems and, at the same time, can give the PV grid-connected system the ability to participate in the frequency and voltage regulation of the power grid independently [8–10]. However, the PV system active power output changes with the external environment (including solar irradiance and temperature) under the effect of maximum power point tracking (MPPT). Since the active output is not controllable, the traditional VSG control strategy cannot be directly used in PV systems. Existing related research in joining the ESSs to the DC or AC side of the PV system [11,12] can effectively solve the above problems. By controlling the charging and discharging of ESSs, the effect of controlling PM output can be simulated in a short time, but it will be affected by the physical constraints of the ESSs. In reference [13], a 50 kW × 30 min lithium-ion battery pack is used to connect a PV array (installed capacity of 500 kW) DC side, and active frequency response is realized by controlling the output of the ESSs. However, any benefit brought by the ESSs to the power grid is based on the economic cost of ESSs [14].

In large-scale PV-PPs such as Hexi New Energy Base in GanSu Province, China, limited by the current energy storage technology level, ESSs cannot be widely used in PV-PPs, and the charging and discharging efficiency is low. Frequent charging and discharging not only makes the energy utilization rate lower but also affects the service life of the ESSs, resulting in greater economic losses. PV-power reserve control (PRC) maintains a part of the power up/down capability by reducing the output of the PV system [15–20] and participates in the power system frequency response in combination with inertial response control and droop control, which is called fast frequency response (FFR) or active power control (APC) [2,21,22]. However, compared with the VSG control method, the VSG is a direct simulation of the internal potential phase motion and its basic inertia and damping characteristics of the SG, and it is the simplest and most effective way to realize the characteristics of the traditional SG. In addition, in the previously published works on PV-PRC, there are different views of whether the voltage operating point should be located to the left or right of the maximum power point voltage (V_{mpp}) in PV-PRC mode. In addition, the PV-PRC model means that a part of the photovoltaic energy is wasted, but there is little work on how to choose the appropriate reserve ratio.

Aiming at the above problems, this paper takes the two-stage PV grid-connected system as the research object. The DC/DC and DC/AC converters implement PV-PRC/MPPT and VSG control respectively. We named this control method as the PV power reserve control type VSG (PV-PRC-VSG) control technology. In this paper, the traditional PV-PRC and PV-VSG are combined and further improved. The main contributions are as follows:

- Propose a new PV-VSG implementation method, which maintains a part of the active power up-regulation capability by operating the PV system in PRC mode and combines the VSG technology to enable the PV system to support sudden power shortages in the power system. The control method is called the PV power reserve control type virtual synchronous generator (PV-PRC-VSG) technology;
- Considering the actual project, in order to ensure the reliable and efficient operation of the inverters, two voltage operating points in PV-PRC mode are analyzed in detail, and the result that the voltage operating point in PRC mode should be located on the right side of the maximum power point voltage is achieved;
- Based on the requirements of the State Grid for wind abandonment and PV energy abandonment and the active support capability of PV-VSG, the upper limit of the reserve ratio in PV-PRC mode is obtained.

In Section 2 of the paper, the traditional PV-VSG technology is introduced in detail. The implementation strategy of PV-PRC and the position of the voltage operating point are elaborated in Section 3. The proposed PV-PRC-VSG control technology and the range of the reserve ratio in PV-PRC mode are described in Section 4. In Section 5, the validity of the proposed method is verified by simulation experiments under various operating conditions.

2. Modeling and Analysis of PV-VSG

2.1. Principle and Embodiment of the VSG

VSG control can be divided into active loop control and reactive loop control from the function and control target. Its active loop includes active frequency droop control and inertial response control, which mainly realizes the function of independent FR. The reactive loop consists of reactive power-voltage droop control and end-voltage closed-loop control, which realizes automatic voltage regulation and voltage amplitude control of the VSG [23]. The basic mathematical model of the VSG is as shown in Equation (1). The VSG control block diagram controlled by the mathematical model is shown in Figure 1.

$$\begin{cases} J\frac{d\omega}{dt} = (P_m - P_e)/\omega \\ P_m = P_{ref} + D_p(\omega_n - \omega) \\ E_m = \frac{1}{Ks}\left(D_q(U_{ref} - U_m) + Q_{ref} - Q_e\right) \end{cases} \tag{1}$$

where J is the moment of inertia, P_{ref} and Q_{ref} are the given values of active power and reactive power, P_e and Q_e are the actual output values of active power and reactive power, ω_n and ω are the rated and actual values of electric angular velocity, D_p and D_q are the droop coefficients of active and reactive loops, U_m and U_{ref} are the actual and given values of grid voltage amplitude, E_m is the internal potential amplitude of the VSG, K is the integral coefficient, and it can be replaced by PI regulator.

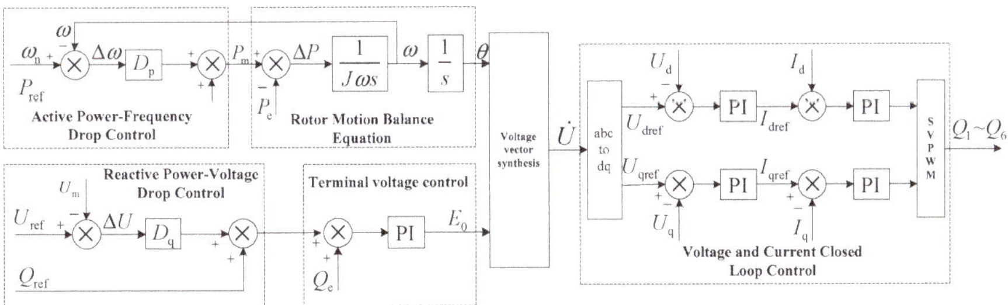

Figure 1. Principle and embodiment of the virtual synchronous generator (VSG).

2.2. Traditional PV-VSG Technology

In this paper, the two-stage grid-connected PV system is taken as the research object, as shown in Appendix A. Because the control objectives of DC/DC and DC/AC converters are different, as shown in Table 1, the control method shown in Figure 1 cannot be applied to the two-stage grid-connected PV system. In addition, if the PV system operates in MPPT mode, the output of the PV system cannot be controlled, and the premise of VSG implementation is that the system must maintain a reserve power, then it cannot achieve autonomous frequency and voltage regulation.

Table 1. Comparison of the control functions of the converter in the two-stage grid-connected photovoltaic virtual (PV) system.

Converter Type	DC/DC Converter	DC/AC Converter
Main Functions	Maximum Power Point Tracking (MPPT) Power Reserve Control Active Power Control	Power Conversion Control DC-Link Voltage Control Reactive Power Control Current Control

Figure 2 shows that for the control effect diagram of above method in reference [13], when the grid frequency changes suddenly, PV output power remains unchanged, and ESSs regulates output power to participate in grid FR. It is noteworthy that the active support provided by PV-ESSs is constrained by the allocation of a capacity cap and support time during the whole frequency decline stage. In addition, the addition of ESSs will greatly increase the construction cost of PV-PPs. The operation of PV-PRC can effectively reduce the above problems.

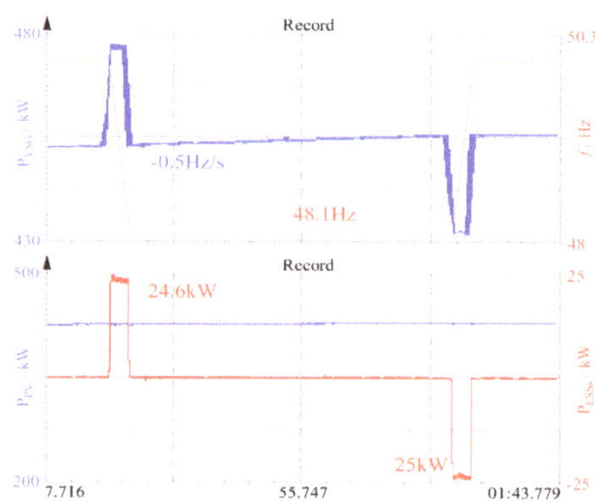

Figure 2. PV-energy storage systems (ESSs)-VSG output wave.

Therefore, in order to apply VSG control technology to the two-stage grid-connected PV system, it is necessary to solve the active standby problem and improve the control algorithm. External ESSs can effectively solve the above problems. The ESSs mainly undertake the reserve capacity required by FR. After adding ESSs, the Equation (1) is converted to:

$$\begin{cases} P_{\mathrm{m}} = P_{\mathrm{mpp}} + P_{\mathrm{ESSs}} \\ P_{\mathrm{ESSs}} = D_{\mathrm{p}}(\omega_{\mathrm{n}} - \omega) \end{cases} \tag{2}$$

3. PV-PRC Principle and Voltage Operating Point Analysis

3.1. Analysis of PV Generator Output Characteristics

The output P–V characteristic curve of the PV module is as shown in Figure 3a. In Figure 3a, there is a unique maximum power value P_{mpp}; P_{deload} is the output power value of the PV system in PRC mode; in PRC mode, the PV system maintains a part of the real-time active power up-regulation capability, as shown in Equation (3), ΔP is called the reserve active power of the PV system.

$$P_{\mathrm{mpp}} = P_{\mathrm{deload}} + \Delta P \tag{3}$$

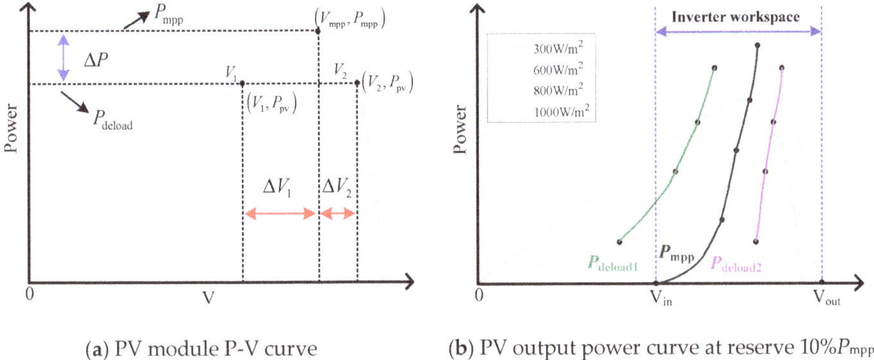

(**a**) PV module P-V curve   (**b**) PV output power curve at reserve 10%P_{mpp}

Figure 3. PV-PRC voltage operating point diagram.

As shown in Figure 2a, the output power of the PV system corresponds to two voltage operating points V_1 and V_2 in the non-MPP. The relationship between V_1, V_2 and V_{mpp} is shown in Equation (4). As shown in Equation (4), the same active reserve is implemented in PRC mode. Since the curve corresponding to the V_2 side has a steeper gradient, the voltage regulation range on the V_2 side is much smaller than that on the V_1 side (i.e., $\Delta V_2 < \Delta V_1$), which will bring the faster response speed.

$$\begin{cases} V_1 < V_{mpp} < V_2 \\ \left|V_{mpp} - V_2\right| = \Delta V_2 < \Delta V_1 = \left|V_{mpp} - V_1\right| \end{cases} \tag{4}$$

where V_{in} is the lowest PV output voltage that can make the inverter work normally; V_{out} is the maximum input voltage that the inverter can withstand; $P_{deload1}$ is the PV output power when PRC mode is operating on the V_1 side; P_{mpp} is the PV output power in MPPT mode; and $P_{deload2}$ is the PV output power when PRC mode is operating on the V_2 side.

In addition, whether operating in MPPT mode or PRC mode, ensuring the safe and reliable operation of the inverter is the necessary prerequisite for realizing grid-connected PV power generation. As shown in Figure 3b, the PV output voltage in the $[V_{in}, V_{out}]$ region can make the inverter work normally. If the output voltage of the PV generator is not in this area, the inverter will go into shutdown or standby state. If PRC mode runs on the V_1 side, when the irradiance changes from 600 to 300 W/m^2, the PV output voltage will not be in the workspace of the inverter. However, when running on the V_2 side, the voltage output is always in the operation area of the inverter.

Therefore, considering the response speed of PRC operation and the adjustable reserve capacity and ensuring the safe and efficient operation of the system, the ideal working area is expected to work on the right side of V_{mpp}, namely the V_2 side.

3.2. PV-PRC Implementation Analysis

Equation (3) shows that the premise of introducing active reserve is that the current maximum output power value P_{mpp} is a known amount. Therefore, the maximum power point estimation (MPPE) link is necessary for the PV-PRC operation, and the active power can be introduced when the MPPE link is found to be P_{mpp}. For the sake of simplicity, the MPPE method in the reference [17] is used, as shown by the red dashed box in Figure 4, taking two sets of PV modules or arrays of the same model and quantity as an example. When operating under the same operating conditions, PV1 operation in MPPT mode, its output power can be used as the available power value of PV2 ($P_{mpp1} = P_{mpp2}$). However, in [17], the PV output power is controlled by controlling the output voltage. The calculation of the voltage command in the process leads to a cumbersome control process. In view of this, this paper improves it by using direct power control to control the output power. In addition, in this kind of control mode, through the lowest voltage limit, the PRC runs with the V_2 side, and the control block diagram is shown in Figure 4.

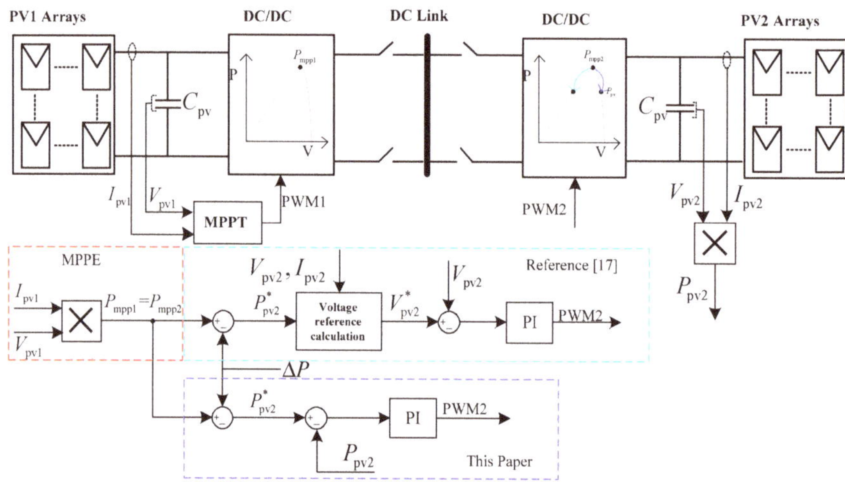

Figure 4. PV-PRC block diagram.

4. PV-PRC-VSG Control Strategy and Reserve Ratio Analysis

In PRC operation mode, a part of the power up-regulation capability is maintained in the PV system, so that the output of the PV system can be adjusted within a certain range, and the regulation of the reserve power has the same effect as the charging and discharging of the energy storage system, and the active output can be adjusted. With the VSG technology, the inertial response and the participating power system primary frequency response can be realized. In this mode, the output power of the virtual prime mover is as shown in Equation (5).

$$\begin{cases} J\omega\frac{d\omega}{dt} = P_m - P_e \\ P_m = P_{deload} + D_p(\omega_n - \omega) \end{cases} \tag{5}$$

Equation (5) can be reduced to:

$$P_e = P_{deload} - J\omega\frac{d\omega}{dt} + D_p(\omega_n - \omega) \tag{6}$$

where from left to right are PV output power, inertial response process demand power and primary frequency modulation demand power.

Define $P_f = -J\omega\frac{d\omega}{dt} + D_p(\omega_n - \omega)$, where P_f is the required power for frequency response.

Assuming that the PV system operates in PRC mode initially and maintains a certain active reserve ΔP, when the frequency change of the grid is detected to exceed the dead zone (± 0.03 Hz), the required power P_f for the PV system to participate in frequency regulation is calculated through the VSG control link. Then, by release or increasing the reserve power, it can participate in the frequency regulation of the power grid to provide power support for the power grid. The control block diagram of the whole system is shown in Figure 5.

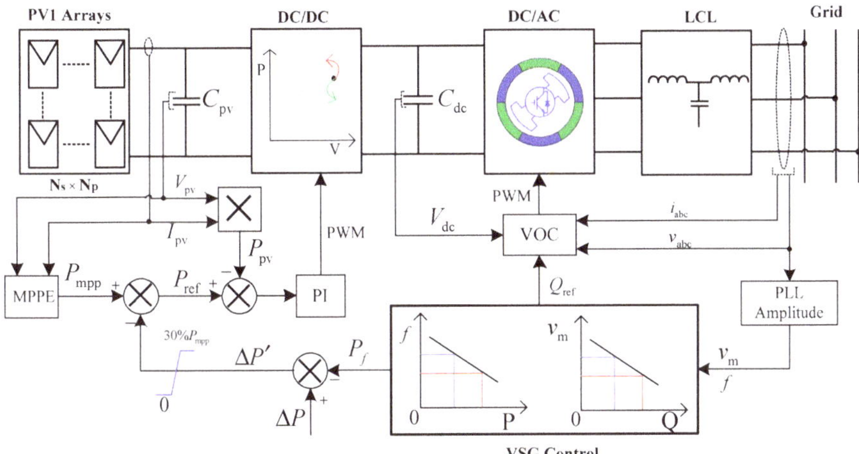

Figure 5. Principle of PRC-VSG implementation.

The PRC operation mode of the PV system essentially abandons part of the PV resources, which is in contradiction with the efficient use of energy and relevant national network codes. However, considering that the realization of PV-PRC-VSG technology can improve the stability of the power system, enhance the acceptance of the power grid to PV energy, and save the investment of ESSs, it is necessary to keep the power reduction within a certain range, and the proportion of the power reserve in the actual operation process is mainly constrained by the following factors:

1. Relevant regulations of the State Energy Administration pointed out that in solving the problem of clean energy consumption, the proportion of PV abandonment and power limitation will be reduced year by year. By the end of 2020, the problem of abandoned wind and PV will be basically solved nationwide (the abandoned water/wind/PV abandonment rate of the three northern regions will remain below 10%, and that of other regions will be below 5%).

2. Analysis of the PRC mode applicable period: Generally, the output of a PV power plants is like a "∩". At the initial and end moments, less PV energy is injected into the power system. The generating units in the power system are mainly undertaken by the SG (thermal power or hydropower). At this time, the equivalent inertia and primary frequency response capability of the system are relatively sufficient, and there is no active reserve ($\Delta P = 0$) in the PV system. In order to maximize energy utilization, the PV system operates in MPPT mode. With the increase of irradiation intensity, the output of the PV system increases gradually. The power system uses thermal power to regulate peak load. With thermal power cut out of the power grid, the equivalent inertia and primary frequency response ability in the power system are greatly reduced. The PV system reduces active output, maintains active reserve ($\Delta P \neq 0$), and participates in frequency response of the power grid.

3. The relevant requirements for PV-VSG to participate in primary frequency response stipulate that when the active power output of PV-VSG is greater than $20\%P_n$, it should have primary frequency response capability; when the frequency deviation exceeds the dead zone (± 0.03 Hz), PV-VSG should adjust the active output to participate in primary frequency response; in the primary frequency modulation process, the upper limit of active power can be increased at least $10\%P_n$, and the upper limit of active power can be reduced at least $20\%P_n$.

To sum up, considering the limitation of abandoning PV energies and the requirement of the VSG for primary frequency modulation, it is more appropriate to reduce power by $10\%P_{mpp}$. During the operation of the power system, when the system frequency increases, the active output of the PV system further reduces the frequency change of the response system by increasing the value of ΔP. When the system frequency decreases, the value of ΔP of the PV system needs to be reduced to release the reserve active power. It is noteworthy that when $P_f > \Delta P$, there is no active power up-regulation ability in the PV system. In addition, ΔP is still limited by the upper limit. When ΔP increases to $30\%P_{\text{mpp}}$, it cannot be increased any more. As shown in Equation (7), because of the limitation of active reserve capacity, the ability of PV-PRC-VSG to participate in primary frequency regulation is limited.

$$\Delta P' = \begin{cases} 30\%P_{\text{mpp}} & ,P_f < -20\%P_{\text{mpp}} \\ \Delta P - P_f & ,-20\%P_{\text{mpp}} \leq P_f < \Delta P \\ 0 & ,P_f \geq \Delta P \end{cases} \tag{7}$$

where $\Delta P'$ is a reference value of reserve power during primary frequency response.

5. Simulation of Proposed Method

To verify the effectiveness of the proposed control strategy, the corresponding simulation model was built in MATLAB/Simulink. The model topology adopts the grid-connected structure of the PV system in Figure 2. The corresponding experimental verification was made for a DC/DC inverter working in PRC mode and DC/DC and DC/AC inverter coordinated control participating in power system frequency response.

5.1. PV-PRC Simulation and Analysis

Firstly, the PRC operation control strategy of the DC/DC side proposed in the second section is simulated and analyzed. Taking the parameters of a single 240-W PV module (JLS60P240W) as an example, the parameters of PV modules are shown in Table 2. The maximum power value of the PV module is calculated in the MPPE process, and then the active reserve is introduced. For the convenience of calculation, the reserve ratio (R) is introduced. The value of R can be calculated by Equation (8). In the operation process control, R is the direct control object.

$$R = \Delta P / P_{\text{mpp}} \times 100\% \tag{8}$$

where: $V_{\text{oc}}(V)$-Open circuit voltage; $I_{\text{sc}}(A)$-short-circuit current; $V_{\text{mpp}}(V)$-Maximum Power Voltage; $I_{\text{mpp}}(A)$-Maximum Power Current; $\beta_{V\text{oc}}(\%/°C)$-temperature coefficient of open circuit voltage; $\alpha_{I\text{sc}}(\%/°C)$-temperature coefficient of short circuit current.

Table 2. PV module parameter data sheet.

$V_{\text{oc}}(\mathbf{V})$	$I_{\text{sc}}(\mathbf{A})$	$V_{\text{mpp}}(\mathbf{V})$	$I_{\text{mpp}}(\mathbf{A})$	$\beta_{V\text{oc}}(\%/°\mathbf{C})$	$\alpha_{I\text{sc}}(\%/°\mathbf{C})$
36.0	8.96	29.4	8.0	−0.37	0.006

The primary task of PRC operation is to calculate the value of P_{mpp} at each moment; that is, the MPPE process. MPPE plays a vital role in the whole PRC operation process, which directly affects the accuracy of the power reserve. In Figure 6a, the red dotted line is the maximum output power (actual P_{mpp} value) of the PV system in MPPT mode, and the black dotted line is the P_{mpp} value calculated by MPPE, which deviates little from the actual value, thus laying a foundation for subsequent experiments.

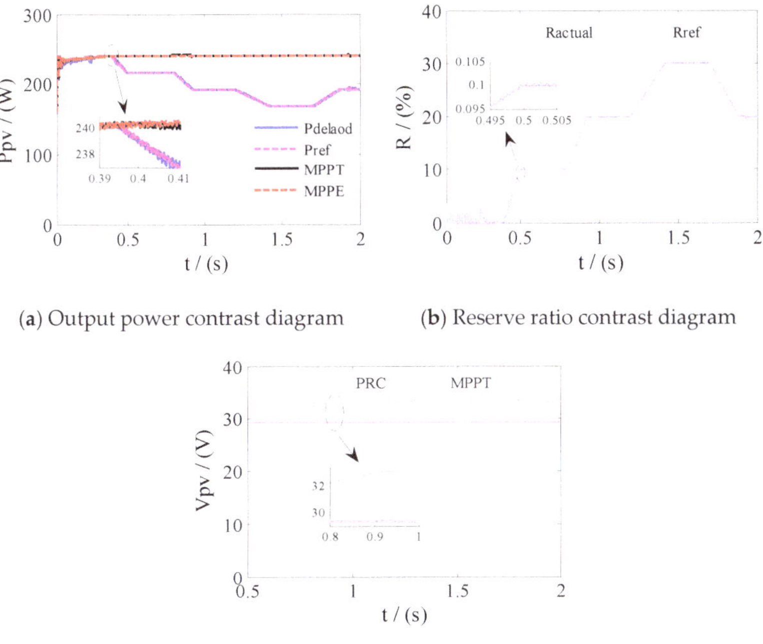

(**a**) Output power contrast diagram (**b**) Reserve ratio contrast diagram

(**c**) Output power contrast diagram

Figure 6. PV-PRC mode simulation waveform.

As shown in Figure 6b, there is no active reserve in the PV system before 0.4 s, and the PV system works in MPPT mode; between 0.4–2 s, the PV system maintains the active reserve in PRC mode. In Figure 5b, the solid line is the given value of the reserve rate, and the dotted line is the actual reserve rate value in the simulation operation. It has good tracking accuracy in the whole operation process. The output power is shown by the blue solid line in Figure 6a. It is noteworthy that the direct power reduction method used in this paper can work in MPPT and PRC modes in a time-sharing manner according to the actual needs, and there is no need to switch control modes, which provides greater convenience for engineering implementation.

Figure 6c is a comparison of output voltage of MPPT mode and PRC mode. It was found that the output voltage in PRC mode is always greater than that in MPPT mode. It is proved that in PRC mode, the output voltage always works on the right side of the P–V curve, which is the same as the expected result in Section 3. It is interesting to find that the DC voltage ripple in PRC mode is much smaller than that in MPPT mode.

5.2. PV-PRC-VSG

In the previous section, the PV-PRC implementation and the voltage operating point were verified. The PV-PRC-VSG technology aims to enable the PV system to provide a certain power support for the power system. This requires DC/DC side and DC/AC side coordinated control. When the system power shortage causes the frequency to change, the DC/DC side provides power support for the power system by adjusting the standby rate R to release or expand the reserve power. The simulation parameters are shown in Table 3.

Table 3. Simulation parameters.

Parameter	Selection	Parameter	Selection
PV module	JLS60P240W	R	$10\%P_{\mathrm{mpp}}$
Ns × Np	12 × 7	P_{mpp}	20 kW
P_{deload}	18 kW	D_{p}	10,000/2 pi

- System frequency rise

The simulation time is 1.5 s, and the system frequency does not change from 0 to 0.7 s. At 0.7 s, the power system has the problem of excess power, which leads to the sudden change of the system frequency from 50 Hz to 50.3 Hz, and it returns to normal at 1.3 s. The frequency variation of the system is shown in Figure 7a. Because of the inertia, the rise curve of the system frequency is smoother. Equation (7) shows that when the system frequency increases, the reserve power rises to 5 kW, and the reserve rate increases to 25% as calculated by Equation (8), and as shown in Figure 7b. The active power of PV output is further reduced to 15 kW, as shown in Figure 7c, and the output voltage and current of the inverter are shown in Figure 7d. The results are in agreement with those of the formulas of Equations (7) and (8). The whole response process essentially maintains power system stability by abandoning a part of PV energy.

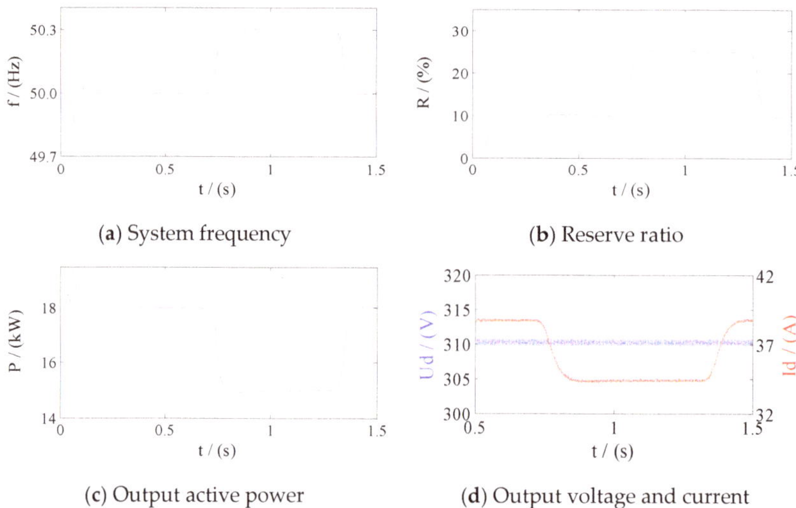

(**a**) System frequency

(**b**) Reserve ratio

(**c**) Output active power

(**d**) Output voltage and current

Figure 7. System output waveform when frequency increases by 0.3 Hz.

- System Frequency Drop in a Small Range

The simulation parameter setting and the initial reserved power amount are the same as above. When the system loses part of the external power due to generator failure at 0.7 s, the system frequency is reduced from 50 to 49.9 Hz, and it returns to normal at 1.3 s, as shown in Figure 8a. When the system frequency drop is detected, the PV system needs to release a part of the reserve active power. According to Equations (7) and (8), the standby rate reference value needs to be changed from 10 to 5%, as shown in Figure 8b. During the frequency drop, a part of the active power is released by the reduction of the reserve ratio to participate in the primary frequency response of the power grid, and the output power is as shown in Figure 8c, and Figure 8d shows the output voltage and current of the inverter.

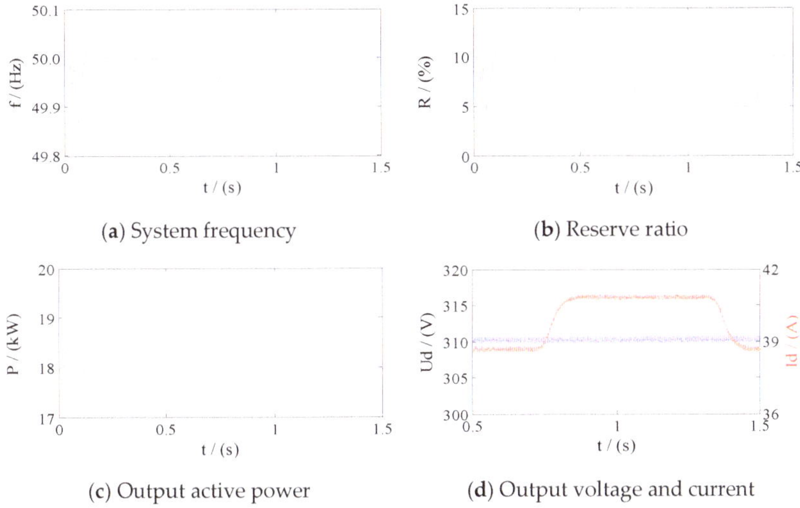

(a) System frequency

(b) Reserve ratio

(c) Output active power

(d) Output voltage and current

Figure 8. System output waveform when frequency decreases by 0.1 Hz.

- Serious Decrease of System Frequency

The initial parameters of the simulation are the same as above. When the large load is suddenly cut in 0.7–1.3 s, the system frequency is reduced from 50 to 49.7 Hz. According to the setting, the PV system reduces the reserve ratio and increases the active output. Due to the limited active reserve of the PV system, the frequency response will be limited. Equation (7) shows that when the frequency drops by more than 0.2 Hz, the PV system will release all reserve power. Therefore, under this condition, the PV system reserve ratio is changed from 10 to 0%, all the active reserve is released, and the grid is integrated into the grid in MPPT mode. The simulation diagram is shown in Figure 9a–d. It can be seen that the frequency modulation effect of the PV-ESSs-VSG system in reference [13] can be achieved without external ESSs, and the active power support time provided is longer.

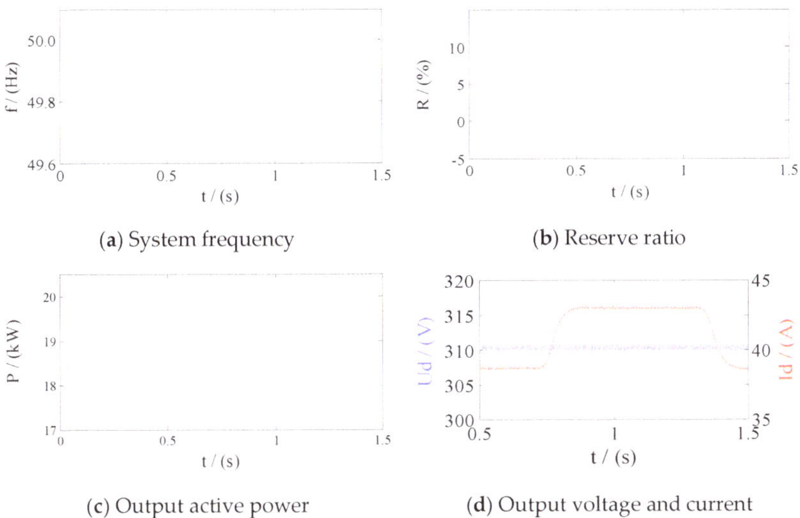

(a) System frequency

(b) Reserve ratio

(c) Output active power

(d) Output voltage and current

Figure 9. System output waveform when frequency decreases by 0.3 Hz.

6. Conclusions

Aiming at the problem that the traditional PV-VSG technology relies on external ESSs, which leads to a large increase in the cost of large-scale PV-PPs construction and limits its application in engineering, this paper puts forward PV-PRC-VSG technology. Boosting the converter realizes the active reserve of the PV system, power output can be adjusted to realize the independent frequency modulation of PV system, and DC/AC inverter to achieve power conversion, and reactive power control. The PV system participates in the frequency response of the power system. Through the analysis and experimental verification, the following conclusions are drawn:

- The PRC method of the PV system in this paper can work in MPPT and PRC modes in a time-sharing manner according to actual needs and does not need to switch the control strategy;
- Considering the efficient and safe operation of the inverters, it is determined that the voltage operating point in PV-PRC mode should be on the right side of V_{mpp}, and the reserve rate should be 10% considering the abandonment of PV energy and frequency modulation ability of the participating system;
- The proposed PV-PRC-VSG control strategy can actively participate in the frequency response of the power system without additional ESSs. It also has a longer power support time, but it is also constrained by the reserve power capacity.

Author Contributions: The research method and control scheme of this paper was put forward by G.B., and H.T. was responsible for the analysis of the experimental results and the writing of the manuscript of the paper. Special thanks were given to K.D., M.M. and N.W. senior engineers of State Grid Gansu Electric Power Corporation Electric Power Research Institute for their practical engineering background and related data.

Funding: This research was supported by State Grid Corporation Science and Technology Project (5227 2217006), and National Natural Science Foundation of China (No. 51267001), and 2018 GanSu university scientific research project - innovation team.

Conflicts of Interest: The authors declare no conflict of interest.

Appendix A

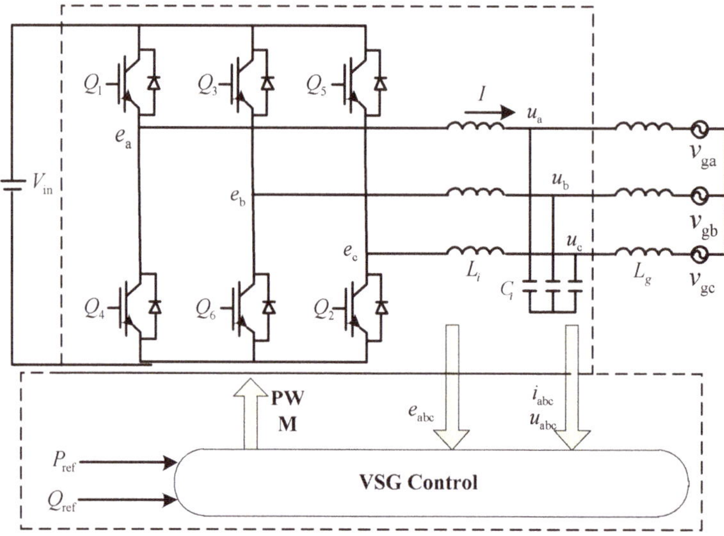

Figure A1. VSG structure in this paper.

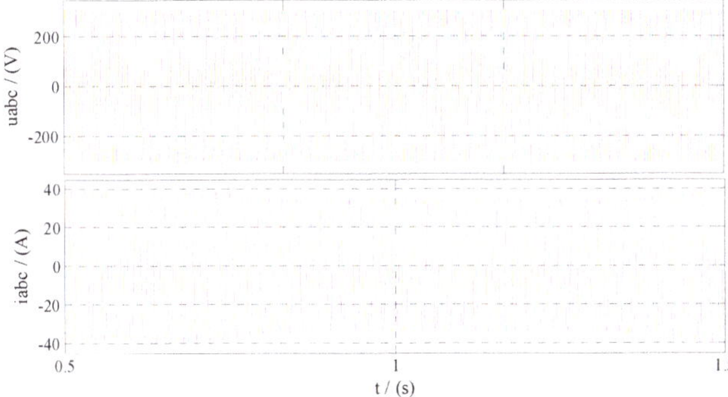

Figure A2. System output voltage and current when the frequency increases by 0.3 Hz.

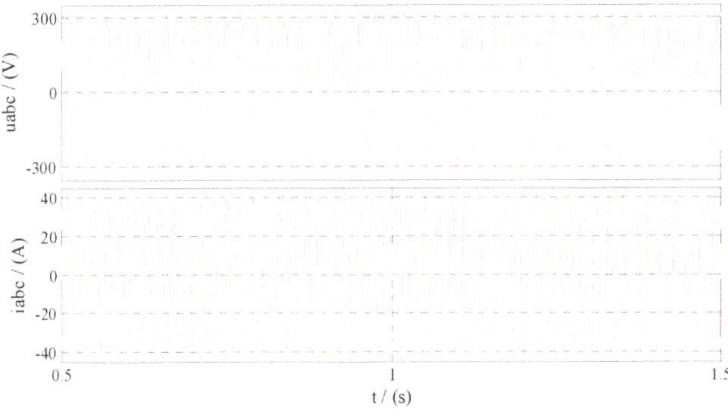

Figure A3. System output voltage and current when the frequency decreases by 0.1 Hz.

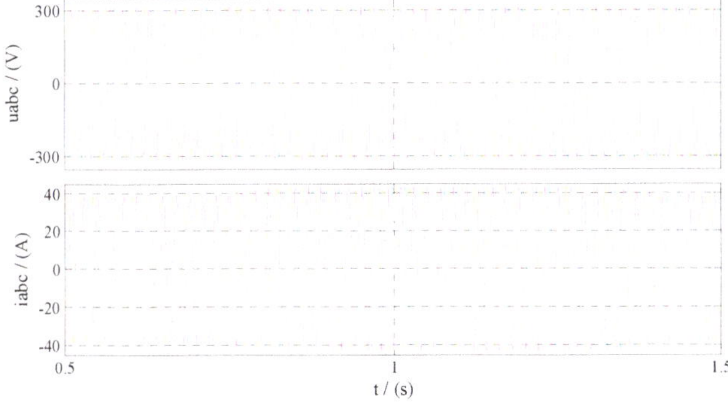

Figure A4. System output voltage and current when the frequency decreases by 0.3 Hz.

References

1. Zhou, X.; Lu, Z.; Liu, Y.; Chen, S. Development models and key technologies of future grid in China. *Proc. CSEE* **2014**, *34*, 4999–5008.
2. Rahmann, C.; Castillo, A. Fast frequency response capability of photovoltaic power plants. The necessity of new grid requirements and definitions. *Energies* **2014**, *7*, 6306–6322. [CrossRef]
3. Liu, Y.; You, S.; Tan, J.; Zhang, Y.; Liu, Y. Frequency response assessment and enhancement of the U.S. Power grids towards extra-high photovoltaic generation penetrations—An industry perspective. *IEEE Trans. Power Syst.* **2018**, *33*, 3438–3449. [CrossRef]

4. Ming, D.; Sheng, W.W.; Wang, X.; Song, Y.T.; Chen, D.Z.; Sun, M. A review on the effect of large-scale PV generation on power systems. *Proc. CSEE* **2014**, *34*, 1–14.

5. Eftekharnejad, S.; Vittal, V.; Heydt, G.T.; Keel, B.; Loehr, J. Impact of increased penetration of photovoltaic generation on power systems. *IEEE Trans. Power Syst.* **2013**, *28*, 893–901. [CrossRef]

6. Tamimi, B.; Cañizares, C.; Bhattacharya, K. System stability impact of large-scale and distributed solar photovoltaic generation: The case of Ontario, Canada. *IEEE Trans. Sustain. Energy* **2013**, *4*, 680–688. [CrossRef]

7. Koran, A.; LaBella, T.; Lai, J.S. High efficiency photovoltaic source simulator with fast response time for solar power conditioning systems evaluation. *IEEE Trans. Power Electron.* **2014**, *29*, 1285–1297. [CrossRef]

8. Zhong, Q.C.; Weiss, G. Synchronverters: Inverters that mimic synchronous generators. *IEEE Trans. Ind. Electron.* **2011**, *58*, 1259–1267. [CrossRef]

9. Zhong, Q.C.; Nguyen, P.L.; Ma, Z.; Sheng, W. Self-synchronized synchronverters: Inverters without a dedicated synchronization unit. *IEEE Trans. Power Electron.* **2014**, *29*, 617–630. [CrossRef]

10. Alipoor, J.; Miura, Y.; Ise, T. Power system stabilization using virtual synchronous generator with alternating moment of inertia. *IEEE J. Emerg. Sel. Top. Power Electron.* **2015**, *3*, 451–458. [CrossRef]

11. Kakimoto, N.; Takayama, S.; Satoh, H.; Nakamura, K. Power modulation of photovoltaic generator for frequency control of power system. *IEEE Trans. Energy Convers.* **2009**, *24*, 943–949. [CrossRef]

12. Hill, C.A.; Such, M.C.; Chen, D.; Gonzalez, J.; Grady, W.M. Battery energy storage for enabling integration of distributed solar power generation. *IEEE Trans. Smart Grid* **2012**, *3*, 850–857. [CrossRef]

13. Yu, G.; Yang, W.; Zhi, L.; Shi, X.; Song, P.; Yang, W.X.; Zheng, W. Engineering application effect analysis and optimization of photovoltaic virtual synchronous generator. *Autom. Elect. Power Syst.* **2018**, *42*, 149–156.

14. Consulting Group of State Grid Corporation of China to Prospects of New Technologies in Power Systems. An analysis of prospects for application of large-scale energy storage technology in power systems. *Autom. Elect. Power Syst.* **2013**, *37*, 3–8.

15. Xin, H.; Liu, Y.; Wang, Z.; Gan, D.; Yang, T. A new frequency regulation strategy for photovoltaic systems without energy storage. *IEEE Trans. Sustain. Energy* **2013**, *4*, 985–993. [CrossRef]

16. Sangwongwanich, A.; Yang, Y.; Blaabjerg, F. A sensorless power reserve control strategy for two-stage grid-connected PV systems. *IEEE Trans. Power Electron.* **2017**, *32*, 8559–8569. [CrossRef]

17. Sangwongwanich, A.; Yang, Y.; Blaabjerg, F.; Sera, D. Delta power control strategy for multistring grid-connected PV inverters. *IEEE Trans. Ind. Appl.* **2017**, *53*, 3862–3870. [CrossRef]

18. Batzelis, E.I.; Kampitsis, G.E.; Papathanassiou, S.A. Power reserves control for PV systems with real-time MPP estimation via curve fitting. *IEEE Trans. Sustain. Energy* **2017**, *8*, 1269–1280. [CrossRef]

19. Batzelis, E.I.; Papathanassiou, S.; Pal, B.C. PV system control to provide active power reserves under partial shading conditions. *IEEE Trans. Power Electron.* **2018**, *33*, 9163–9175. [CrossRef]

20. Li, X.; Wen, H.; Zhu, Y.; Jiang, L.; Hu, Y.; Xiao, W. A novel sensorless photovoltaic power reserve control with simple real-time MPP estimation. *IEEE Trans. Power Electron.* **2019**, *34*, 7521–7531. [CrossRef]

21. Hoke, A.; Maksimović, D. Active power control of photovoltaic power systems. In Proceedings of the 2013 1st IEEE Conference on Technologies for Sustainability (SusTech), Portland, OR, USA, 1–2 August 2013; pp. 70–77.

22. Hoke, A.F.; Shirazi, M.; Chakraborty, S.; Muljadi, E.; Maksimovic, D. Rapid active power control of photovoltaic systems for grid frequency support. *IEEE J. Emerg. Sel. Top. Power Electron.* **2017**, *5*, 1154–1163. [CrossRef]

23. Bevrani, H. Virtual synchronous generators: A survey and new perspectives. *Int. J. Electr. Power Energy Syst.* **2014**, *54*, 244–254. [CrossRef]

Article

Research on the Filters for Dual-Inverter Fed Open-End Winding Transformer Topology in Photovoltaic Grid-Tied Applications

Baoji Wang [1], Xing Zhang [1,*], Chao Song [1] and Renxian Cao [2]

[1] School of Electrical Engineering and Automation, Hefei University of Technology, Hefei 230009, China; mail.wbj@gmail.com (B.W.); songchaocumt@gmail.com (C.S.)

[2] Sungrow Power Supply Co., Ltd., Hefei 230088, China; crx@sungrow.cn

* Correspondence: honglf@ustc.edu.cn; Tel.: +86-136-0560-1932

Received: 3 June 2019; Accepted: 17 June 2019; Published: 18 June 2019

Abstract: Owing to the necessity of the transformer for the multi-parallel inverters connected to the medium-voltage (MV) grid, the conventional multi-parallel inverter topology can be reconfigured to the dual-inverter fed open-end winding transformer (DI-OEWT) topology to obtain lower output voltage harmonics, which can reduce the requirement of the filter inductance. However, due to the special structure of the DI-OEWT topology, the arrangement scheme of the filter can be more than one kind, and different schemes may affect the filter performance. In this paper, research on the existing two kinds of filters, as well as a proposed one, for the DI-OEWT topology used in photovoltaic grid-tied applications is presented. The equivalent circuits of these filters are derived, and based on this, the harmonic suppression capability of these filters is analyzed and compared. Furthermore, a brief parameter design method of these filters is also introduced, and based on the design examples, the inductance and capacitance requirements of these filters are compared. In addition, these filters are also evaluated in terms of the applicability for fault tolerance. At last, the analysis is verified through an experiment on a 30 kW dual-three-level inverter prototype.

Keywords: dual inverter; open-end winding transformer; photovoltaic application; filter

---

1. Introduction

In recent years, the increasing exhaustion of traditional fossil energy has resulted in an emerging interest in the development of renewable energies, one of which is solar energy that has already obtained widespread applications. Among all types of commercial solar energy applications, photovoltaic grid-tied system plays an important role, and large-scale photovoltaic power plants have become dominant [1].

In the large-scale photovoltaic power plants, three-phase single-stage central inverters are widely used because of their many advantages, such as cost-effectiveness, simplicity in hardware design and easy maintenance. The two-level, three-level T-type, and three-level neutral point clamp (NPC) voltage source inverters (VSI) are the most widely used topologies among the current commercial central inverters [2]. Furthermore, to increase the transmission efficiency of the overall electrical equipment, the inverters are generally connected to a medium-voltage (MV) grid of a voltage level from 10 kV to 35 kV. Additionally, multi-parallel inverter topology, in which the inverters are connected in parallel through step-up MV transformers, are commonly used in these systems. A typical commercial application example is 1 MW MV turnkey station, which is composed of two 500 kW central inverters and one 1 MVA MV transformer [3]. Since the transformer is essential for the inverters connected to the MV grid, the multi-parallel inverter topology has the possibility to be reconfigured to the dual-inverter fed open-end winding transformer (denominated here as DI-OEWT) topology.

The dual-inverter topology was first proposed for the motor drive application in [4]. Since then, it has already been extended to many applications, such as STATCOM [5], active power filter [6], dynamic voltage restorer [7], photovoltaic grid-tied inverter [8]. It utilizes dual-inverter structure connected to the open-end windings of an induction motor or a three-phase transformer. Through proper modulation strategy, the harmonic cancellation effect can be realized among the two inverters, the dual-inverter topology with two N-level inverters can have the same output voltage levels as a (2N-1)-level inverter [9]. Therefore, lower dv/dt and lower harmonics in the output voltage can be obtained, which can reduce the filter requirement. It can also double the DC voltage utilization. A tdual-N-level inverter with half the DC link voltage (compared to a conventional single inverter scheme) is capable of producing the same AC voltage as a single (2N-1) level inverter, which will reduce the voltage capacity of the power switch devices and decrease the switching losses [10]. What is more, the DC sources requirement of the dual-inverter is minimal over other multilevel topologies [11–13]. The two inverters of the dual-inverter can also be supplied by one single DC source for cost saving [14]. Another merit of the dual-inverter topology is the availability of higher redundant switching state combinations compared to the single inverter, which can be used to achieve switching frequency reduction [15], common-mode voltages suppression [16], capacitor voltage balance [17]. Furthermore, it also offers the advantage of fault tolerance. In case of a fault in the inverter, the inverter can still work with some adjustment [18–20].

However, the advantages of the dual-inverter topology described above are not all applicable to the topology used in photovoltaic grid-tied applications. Firstly, the DC bus voltage is limited to the photovoltaic array voltage, which is usually 1000 V or 1500 V (open-circuit voltage). It is uneconomic to reduce this voltage because that will increase the system installation costs and narrow the maximum power point tracking (MPPT) voltage operation range [2]. In addition, it will be better for the dual-inverter to be supplied by two separate arrays (two DC sources), which can not only achieve multiple MPPTs, but also suppress the circulating current among the two inverters [21,22]. The fault tolerance capability cannot be a special advantage of the dual-inverter because the multi-parallel inverter topology has the same capability as well. To sum up, compared with the conventional multi-parallel inverter topology, the most attractive advantage of the DI-OEWT topology is the improvement of the harmonic quality in the output voltages, thus can reduce the filter requirement and save the filter costs.

However, there are few papers considering the selection or design of the filters used in DI-OEWT-based grid-tied applications at present. In [23–28], single inductor filter was adopted in these DI-OEWT based grid-tied systems, which is obviously not a good choice. Owing to the weak harmonic suppression capability, it needs a large inductance value to meet the grid standard. To reduce the inductance cost, high order filter is preferred [29]. In [30], two kinds of high order filters for DI-OEWT topology were proposed. One is the "individual capacitor type filter", that the two inverters of the DI-OEWT topology are connected to the open-end windings of the transformer through two individual inductor-capacitor filters. The other one is the "common capacitor type filter", which is also presented in [21,22], that the two inverters are connected to the open-end windings of the transformer through two individual inductors but one common capacitor. However, the authors in [30] were mainly focused on active damping methods of the two different filters, and the authors in [21,22] were mainly focused on the magnetic integration design of the filter inductors, none of these papers analyzed the harmonic suppression capability or the parameter design method of these filters.

In addition, the transformer's leakage inductance is a significant component of the filter, and for the transformer used in the high power MV grid-tied system, the value is usually no less than 6% [31]. This is a relatively big value and should be used properly in the selection and design of the filter, but none of the present literatures has paid attention to this. Another point deserves to be considered is the fault tolerance scheme of the DI-OEWT topology, which is very important for such multi-inverter type topology.

In light of the above, this paper aims to research the filters for the DI-OEWT topology used in photovoltaic grid-tied applications. First, the equivalent circuits of the existing two kinds of high order filters presented in [21,22,30] are derived, and based on this, the harmonic suppression capabilities of these filters are analyzed and compared. According to the analysis results, a new high order filter for DI-OEWT topology is also proposed. Furthermore, a brief parameter design method of the existing and the proposed filters for DI-OEWT is introduced and based on the design examples, the inductance and capacitance requirements of these filters are compared. Besides, these filters are also evaluated in terms of the applicability for fault tolerance.

The rest of the paper is organized as follows: The inductor-capacitor-inductor (LCL) filter used in multi-parallel inverters is presented in Section 2 as the comparison object. The model and harmonic suppression capability analysis of the existing and the proposed filters for DI-OEWT topology are shown in Section 3. The filter design method, together with the design examples, as well as the fault-tolerant scheme, are presented in Section 4. Experimental results are given in Section 5 to validate the filter parameters. Finally, the conclusions are summarized in Section 6.

2. LCL Filter for the Conventional Multi-Parallel Inverters

The system configuration of the conventional multi-parallel inverters used in photovoltaic applications is illustrated in Figure 1. The two inverters are connected in parallel to the low voltage side of the MV transformer through individual LC filters. The leakage inductance L_t' of the transformer and the grid inductance L_g' play the role of the grid-side filter inductor, so the two inverters can also be viewed as connecting in parallel through individual LCL filters.

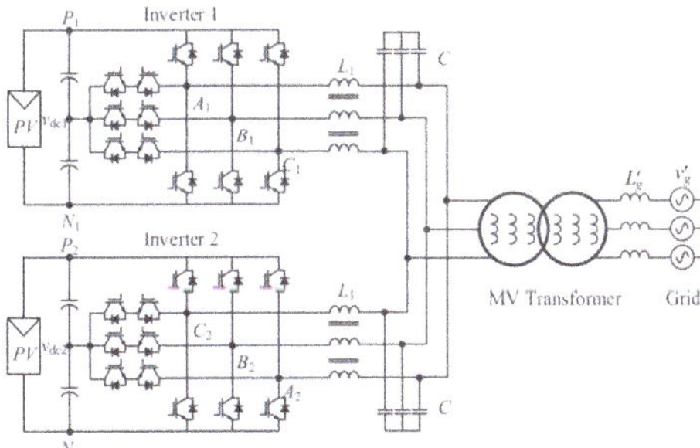

Figure 1. System configuration of the conventional multi-parallel inverters.

It is essential to derive the equivalent circuit of the filter based VSI for the filter design and harmonic suppression capability analysis. Since the LCL filter used in parallel inverters has no difference with that used in single inverters, the single-phase equivalent circuit of the LCL filter can be directly obtained from many existing papers [32,33], as shown in Figure 2. In the figure, the equivalent circuit is drawn referred to the low-voltage side of the transformer, v_{inv1} and i_{inv1} (take inverter 1 as an example) are the inverter output phase voltage and current, respectively. v_g and i_g are the grid voltage and current, respectively. The inverter-side inductor is represented as L_1, the combined inductance of the transformer leakage inductance L_t and the grid inductance L_g are represented as L_2, and the filter capacitor is represented as C.

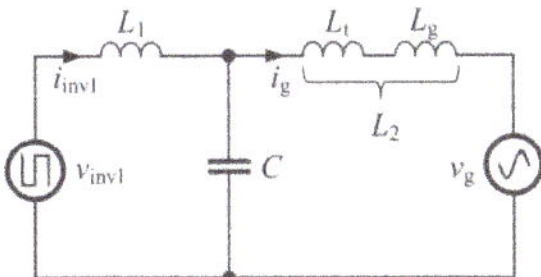

Figure 2. Single-phase equivalent circuit of the LCL filter.

While the inverter is working, the harmonics in i_{inv1} should be suppressed by the filter to limit the current ripple, and the harmonics in i_g should be suppressed to meet the grid standard [34]. Assuming that no harmonics exist in the grid voltage, the inverter output voltage v_{inv1} is the only harmonic source of the system, then the transfer functions $v_{inv1}(s)$ to $i_g(s)$ and $i_{inv1}(s)$ can be used to reflect the harmonic suppression capability of the LCL filter, as shown in Equations (1) and (2), respectively

$$i_g(s) = \frac{1}{L_1 L_2 C s^3 + (L_1 + L_2)s} v_{inv1}(s) \tag{1}$$

$$i_{inv1}(s) = \frac{L_2 C s^2 + 1}{L_1 L_2 C s^3 + (L_1 + L_2)s} v_{inv1}(s) \tag{2}$$

where $L_2 = L_t + L_g$. The value of L_t and L_g is related to the transformer impedance voltage V_k and the grid short-circuit ratio (SCR), respectively [35]. Therefore, L_2 can be calculated as

$$L_2 = \frac{3v_g^2}{2\pi f_0 P_{rated}} V_k + \frac{3v_g^2}{2\pi f_0 P_{rated} SCR} = \frac{3v_g^2}{2\pi f_0 P_{rated}} \left(V_k + \frac{1}{SCR} \right) \tag{3}$$

where P_{rated} is the system rated power, f_0 is the fundamental frequency.

3. Existing and Proposed Filters for the DI-OEWT Topology

3.1. Type-1 Filter for the DI-OEWT Topology

Direct replacing of the two-winding transformer in the multi-parallel inverter topology (as shown in Figure 1) with the open-end winding (OEW) transformer, can obtain the DI-OEWT topology with the "individual capacitor type filter" (here we call it "Type-1 filter"). The filter was proposed in [30], as illustrated in Figure 3. It is worth noting that the voltage of the low-voltage side of the OEW transformer is twice the voltage of the replaced transformer, since the AC sides of the two inverters are connected in series through the low-voltage side of the transformer [8]. Accordingly, the current level of the inverter remains unchanged, so the multi-parallel inverter topology can be easily reconfigured to the DI-OEWT topology.

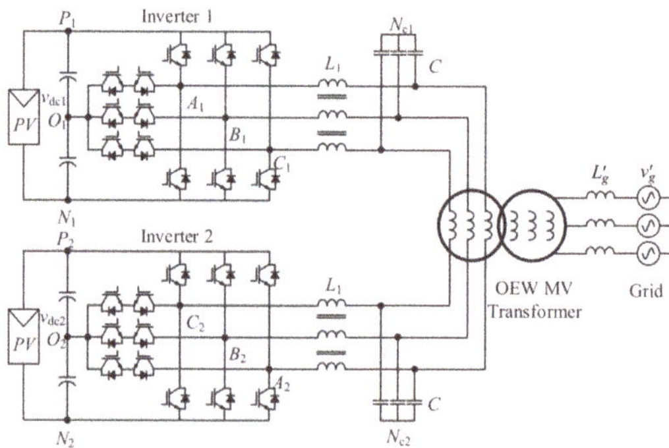

Figure 3. System configuration of the dual-inverter fed open-end winding transformer (DI-OEWT) topology with the Type-1 filter.

To analyze the Type-1 filter, firstly, the three-phase equivalent circuit of the DI-OEWT topology with the Type-1 filter is derived in Figure 4. The voltage source V_{X1O1} and V_{X2O2} represent the pole voltage of phase X_1 of inverter 1 and the pole voltage of phase X_2 of inverter 2, respectively ($X = A$, B, C). In a balanced and symmetrical three-phase system, for inverter 1, a common-mode (CM) voltage exists between the neutral point O_1 and the capacitor common point N_{c1}, this CM voltage is expressed as $V_{O1Nc1} = -1/3(V_{A1O1} + V_{B1O1} + V_{C1O1})$. Similarly, the corresponding CM voltage of inverter 2 is expressed as $V_{O2Nc2} = -1/3(V_{A2O2} + V_{B2O2} + V_{C2O2})$. According to Kirchhoff's voltage law (KVL), the voltage across the two capacitor common points N_{c1}, N_{c2} can be derived as

$$V_{Nc1Nc2} = \frac{1}{3}\left[\begin{array}{l} -(V_{CA1} + V_{CB1} + V_{CC1}) + (V_{CA2} + V_{CB2} + V_{CC2}) \\ +(V_{L\sigma1A} + V_{L\sigma1B} + V_{L\sigma1C}) + (V_{tA1} + V_{tB1} + V_{tC1}) \end{array} \right] \tag{4}$$

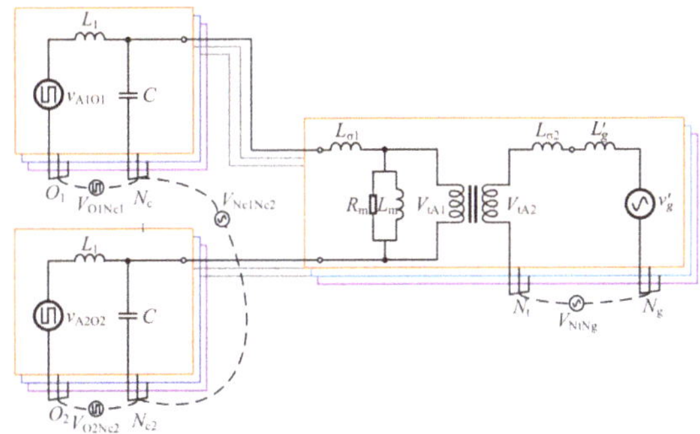

Figure 4. Three-phase equivalent circuit of the DI-OEWT topology with the Type-1 filter.

The voltage across the neutral point N_t of the transformer high voltage side and the neutral point N_g of the three-phase grid can be derived as

$$V_{NtNg} = \frac{1}{3}\left[\begin{array}{l} (V_{tA2} + V_{tB2} + V_{tC2}) + (V_{L\sigma2A} + V_{L\sigma2B} + V_{L\sigma2C}) \\ +\left(V_{LgA} + V_{LgB} + V_{LgC}\right) + \left(v'_{gA} + v'_{gB} + v'_{gC}\right) \end{array} \right] \tag{5}$$

In Equations (4) and (5), V_{CX1} and V_{CX2} represent the voltage on the filter capacitor of phase X_1 and phase X_2, respectively. $V_{L\sigma1X}$ and $V_{L\sigma2X}$ represent the phase X voltage on the leakage inductance at low voltage side ($L_{\sigma1}$) and high voltage side ($L_{\sigma2}$) of the MV transformer, respectively. V_{LgX} represents the phase X voltage on the grid inductance L'_g. V_{tX1} and V_{tX2} represent the voltage of the low and high voltage side of the MV transformer, respectively. v'_{gX} is the phase X voltage of the MV grid. Considering only the line frequency component and the balanced three-phase system, it is easy to deduce that $V_{Nc1Nc2} = 0$ and $V_{NtNg} = 0$. Thus, N_{c1} can be connected to N_{c2} and N_t can be connected to N_g. Then, the single-phase equivalent circuit of the DI-OEWT topology with the Type-1 filter can be derived as shown in Figure 5 with further simplification. Where the inverter 1 output phase voltage $v_{inv1} = V_{A1O1} + V_{O1Nc1}$, the inverter 2 output phase voltage $v_{inv2} = V_{A2O2} + V_{O2Nc2}$, i_{inv1} and i_{inv2} represent the output phase current of inverter 1 and inverter 2, respectively.

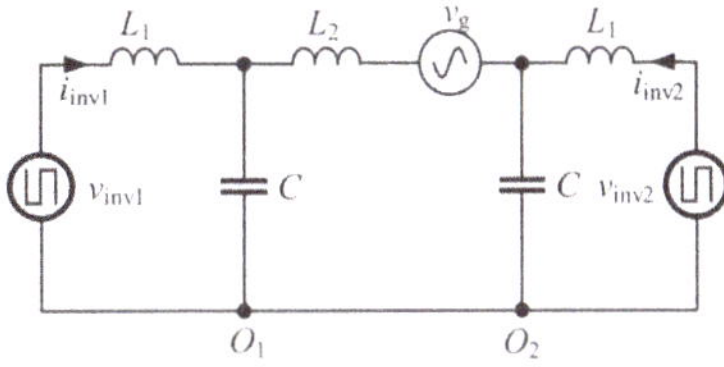

Figure 5. Single-phase equivalent circuit of the DI-OEWT topology with the Type-1 filter.

From Figure 5, the transfer functions $v_{inv1}(s)$, $v_{inv2}(s)$ to $i_g(s)$ and $i_{inv1}(s)$ can be, respectively, calculated as

$$i_g(s) = \frac{1}{L_1 L_2 C s^3 + (2L_1 + L_2)s}[v_{inv1}(s) - v_{inv2}(s)] \tag{6}$$

$$i_{inv1}(s) = G_1(s)v_{inv1}(s) - G_2(s)v_{inv2}(s) \tag{7}$$

where

$$i_{\text{inv1}}(s) = G_1(s)v_{\text{inv1}}(s) - G_2(s)v_{\text{inv2}}(s)$$

$$G_2(s) = \frac{1}{L_1^2 L_2 C^2 s^5 + \left(2L_1^2 C + 2L_1 L_2 C\right)s^3 + (2L_1 + L_2)s}$$

Figure 6a shows Bode plots of the transfer function $v_{\text{inv1}}(s)$, $v_{\text{inv2}}(s)$ to $i_g(s)$ of both the LCL filter and the Type-1 filter, while the inverter-side inductor L_1, capacitor C, impedance voltage V_k and SCR are all the same in both filters. Figure 6b presents Bode plots of the transfer function $v_{\text{inv1}}(s)$, $v_{\text{inv2}}(s)$ to $i_{\text{inv}}(s)$ with the aforementioned parameters. According to Figure 6 and Equations (1), (2), (6), (7), the following can be obtained.

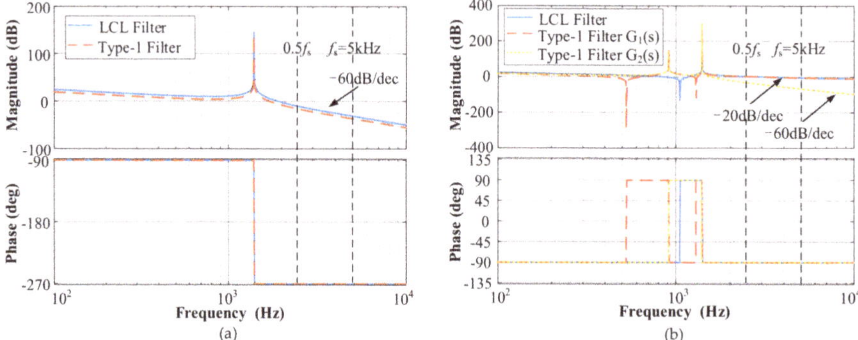

Figure 6. Bode plots of transfer function $v_{\text{inv1}}(s)$, $v_{\text{inv2}}(s)$ to $i_g(s)$ and $i_{\text{inv1}}(s)$ in LCL and Type-1 filters (**a**) $v_{\text{inv1}}(s)$, $v_{\text{inv2}}(s)$ to $i_g(s)$, (**b**) $v_{\text{inv1}}(s)$, $v_{\text{inv2}}(s)$ to $i_{\text{inv1}}(s)$.

(1) Figure 6a suggests that, for i_g, Type-1 filter has superior high-frequency harmonic suppression capabilities than the LCL filter. Meanwhile, compare Equation (6) with (1), the harmonic source of i_g in DI-OEWT topology is $v_{\text{inv1}} - v_{\text{inv2}}$, while in multi-parallel inverter topology is v_{inv1}. Owing to the harmonic cancellation effect in the output voltage of DI-OEWT, $v_{\text{inv1}} - v_{\text{inv2}}$ has lower harmonics than v_{inv1}. Thus, i_g in DI-OEWT can achieve a much better current quality than the multi-parallel inverter topology. Conversely, Type-1 filter can reduce filter requirement than LCL filter. Since L_2 is limit by the V_k and SCR, which cannot be reduced, so L_1 or C may be reduced.

(2) From Equation (7), for the inverter output phase current i_{inv1}, $G_1(s)$ and $G_2(s)$ can reflect the suppression capability of the Type-1 filter on v_{inv1} and v_{inv2}, respectively. From Figure 6b, in the high-frequency band, the harmonics attenuation rate of $G_2(s)$ is −60 dB/dec, way above the harmonics attenuation rate of $G_1(s)$, which is −20 dB/dec. Therefore, the harmonic source of i_{inv1} in DI-OEWT topology is mainly the v_{inv1}, the same as the voltage in multi-parallel inverter topology according to Equation (2). Meanwhile, the harmonics attenuation rate of LCL filter is also −20 dB/dec, thus, i_{inv1} of both the scheme has almost the same current quality. Because the current ripple of i_{inv1}, which is mainly suppressed by inverter-side inductor L_1, is strictly limited by the semiconductor current rating and loss requirement. Therefore, L_1 of Type-1 filter cannot be reduced.

In summary, compared with the multi-parallel inverter topology, the DI-OEWT topology with Type-1 filter can only reduce the value of filter capacitor C, showing that the Type-1 filter maybe not a better solution for the DI-OEWT topology.

Another way to explain why the Type-1 filter has the above characteristics can be shown as follows. Firstly, the reason why the DI-OEWT topology can achieve multilevel output and lower harmonics in output voltages is the harmonic cancellation effect between the two inverters. However, the two individual sets of shunt capacitors of the Type-1 filter break this cancellation loop. Some high-frequency harmonics, which should have been canceled, flow through these capacitors, leading to the increase of filter burden. In other words, Type-1 filter does not make judicious use of the merit of the DI-OEWT topology.

3.2. Type-2 Filter for the DI-OEWT Topology

The above analysis suggests that the special working characteristic (the harmonic cancellation characteristic) of the DI-OEWT topology should be considered when designing the filter. The "common capacitor type filter" (here we call it "Type-2 filter") for DI-OEWT topology, which is presented in [21,22], seems to be a reasonable one, as illustrated in Figure 7. Two inverters are connected to a common shunt capacitor branch of the filter through the two inverter-side inductors, respectively. Then, the harmonic cancellation of the dual-inverter can be realized through the common capacitor branch. This common capacitor branch can be seen as the series connection of the two individual capacitor branches of the Type-1 filter, and the voltage rating of the former is the twice of the latter, so the actual value of the capacitor in Type-2 filter is half the value in Type-1 filter when the two has the same per unit (p.u.) value. Therefore, the capacitor here is represented as $C/2$.

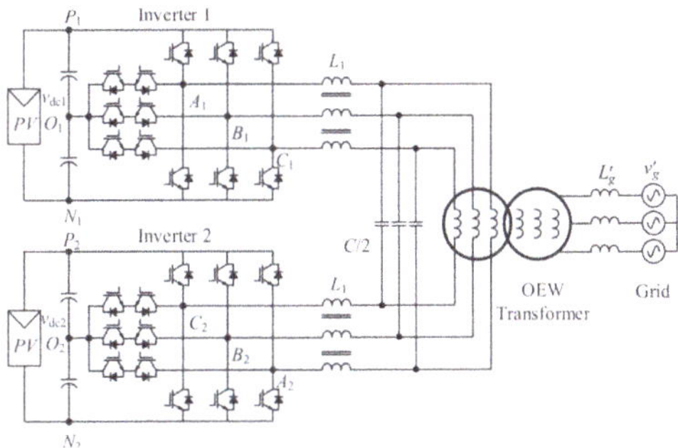

Figure 7. System configuration of the DI-OEWT topology with the Type-2 filter.

Figure 8 shows the three-phase equivalent circuit of the DI-OEWT topology with the Type-2 filter. With the same derivation method as described in Section 3.1, we can get the corresponding single-phase equivalent circuit as illustrated in Figure 9. From the figure, the transfer functions $v_{\text{inv1}}(s)$, $v_{\text{inv2}}(s)$ to $i_g(s)$ and $i_{\text{inv1}}(s)$ can be, respectively, calculated as

$$i_g(s) = \frac{1}{L_1 L_2 C s^3 + (2L_1 + L_2)s}[v_{\text{inv1}}(s) - v_{\text{inv2}}(s)] \tag{8}$$

$$i_{\text{inv1}}(s) = \frac{(L_2 C/2)s^2 + 1}{L_1 L_2 C s^3 + (2L_1 + L_2)s}[v_{\text{inv1}}(s) - v_{\text{inv2}}(s)] \tag{9}$$

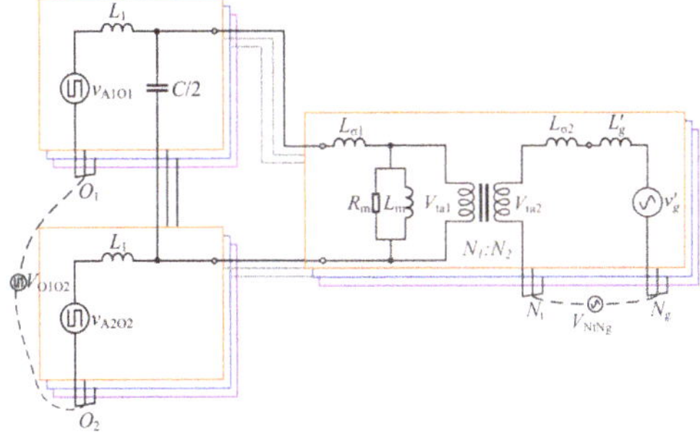

Figure 8. Three-phase equivalent circuit of the DI-OEWT topology with the Type-2 filter.

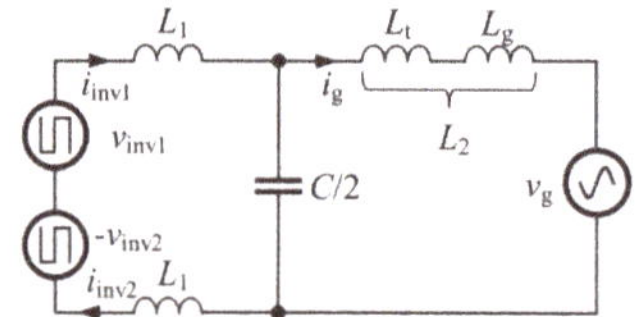

Figure 9. Single-phase equivalent circuit of the DI-OEWT topology with the Type-2 filter.

It can be found from Equations (8) and (9) that the harmonic source of i_g and i_{inv1} are all $v_{inv1} - v_{inv2}$, proves once again that the Type-2 filter does make judicious use of the merit of the DI-OEWT topology.

Figure 10 shows the bode plots of the transfer functions $v_{inv1}(s)$, $v_{inv2}(s)$ to $i_g(s)$ and $i_{inv1}(s)$ of both the LCL filter and the Type-2 filter (with the same corresponding parameters), respectively. The figures suggest that, for both i_g and i_{inv1}, Type-2 filter has superior high-frequency harmonic suppression capabilities than the LCL filter. That can lead to a decreasing in the total inductance, capacitance, and volume. Therefore, the Type-2 filter can be an option for the DI-OEWT topology.

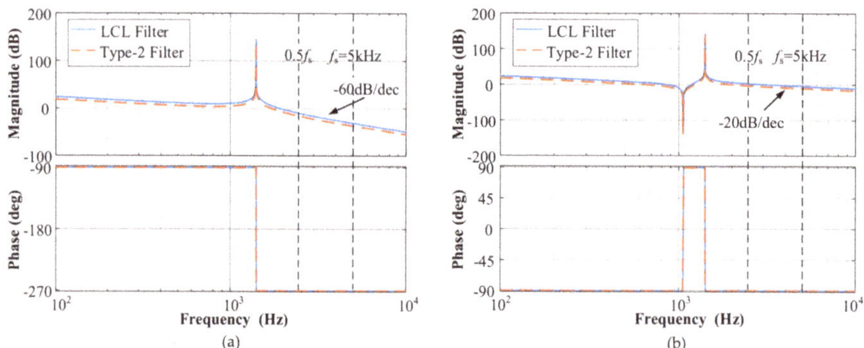

Figure 10. Bode plots of transfer functions $v_{inv1}(s)$, $v_{inv2}(s)$ to $i_g(s)$ and $i_{inv1}(s)$ in LCL filter and Type-2 filter. (**a**) $v_{inv1}(s)$, $v_{inv2}(s)$ to $i_g(s)$, (**b**) $v_{inv1}(s)$, $v_{inv2}(s)$ to $i_{inv1}(s)$.

Furthermore, look again at Figure 9, it shows that the two inverter-side inductors L_1 of the two inverters are actually connected in series, so these two inductors can merge into one single inductor, further reducing the volume and cost due to the saving of inductor magnetic cores. In this case, the complex magnetic integration method for reducing the inverter-side inductors' magnetic component size, which was proposed in [22], is actually unnecessary.

3.3. The Proposed Type-3 Filter for the DI-OEWT Topology

According to the above analysis of the existing filters for DI-OEWT topology, only Type-2 filter is suitable because its structure is fit for the working characteristic of the DI-OEWT topology. However, the leakage inductance of the transformer, which is a relatively large value in a high-power MV transformer, has not got special attention and rational utilization.

In Section 3.1, we have mentioned that the voltage of the low-voltage side of the OEW transformer is the twice of the replaced transformer. Then, it can be easy to deduce from Equation (3) that the leakage inductance of OEW transformer is four times as the replaced one (referred to the low-voltage side). Setting such a large inductance as the grid-side inductor may be not a cost-saving solution.

Therefore, in this paper, a new high order filter (here we call it "Type-3 filter") for DI-OEWT topology is proposed, as illustrated in Figure 11. The two inverters are directly connected to the low-voltage side of the OEW transformer, the shunt capacitor C'_H and grid-side inductor L'_H are set at the high voltage side of the OEW transformer. In this filter, the harmonic cancellation effect of the

DI-OEWT can be realized through the transformer, without any other shunt branch. The transformer leakage inductance L_t is used as the inverter-side inductor of the filter. If the initial value of L_t is not enough to suppress the inverter-side current ripple, it can be increased just by increasing the impedance voltage V_k of the transformer.

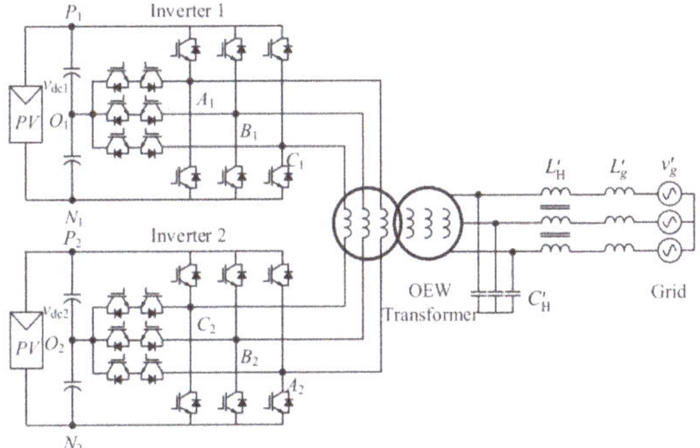

Figure 11. System configuration of the DI-OEWT topology with the Type-3 filter.

With the similar derivation method described above, here we directly give the single-phase equivalent circuit of the DI-OEWT topology with the Type-3 filter, as shown in Figure 12, where C_H and L_H represent the shunt capacitor and grid-side inductor referred to the low-voltage side, respectively.

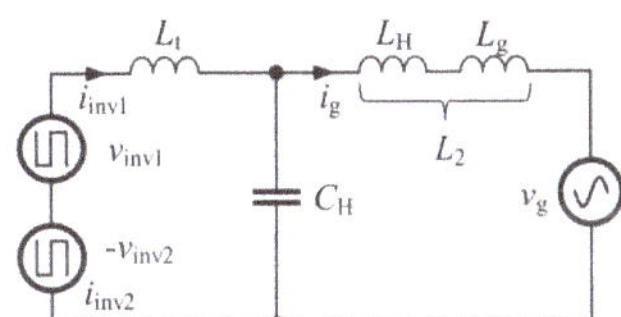

Figure 12. Single-phase equivalent circuit of the DI-OEWT topology with the Type-3 filter.

From Figure 12, the transfer functions $v_{inv1}(s)$, $v_{inv2}(s)$ to $i_g(s)$ and $i_{inv1}(s)$ can be, respectively, calculated as

$$i_g(s) = \frac{1}{(L_t L_2 C_H)s^3 + (L_t + L_2)s}[v_{inv1}(s) - v_{inv2}(s)] \tag{10}$$

$$i_{inv1}(s) = \frac{L_2 C_H s^2 + 1}{L_t L_2 C_H s^3 + (L_t + L_2)s}[v_{inv1}(s) - v_{inv2}(s)] \tag{11}$$

As can be seen from comparing Figures 9 and 12, the equivalent circuit of the Type-3 and Type-2 filter are actually the same, while the difference between them is the role of the leakage inductance of the OEW transformer. Due to this difference, the extra required capacitance and the inductance value of the two filters may be different. This different value should be compared with specific examples, which will be presented in the following section.

4. Parameter Design and Evaluations of the Filters

4.1. Parameter Design of the Filters

The design procedure of the Type-3 filter, together with the Type-1 and Type-2 filters, for the DI-OEWT topology is covered in the following. First, a 30 kW OEW dual-three-level (D3L) inverter, with 5 kHz switching frequency, 460 V~850 V (MPPT lower and upper limit voltage) DC voltage, 380 V line-to-line grid voltage is adopted. The primary side of the transformer is set as the OEW structure, and the secondary side of the transformer is connected in delta. It is worth noting that the primary side line-to-line voltage of the transformer in conventional multi-parallel inverter topology is usually 315 V, so the phase voltage ratio of the OEW transformer used here is 364 V/380 V, where 364 V = 2 × 315 V/1.732. In the following, the filter components calculations are presented on a per-unit basis, and the corresponding base values are listed in Table 1.

Table 1. Per-unit base values of the system.

Symbol	Parameter	Formula	Value
P_B	Rated base power	-	30 kW
V_B	Ac base voltage	-	364 V
I_B	Ac base current	$P_B/(3V_B)$	27.5 A
f_B	Base frequency	-	50 Hz
Z_B	Base impedance	$3V_B^2/P_B$	13.2496 Ω
L_B	Base inductance	$Z_B/(2\pi f_B)$	42.17 mH
C_B	Base capacitance	$1/(2\pi f_B Z_B)$	240 μF

(1) Inverter-side Inductor: The inverter-side inductor value is determined by the requirement of the inverter-side current ripple, owing to the semiconductor current rating and efficiency requirement, this current ripple must be limited within a certain range. Besides, the equation for calculating the inductor value is related to the inverter topology as well as the modulation strategy, here the D3L inverter is modulated by the decoupled SVPWM strategy [17].

The decoupled SVPWM strategy has the characteristic that the total reference voltage signal is divided into two opposite parts for the two constituent inverters, and each of the inverter is switched independently of the other with the standard SVPWM strategy. Such characteristic is very appropriate for the two inverters tracking the individual MPPs. The space vector diagram of the individual three-level inverters is shown in Figure 13. In Figure 13, under the decoupled SVPWM strategy, the reference voltage space vector of the two inverters V_{r1} and V_{r2} are synthesized by the nearest three voltage vectors $\{V_0, V_1, V_2\}$ and $\{V'_0, V'_3, V'_4\}$, respectively. The pulse sequence is symmetrical with seven pieces of segments in each switching cycle T_s and the dwell times for each of the voltage vectors satisfy the following expression.

$$\begin{cases} V_0 T_0 + V_1 T_1 + V_2 T_2 = V_{r1} T_s \\ V'_0 T'_0 + V'_3 T'_3 + V'_4 T'_4 = V_{r2} T_s \end{cases} \tag{12}$$

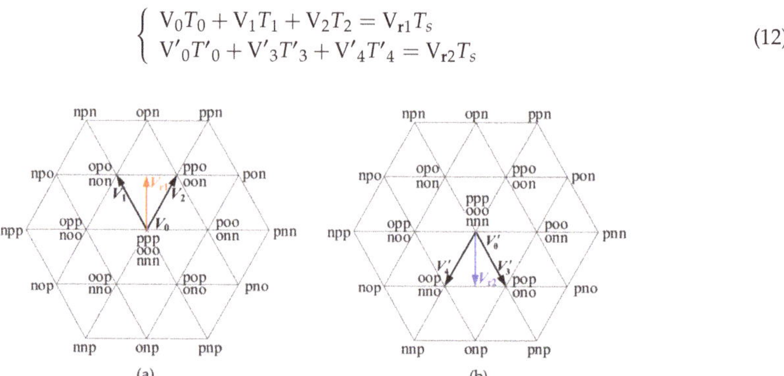

Figure 13. Space vector diagram of the individual three-level inverters. (**a**) Inverter 1, (**b**) inverter 2.

The peak to peak value of the inductor current ripple is defined by volt-seconds applied to the inductor over the switching period [32,35]. For Type-1 filter, as mentioned in Section 3.1, the harmonic source of the inverter-side current is mainly the output phase voltage of one inverter. Take inverter 1 as an instance, the maximum current ripple occurs when the zero vector V_0 dwell time $T_0 = 0$, and the other two vectors V_1 and V_2 equally divide the switching period, $T_1 = T_2 = T_s/2$. In such case, the modulation index M = 1/ $\sqrt{3}$ and the DC voltage of the inverter is slightly larger than the MPPT upper limit voltage, the reference voltage vector V_{r1} is in the midway between V_1 and V_2. The pulse sequence is degenerated into five pieces of segments and the corresponding inductor voltage and current waveform (take phase A1 as an example and $v_{L1} \approx v_{inv1}$) are shown in Figure 14.

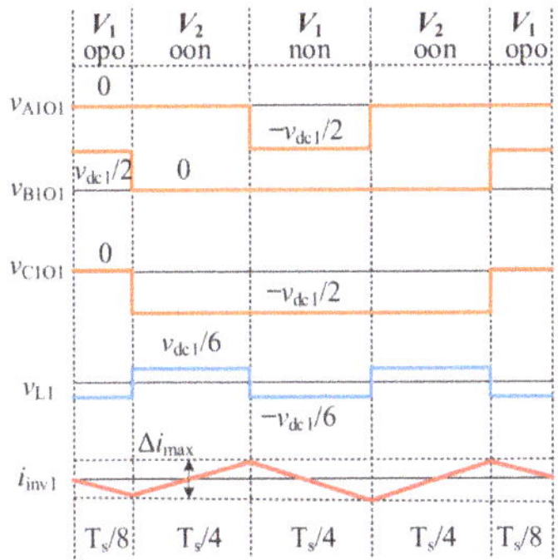

Figure 14. Phase A1 inductor voltage and current waveform.

According to the volt-second balance principle, the following expression can be got as:

$$L_1 \Delta i_{max} = \frac{v_{dc1}}{6} \frac{T_s}{4} \tag{13}$$

where Δi_{max} is the maximum current ripple.

From Equation (13), the minimum inverter-side inductor value of the Type-1 filter can be estimated by

$$L_{1min} = \frac{v_{dc1}}{24 \Delta i_{max} f_s} \tag{14}$$

where $f_s = 1/T_s$ is the switching frequency.

For Type-2 and Type-3 filter, the harmonic source of the inverter-side current is the difference between the output phase voltage of the two inverters ($v_{inv1} - v_{inv2}$). The maximum current ripple will occur in the case that the reference voltage vector V_{r1} is in the midway between V_1 and V_2, V_{r2} is in the midway between V'_3 and V'_4. In this case, $T_0 = T'_0 = 0$, $T_1 = T_2 = T'_3 = T'_4 = T_s/2$. For the convenience of analysis, assuming $v_{dc1} = v_{dc2} = v_{dc}$, then the inverter-side inductor voltage and current waveform of the Type-2 and Type-3 filter can be roughly derived as shown in Figure 15 (take phase A1-A2 as an example and $v_L \approx v_{inv1} - v_{inv2}$).

Accordingly, the minimum inverter-side inductor value of the Type-2 and Type-3 filter can be derived based on the volt-second balance across the inductor as

$$L_{min} = \frac{v_{dc}}{12 \Delta i_{max} f_s} \tag{15}$$

where $L = 2L_1$ for Type-2 filter and $L = L_t$ for Type-3 filter.

For the values $v_{dc} = 850$ V, $f_s = 5000$ Hz, and about 15% current ripple, the estimated minimum inverter-side inductor value for each filter can be calculated according to Equations (14) and (15) as follows:

Type-1 filter: two 0.02845 p.u. inductor;

Type-2 filter: one 0.0569 p.u. inductor;

Type-3 filter: transformer leakage inductance, 0.06 p.u..

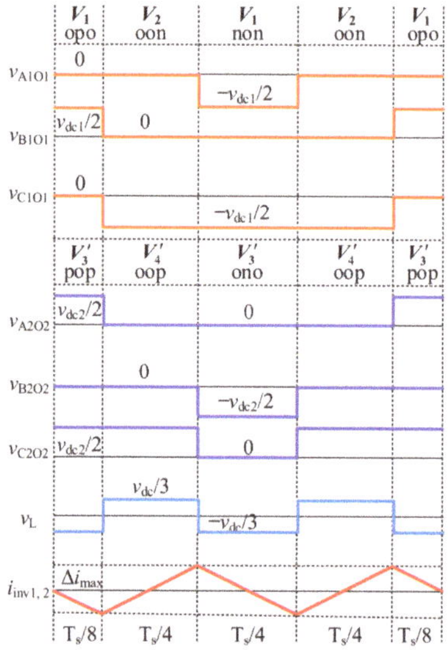

Figure 15. Phase A1-A2 inductor voltage and current waveform.

(2) Grid-side Inductor and Shunt Capacitor: The grid-side inductor and shunt capacitor together constitute a shunt network to suppress the grid current harmonics, so as to satisfy the standard. For example, recent published Chinese standard NB/T 32004-2018 [34] requires the harmonics greater than 35th should be less than 0.3% of the 30% of the rated fundamental current. Besides, the capacitor value is limited to the maximum absorbed reactive power at rated load and typically less than 5%. Moreover, the resonant frequency ω_{res} of the filter is often chosen as Equation (16) with the intention of not creating resonance problem in the lower and higher parts of the harmonic spectrum.

$$10 \times 2\pi f_B \leq \omega_{res} \leq 0.5 \times 2\pi f_s \tag{16}$$

where the value of ω_{res} is $\sqrt{(2L_1 + L_2)/L_1 L_2 C}$ for Type-1, Type-2 filter and $\sqrt{(L_t + L_2)/L_t L_2 C_H}$ for Type-3 filter.

For Type-1 and Type-2 filter, the grid-side inductor value is limited by the transformer leakage inductance, and its value is 0.06 p.u. (neglect the grid inductance). The capacitor values of these two filters are computed considering the attenuation of the filter. For instance, the maximum harmonic of the grid current greater than 35th usually occurs around the switching frequency. To limit the maximum current harmonic lower than 0.3%, the capacitor value of these two filters can be computed from Equations (6) or (8) considering the most dominant harmonic $V(h)$ around the switching frequency in the inverter output voltage spectrum as

$$C \geq \frac{2\pi f_B h (2L_1 + L_2) \dfrac{I_B \times \sqrt{2} \times 30\% \times 0.3\%}{V(h)} + 1}{L_1 L_2 (2\pi f_B h)^3 \dfrac{I_B \times \sqrt{2} \times 30\% \times 0.3\%}{V(h)}} \tag{17}$$

where h is the most dominant harmonic order. For D3L inverter with the decoupled SVPWM, if the DC voltage and the power (or the modulation index) of the two inverters are all the same, the harmonic around the switching frequency can be totally cancelled. But when the two inverters are tracking the individual MPPs, the DC voltage and the power (or the modulation index) of the two inverters may be different, this total cancellation cannot be realized. Assuming the worst case that no harmonic cancellation occurs around the switching frequency harmonics, the most dominant harmonic order $h = m_f - 2$, the value of the ($m_f - 2$) order harmonic can be got from simulation on a single three-level inverter as $V(m_f - 2) \approx 0.055$ p.u. (20.2 V), where m_f is the modulation frequency index, and the value here is 100 (5000 Hz/50 Hz). Then from Equation (17) a minimum value of C can be calculated as 0.0175 p.u. Considering the constraints illustrated in Equation (16) and taking a certain margin, the final capacitor value of Type-1 and Type-2 filter is chosen as 0.0208 p.u.

For Type-3 filter, both the grid-side inductor and the shunt capacitor need to be designed. Usually, we choose a larger capacitor to reduce the requirement of the grid-side inductor for saving cost. Here the capacitor is starting with the value as 0.0416 p.u. With the same methodology as above, to ensure the grid current to satisfy the harmonic requirement, the grid-side inductor value can be computed from Equation (10) as

$$
L_2 \geq \frac{2\pi f_B h L_t \dfrac{I_B \times \sqrt{2} \times 30\% \times 0.3\%}{V(h)} + 1}{2\pi f_B h \left[L_t C [2\pi f_B h]^2 - 1\right] \dfrac{I_B \times \sqrt{2} \times 30\% \times 0.3\%}{V(h)}}
\tag{18}
$$

Also, considering the value of the most dominant harmonic order $V(m_f - 2) \approx 0.055$ p.u., a minimum value of L_2 can be calculated as 0.0218 p.u.. Considering (16) and taking a certain margin, the final value of L_2 is selected as 0.0237 p.u.

4.2. Fault-Tolerant Configuration for DI-OEWT Topology

Like the multi-parallel inverter topology, one of the advantages of the DI-OEWT topology is the availability of fault-tolerant operation. Many papers have investigated the fault-tolerant method for the dual-inverter topology used in motor drives, such as hybrid modulation strategies [18], dual-inverter topology reconfiguring method [19], fault-tolerant direct thrust force control method [20]. However, these methods are not suitable for the DI-OEWT topology used in photovoltaic applications, due to the DC voltages and power of the two inverters in such case are often different for the separate MPPT purpose.

To realize fault-tolerant operation in these systems, a more reasonable way is to cut off the fault inverter while allowing the healthy inverter to continue working. However, only one inverter cannot supply the whole voltage of the transformer due to the DC voltage limit, so the OEW transformer need to be reconfigured. Figure 16 presents a possible configuration to achieve the above purpose. As shown in the figure, a center-tapped (at the low voltage side) OEW transformer is used here, S_1 and S_2 are the contactors of the respective inverters connecting to the open-end windings of the transformer, S_3 is the contactor to short the three-phase center-taps. In case of a fault in one inverter, the corresponding contactor (S_1 or S_2) of the fault inverter is switched off, and contactor S_3 is switched on, then half of the primary windings are connected in star across the healthy inverter. The healthy inverter only needs to supply half of the original AC voltage, so that it can still work properly without any change.

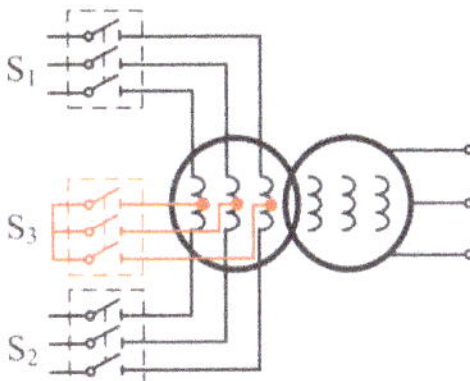

Figure 16. Fault-tolerant configuration of the DI-OEWT topology.

Nevertheless, considering the configurations of the filters described above, not all types of filters are applicable for the presented fault-tolerant scheme. For DI-OEWT topology with the Type-1 and the proposed Type-3 filter, when a fault occurs in one inverter, the healthy inverter can still work through the modified filter. While for DI-OEWT topology with the Type-2 filter, fault-tolerant operation cannot be realized because the shunt capacitor cannot decouple from the faulty inverter.

4.3. Summary and Discussion

The designed parameters of the three kinds of filters, as well as the applicability of the presented fault-tolerant scheme on the filters, are summarized in Table 2.

Table 2. Comparison of the three kinds of filters.

Filters	Type-1 Filter	Type-2 Filter	Type-3 Filter
Inverter-side Inductor	0.02845 p.u. × 2 with 2 magnetic cores	0.0569 p.u. with 1 magnetic core	Tr leakage inductance 0.06 p.u.
Grid-side Inductor	Tr leakage inductance 0.06 p.u.	Tr leakage inductance 0.06 p.u.	0.0237 p.u.
Shunt CapacitoR	0.0208 p.u.	0.0208 p.u.	0.0416 p.u.
Total Inductance	0.1169 p.u.	0.1169 p.u.	0.0837 p.u.
Fault-tolerant applicability	Yes	No	Yes

Note: Tr represents the OEW transformer.

From analyzing the table, the followings can be seen.

(1) The Type-1 and Type-2 filter have the same filter parameters, but the inverter-side inductor of the Type-1 filter needs two magnetic cores while the Type-2 filter only needs one, which means the Type-2 filter has a smaller size and cost than the Type-1 filter.

(2) Besides the transformer leakage inductance, the value of the extra inductance requirement of the Type-3 filter is reduced by a factor of 58.35%, while the capacitance requirement is increased about 50%, compared to the Type-1 and Type-2 filter.

(3) The total inductance of the Type-1 and Type-2 filter is more than 10% of the system base inductance, which may cause a larger fundamental voltage drop across the inductor and increase the DC voltage lower limit value.

(4) Unlike the Type-1 and Type-3 filter, the DI-OEWT topology with the Type-2 filter cannot realize fault-tolerant operations with the presented fault tolerate scheme.

(5) In terms of the inductance requirement, magnetic cores numbers and the fault-tolerant applicability, the proposed Type-3 filter has a certain advantage over the existing Type-1 and Type-2 filter.

5. Experiment Results

5.1. Experiment Setup Description

To verify the above analysis and design methodology, a 30 kW three-phase D3L inverter fed open-end winding transformer prototype based on DSP TMS320F28377D controller is constructed, as shown in Figure 17. The D3L inverter is supplied by two rectifiers (DC Source 1 and 2) with the DC voltage $v_{dc} = 850$ V. The switching frequency of the inverter is 5 kHz. For the OEW transformer, the phase voltage ratio is 364 V/380 V, the leakage inductance is about 0.025 p.u., we compensate it to 0.06 p.u. through adding extra inductors.

The designed parameters of the three kinds of filters tabulated in Table 2 are tested by experiment. Here the D3L inverter with different filters are divided into three cases, where Case-1 represents the D3L inverter with the Type-1 filter, Case-2 represents the D3L inverter with the Type-2 filter, and Case-3 represents the D3L inverter with the Type-3 filter. In addition, the experimental analysis is carried out in two operating modes. One is the two inverters of the D3L inverter working at the power balance mode. Another one is the two inverters of the D3L inverter working at the power unbalance mode. Here, the power ratio of the two inverters is chosen as 3/2, corresponding to a relatively large imbalance ratio of 33.33%. Besides, to suppress the resonance caused by the high order filter, the filter based active damping method [36] is used here.

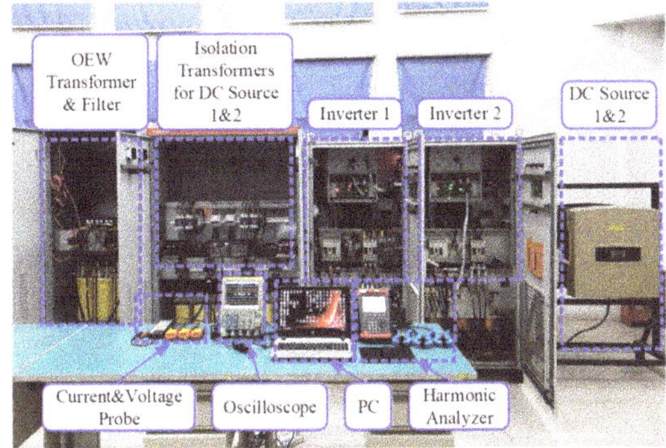

Figure 17. Experimental platform of the DI-OEWT grid-tied system.

5.2. Inverter-Side Current Analysis

Figures 18–20 show the experimental measured inverter-side currents and the corresponding fast Fourier transformation (FFT) waveforms of the three cases in power balance mode, respectively. From Figure 18a, Figure 19a, and Figure 20a, the inverter-side current ripple at the rated load in the three cases are around 14.9%, 14.6%, and 14.1%, respectively, which are roughly in agreement with the value calculated by Equations (14) and (15). From the FFT waveforms of the inverter-side currents as illustrated in Figure 18b, Figure 19b, and Figure 20b, it can be noticed that the current harmonics around the switching frequency are roughly canceled out in Case-2 and Case-3, but still exist in Case-1. The results demonstrate the analysis in Section 3 that the two individual sets of shunt capacitors of the Type-1 filter break the harmonic cancelation loop, causing the harmonics which should have been canceled still exist in the inverter-side current.

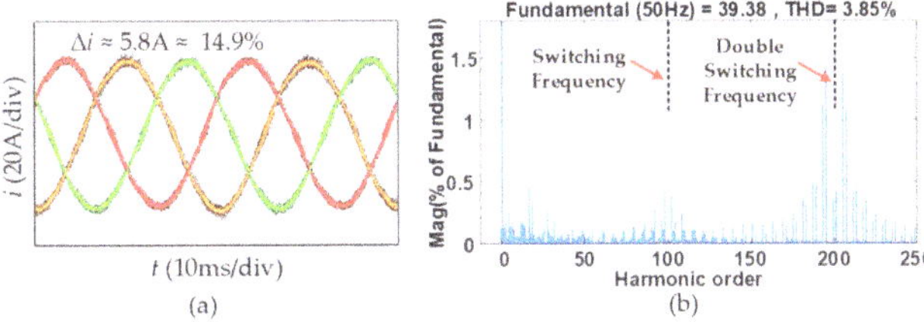

Figure 18. Inverter-side current experimental waveforms of Case-1 in power balance mode. (**a**) Inverter-side current and current ripples. (**b**) FFT waveforms of the inverter-side current.

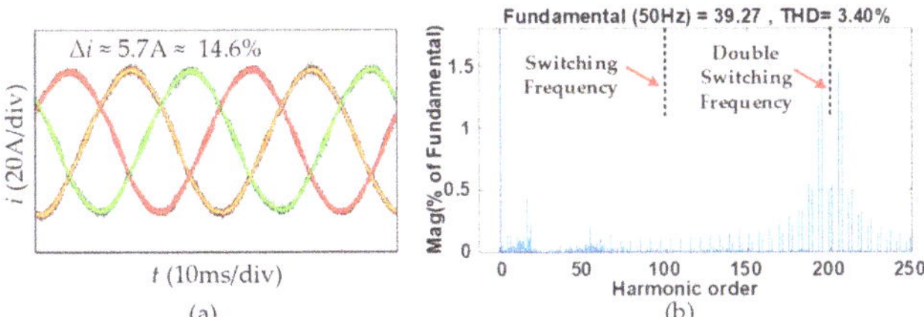

Figure 19. Inverter-side current experimental waveforms of Case-2 in power balance mode. (**a**) Inverter-side current and current ripples. (**b**) FFT waveforms of the inverter-side current.

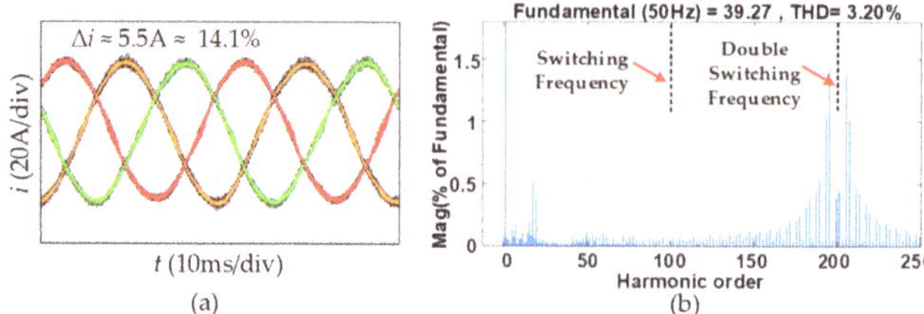

Figure 20. Inverter-side current experimental waveforms of Case-3 in power balance mode. (**a**) Inverter-side current and current ripples. (**b**) FFT waveforms of the inverter-side current.

The experimental measured inverter-side currents and the corresponding FFT waveforms of the three cases in power unbalance mode are shown in Figures 21–23, respectively. Observing these figures, when the D3L inverter is working in power unbalance mode, the inverter-side current ripple of the three cases are around 14.6%, 14.4%, and 13.8%, respectively, which are still within the design limit. Besides, as observed from Figures 22b and 23b, the harmonics around the switching frequency also exist in the inverter-side currents in Case-2 and Case-3, which means that the total harmonic cancellation around the switching frequency cannot be realized in power unbalance mode.

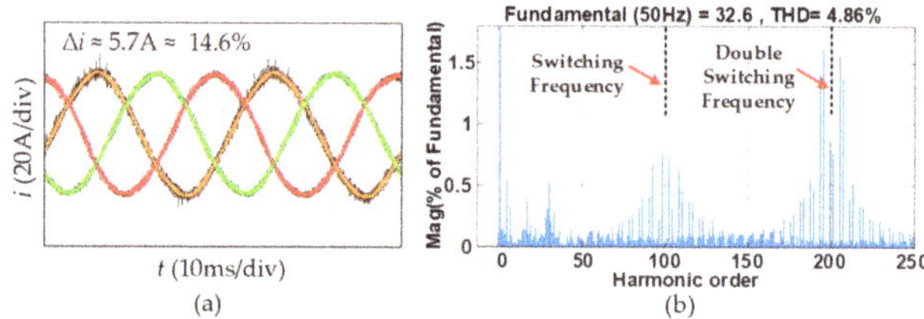

Figure 21. Inverter-side current experimental waveforms of Case-1 in power unbalance mode. (**a**) Inverter-side current and current ripples. (**b**) FFT waveforms of the inverter-side current.

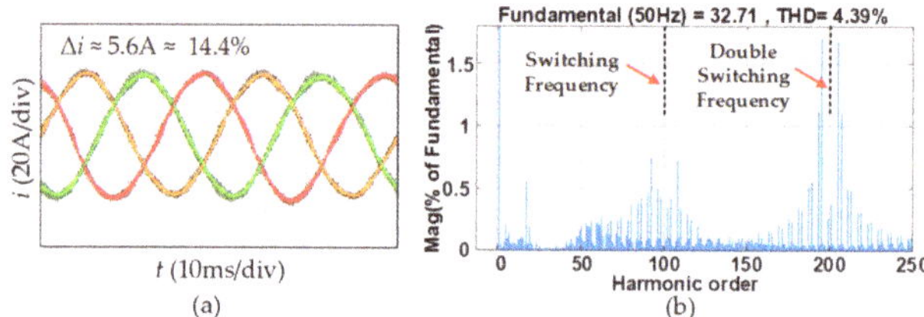

Figure 22. Inverter-side current experimental waveforms of Case-2 in power unbalance mode. (**a**) Inverter-side current and current ripples. (**b**) FFT waveforms of the inverter-side current.

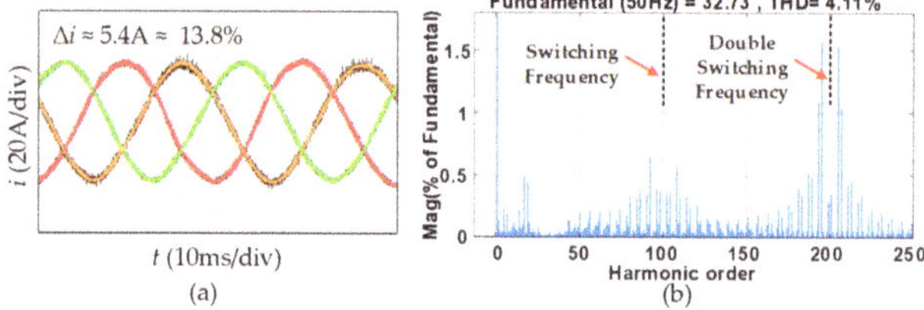

Figure 23. Inverter-side current experimental waveforms of Case-3 in power unbalance mode. (**a**) Inverter-side current and current ripples. (**b**) FFT waveforms of the inverter-side current.

5.3. Grid Current Analysis

The experimental tests for the grid current are conducted at 30% load to verify whether the current quality satisfies the grid standard [34]. Figures 24 and 25 show the grid current waveforms and the corresponding harmonic spectrums of the three cases in power balance mode, respectively. Figures 26 and 27 show the grid current waveforms and the corresponding harmonic spectrums of the three cases in power unbalance mode, respectively. The test cases along with the grid current THD and the dominating current harmonic amplitude percentage (higher than 35th) are summarized in Table 3. As observed from the table, even though the inverter is working at 30% load, the current THD are all still below the 5% [34] limit, and the dominant harmonics (higher than 35th) are all within the 0.3% limit as well.

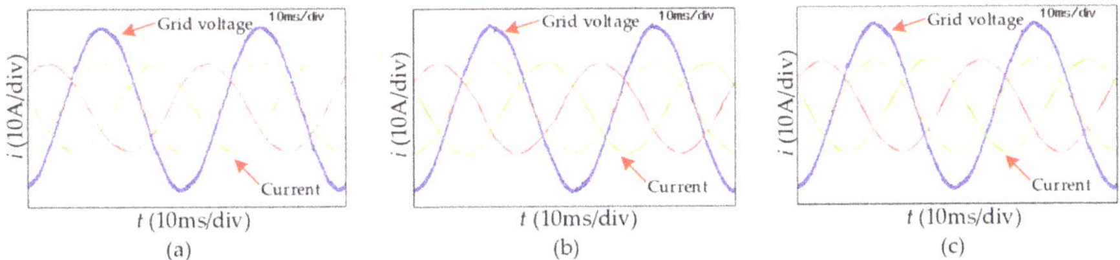

Figure 24. Grid current experimental waveforms of the three cases in power balance mode. (**a**) Case-1. (**b**) Case-2. (**c**) Case-3.

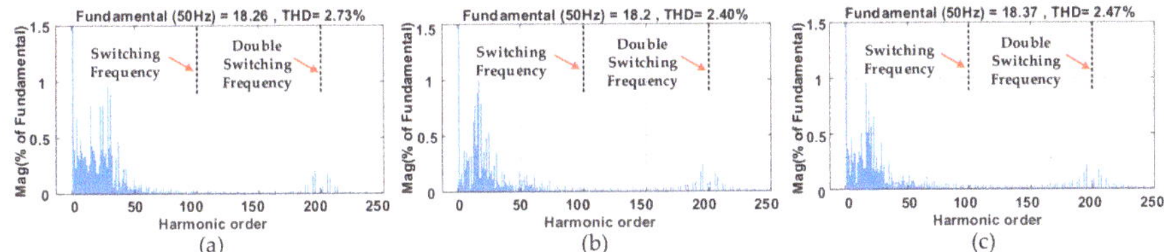

Figure 25. Grid current FFT waveforms of the three cases in power balance mode. (**a**) Case-1, (**b**) Case-2, (**c**) Case-3.

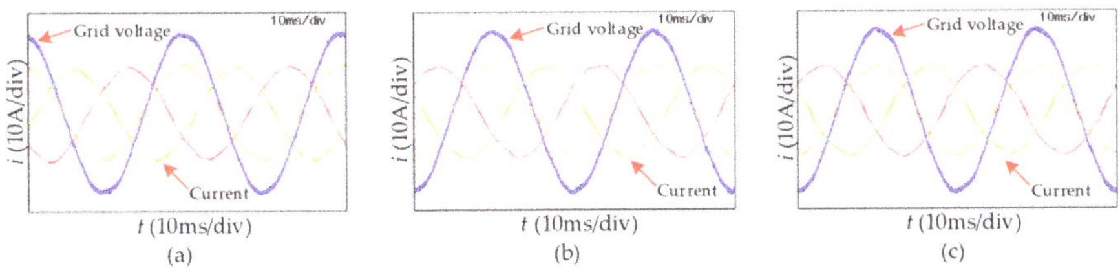

Figure 26. Grid current experimental waveforms of the three cases in power unbalance mode. (**a**) Case-1, (**b**) Case-2, (**c**) Case-3.

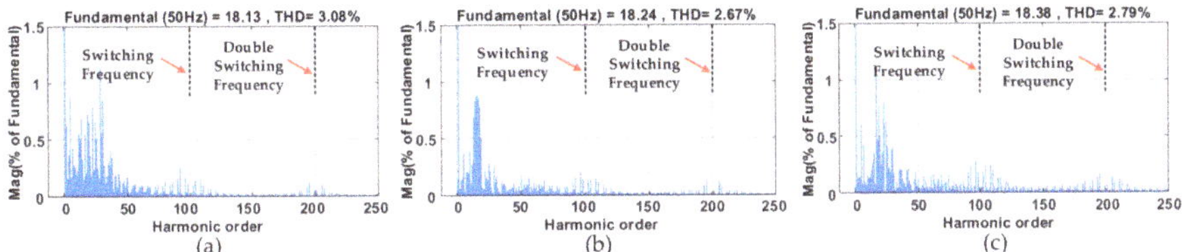

Figure 27. Grid current FFT waveforms of the three cases in power unbalance mode. (**a**) Case-1, (**b**) Case-2, (**c**) Case-3.

Table 3. THD and dominant harmonic percentage of grid current in different cases and modes.

Cases	Case-1		Case-2		Case-3	
Working Modes	Balance	Unbalance	Balance	Unbalance	Balance	Unbalance
THD	2.73%	3.08%	2.4%	2.67%	2.47%	2.79%
Dominant Harmonic Percentage	0.2%	0.25%	0.22%	0.2%	0.21%	0.28%

In summary, the analysis of the experimental inverter-side currents and the grid currents of all cases confirms the effectiveness of the filters' parameters designed using the proposed method. Consequently, the discussions in Section 4.3 can be confirmed as well.

6. Conclusions

This paper investigates the filters for the DI-OEWT topology used in the photovoltaic grid-tied applications. The equivalent circuits of the existing Type-1 and Type-2 filter are derived in detail, and the corresponding harmonic suppression transfer functions are given. Some findings can be obtained from analyzing these equivalent circuits and the transfer functions, as well as comparing with the LCL filter used in the multi-parallel inverter topology. It can be found that Type-1 filter does not make judicious use of the merit (output harmonic cancellation) of the DI-OEWT topology, because its two individual sets of shunt capacitors break the cancellation loop. Another finding is that the two inverter-side inductors of the Type-2 filter are actually connected in series and can be merged into one single inductor to further reduce the filter volume and cost.

From the analysis of the existing filters and with consideration of the large leakage inductance value of the high power MV transformer, a new high order filter (named Type-3 filter) for DI-OEWT topology is also proposed. Unlike the existing filters, in which the transformer leakage inductance is set as the grid-side inductor, the Type-3 filter set the leakage inductance as the inverter-side inductor. Such an arrangement can maximize the transformer leakage inductance utilization, so as to reduce the requirement of the extra inductance.

A brief parameter design method for the three kinds of filters is also introduced. With the design examples of the three filters, the extra inductance requirement of the Type-3 filter can have a 58.35% higher reduction than the Type-1 and Type-2 filters. In addition, a fault tolerance scheme for DI-OEWT topology is also presented in the paper, but only Type-1 and Type-3 filters are applicable for the scheme. Finally, through evaluating the three filters in terms of the inductance requirement, magnetic core number and applicability for the fault tolerance scheme, the proposed Type-3 filter has a certain advantage over the existing Type-1 and Type-2 filters.

Experimental results of a 30 kW D3L inverter prototype verify the effectiveness of the proposed filter design method and validity of the analysis.

Besides the stability analysis, damping methods of the dual-inverter with the filters are also important issues, we will focus on these topics in future research.

Author Contributions: Conceptualization, B.W. and X.Z.; data curation, B.W.; formal analysis, B.W.; funding acquisition, X.Z. and R.C.; methodology, B.W.; experiment, B.W. and C.S.; writing—original draft, B.W.; writing—review & editing, B.W.

Funding: This work was funded by the National Natural Science Foundation of China (NO. U1766207).

Acknowledgments: The authors declare no acknowledgments.

Nomenclature

List of Abbreviations

LCL	inductor-capacitor-inductor
MV	medium-voltage
DI-OEWT	dual-inverter fed open-end winding transformer
NPC	neutral point clamp
VSI	voltage source inverter
MPPT	maximum power point tracking
SCR	short-circuit ratio
OEW	open-end winding
CM	common-mode
KVL	Kirchhoff's voltage law
FFT	fast Fourier transformation
p.u.	per unit

List of Symbols

$v_{\mathrm{inv}1}$	inverter 1 output phase voltage
$v_{\mathrm{inv}2}$	inverter 2 output phase voltage
$i_{\mathrm{inv}1}$	inverter 1 output phase current
$i_{\mathrm{inv}2}$	inverter 2 output phase current
v_g	grid voltage
i_g	grid current
L_1	inverter-side inductor
L_t	transformer leakage inductance
L_g	grid inductance
L_2	combined inductance of L_t and L_g
C	filter capacitor
C'_H	shunt capacitor of Type-3 filter
L'_H	grid-side inductor of Type-3 filter
C_H	shunt capacitor of Type-3 filter referred to the low-voltage side
L_H	grid-side inductor of Type-3 filter referred to the low-voltage side
V_k	impedance voltage
P_{rated}	system rated power
f_0	fundamental frequency
V_{X1O1}	pole voltage of phase X_1 of inverter 1, $X = A,B,C$
V_{X2O2}	pole voltage of phase X_2 of inverter 2
V_{O1Nc1}	CM voltage of inverter 1
V_{O2Nc2}	CM voltage of inverter 2
V_{Nc1Nc2}	voltage across the two capacitor common points N_{c1}, N_{c2}
V_{NtNg}	voltage across N_t and N_g
V_{CX1}	voltage on filter capacitor of phase X_1
V_{CX2}	voltage on filter capacitor of phase X_2
$V_{L\sigma1X}$	phase X voltage on the leakage inductance at transformer low voltage side ($L_{\sigma1}$)
$V_{L\sigma2X}$	phase X voltage on the leakage inductance at transformer high voltage side ($L_{\sigma2}$)
V_{LgX}	phase X voltage on the grid inductance
V_{tX1}	voltage of the low voltage side of the MV transformer
V_{tX2}	voltage of the high voltage side of the MV transformer
v'_{gX}	phase X voltage of the MV grid
$v_{\mathrm{dc}1}$	DC voltage of inverter 1
$v_{\mathrm{dc}2}$	DC voltage of inverter 2
ω_{res}	filter resonant frequency
T_s	switching period
f_s	switching frequency
m_f	modulation frequency index
h	dominant harmonic order

References

1. Zhao, Z.; Lei, Y.; He, F.; Lu, Z. Overview of large-scale grid-connected photovoltaic power plants. *Autom. Electr. Power Syst.* **2011**, *12*, 101–107.
2. Serban, E.; Ordonez, M.; Pondiche, C. DC-bus voltage range extension in 1500 V photovoltaic inverters. *IEEE J. Emerg. Sel. Top. Power Electron.* **2015**, *3*, 901–917. [CrossRef]
3. Sungrow. SG1250UD Inverter Product Datasheet. Available online: http://en.sungrowpower.com/product_view?id=146.html (accessed on 18 June 2019).
4. Takahashi, I.; Ohmori, Y. High-performance direct torque control of an induction motor. *IEEE Trans. Ind. Appl.* **1989**, *25*, 257–264. [CrossRef]
5. Anand, S.; Fernandes, B.G.; Chatterjee, K. DC voltage controller for asymmetric-twin-converter-topology-based high-power STATCOM. *IEEE Trans. Ind. Electron.* **2013**, *60*, 11–19. [CrossRef]
6. De Almeida Carlos, G.A.; Jacobina, C.B.; dos Santos, E.C.; Fabrício, E.L.L.; dos Santos Rocha, N. Shunt active power filter with open-end winding transformer and series-connected converters. *IEEE Trans. Ind. Appl.* **2015**, *51*, 3273–3283. [CrossRef]
7. De Almeida Carlos, G.A.; dos Santos, E.C.; Jacobina, C.B.; Mello, J.P.R.A. Dynamic voltage restorer based on three-phase inverters cascaded through an open-end winding transformer. *IEEE Trans. Power Electron.* **2016**, *31*, 188–199. [CrossRef]
8. Grandi, G.; Rossi, C.; Ostojic, D.; Casadei, D. A new multilevel conversion structure for grid-connected PV applications. *IEEE Trans. Ind. Electron.* **2009**, *56*, 4416–4426. [CrossRef]
9. Corzine, K.A.; Wielebski, M.W.; Peng, F.Z.; Wang, J. Control of cascaded multilevel inverters. *IEEE Trans. Power Electron.* **2004**, *19*, 732–738. [CrossRef]
10. Edpuganti, A.; Rathore, A.K. New optimal pulsewidth modulation for single dc-link dual-inverter fed open-end stator winding induction motor drive. *IEEE Trans. Power Electron.* **2015**, *30*, 4386–4393. [CrossRef]
11. Kubasiak, N.; Mercorelli, P.; Liu, S. Model predictive control of transistor pulse converter for feeding electromagnetic valve actuator with energy storage. In Proceedings of the 44th IEEE Conference on Decision and Control, Seville, Spain, 15 December 2005.
12. Mercorelli, P. A multilevel inverter bridge control structure with energy storage using model predictive control for flat systems. *J. Eng.* **2013**, *2013*, 750190. [CrossRef]
13. Mercorelli, P.; Kubasiak, N.; Liu, S. Multilevel bridge governor by using model predictive control in wavelet packets for tracking trajectories. In Proceedings of the IEEE International Conference on Robotics and Automation (ICRA), New Orleans, LA, USA, 26 April–1 May 2004; pp. 4079–4084.
14. Kiadehi, A.D.; Drissi, K.E.K.; Pasquier, C. Adapted NSPWM for single DC-link dual-inverter fed open-end motor with negligible low-order harmonics and efficiency enhancement. *IEEE Trans. Power Electron.* **2016**, *31*, 8271–8281.
15. Srinivas, S.; Ramachandra Sekhar, K. Theoretical and experimental analysis for current in a dual-inverter-fed open-end winding induction motor drive with reduced switching PWM. *IEEE Trans. Ind. Electron.* **2013**, *60*, 4318–4328. [CrossRef]
16. Somasekhar, V.T.; Srinivas, S.; Kumar, K.K. Effect of zero-vector placement in a dual-inverter fed open-end winding induction motor drive with alternate sub-hexagonal center PWM switching scheme. *IEEE Trans. Power. Electron.* **2008**, *23*, 1584–1591. [CrossRef]
17. Wu, D.; Wu, X.; Su, L.; Yuan, X.; Xu, J. A dual three-level inverter based open-end winding induction motor drive with averaged zero-sequence voltage elimination and neutral-point voltage balance. *IEEE Trans. Ind. Electron.* **2016**, *63*, 4783–4795. [CrossRef]
18. Zhao, W.; Zhao, P.; Xu, D.; Chen, Z.; Zhu, J. Hybrid modulation fault-tolerant control of open-end windings linear vernier permanent-magnet motor with floating capacitor inverter. *IEEE Trans. Power Electron.* **2018**, *34*, 2563–2572. [CrossRef]
19. Yu, Z.; Kong, W.; Jiang, D.; Qu, R.; Li, D.; Jia, S.; Sun, J.; Gan, C. Fault-tolerant control strategy of the open-winding inverter for DC-biased vernier reluctance machines. *IEEE Trans. Power Electron.* **2019**, *34*, 1658–1671. [CrossRef]
20. Zhao, W.; Wu, B.; Chen, Q.; Zhu, J. Fault-tolerant direct thrust force control for a dual inverter fed open-end winding linear vernier permanent-magnet motor using improved SVPWM. *IEEE Trans. Ind. Electron.* **2018**, *65*, 7458–7467. [CrossRef]

21. Gohil, G.; Bede, L.; Teodorescu, R.; Kerekes, T.; Blaabjerg, F. Dual-converter-fed open-end transformer topology with parallel converters and integrated magnetics. *IEEE Trans. Ind. Electron.* **2016**, *63*, 4929–4941. [CrossRef]

22. Gohil, G.; Bede, L.; Teodorescu, R.; Kerekes, T.; Blaabjerg, F. Optimized integrated harmonic filter inductor for dual-converter-fed open-end transformer topology. *IEEE Trans. Power Electron.* **2017**, *32*, 1818–1831. [CrossRef]

23. Kumar, N.; Saha, T.; Dey, J. Sliding-mode control of PWM dual inverter-based grid-connected PV system: Modeling and performance analysis. *IEEE J. Emerg. Sel. Top. Power Electron.* **2016**, *4*, 435–444. [CrossRef]

24. Kumar, N.; Saha, T.K.; Dey, J. Control, implementation, and analysis of a dual two-level photovoltaic inverter based on modified proportional resonant controller. *IET Renew. Power Gener.* **2018**, *12*, 598–604. [CrossRef]

25. Pires, V.F.; Foito, D.; Sousa, D.M. Conversion structure based on a dual T-type three-level inverter for grid connected photovoltaic applications. In Proceedings of the 2014 IEEE 5th International Symposium on Power Electronics for Distributed Generation Systems (PEDG), Galway, Ireland, 24–27 June 2014; pp. 1–7.

26. Oleschuk, V.; Ermuratskii, V. Dual-inverter-based photovoltaic system with discontinuous synchronized PWM. In Proceedings of the 2014 IEEE International Conference on Intelligent Energy and Power Systems (IEPS), Kiev, Ukraine, 2–6 June 2014; pp. 86–89.

27. Pires, V.F.; Martins, J.F.; Hao, C. Dual-inverter for grid connected photovoltaic system: modeling and sliding mode control. *Sol. Energy* **2012**, *86*, 2106–2115. [CrossRef]

28. Anand, S.; Fernandes, B.G. Multilevel open-ended transformer based grid feeding inverter for solar photovoltaic application. In Proceedings of the IECON 2012-38th Annual Conference on IEEE Industrial Electronics Society, Montreal, QC, Canada, 25–28 October 2012; pp. 5738–5743.

29. Beres, R.N.; Wang, X.; Blaabjerg, F.; Liserre, M.; Bak, C.L. Optimal design of high-order passive-damped filters for grid-connected applications. *IEEE Trans. Power Electron.* **2016**, *31*, 2083–2098. [CrossRef]

30. Yin, J. Research on the Control Strategies of Dual Converter Based on Open-End Winding Topology. Ph.D. Dissertation, Beijing Jiaotong University, Beijing, China, April 2015; pp. 92–93.

31. Siemens, A.G. GEAFOL Cast-Resin Transformer 100 to 16000 kVA. In *Cast-Resin Transformers Catalog TV1*; Siemens AG: München, Germany, 2007.

32. Jiao, Y.; Lee, F.C. LCL filter design and inductor current ripple analysis for a three-level NPC grid interface converter. *IEEE Trans. Power Electron.* **2015**, *30*, 4659–4668. [CrossRef]

33. Juntunen, R.; Korhonen, J.; Musikka, T.; Smirnova, L.; Pyrhönen, O.; Silventoinen, P. Comparative analysis of LCL-filter designs for paralleled inverters. In Proceedings of the 2015 IEEE Energy Conversion Congress and Exposition (ECCE), Montreal, QC, Canada, 20–24 September 2015; pp. 2664–2672.

34. National Energy Administration. *Technical Specification of PV Grid-Connected Inverter*; NB/T 32004-2018; China, July 2018. Available online: https://www.chinesestandard.net/PDF/English.aspx/NBT32004-2018 (accessed on 18 June 2019).

35. Rockhill, A.A.; Liserre, M.; Teodorescu, R.; Rodriguez, P. Grid-filter design for a multimegawatt medium-voltage voltage-source inverter. *IEEE Trans. Ind. Electron.* **2011**, *58*, 1205–1217. [CrossRef]

36. Dannehl, J.; Liserre, M.; Fuchs, F.W. Filter-based active damping of voltage source converters with LCL filter. *IEEE Trans. Ind. Electron.* **2011**, *58*, 3623–3633. [CrossRef]

Article

Experimental Comparison of Two-Level Full-SiC and Three-Level Si–SiC Quasi-Z-Source Inverters for PV Applications

Serhii Stepenko [1,2,*], **Oleksandr Husev** [1,3], **Dmitri Vinnikov** [1], **Carlos Roncero-Clemente** [4], **Sergio Pires Pimentel** [1,5] **and Elena Santasheva** [1,4]

[1] Department of Electrical Power Engineering and Mechatronics, Tallinn University of Technology, 19086 Tallinn, Estonia
[2] Department of Information Measuring Technologies, Metrology and Physics, Chernihiv National University of Technology, 14027 Chernihiv, Ukraine
[3] Biomedical Radioelectronic Apparatus and Systems Department, Chernihiv National University of Technology, 14027 Chernihiv, Ukraine
[4] Department of Electrical, Electronic and Control Engineering, School of Industrial Engineering, University of Extremadura, 06006 Badajoz, Spain
[5] School of Electrical, Mechanical, and Computer Engineering, Federal University of Goias (UFG), Goiania 74690-900, Brazil
* Correspondence: serhii.stepenko@taltech.ee

Received: 6 June 2019; Accepted: 25 June 2019; Published: 28 June 2019

Abstract: The paper presents a comparative study of two solar string inverters based on the Quasi-Z-Source (QZS) network. The first solution comprises a full-SiC two-level QZS inverter, while the second design was built based on a three-level neutral-point-clamped QZS inverter with Silicon based Metal–Oxide–Semiconductor Field-Effect Transistors (Si MOSFETs). Several criteria were taken into consideration: the size of passive elements, thermal design and size of heatsinks, voltage stress across semiconductors, and efficiency investigation. The Photovoltaic (PV)-string rated at 1.8 kW power was selected as a case study system. The advantages and drawbacks of both solutions are presented along with conclusions.

Keywords: DC–AC converters; efficiency; neutral-point-clamped inverter; PV applications; PV inverters; PV systems; quasi-z-source; two-level inverter; three-level inverter; converter topologies

---

1. Introduction

Continuous development and improvements of Photovoltaic (PV) system designs along with related technologies, such as Wide Bandgap (WBG) GaN/SiC devices, Digital Signal Processor (DSP)- and Field-Programmable Gate Array (FPGA)-based control units have gradually decreased their costs. This allows new solutions featuring high efficiency and easy implementation which make them commercially attractive. At the same time, power rates and voltage operation ranges determine the availability of certain PV applications, especially in small-scale installations. In addition to efficiency and power density, the reliability of PV inverters is the key factor influencing the feasibility of single-phase industrial implementations [1,2], where Full-Bridge (FB) Voltage-Source Inverters (VSIs) are mostly used. Many DC–AC solutions for connecting PV modules to a single-phase grid are discussed in Reference [3]. The relative costs assessed based on the calculated ratings, component surveys at different vendors, and linear regression analysis were also taken into account in the evaluation.

The Z-Source Inverter (ZSI) [4] is an alternative to VSIs and Current-Source Inverters (CSIs) due to its ability to provide buck–boost operation within the single stage and its improved reliability based on its natural immunity against short-circuit. Its benefits have made it a promising solution for PV systems and have urged investigations in this area, which has resulted in many DC–DC and DC–AC topologies for single-phase and three-phase applications [5–15].

The Quasi-Z-Source Inverter (QZSI) was derived from the ZSI and has become a desirable topology for PV applications [5] due to its inheritance of all the advantages of ZSI enhanced by lower component ratings and continuous input current. The application of multilevel inverters has advantages in higher power designs, where the high voltage stress on the inverter's switches can be avoided [16–19]. The combination of the QZSI with the Three-Level (3L) Neutral-Point-Clamped (NPC) inverter has created a new promising topology, described in detail in Reference [6]. It features certain advantages such as low voltage stress on the power switches, single-stage buck–boost operation, continuous input current, short-circuit immunity, and low total harmonic distortion of the output voltage and current.

A detailed comparative study of basic and derived impedance-source networks for buck–boost inverter applications is provided in Reference [7], mostly for three-phase applications. The investigation of loss distribution was addressed recently in References [8,9] for QZSI-based topologies along with methods for their reduction and efficiency improvement.

Many publications devoted to the ZSI- and QZSI-derived solutions for PV, wind, and Microgrids applications have appeared recently [10–14]. They address certain issues, such as current harmonics reduction, voltage gain improvement, leakage current reduction, etc. The authors of Reference [11] emphasize the use of the coupled-inductor and SiC devices to optimize power density. A good comparison of impedance-source networks suitable for DC and AC applications by means of the passive components' number and size, semiconductor devices stress, and range of the input voltage variation is provided in Reference [15]. The increased voltage stress across semiconductors was reported as the main drawback of ZSI/QZSI. High-voltage gain solutions with additional magnetics may mitigate this.

An extreme high efficiency of 99.4% was reported for a three-phase 50 kW full-SiC PV string inverter in Reference [20]. Another full-SiC solution for a 25 kW three-phase PV string inverter demonstrated 97.7% peak efficiency [21]. These are examples of extra high efficiency, which, however, can be achieved much easier in high-power systems. The latter includes a detailed step-by-step explanation and design guidelines for all the components of the system.

Some low-power low-voltage designs are presented in References [22–26]. An example of an efficient converter based on the zeta inverter topology using 300 V Si + 1200 V SiC Metal–Oxide–Semiconductor Field-Effect Transistors (MOSFETs) is provided in Reference [22], with efficiency up to 95%; however, the nominal power was 220 W and the maximal was 440 W. A CSI-based single-phase solution for leakage current reduction is shown in Reference [23], where Insulated-Gate Bipolar Transistors (IGBTs) were used.

Several 350–400 W designs based on a quasi-switched-boost inverter with an efficiency of 91.3–94% are reported in References [24,25]. A good analysis of power losses, efficiency, and temperature is provided in Reference [26] for a CSI-based solution with SiC MOSFETs; additionally, power losses for all-SiC and hybrid approaches were analyzed, but the experimental results are not shown in the paper.

A valuable and interesting experimental comparison presented in Reference [27] is devoted to three topologies of a three-phase Two-Level (2L) inverter: a QZSI, a VSI with a boost converter, and a VSI with an interleaved boost converter. A detailed description of the methodology for comparison could be a very good reference for such an analysis. However, since the investigated input voltage range was 400–600 V, the operation of the QZSI was not assessed completely by means of the boost mode and the California Energy Commission (CEC) efficiency was not reported.

The most relevant solutions reported for single-phase and three-phase PV applications and supported by experimental verification are listed in Table 1. It should be mentioned that different solutions have been used to reach certain installed goals and satisfy some specific requirements.

Table 1. Parameters and characteristics of existing solutions.

References	Inverter Topology	Rated Power, kW	Input Voltage, V	Semiconductor Devices and Switching Frequency	Output Rms Voltage and Frequency	Peak/CEC Efficiency, %
[1]	SAF with FB VSI	2	450	900 V SiC MOSFET/45 kHz 650V Si IGBT/45 kHz	240 V, 60 Hz	97.75/97.2 97.0/96.4
[2]	FB VSI	0.5	280	Si MOSFET/19.2 kHz	110 V, 60 Hz	96/-
[4]	3-Phase ZSI	4.5	150	IGBT/10 kHz	208 V, 60 Hz	-/-
[5]	3-Phase QZSI	4.3	189–400	600 V diode & 600 V IGBT/10 kHz	208 V, 60 Hz	-/-
[6]	3l NPC QZSI	1	220–325	600 V diode and 600 V Si MOSFET/100 kHz	230 V, 50 Hz	94/-
[11]	CUK-based ZSI	0.4	90	1200 V IGBT K40T1202/20 kHz	110 V, 50 Hz	-/-
[13]	3-switch ZSI SEPIC	0.5	100	1200 V IGBT K40T1202/20 kHz	124 V, 50 Hz	91.7/-
[14]	QZSI +2 bi- directional switches	1	250	SiC diode C4D20120D and SiC MOSFET C2M0080120D/10 kHz 1200 V IGBT IKW25T120/10 kHz	220 V, 60 Hz	95.1/- 92.9/-
[20]	Interleaved boost DC–DC + T-type 3L 3-phase DC–AC	50	450-800	1.2kV SiC C4D20120D + 1.2kV SiC MOSFET C2M0025120D + 600V SiC C3D16060D + 1.2kV SiC MOSFET C2M0025120D/ 75 kHz	480 V, 50 Hz	99.4/-
[21]	HF link DC–AC–DC + 3-phase 2L VSI	25	533	SiC HB module CAS120M12BM2 + SiC diodes C4D40120D and 3 Phase SiC module CCS050M12CM2	400 V, 50 Hz	98.5/-
[22]	Zeta inverter	0.22	48	300 V MOSFET IXFK150N30P3 + 1200 V SiC MOSFET UJC1206k/50 kHz	220 V, 60 Hz	95/-
[24]	QSBI	0.35	50–72	Diodes STPS60SM200C and IXYS30-60A + MOSFETs IRFP4668 and IRFP460/20 kHz	110 V, 50 Hz	91.3/-
[25]	QSBI	0.4	58–100	Diodes DSEP 30–06A + MOSFETs IRFP460/10 and 20 kHz	110 V, 50 Hz	94/-
[27]	QZSI BC+VSI IBC+VSI	6	400–600	1 × C4D20120D diodes + 6 × C2M0080120D/100 kHz 1 × C4D20120D diodes + 1 × C2M0080120D + 6 × C2M0080120D/100 kHz 2 × C4D20120D diodes + 1 × C2M0080120D + 6 × C2M0080120D/100 kHz	220 V, 50 Hz	95.97/- 95.96/- 96.11/-

In some cases, different semiconductor technologies were tested. Thus, in Reference [1], different WBG and Si devices were investigated and evaluated (650 V GaN switches by Transphorm, RFMD and GaN Systems, 650 V SiC switches by RoHM, 900 V SiC by Wolfspeed and F5 series IGBT switches by Infineon). The final choice was to use 900 V SiC devices due to the voltage margin of 200% over the maximum DC bus voltage. The power levels of different PV applications could vary significantly. Particularly, the topologies discussed in Table 1, have been verified by experimental prototypes in the range from 220 W to 50 kW.

A 1800 W single-stage distributed PV plant was taken as a case study in Reference [28]. The experimental results of the developed 1 kW two-string prototype with different PV strings at various PV conditions are shown in Reference [29]. The industrial PV-string inverter SMA Sunny Boy 1600TL with a maximum input power of 1700 W was investigated in Reference [30].

The main requirements for off-grid and grid-connected PV systems include efficiency, reliability, and high-power density. These features could be available by providing low-input current ripple as well as low DC-link voltage ripple. This results in high-output current quality with the minimal possible requirements to the output filter. The importance of the power decoupling between the modules and the grid is discussed in Reference [3]. Some theoretical and simulation results for the 2L QZSI and the 3L NPC QZSI are reported in References [31–33].

To improve the reliability of the system and achieve higher power density by the reduction of redundant passive components, the approach of interleaving is often used in VSI. It enables significant reduction of the current ripple in QZS-stage inductors and the voltage ripple at the DC-link [27,31,33–35]. A topology of the Interleaved QZSI (IQZSI) under the Simple Boost Control (SBC) was proposed in Reference [34] for PV applications. Its certain benefits, including the reduced output THD and QZS-stage passive elements, potentially lead to higher power density of the system. To improve utilization of the DC-link and achieve higher gain, the Maximum Boost Control (MBC) [36] with appropriate modification was required. It smoothes out variation in the Shoot-Through (ST) duty cycle. The operation of MBC in IQZSI revealed the importance and proved the necessity of power decoupling in such PV systems [35]. Additionally, some control approaches for 3l NPC QZSI are proposed in [37–39].

Although many solutions were claimed as suitable for PV applications, in most of the listed studies, the case study tasks for PV applications are not positioned in detail. Moreover, there are numerous works that present SiC-based solutions, including those built on QZS network, however, the discussions on the feasibility and experimental investigations of the alternative approaches for 2L and multi-level approaches based on Si, SiC, and Si+SiC designs are absent. The peak and especially the CEC efficiency [40] of the proposed PV solutions are often not analyzed in the papers. The calculation for the passive components is usually significantly simplified and in practical experience, some capacitors or inductors can be smaller or with an increased ripple [15]. Thus, our study aimed to discuss the most urgent peculiarities in the implementation of the 2L full-SiC and the 3L Si–SiC inverters based on the QZS network and to share our experiences to advance the application of these solutions in PV systems.

The paper is organized as follows. Section 2 outlines the main specifications of the case study system, provides the system parameters, and explains both of the converters with the control approach. Section 3 presents the design guidelines for element selection. Section 4 describes the experimental prototypes built based on the 2L QZSI and the 3L NPC QZSI topologies, explains the structure of the experimental setup along with the equipment used, and demonstrates the obtained results, including operation waveforms, measured efficiency, and temperature dependencies. Section 5 presents a comparative evaluation of both topologies followed by the conclusions provided in Section 6.

2. Case Study System

2.1. System Parameters and Specifications

The PV system being considered for PV string application which could comprise 5 ... 10 PV panels with total power up to 1800 W. The PV panel SPR-200-BLK from SunPower was selected for the case study [41]. The main system and PV panels parameters are provided in Table 2 and typical P-V and I-V dependencies are shown in Figure 1. The operating power profile of the design solution according to the case study PV string is also depicted in Figure 1. In the input voltage range from 200 V to 400 V, the converter was assumed to operate with the rated input current of 5 A.

Depending on the operating conditions and the type of panels, the power conversion efficiency can vary, but is aimed to be in the range from 92% to 96%. The converter was aimed to operate at the rated (nominal) power of 1800 W, with its maximum efficiency of 97.1% in the nominal mode, which corresponds to the input voltage of 360 V. In this operating point, the converter has its highest CEC efficiency, which is over 96% for both topologies (2L QZSI and 3L NPC QZSI).

Table 2. Photovoltaic (PV) panel and system parameters.

PV panel parameters	Values	System parameters	Values
Standard Test Conditions (STCs): Air Mass (AM) 1.5, Irradiance 1000 W/m^2, cell temperature 25 °C		Nominal power Nominal voltage	P_{nom} = 1800 W V_{nom} = 200–400 V
Nominal power (+/−5%) Rated voltage Rated current Open circuit voltage Short circuit current	P_{nom} = 200 W V_{mpp} = 40.0 V I_{mpp} = 5.00 A V_{oc} = 47.8 V I_{sc} = 5.40 A	Nominal current Load RMS voltage Output current THD Min operating power Max operating power	I_{nom} = 5 A V_{load} = 230 V THD_I < 3% P_{min} = 90 W P_{max} = 2000 W
Nominal Operating Cell Temperature (NOCT): Air Mass (AM) 1.5, Irradiance 800 W/m^2, cell temperature 46 °C +/−2 °C		Min operating voltage Max operating voltage	V_{min} = 180 V V_{max} = 480 V
Nominal power Rated voltage Rated current Open circuit voltage Short circuit current	P_{nom} = 146 W V_{mpp} = 36.5 V I_{mpp} = 4.01 A V_{oc} = 44.5 V I_{sc} = 4.38 A	Voltage ripple Max input current Current ripple Number of PV panels Case study PV panels	ΔV < 5% I_{max} = 10 A ΔI < 10% N = 5 … 10 SPR-200-BLK

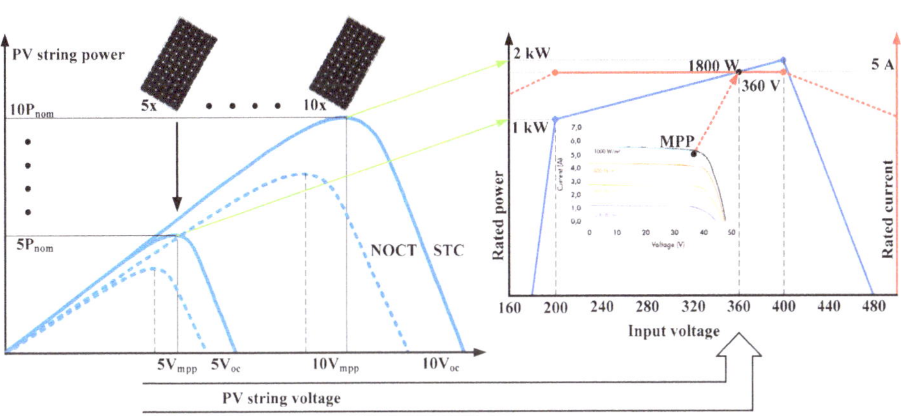

Figure 1. PV string characteristics and system power profile.

2.2. Description of Topologies

The PV system considered was built based on two different approaches: on the 2L QZSI (Figure 2) and on the 3L NPC QZSI (Figure 3). The 2L QZSI proposed in Reference [5] is described in detail as a three-phase application for PV systems. The main parts of the topology include the QZS network represented by L_1, D_1, C_1, L_2, and C_2; the FB 2L inverter based on MOSFET switches S_1, S_2, S_3, and S_4; and the output filter L_{F1}, C_F, and L_{F2} feeding the load or connected to the grid. Detailed discussions and explanations on the 2L QZSI for a single-phase PV application as well as the control approaches, including SBC, MBC, constant boost control and their modifications, are provided in [31,33–35]. In this study, SBC was used for generating the ST states.

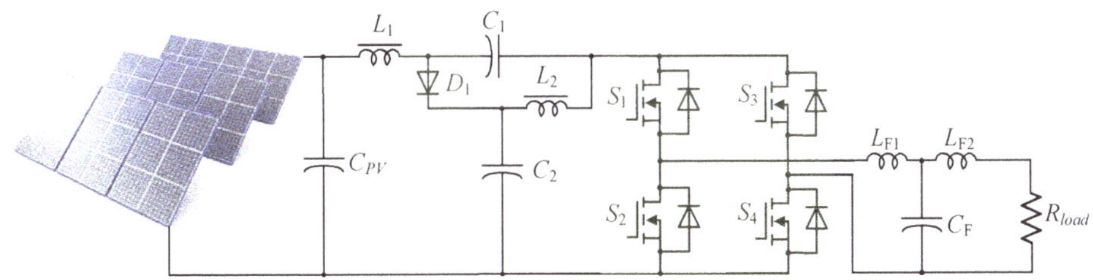

Figure 2. The 2L QZS inverter.

The 3L NPC QZSI (Figure 3) was proposed and discussed in detail as a single-phase application in Referene [6]. The study also provides the main design guidelines and the experimental results. The main parts of the topology include the QZS network, which in this case was divided by a neutral point into two symmetrical parts, represented by L_1, C_1, D_1, L_2, C_2 and L_3, C_3, D_2, L_4, and C_4; an FB 3L inverter with switches S_1, S_2, S_3, S_4, S_5, S_6, S_7, and S_8; clamping diodes D_3, D_4, D_5, and D_6; and an output filter L_{F1}, C_F, and L_{F2} feeding the load or connected to the grid.

The topology was proved as an efficient PV converter [37–39], including maximum power point tracking (MPPT) implementation along with continuous input current [37] and operation in the grid-connected mode [38,39]. The implementation of this topology under different control approaches is discussed in detail in References [6,7,32,38,39]. In our study, the SBC approach was used for generating the ST states.

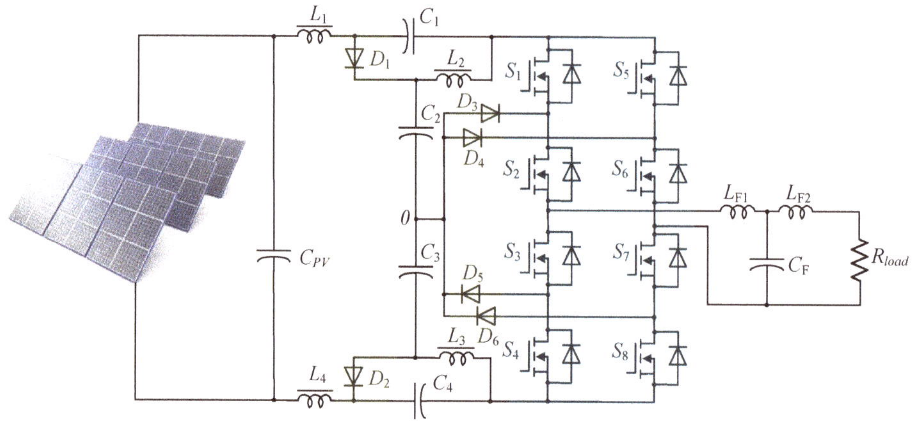

Figure 3. The 3l NPC QZS inverter.

3. General Design Guidelines

3.1. Selection of Passive Components

The passive element values of the QZS network for both cases were estimated according to the guidelines [6] based on the same approach that includes High-Frequency (HF) and Low-Frequency (LF) ripple analysis. The HF ripple of the input current was taken into account as follows:

$$L_1 \geq \frac{V_{OUT}^2 \cdot (1 - 2 \cdot D_s)}{2 \cdot (1 - D_s) \cdot K_{LH1} \cdot P_{OUT}} \cdot T_s \cdot D_s, \tag{1}$$

where L_1 is the value of QZS network inductance, V_{OUT} is the output voltage, D_S is the duration of ST state, T_S is the switching period, K_{LH1} is an assumed HF ripple of input current, and P_{OUT} is the output power. The main peculiarities of the calculation and selection process are as follows.

For the 3L NPC QZSI for appropriate inductances L_1, L_2, L_3, L_4 chosen according to Equation (1), we assumed the max HF ripple to be limited to 10%, which means $K_{LH1} = 0.1$. For the output voltage $V_{OUT} = 230$ V and max $D_S = 0.225$, the switching period $T_s = 1/f_s = 1/65$ kHz and output power $P_{out} = 900$ W. According to Equation (1), it gives us the minimal value of $L_1 = 0.72$ mH.

Since it is a minimal possible value (which provides boundary conduction mode), and assuming possible variation of the inductance under the temperature and other impacts, the value of 0.9 mH was chosen to assure the Continuous Conduction Mode (CCM).

For this value, according to Equation (2), the HF current ripple should be 8%.

$$K_{LH1} = \frac{\Delta I_{L1}}{2 \cdot I_{IN}} \approx \frac{V_{OUT}^2 \cdot (1 - 2 \cdot D_S)}{2 \cdot (1 - D_S) \cdot L_1 \cdot P_{OUT}} \cdot T_S \cdot D_S, \tag{2}$$

Since inductances in the 3L NPC QZSI are connected in series, the equivalent inductances for the 2L QZSI could be assumed as $L_1 = L_1 + L_4$, $L_2 = L_2 + L_3$. Thus, equivalent inductances of 1.8 mH were chosen for the 2L QZSI. For QZS capacitances C_1 and C_2, we assumed the voltage ripple to be limited to 2% and 1% correspondingly:

$$K_{CL1} = \frac{\bar{v}_{C1}}{V_{C1}} = \frac{8 \cdot P_{OUT} \cdot (1 - D_S) \cdot (4\pi \cdot T \cdot L_2 + R \cdot T^2)}{3\pi \cdot V_{OUT}^2 \cdot D_S \cdot \sqrt{16\pi^2 \cdot C_1^2 \cdot R^2 + (16\pi^2 \cdot C_1 \cdot L_2 - T^2)^2}}, \tag{3}$$

$$K_{CL2} = \frac{\bar{v}_{C2}}{V_{C2}} = \frac{8 \cdot P_{OUT} \cdot (4\pi \cdot T \cdot L_1 + R \cdot T^2)}{3\pi \cdot V_{OUT}^2 \cdot \sqrt{16\pi^2 \cdot C_2^2 \cdot R^2 + (16\pi^2 \cdot C_2 \cdot L_1 - T^2)^2}}, \tag{4}$$

According to Equations (3) and (4), the minimal values $C_1 = 1000\ \mu F$ and $C_2 = 233\ \mu F$ were selected. Taking into account the maximal RMS current of the capacitors and decreasing the capacitance under the voltage near to the maximal rated level and the temperature impact, the electrolytic capacitances were chosen as $C_1 = 2700\ \mu F$ and $C_2 = 860\ \mu F$.

At the same time, one can assess the LF ripple according to Equation (5), which for the chosen value of 0.9 mH, $C_2 = 860\ \mu F$ gives us the level of 25%:

$$K_{LL1} = \frac{\Delta I_{L1}}{I_{IN}} \approx \frac{\Delta I_{L1} \cdot V_{IN}}{P_{OUT}} \approx \frac{8 \cdot (1 - 2 \cdot D_S) \cdot T^2}{2\pi \cdot (1 - D_S) \cdot \sqrt{16\pi^2 \cdot C_2^2 \cdot R^2 + (16\pi^2 \cdot C_2 \cdot L_1 - T^2)^2}}, \tag{5}$$

Since the capacitances C_1, C_4, and C_2, C_3 in the 3l NPC QZSI are connected in the series under ST, the equivalent capacitance of the asymmetrical QZS network will be twice lower. Thus, taking into account maximal possible voltages, the electrolytic capacitances of $C_1 = 1200\ \mu F$ and $C_2 = 680\ \mu F$ were chosen for the 2L QZSI topology, which is summed up in Table 3.

It should also be mentioned, that in the 3l NPC QZSI prototype, capacitances C_1 and C_4 were physically installed as a combination of parallel connection of 1200 μF and 1500 μF, while capacitances C_2 and C_3 were combined as 390 μF and 470 μF in parallel connection.

For the standalone application (off-grid), the simplest L or LC filter could be used. The application of an LC filter could also provide better efficiency due to the fewer losses. However, since the case study of the PV system is considered for grid-connected applications, we used the LCL filter in both cases. This provides better stability in the grid-connected mode.

The values of the passive components of the output filter were assessed and chosen based on the classical approach, which is reported in Reference [42]. Thus, for both converters the same output LCL filters were chosen with $L_{F1} = 560\ \mu H$, $C_F = 15\ \mu F$, and $L_{F2} = 200\ \mu H$.

3.2. Selection of Semiconductor Devices and Heatsinks

The main difference in the proposed solutions was observed during the selection of semiconductor devices. The peak voltage across the QZSI bridge is increasing with the input voltage decreasing. It is explained by the necessity of ST implementation that deteriorates the DC-link voltage utilization. Thus, 1200 V SiC power switches should be used in the 2L QZSI solution for the case study system. To overcome this limitation, the 3l NPC QZSI is considered as an alternative approach. Eight 650 V Si MOSFETs with a fast body diode were used in it. Also, six 650 V SiC diodes (2 in QZS network + 4 as clamping diodes) were used, representing the Si–SiC approach. The SiC diode and four 1200 V SiC

MOSFETs were used in the 2L QZSI representing the full-SiC approach. All semiconductor devices along with chosen passive elements are provided in Table 3.

Table 3. Selected elements.

Components	2L QZSI	3l NPC QZSI
QZS-stage inductors	$L_1 = L_2 = 1.8$ mH	$L_1 = L_2 = L_3 = L_4 = 0.9$ mH
QZS-stage capacitors	EKMS3B1VSN122MA50S, $C_1 = 1200$ μF, 105 °C, 315 V, 3000 Hrs, 3.25 A, ESR 100 mΩ; ALC10(1)681DL500, $C_2 = 680$ μF, 85 °C, 500 V, 2000 Hrs, 3.65 A, ESR 244 mΩ	ESMQ201VSN122MQ40S, $C_1 = C_4 = 1200$ μF, 85 °C, 200 V, 2000 Hrs, 3.5 A, ESR 166 mΩ; B43504G2158M80, $C_1 = C_4 = 1500$ μF, 105 °C, 200 V, 3000 Hrs, 3.4 A, ESR 100 mΩ; B43545C9397M000, $C_2 = C_3 = 390$ μF, 105 °C, 400 V, 5000 Hrs, 2.3 A, ESR 150 mΩ; LPW471M2GQ45M, $C_2 = C_3 = 470$ μF, 85 °C, 400 V, 2000 Hrs, ESR 420 mΩ
Output filter	$L_{F1} = 0.56$ mH, $L_{F2} = 0.2$ mH, $C_F = 15$ μF	
QZS-stage and clamping diodes	D_1: SiC C4D02120A, $V_{RRM} = 1200$ V	D_1–D_6: SiC C3D10065A, $V_{RRM} = 650$ V
Inverter bridge switches	S_1–S_4: C2M0080120D MOSFETs SiC, $V_{DS} = 1200$ V, $R_{DS} = 80$ mΩ	S_1–S_8: IPW65R041CFD MOSFETs Si , $V_{DS} = 650$ V, $R_{DS} = 41$ mΩ
Gate drivers	ACPL-H342 (2.5 A max peak output current)	

The selected semiconductors are equivalent by means of conduction losses. At the same time, the main differences between the Si and the SiC technology lie in the switching losses and maximum operation temperature. On the one hand, the full-SiC design may provide lower switching losses, on the other hand, the operation temperature can be higher. The practical benefit of the higher semiconductor temperature limit lies in the reduced size of heatsink required. In the heatsink design, our approach was to select the type and volume that can provide the required operation temperature. The heatsink was collected of several items, whereas the thermal resistance of each was equal to 2.8 °C/W. Taking into account the higher operation temperature of SiC devices, the volume of heatsink for the 2L solution with the full-SiC approach was twice as small. Thus, for the nominal input voltage, the expected maximal temperature of the heatsink in the 2L solution was about 90 °C, while in the case of conventional Si MOSFETs, it was expected up to 70 °C.

4. Experimental Study

4.1. Experimental Setup and Tested Prototypes Description

The general approach to the experimental verification is shown by the structure of the experimental setup in Figure 4, which was intended to facilitate a comparison of the proposed solutions by means of an efficiency study and verification of theoretical statements. The experimental setup includes the following equipment: programmable DC power supply (PV array simulator) Chroma 62150H-1000S [43]; high-performance power analyzer YOKOGAVA WT1800 [44]; oscilloscope Tektronix MSO 4034B [45]; 2L QZSI or 3l NPC QZSI as a PV inverter incorporating output LCL filter (Figure 5); resistive load for up to 3 kW output power; both converters were controlled by FPGA Cyclone IV EP4CE22E22C8 [46]; the temperatures of the inverter switches (T_{sw}) and heatsinks (T_{swh}), QZS-stage diodes (T_{qzsD}) and heatsinks (T_{qzsDh}), clamping diodes (T_{clD}) and heatsinks (T_{clDh}) in 3l NPC QZSI, were measured using an infrared thermal camera Fluke Ti10 [47]; passive element values were measured using HM8118 LCR Bridge/Meter [48].

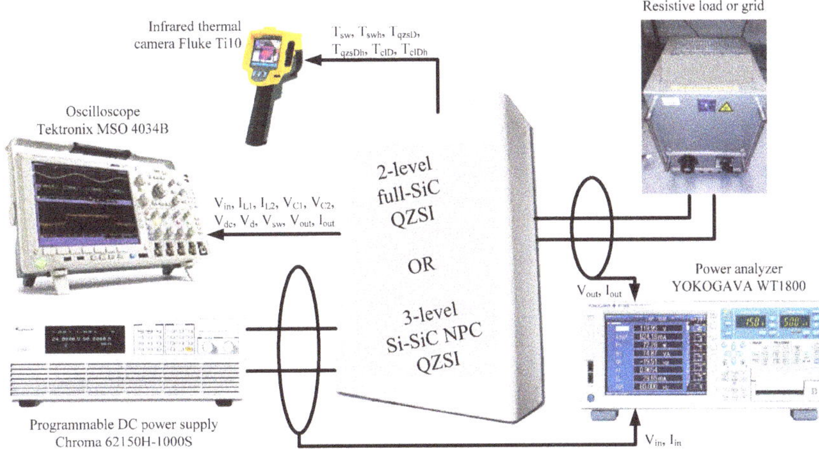

Figure 4. The structure of the experimental verification setup.

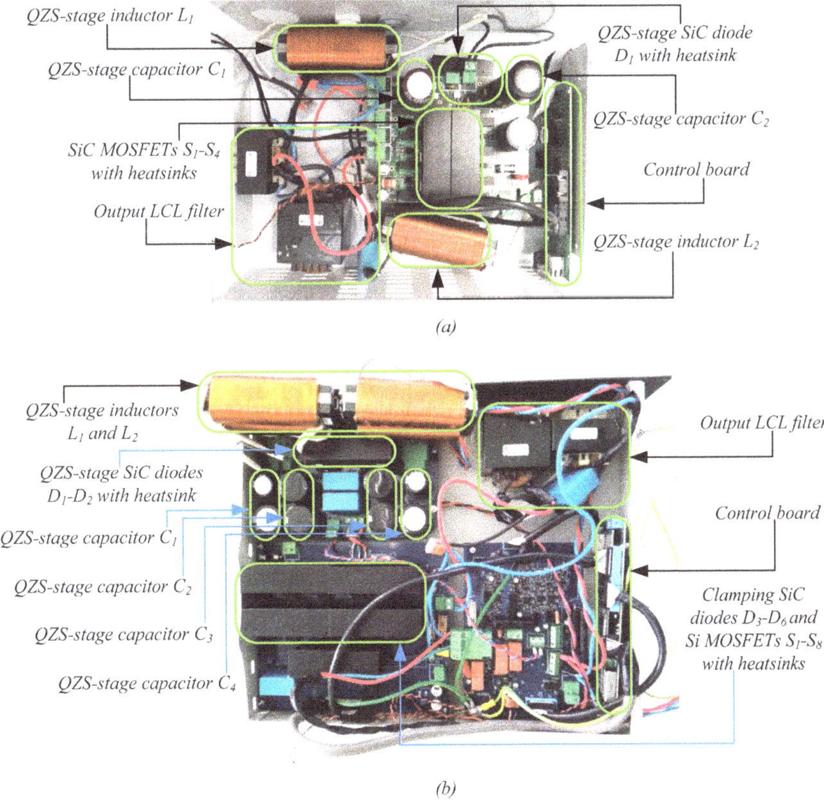

Figure 5. Experimental prototypes of the 2L QZSI (**a**) and the 3L NPC QZSI (**b**).

4.2. Operation Waveforms and Characteristics of the Prototypes

The experimental waveforms are shown below in Figures 6 and 7 for both the 2L QZSI and the 3L NPC QZSI. Figure 6 shows the experimental results in the open loop mode of the 2L QZSI. Figure 6a,b correspond to the nominal power point 1800 W with the input voltage 360 V. The operation in the boost mode (power point of 1 kW) is presented in Figure 6c,d. To boost input voltage of 200 V up to 360 V at DC-link, the duty cycle D_S = 0.225 was applied. The impact of the ST states on the ripples can be observed.

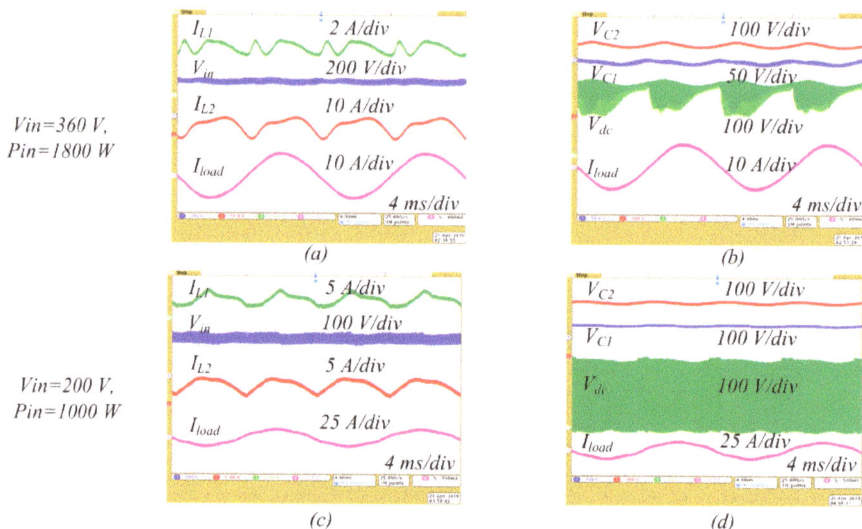

Figure 6. Operation waveforms of 2L QZSI at a nominal power level of 1800 W (**a,b**) and in boost voltage mode at a reduced power level of 1000 W (**c,d**).

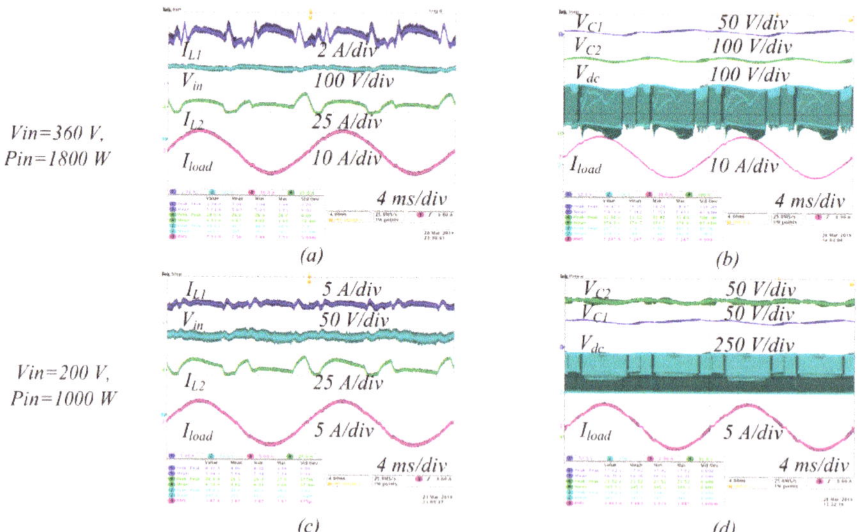

Figure 7. Operation waveforms of 3L NPC QZSI at a nominal power level of 1800 W (**a,b**) and in boost voltage mode at a reduced power level of 1000 W (**c,d**).

The appropriate waveforms for 3l NPC QZSI are depicted in Figure 7.

The diagrams presented are similar. The LF ripple of the input current was about 30% for both solutions that slightly exceeded the analytically calculated level of 25%. However, the input currents for both converters were in the CCM and the HF ripples for the 2L QZSI and the 3L NPC QZSI corresponded to the calculated value of 8%. The voltage ripples at C_1 are 1.5–1.6% and ripples at C_2 were 1.2–1.4%, which completely satisfied or nearly the calculated values of 2% and 1%, correspondingly.

In addition, Figure 8 shows the experimental diagrams of the PV-inverter operation in the closed-loop grid-connected mode for the nominal and reduced input power. It can be observed that the input current ripples had an LF ripple that corresponded to the double-frequency ripple. In the second case (Figure 8b), the current ripple had an HF component. It should be noted that a closed-loop system stabilizes the behavior of the converter. The observed current ripples completely corresponded to the theoretical expectation.

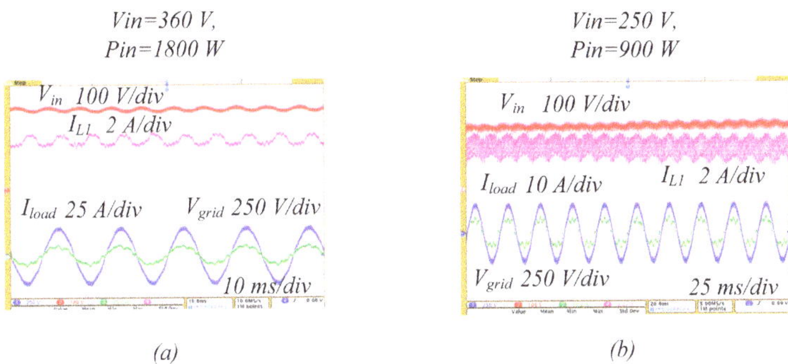

Figure 8. Grid-connected operation of 3L NPC QZSI at a nominal power level of 1800 W (**a**) and in boost voltage mode at a reduced power level of 900 W (**b**).

4.3. Efficiency Evaluation

Figure 9a shows the efficiency dependencies versus the input power. It corresponded to the different irradiance levels of the PV-string. It means that different operation points corresponded to the different input voltages and input currents. Since PV inverters are not operating at a nominal power point constantly, the important characteristics of the PV inverter performance is the weighted CEC efficiency [46]. The value of the weighted CEC efficiency was obtained by assigning a probable percentage of time the inverter resides at a certain operating point. If we denote the efficiency at 50% of the nominal power by "Eff50%", the average EU (European) and CEC efficiency values weighted appropriately are defined as:

$$\eta_{EU} = 0.03 \cdot \text{Eff5\%} + 0.06 \cdot \text{Eff10\%} + 0.13 \cdot \text{Eff20\%} + 0.10 \cdot \text{Eff30\%} + 0.48 \cdot \text{Eff50\%} + 0.20 \cdot \text{Eff100\%}, \quad (6)$$

$$\eta_{CEC} = 0.04 \cdot \text{Eff10\%} + 0.05 \cdot \text{Eff20\%} + 0.12 \cdot \text{Eff30\%} + 0.21 \cdot \text{Eff50\%} + 0.53 \cdot \text{Eff75\%} + 0.05 \cdot \text{Eff100\%}, \quad (7)$$

The CEC efficiency measurement and calculation results corresponding to Figure 9a are shown in Table 4. For both inverters, it exceeds 96%. However, for the 2L QZSI it supersedes by 0.4%. The EU efficiency was also measured and for both converters it exceeded 95%.

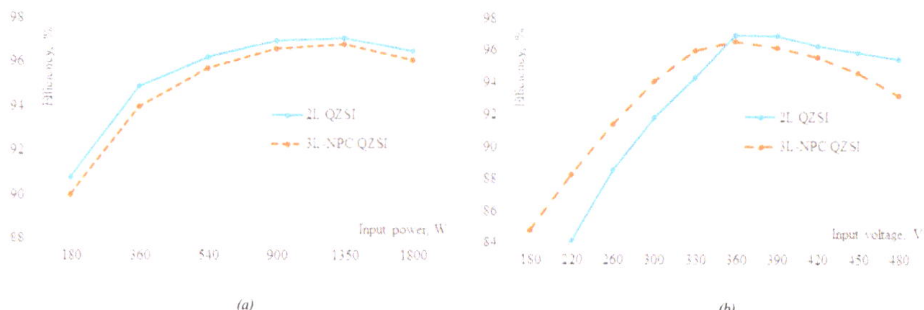

Figure 9. Efficiency evaluation of the 2L QZSI and the 3L NPC QZSI under different solar irradiance (power) levels (**a**) and within input voltage operating range (**b**).

Table 4. The CEC efficiency of 2L QZSI and 3L NPC QZSI.

Power Checkpoints, % of Nominal	Input Power, W	2L QZSI Efficiency, %	3l NPC QZSI Efficiency, %
10	180	90.8	90.0
20	360	94.9	94.0
30	540	96.2	95.7
50	900	97.0	96.6
75	1350	97.1	96.8
100	1800	96.5	96.1
-	CEC efficiency	96.6	96.2

It should be mentioned that efficiency curves for the 2L QZSI and the 3L NPC QZSI are characterized by different shapes. Figure 9b shows efficiency dependence versus different input voltage. This case illustrates the efficiency of the converters with different numbers of PV panels or at shadowed conditions. It can be seen that the 2L solution had higher peak efficiency, which corresponds to the nominal operation point, but more significant efficiency decrease occurred at low input voltage. In general, for the zero ST duty cycle, the 2L QZSI demonstrated 0.4 ... 2.3% higher efficiency than the 3L NPC QZSI. The situation changed when a non-zero ST duty cycle was utilized. It can be explained by higher conduction losses in SiC transistors, which take force under higher current rates in ST mode.

4.4. Evaluation of Temperature Behavior of Semiconductors and Heatsinks

The temperature of semiconductor devices and heatsinks was controlled by an infrared thermal camera Fluke Ti10. The results are presented in Figures 10 and 11. It should be mentioned that clamping diodes and inverter power switches in the 3L NPC QZSI had four common heatsinks, each one intended for two MOSFETs and one clamping diode. One more heatsink was used for two QZS-stage diodes, as can be seen from Figure 5. At the same time, two heatsinks were used in the 2L QZSI for inverter switches and one heatsink for the QZS-stage diode.

Figure 10a shows the temperature of the QZS-stage diode and its heatsink in the 2L QZSI under different power levels. These points correspond to Figure 9a. It can be seen that the diode and the heatsink temperature rose under the power increase. The maximum temperature corresponded to the maximum power and fully corresponded to the theoretical assumptions. Figure 10b shows the temperature of the QZS-stage diodes and their heatsink in the 3L NPC QZSI under different power levels. It can be seen that the diode and heatsink temperature was slightly higher than in the 2L solution. The thermal images of the QZS-stage diode D_1 at close to nominal power level (1850 W) are shown in Figure 11a,c. For the 2L QZSI and 3L NPC QZSI, the hottest points were the temperatures 130 °C and 140 °C, respectively.

Figure 10c,d show similar diagrams for SiC power switches in the 2L QZSI and Si power switches in the 3L NPC QZSI. The maximum temperature of the SiC power switches in the nominal mode was not higher than 95 °C in the 2L QZSI, while the maximum temperature of the Si power switches in the

3L NPC QZSI was much lower and did not exceed 75 °C. It should be mentioned, that under the ST states application in the boost mode, the temperature of the bridge power switches rose significantly in both cases.

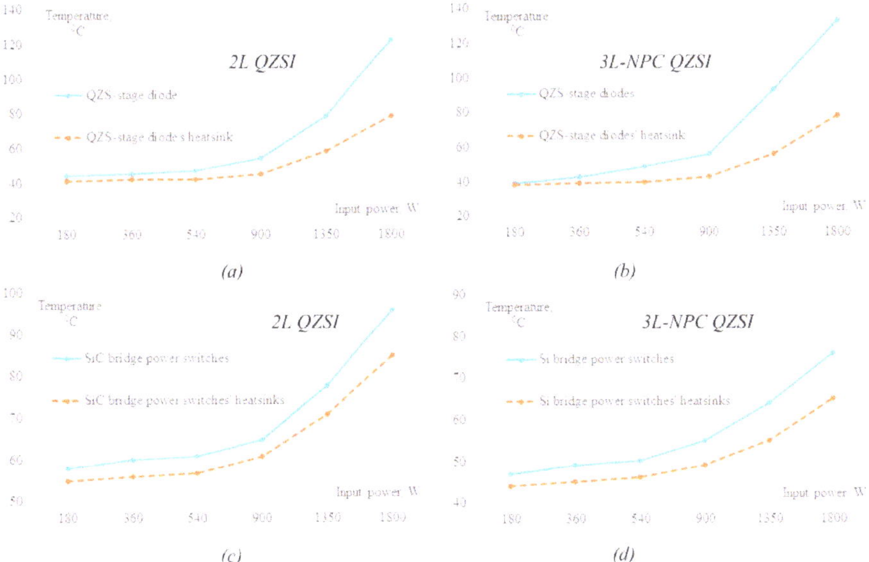

Figure 10. Temperatures of the QZS-stage diode and heatsink in the 2L QZSI (**a**) and 3L NPC QZSI (**b**), temperatures of the SiC bridge power switches and heatsinks in the 2L QZSI (**c**), and temperatures of the Si bridge power switches and heatsinks in the 3L NPC QZSI (**d**).

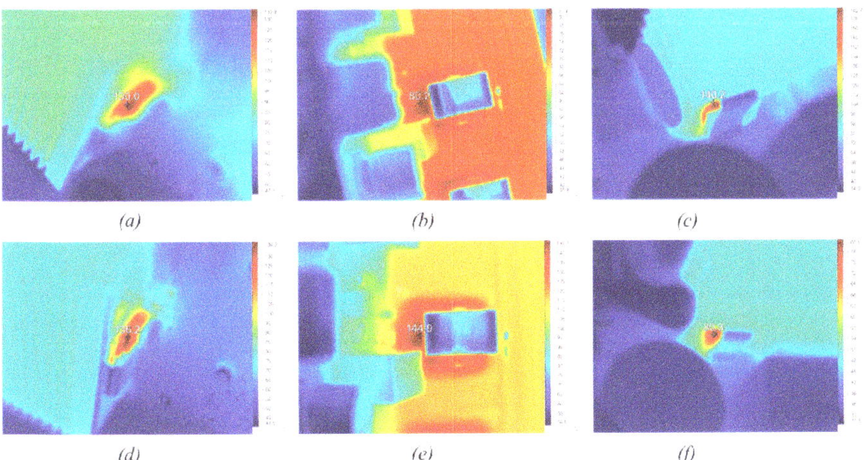

Figure 11. Thermal images in the nominal operation mode: QZS-stage diode (**a**) and SiC bridge power switches (**b**) in the 2L QZSI and QZS-stage diode in the 3L NPC QZSI (**c**); thermal images in the boost mode under ST states application: QZS-stage diode (**d**) and SiC bridge power switches (**e**) in the 2L QZSI and QZS-stage diode in the 3L NPC QZSI (**f**).

In the case of the 2L solution, the SiC temperature reaches 160 °C, while in the 3L NPC solution the Si temperature did not exceed 120 °C at the maximum power points. Figure 11b,e show the SiC thermal pictures for two modes at the same power point of 1850 W. It can be seen that under the transient from nominal to boost mode, the temperature of the QZS-stage diodes changed insignificantly in both solutions.

The SiC power switches temperature in the 2L solution rose from 90 … 110 °C to 130 … 160 °C, while in the 3L solution based on the Si power switches, the temperature rose from 70 … 80 °C to 100 … 110 °C only. Finally, the SiC semiconductor devices can safely operate with higher temperature. It means that the size of the heatsinks in the case of full-SiC design can be smaller.

5. Comparative Analysis

This section presents the results of the comparison of different characteristics discussed and verified above. Figure 12 shows a diagram that includes several parameters for the 2L QZSI and the 3L NPC QZSI: volume of capacitors, volume of inductors, summarized voltage stress across semiconductors, heatsink volume, and CEC efficiency.

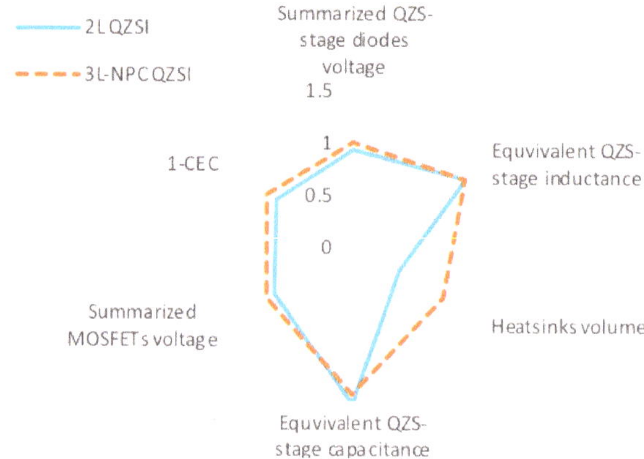

Figure 12. Comparative diagram for the 2L QZSI and 3L NPC QZSI.

As can be seen, the equivalent inductances and capacitances of the QZS-stage are practically equal. It is explained by the same operation conditions of the proposed solutions and the same switching frequency. In the case of the 3L NPC, the capacitors and inductors are split, but the overall size remains the same.

The main difference concerns semiconductors. Overall voltage stress across full bridge transistors remains the same, but the 3L NPC requires additional clamping diodes. It was also clearly shown, that due to the higher operation temperature of the SiC devices in 2L QZSI, the heatsink can be selected significantly smaller.

Finally, the diagram also shows that power losses (1-CEC) are slightly smaller in the 2L QZSI than in the 3L NPC QZSI. It was achieved even under the higher operation temperature of SiC semiconductors.

6. Conclusions

In this study, a PV-string with a nominal power level of 1800 W was chosen as a case study for the evaluation of two PV-inverters based on the 2L QZSI full-SiC solution and the 3L NPC QZSI solution with Si MOSFETs. However, both investigated topologies could be easily up-scaled (with appropriately selected passive components) and safely operated up to twice the higher power level. The main conclusion from our analysis is that the full-SiC 2L QZSI solution has a clear advantage over the 3l NPC QZSI solution with Si MOSFETs. First of all, it simplifies the Printed Circuit Board (PCB) and reduces the number of auxiliary components around switches. Secondly, it may provide higher efficiency along with the lower volume of heatsink. It should be mentioned that efficiency and heatsink volume can be used as trade-off parameters for industrial optimization. The heatsink volume increase will lead to the temperature decreasing and to efficiency improvements correspondingly. On the other hand, the reliability and longtime operation of the full-SiC solution is an open question for discussion and can be a limiting factor in industrial implementation.

Author Contributions: Conceptualization, S.S. and O.H.; methodology, S.S. and C.R.-C.; software, O.H. and S.S.; validation, S.S. and S.P.P.; formal analysis, S.S. and C.R.-C.; investigation, S.S. and E.S.; resources, D.V., O.H. and S.S.; data curation, S.S. and O.H.; writing—original draft preparation, S.S.; writing—review and editing, S.S., O.H. and D.V.; visualization, S.S.; supervision, D.V. and O.H.; project administration, O.H. and S.S.; funding acquisition, S.S.

Funding: This research was carried out with the support of the European Regional Development Fund and the programme Mobilitas Pluss under the project MOBJD126 awarded by the Estonian Research Council. This work was also supported by the Estonian Centre of Excellence in Zero Energy and Resource Efficient Smart Buildings and Districts (ZEBE), grant 2014-2020.4.01.15-0016 funded by the European Regional Development Fund, and by the Estonian Research Council grant PUT1443.

Conflicts of Interest: The authors declare no conflict of interest.

References

1. Ghosh, R.; Wang, M.; Mudiyula, S.; Mhaskar, U.; Mitova, R.; Reilly, D.; Klikic, D. Industrial Approach to Design a 2-kVa Inverter for Google Little Box Challenge. *IEEE Trans. Ind. Electron.* **2018**, *65*, 5539–5549. [CrossRef]
2. Tang, Y.; Yao, W.; Loh, P.C.; Blaabjerg, F. Highly reliable transformerless photovoltaic inverters with leakage current and pulsating power elimination. *IEEE Trans. Ind. Electron.* **2016**, *63*, 1016–1026. [CrossRef]
3. Kjaer, S.B.; Pedersen, J.K.; Blaabjerg, F. A Review of Single-Phase Grid-Connected Inverters for Photovoltaic Modules. *IEEE Trans. Ind. Appl.* **2005**, *41*, 1292–1306. [CrossRef]
4. Peng, F.Z. Z-Source Inverter. *IEEE Trans. Ind. Appl.* **2003**, *39*, 504–510. [CrossRef]
5. Li, Y.; Anderson, J.; Peng, F.Z.; Liu, D. Quasi-Z-Source Inverter for Photovoltaic Power Generation Systems. In Proceedings of the 2009 Twenty-Fourth Annual IEEE Applied Power Electronics Conference and Exposition, Washington, DC, 15–19 February 2009; pp. 918–924. [CrossRef]
6. Husev, O.; Roncero-Clemente, C.; Romero-Cadaval, E.; Vinnikov, D.; Stepenko, S. Single phase three-level neutral-point-clamped quasi-Z-source inverter. *IET Power Electron.* **2015**, *8*, 1 10. [CrossRef]
7. Husev, O.; Blaabjerg, F.; Roncero-Clemente, C.; Romero-Cadaval, E.; Vinnikov, D.; Siwakoti, Y.P.; Strzelecki, R. Comparison of Impedance-Source Networks for Two and Multilevel Buck–Boost Inverter Applications. *IEEE Trans. Power Electron.* **2016**, *31*, 7564–7579. [CrossRef]
8. Abdelhakim, A.; Davari, P.; Blaabjerg, F.; Mattavelli, P. Switching Loss Reduction in the Three-Phase Quasi-Z-Source Inverters Utilizing Modified Space Vector Modulation Strategies. *IEEE Trans. Power Electron.* **2017**, *33*, 4045–4060. [CrossRef]
9. Roncero-Clemente, C.; Romero-Cadaval, E.; Pires, V.F.; Martins, J.F.; Vilhena, N.; Husev, O. Efficiency and Loss Distribution Analysis of the 3L Active NPC qZS Inverter. In Proceedings of the 2018 IEEE 12th International Conference on Compatibility, Power Electronics and Power Engineering (CPE-POWERENG 2018), Doha, Qatar, 10–12 April 2018; pp. 1–6. [CrossRef]
10. Huang, S.; Zhang, Y.; Hu, S. Stator Current Harmonic Reduction in a Novel Half Quasi-Z-Source Wind Power Generation System. *Energies* **2016**, *9*, 770. [CrossRef]
11. Wang, B.; Tang, W. A New CUK-Based Z-Source Inverter. *Electronics* **2018**, *7*, 313. [CrossRef]
12. Priyadarshi, N.; Padmanaban, S.; Ionel, D.M.; Mihet-Popa, L.; Azam, F. Hybrid PV-Wind, Micro-Grid Development Using Quasi-Z-Source Inverter Modeling and Control—Experimental Investigation. *Energies* **2018**, *11*, 2277. [CrossRef]
13. Wang, B.; Tang, W. A Novel Three-Switch Z-Source SEPIC Inverter. *Electronics* **2019**, *8*, 247. [CrossRef]
14. Choi, W.-Y.; Yang, M.-K. Transformerless Quasi-Z-Source Inverter to Reduce Leakage Current for Single-Phase Grid-Tied Applications. *Electronics* **2019**, *8*, 312. [CrossRef]
15. Husev, O.; Shults, T.; Vinnikov, D.; Roncero-Clemente, C.; Romero-Cadaval, E.; Chub, A. Comprehensive Comparative Analysis of Impedance-Source Networks for DC and AC Application. *Electronics* **2019**, *8*, 405. [CrossRef]

16. Noman, A.; A Al-Shamma'a, A.; Addoweesh, K.; Alabduljabbar, A.; Alolah, A. Cascaded Multilevel Inverter Topology Based on Cascaded H-Bridge Multilevel Inverter. *Energies* **2018**, *11*, 895. [CrossRef]

17. Hammami, M.; Rizzoli, G.; Mandrioli, R.; Grandi, G. Capacitors Voltage Switching Ripple in Three-Phase Three-Level Neutral Point Clamped Inverters with Self-Balancing Carrier-Based Modulation. *Energies* **2018**, *11*, 3244. [CrossRef]

18. Wu, M.; Song, Z.; Lv, Z.; Zhou, K.; Cui, Q. A Method for the Simultaneous Suppression of DC Capacitor Fluctuations and Common-Mode Voltage in a Five-Level NPC/H Bridge Inverter. *Energies* **2019**, *12*, 779. [CrossRef]

19. Tran, V.-T.; Nguyen, M.-K.; Ngo, C.-C.; Choi, Y.-O. Three-Phase Five-Level Cascade Quasi-Switched Boost Inverter. *Electronics* **2019**, *8*, 296. [CrossRef]

20. Mookken, J.; Agrawal, B.; Liu, J. Efficient and Compact 50kW Gen2 SiC Device Based PV String Inverter. In Proceedings of the PCIM Europe 2014; International Exhibition and Conference for Power Electronics, Intelligent Motion, Renewable Energy and Energy Management, Nuremberg, Germany, 20–22 May 2014; pp. 1–7.

21. Öztürk, S.; Canver, M.; Çadırcı, I.; Ermiş, M. All SiC Grid-Connected PV Supply with HF Link MPPT Converter: System Design Methodology and Development of a 20 kHz, 25 kVA Prototype. *Electronics* **2018**, *7*, 85. [CrossRef]

22. Choi, W.-Y.; Yang, M.-K. High-Efficiency Design and Control of Zeta Inverter for Single-Phase Grid-Connected Applications. *Energies* **2019**, *12*, 974. [CrossRef]

23. Guo, X.; Zhang, J.; Zhou, J.; Wang, B. A New Single-Phase Transformerless Current Source Inverter for Leakage Current Reduction. *Energies* **2018**, *11*, 1633. [CrossRef]

24. Nguyen, M.-K.; Choi, Y.-O. Voltage Multiplier Cell-Based Quasi-Switched Boost Inverter with Low Input Current Ripple. *Electronics* **2019**, *8*, 227. [CrossRef]

25. Nguyen, M.; Choi, Y. PWM Control Scheme for Quasi-Switched-Boost Inverter to Improve Modulation Index. *IEEE Trans. on Power Electron.* **2018**, *33*(5), 4037–4044. [CrossRef]

26. Fernández, E.; Paredes, A.; Sala, V.; Romeral, L. A Simple Method for Reducing THD and Improving the Efficiency in CSI Topology Based on SiC Power Devices. *Energies* **2018**, *11*, 2798. [CrossRef]

27. Wolski, K.; Zdanowski, M.; Rabkowski, J. High-frequency SiC-based inverters with input stages based on quasi-Z-source and boost topologies—Experimental comparison. *IEEE Trans. Power Electron.* **2019**. [CrossRef]

28. Tiku, D. Modular Multilevel MMI(HB) Topology for Single-Stage Grid Connected PV Plant. In Proceedings of the 11th IET International Conference on AC and DC Power Transmission, Birmingham, UK, 2015; pp. 1–8. [CrossRef]

29. Keyhani, H.; Toliyat, H.A. Single-Stage Multistring PV Inverter With an Isolated High-Frequency Link and Soft-Switching Operation. *IEEE Trans. Power Electron.* **2014**, *29*, 3919–3929. [CrossRef]

30. Strache, S.; Wunderlich, R.; Heinen, S. A Comprehensive, Quantitative Comparison of Inverter Architectures for Various PV Systems, PV Cells, and Irradiance Profiles. *IEEE Trans. Sustain. Energy* **2014**, *5*, 813–822. [CrossRef]

31. Stepenko, S.; Roncero-Clemente, C.; Husev, O.; Makovenko, E.; Pimentel, S.P.; Vinnikov, D. New interleaved single-phase quasi-Z-source inverter with active power decoupling. In Proceedings of the 2018 IEEE 12th International Conference on Compatibility, Power Electronics and Power Engineering (CPE-POWERENG 2018), Doha, Qatar, 10–12 April 2018; pp. 1–6. [CrossRef]

32. Makovenko, E.; Husev, O.; Romero-Cadaval, E.; Vinnikov, D.; Stepenko, S. Single-Phase Three-Level qZ-Source Inverter Connected to the Grid with Battery Storage and Active Power Decoupling Function. In Proceedings of the 2018 IEEE 59th International Scientific Conference on Power and Electrical Engineering of Riga Technical University (RTUCON), Riga, Latvia, 12–13 November 2018; pp. 1–6. [CrossRef]

33. Stepenko, S.; Husev, O.; Pimentel, S.P.; Makovenko, E.; Vinnikov, D. Small Signal Modeling of Interleaved Quasi-Z-Source Inverter with Active Power Decoupling Circuit. In Proceedings of the 2018 IEEE 59th International Scientific Conference on Power and Electrical Engineering of Riga Technical University (RTUCON), Riga, Latvia, 12–13 November 2018; pp. 1–6. [CrossRef]

34. Roncero-Clemente, C.; Husev, O.; Stepenko, S.; Romero-Cadaval, E.; Vinnikov, D. Interleaved single-phase quasi-Z-source inverter with special modulation technique. In Proceedings of the 2017 IEEE First Ukraine Conference on Electrical and Computer Engineering (UKRCON), Kiev, Ukraine, 29 May–2 June 2017; pp. 593–598. [CrossRef]
35. Roncero-Clemente, C.; Stepenko, S.; Husev, O.; Romero-Cadaval, E.; Vinnikov, D. Maximum boost control for interleaved single-phase Quasi-Z-Source inverter. In Proceedings of the IECON 2017-43rd Annual Conference of the IEEE Industrial Electronics Society, Beijing, China, 29 October–1 November 2017; pp. 7698–7703. [CrossRef]
36. Nguyen, M.-K.; Choi, Y.-O. Maximum Boost Control Method for Single-Phase Quasi-Switched-Boost and Quasi-Z-Source Inverters. *Energies* **2017**, *10*, 553. [CrossRef]
37. Roncero-Clemente, C.; Stepenko, S.; Husev, O.; Miñambres-Marcos, V.; Romero-Cadaval, E.; Vinnikov, D. Three-Level Neutral-Point-Clamped Quasi-Z-Source Inverter with Maximum Power Point Tracking for Photovoltaic Systems. In *Technological Innovation for the Internet of Things. DoCEIS 2013. IFIP Advances in Information and Communication Technology*; Camarinha-Matos, L.M., Tomic, S., Graça, P., Eds.; Springer: Berlin/Heidelberg, Germany, 2013; Volume 394, pp. 334–342. [CrossRef]
38. Pimentel, S.P.; Husev, O.; Stepenko, S.; Vinnikov, D.; Roncero-Clemente, C.; Makovenko, E. Voltage Control Tuning of a Single-Phase Grid-Connected 3L qZS-Based Inverter for PV Application. In Proceedings of the 2018 IEEE 38th International Conference on Electronics and Nanotechnology (ELNANO), Kiev, Ukraine, 24–26 April 2018; pp. 692–698. [CrossRef]
39. Pimentel, S.P.; Husev, O.; Vinnikov, D.; Roncero-Clemente, C.; Stepenko, S. An Indirect Model Predictive Current Control (CCS-MPC) for Grid-Connected Single-Phase Three-Level NPC Quasi-Z-Source PV Inverter. In Proceedings of the 2018 IEEE 59th International Scientific Conference on Power and Electrical Engineering of Riga Technical University (RTUCON), Riga, Latvia, 12–13 November 2018; pp. 1–6. [CrossRef]
40. Types of Solar Inverter Efficiency - Peak, Euro, and CEC-Solar Selections. Available online: http://www.solarselections.co.uk/blog/types-of-solar-inverter-efficiency-peak-vs-euro (accessed on 10 May 2019).
41. SolarHub-PV Module Details: SPR-200-BLK-U-by SunPower. Available online: http://www.solarhub.com/product-catalog/pv-modules/764-SPR-200-BLK-U-SunPower (accessed on 13 May 2019).
42. Husev, O.; Chub, A.; Romero-Cadaval, E.; Roncero-Clemente, C.; Vinnikov, D. Voltage Distortion Approach for Output Filter Design for Off-Grid and Grid-Connected PWM Inverters. *J. Power Electron.* **2015**, *15*, 278 287. [CrossRef]
43. Model 62150h 600s/1000s-Chroma Systems Solutions. Available online: http://www.chromausa.com/pdf/Br-62150H-600S+1000S-dcsupply-112011.pdf (accessed on 24 May 2019).
44. WT1800 High Performance Power Analyzer | Yokogawa Test & Measurement Corporation. Available online: https://tmi.yokogawa.com/us/solutions/products/power-analyzers/wt1800-high-performance-power-analyzer/ (accessed on 24 May 2019).
45. MSO/DPO4000B Mixed Signal Oscilloscope | Tektronix. Available online: https://www.tek.com/oscilloscope/mso4000-dpo4000 (accessed on 24 May 2019).
46. Cyclone IV Device Handbook-Intel. Available online: https://www.intel.com/content/dam/www/programmable/us/en/pdfs/literature/hb/cyclone-iv/cyclone4-handbook.pdf (accessed on 24 May 2019).
47. Fluke Ti10 Infrared Camera | Fluke. Available online: https://www.fluke.com/en-us/product/thermal-cameras/ti10 (accessed on 27 May 2019).
48. HM8118 LCR Bridge/Meter | Overview | Rohde & Schwarz. Available online: https://www.rohde-schwarz.com/pk/product/hm8118-productstartpage_63493-44101.html (accessed on 28 May 2019).

Article

Evaluation of Interconnection Configuration Schemes for PV Modules with Switched-Inductor Converters under Partial Shading Conditions

Kamran Ali Khan Niazi [1,*], **Yongheng Yang** [1], **Mashood Nasir** [2] **and Dezso Sera** [1]

1 Department of Energy Technology, Aalborg University, 9220 Aalborg, Denmark
2 Electrical Engineering Department, Lahore University of Management Sciences, Lahore 54792, Pakistan
* Correspondence: kkn@et.aau.dk

Received: 6 June 2019; Accepted: 19 July 2019; Published: 21 July 2019

Abstract: Partial shading on photovoltaic (PV) arrays reduces the overall output power and causes multiple maximas on the output power characteristics. Due to the introduction of multiple maximas, mismatch power losses become apparent among multiple PV modules. These mismatch power losses are not only a function of shading characteristics, but also depend on the placement and interconnection patterns of the shaded modules within the array. This research work is aimed to assess the performance of 4×4 PV array under different shading conditions. The desired objective is to attain the maximum output power from PV modules at different possible shading patterns by using power electronic-based differential power processing (DPP) techniques. Various PV array interconnection configurations, including the series-parallel (SP), total-cross-tied (TCT), bridge-linked (BL), and center-cross-tied (CCT) are considered under the designed shading patterns. A comparative performance analysis is carried out by analyzing the output power from the DPP-based architecture and the traditional Schottky diode-based architecture. Simulation results show the gain in the output power by using the DPP-based architecture in comparison to the traditional bypassing diode method.

Keywords: partial shading; photovoltaic (PV) arrays; multiple maximas; mismatch; differential power processing (DPP); series-parallel (SP); total-cross-tied (TCT); bridge-linked (BL); center-cross-tied (CCT)

---

1. Introduction

Solar energy is free and abundant [1]. Environmental concerns are widely reduced by using solar energy for power generation. Therefore, the photovoltaic (PV) energy is becoming the most emerging and promising solution to address environmental problems. Generally, the efficiency of solar PV energy conversion using PV panels is low [2], and therefore, many researchers are working on improving the efficiency and output energy yield [3]. Factors that may affect the conversion efficiency generally include the effect of soiling, dirt and dust, elevated temperature, and sudden irradiance changes [4]. Similarly, the output power produced by PV arrays is remarkably reduced due to partial shading conditions [5,6]. Partial shading is generally induced over a PV module, string, or on a whole small PV system. It is due to cloud shadows, dust, permanent cracks on shields or surfaces; as well as shade due to various structures including trees, leaves, and buildings or towers [7]. Partial shading causes a reduction in the irradiance, and also distributes irradiance in a non-uniform pattern over the surface of various PV modules in an array [8,9]. Hence, the current from the PV array is constrained by the shaded PV modules, which in turn is detrimental for the other healthy PV modules connected in the series [10]. Consequently, in practice, a parallel-connected diode termed as a bypass diode (D_1 and D_2) is installed across it to minimize the effects of mismatching, as shown in Figure 1a. During mismatching, this bypass diode will be ON and the current starts flowing through it (as shown in Figure 1b). In this case, various maximas appear on the power–voltage (P–V) characteristics. These multiple peaks are known as local maxima's, as examplified in Figue 1c. When multiple peaks are present, the conventional maximum power point techniques (MPPTs) may not work accurately. Therefore, a global maximum power point technique (GMPPT), capable to distinguish between local and global maxima is generally needed to maximize the overall output power from the array. In the literature, many conventional MPPT techniques have been presented and their behavior on partial shading conditions is analyzed [11–13]. Various artificial intelligence techniques including fuzzy logic [14,15], neural networks [16], and genetic algorithms [17] are generally employed to track the global maximum power point (GMPP). However, these techniques have limitations and exhibit false tracking over varying conditions of irradiance and temprature. Furthermore, the convential MPPT methods should be retrofitted with more sensing and control requirements [18,19].

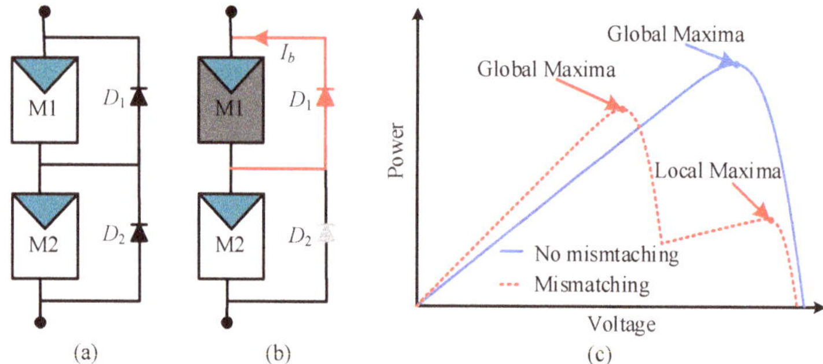

Figure 1. A PV module (M1 and M2) with parallel-connected bypass diodes D_1 and D_2: (**a**) schematic diagram, (**b**) schematic showing the current flow direction while shaded and bypassed [19], (**c**) P–V characteristic of series-connected two PV modules while one is shaded (mismatching occurs). Here, I_b is a bypass current through D_1.

In addition to developing advanced MPPT algorithms, an alternative is to directly mitigate the local peaks under partial shading. Differential power processing (DPP) converters [20–25] are typical representatives, that enable each PV module to produce maximum output power. DPP converters eliminate the problem of multiple maximas in the PV string, as highlighted in Figure 2. In addition, Figure 3a further exemplifies one DPP configuration known as the PV–PV voltage balance converter [20]. This PV–PV converter only processes the mismatched power and thus, it is used in this paper. The working principle of the PV–PV DPP is shown in Figure 3b,c, where the PV module M1 is shaded and the PV module M2 in the non-shaded mode. I_L is a mismatch current, which passes through the inductor L. The transistors Q_1 and Q_2 operate complementarily to each other. Nevertheless, the switching will induce power losses that can be found as [20]

$$P_{SWLOSS} = 2k_0 \left[V_G \sqrt{\frac{V_G}{2V_B}} + 4V_D I_L \sqrt{\frac{V_D}{V_B}} \right] f_{SW} \tag{1}$$

in which k_0 is a material property dependant device constant, V_B is the breakdown voltage of the device, V_D is the voltage at the device terminal, V_G is the voltage at the gate terminal, I_L is the current passing through the inductor, and f_{sw} is the switching frequency of the power device. The PV–PV DPP topology is based on the switched-inductor between two PV modules. Therefore, it is named as switched-inductor (SL)-based topology. The SL-based topology can be represented by a simplified model as illustrated in Figure 4. To have the same voltage (i.e., V_1 and V_2) across both PV modules, the value of effective impedance Z_{EFF} should be minimum. However, practically it is unavoidable to have zero value of Z_{EFF} but since the value of Z_{EFF} is frequency dependent. Therefore, it is possible to achieve the minimum value by operating the converter at frequencies near to the resonant frequency.

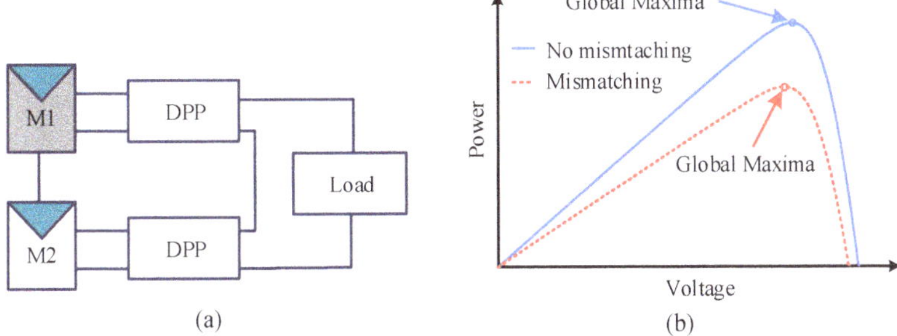

Figure 2. PV modules (M1 and M2) with parallel-connected DPP topologies: (**a**) schematic diagram when M1 is shaded and (**b**) P–V characteristic of series-connected two PV modules during mismatching and no mismatching.

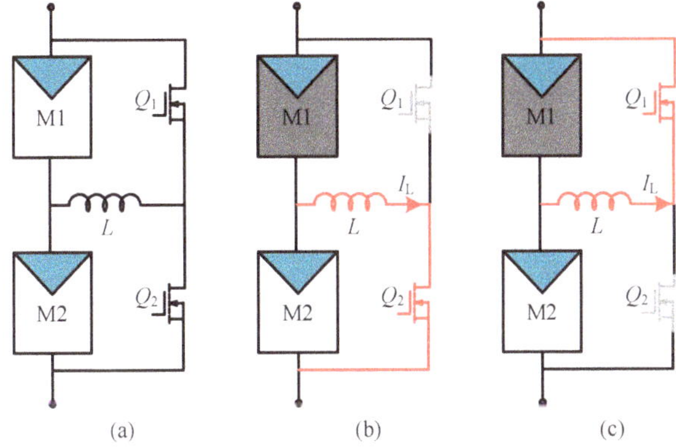

Figure 3. Switched-inductor (SL)-based PV–PV voltage balance converter [20]: (**a**) schematic diagram containing two series-connected PV modules M1 and M2 without shading; (**b**) M1 is shaded, where Q_1 is ON and Q_2 is OFF; and (**c**) M1 is shaded and Q_1 is OFF and Q_2 is ON. Here, I_L is a current passing through an inductor L.

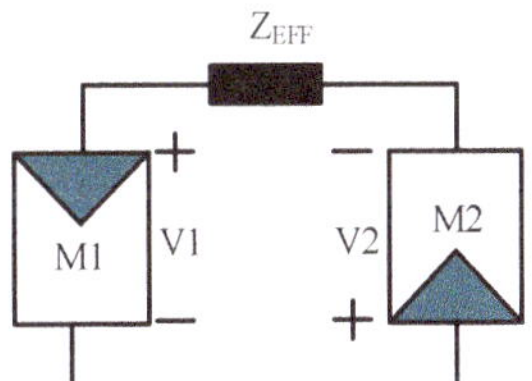

Figure 4. Equivalent circuit for the SL-based DPP technique. Here, Z_{EFF} is the effective impedance, V_1, and V_2 are the voltages across the PV module M1 and M2, respectively.

Moreover, another important factor to consider in the design of the SL-based topology is the quality factor Q, which is given by expression (2)

$$Q = (1/R_{ESR})\sqrt{L/C} \tag{2}$$

where L is the inductance shown in Figure 3 and C is the stray capacitance, which can be neglected as its value is very small. For practical reasons and to have adequate voltage stress across inductor L, the value of Q is selected in the range of 1–10 and the switching frequency is selected as 50 kHz.

The value of switching frequency is selected closer to the resonant frequency in order to have a minimum value of Z_{EFF}. For better understanding, Z_{EFF} can be considered as a series LC network whose impedance versus frequency curve is shown in Figure 5, which can be calculated by using (3). At lower frequencies (below resonant frequency), the topology behaves as overall capacitive. However, as the frequency increases, the value of Z_{EFF} decreases to its minimum value at the resonant frequency where the impedances of L and C cancels each other. Similarly, beyond the resonant frequency and at higher frequencies, the nature of Z_{EFF} is overall inductive. Therefore, the value of switching frequency is selected in the vicinity of resonant frequency to achieve a minimum value of Z_{EFF}, which will, in turn, equalizes the voltages of PV modules in a system. Moreover, the value of switching frequency is selected to be higher than the resonant frequency to achieve soft-switching operation for all the switches, which can minimize the switching power loss.

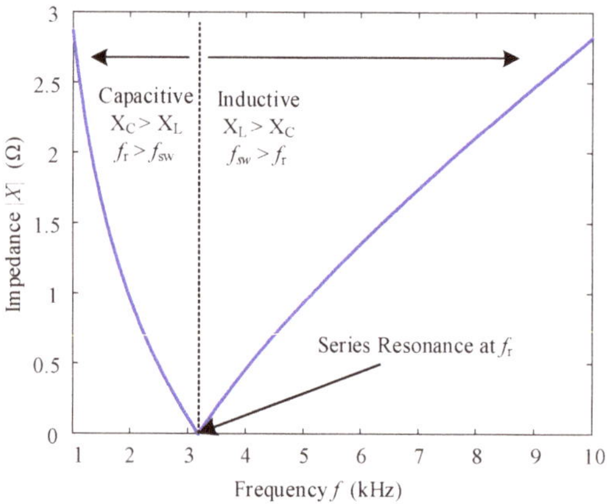

Figure 5. A curve showing the variation of impedance with frequency for a series resonant circuit.

$$|X| = |X_L| - |X_C| \tag{3}$$

where X_L is the inductive impedance and X_C is the capacitive impedance.

The power losses during partial shading are dependent upon the patterns of shading. Various schemes have been presented in the literature to minimize the detrimental effects caused by non-uniform shading [26]. One possibility is to reconfigure the interconnection PV modules in case if there is partial shading. The most commonly used interconnection scheme for PV arrays is series-parallel (SP). Configurations like the bridge-linked (BL), central-cross-tied (CCT), and total-cross-tied (TCT) (see Figure 6) can also be adopted to minimize the mismatching effect due to partial shading [27]. Although, it has been reveailed that the TCT configuration yields maximum power for conventional bypass technique [28], more attempts have been made to mitigate the effects of partial shading using SP configurations due to its simplicity [29]. Using TCT enhances the longevity of the PV module with an estimated increase in a lifetime by 30% [30]. Alternately an electronic array reconfiguration scheme abbreviated as EAR has been proposed, which may modify the interconnection pattern of modules using electronic switches. In EAR, during operation and the decision of reconfiguration is based upon the pattern of shading, while control is achieved using a switch matrix [31]. The electrical reconfiguration using switches and relays, which can be effectively realized for small systems. However, for large PV arrays in solar parks, etc., the electronic switches, their interaction, and controllability becomes complex and difficult to handle due to the constraints of switching [32]. Similarly, another technique termed as disperse interconnection scheme (SDS) for multiple PV modules in an array has been presented in [33]. This technique is also based on changing the electrical configuration of the modules in the PV array. This technique has superior output yield in comparison to other interconnection schemes as discussed in [29]. However, cost and complexity associated with the changing connections in large PV arrays can be cumbersome, and the overall system yield may become infeasible [29]. Therefore, various aspects by including efficiency and cost must be carefully considered for the optimal system design. A static reconfiguration is considered effective over other dynamic interconnection techniques [23]. The static reconfiguration technique does not involve the dynamic change of interconnection. It is based upon the one-time constant arrangement of PV modules with predefined interconnection settings in an array under different partial shading conditions [34,35]. Different interconnection topologies, which are discussed above—namely, SP, BL, and TCT for PV arrays—have been proposed. These interconnection schemes are tested by using intelligent power electronics to minimize power losses due to mismatch. The power electronic-based technique replaces the shunting bypass diodes, which are generally connected in parallel to PV modules in the array.

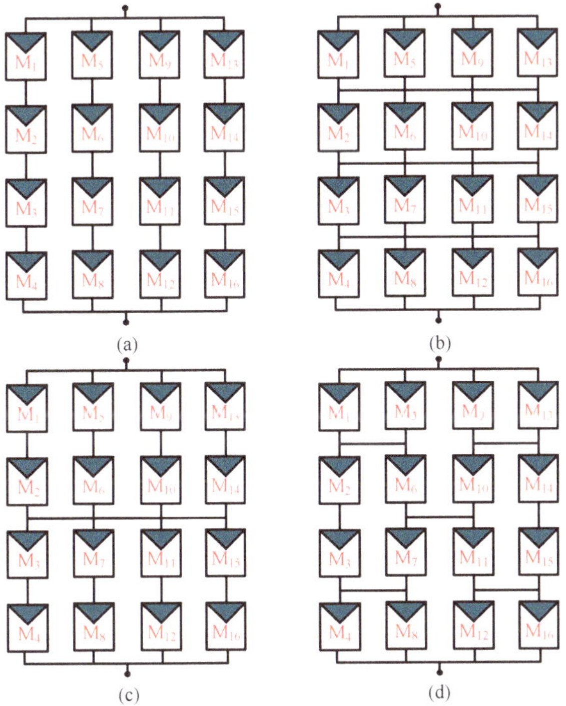

Figure 6. PV array topologies: (**a**) series-parallel (SP), (**b**) total-cross-tied (TCT), (**c**) central-cross-tied (CCT), and (**d**) bridge-linked (BL).

The performance analysis of above-mentioned interconnection schemes has been widely discussed with traditional bypass diodes in the literature so far. However, when DPP converters have adopted the performance of those configurations has not yet been explored. Therefore, in this paper, the four array interconnection schemes—i.e., the SP, TCT, CCT, and BL configurations with SL-DPP—are tested under different patterns of shading and irradiance. The performance of proposed schemes (SL-based interconnection schemes) is evaluated by introducing different shading patterns and its comparison with state of the art topology, i.e., bypass diode (for comparative analysis). This study uses a 4 × 4 PV array of system, which has a size of 968 W. The output power and mismatch losses under various shading patterns with SL-DPP and bypass diodes are presented. The organization of the rest of the paper is given below. Section 2 introduces the different interconnection configurations and shading pattern designs. Section 3 outlines various important results and based upon the highlighted results conclusions are outlined in Section 4.

2. Configurations and Shading Pattern Designs

The four interconnection schemes—i.e., the SP, TCT, CCT, and BL configurations—are compared at different shading patterns for the 4 × 4 PV array. The PV array with these configurations using DPP converters and bypass diodes are evaluated. The rating of PV modules is given in Table 1. The rated maximum output power from the 4 × 4 PV system is 968 Watt at 1000 W/m^2 and STP. Different types of shading patterns including, one module shading, short wide, long narrow shading, central shading, and diagonal shading are applied at all above discussed interconnection schemes on 4 × 4 PV array at different irradiances, as shown in Figure 7. The variations in irradiance from 200 W/m^2 to 800 W/m^2 with a difference of 100 W/m^2 is considered to compare the performance of SL-based DPPs with conventional parallel-connected bypass diodes. Various shading pattern designs for PV array configuration schemes are as follows.

Table 1. Specification of the PV module at STC (1000 W/m^2 and 25 °C).

Parameters	Rating
Rated Peak Power (P_{max})	60 W
Voltage at MPP (V_{mp})	17.10 V
Current at MPP (I_{mp})	3.50 A
Open Circuit Voltage (V_{oc})	21.10 V
Short Circuit Current (I_{sc})	3.80 A

2.1. One Module Shading

In the condition of one module shading, one module from first row and first column is shaded from 4 × 4 PV array, i.e., PV module M1, as shown in Figure 7a.

2.2. Short Wide Shading

For short wide shading, four PV modules from the first two rows and columns are shaded, i.e., M1, M2, M5, and M6, as shown in Figure 7b.

2.3. Long Narrow Shading

In long narrow shading, PV modules placed at the last column of PV array, which includes PV modules, M13, M14, M15, and M16 along with the last row, which are M4, M8, and M12 are not shaded, as shown in Figure 7c. Rest of all PV modules are shaded from 4 × 4 PV array.

2.4. Central Shading

Four PV modules from the center are shaded—i.e., M6, M7, M10, and M11—while all other remain unshaded, as shown in Figure 7d.

2.5. Diagonal Shading

For diagonal shading M4, M7, M10, and M13 are shaded, as shown in Figure 7e. In diagonal shading, one module gets shaded from each row and column of a 4 × 4 PV array for all configuration schemes.

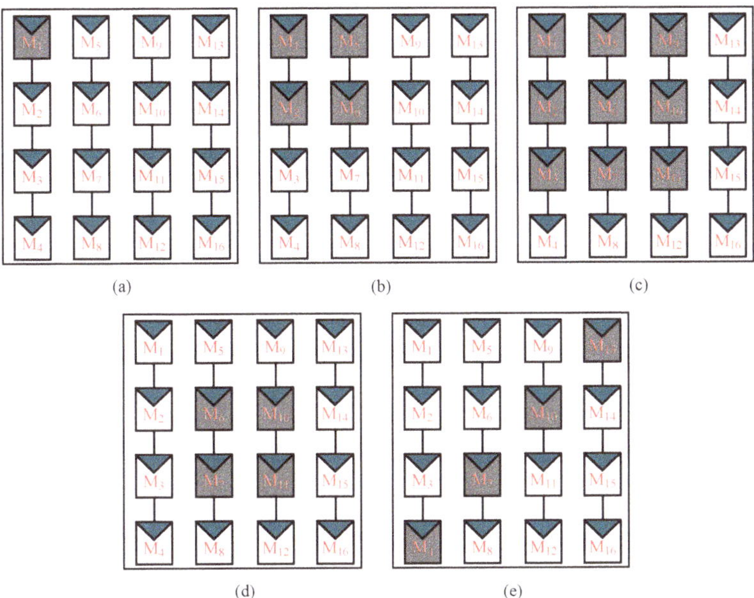

(a)   (b)   (c)

(d)   (e)

Figure 7. Shading pattern designs for PV array configuration schemes: (**a**) one module, (**b**) short wide, (**c**) long narrow, (**d**) central, and (**e**) diagonal.

3. Results and Discussion

The simulation results for the configurations with traditional bypass diodes and power electronic-based DPP architectures are shown in Figures 8–11. The systems are evaluated at 1000 W/m^2, 800 W/m^2, 600 W/m^2, 500 W/m^2, 400 W/m^2, and 200 W/m^2 by using a power sim (PSIM).

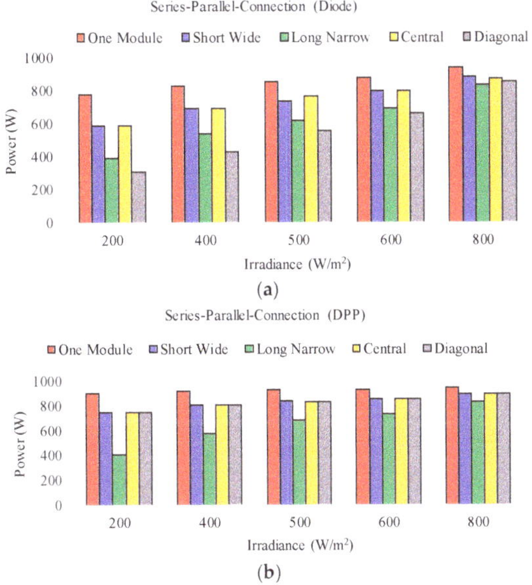

Figure 8. Output for series-parallel (SP) connection under various shading conditions with: (**a**) diode connection and (**b**) DPP connection.

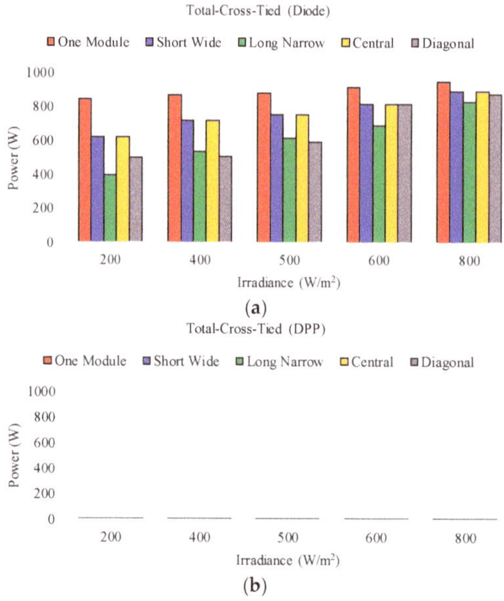

Figure 9. Output for total-cross-tied (TCT) connection under various shading conditions with: (**a**) diode connection and (**b**) DPP connection (zero output).

The unshaded modules in Figure 7 experience an irradiance of 1000 W/m^2. It is seen in Figures 8 and 10 that PV strings with DPP architectures have higher output power for the SP and CCT configurations than the systems with traditional bypass diodes. However, when DPP converters are adopted, there is almost zero or below 1 W of output power for all the shading scenarios with the TCT and BL configurations, as shown in Figures 9b and 11b. As the SL-based DPP architecture requires at least two series-connected PV modules for the converter to work properly, as shown in Figure 3. However, in TCT and BL architectures, their interconnections with other parallel PV strings affect the working principle of the used SL-based DPP topology.

In SP configuration, the output power for bypass diode and DPP under all shading patterns is shown in Figure 8a,b. The output power is 778 W, 826 W, 847 W, 872 W, and 933 W with bypass diode for one module, short wide, long narrow, central, and diagonal shading patterns, respectively, as shown in Figure 8a. In SP, output power with proposed SL-based DPP architecture is 905 W, 918 W, 924 W, 930 W, and 942 W at 200 W/m^2, 400 W/m^2, 500 W/m^2, 600 W/m^2, and 800 W/m^2, respectively during one module, short wide, long narrow, central, and diagonal shading, respectively (see Figure 8b). For bypass diode technique, the power output in SP using short wide and central shading is almost similar as four PV modules are shaded under both schemes, two from each PV strings. Whereas, in the diagonal shading, each module is shaded from each parallel-connected PV string. Therefore, it has a more severe effect on output power, especially for diode bypass architecture, which can be seen from Figure 8a. DPP with SP interconnection, which is shown in Figure 8b has almost the same output power under given irradiances.

The power output received from CCT interconnection is shown in Figure 10. Similar to SP, in CCT one module shading has maximum power for all shading patterns, i.e., 821 W, 857 W, 870 W, 900 W, and 945 W at 200 W/m^2, 400 W/m^2, 500 W/m^2, 600 W/m^2, and 800 W/m^2, respectively with bypass diode architecture. Similarly, 905 W, 918 W, 925 W, 930 W, and 942 for SL-based DPP architecture at 200 W/m^2, 400 W/m^2, 500 W/m^2, 600 W/m^2, and 800 W/m^2, respectively under one module shading scheme.

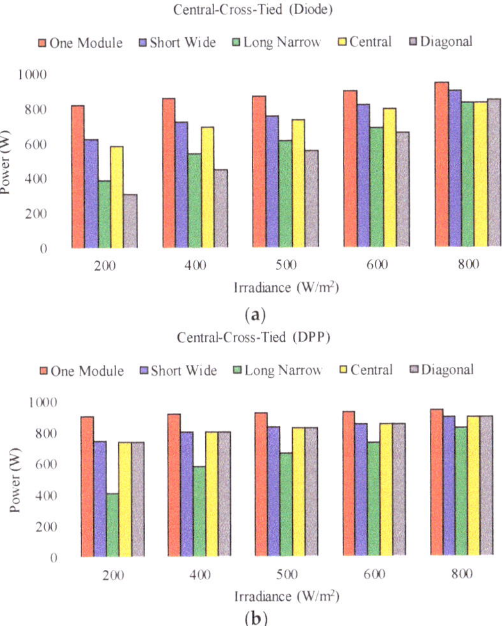

Figure 10. Output for central-cross-tied (CCT) connection under various shading conditions with: (**a**) diode connection and (**b**) DPP connection.

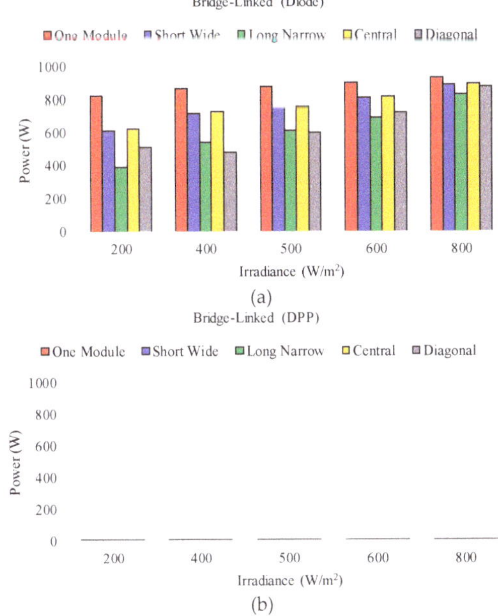

Figure 11. Output for bridge-linked (BL) connection under various shading conditions with: (**a**) diode connection and (**b**) DPP connection (zero output).

For short wide, long narrow, central, and diagonal shading the output power is given in Figure 10a,b with bypass diode and DPP, respectively. Figures 9 and 11 show the output power from TCT and

BL interconnections for the traditional diode. In TCT and BL connections, DPP architecture is not applicable as discussed before. Therefore, the output power is almost 0 W, as shown in Figures 9b and 11b. Power losses are also calculated from Figures 8 and 10 during one module, short wide, long narrow, central, and diagonal shading for SP and CCT traditional bypass diodes and DPP converters. These losses are calculated only for the SP and CCT interconnections because TCT and BL interconnection schemes are not applicable on the SL-based DPP converter. For instance, power losses during one module shading for the worst case—i.e., 200 W/m^2 are 15.34%, 10.66% for SP and CCT, respectively by using the bypass diode. It is only 1.52% for the DPP architecture by using SP and CCT interconnections during one module shading. The power losses decrease with an increase in irradiance. For short wide and long narrow shading at 200 W/m^2, the power losses for traditional bypass diode are 24.19% and 19.12% during short wide and 40.42% and 40.73% during long narrow shading for SP and CCT, respectively. Similarly, at 200 W/m^2, DPP architecture has 3.66% and 3.40% power losses during short wide shading for SP and CCT, respectively. Power loss for DPP during long wide shading is 37.51% and 40.58% for SP and CCT at 200 W/m^2. For the rest of the irradiances, the power loss decreases as the irradiance increases both for diode and DPP.

The power losses for SP and CCT string interconnections during central and diagonal shading for diode at 200 W/m^2 is 24.19% both for SP and CCT while 4.11% by using DPP. During diagonal shading, it has similar power losses for SP and CCT interconnections, which is 60.85% for bypass diode and 4.05% for DPP at 200 W/m^2. Similarly, these power losses decrease with an increase of irradiance in a diagonal shading pattern also. Overall, PV strings with SP and CCT interconnections with DPP architecture have more output power than a traditional diode. For short wide, central, and diagonal shading, PV strings with DPP architecture are producing almost the same output power because four PV modules are shaded for all of them. DPP architecture is not applicable to TCT and BL interconnections. In all, DPP extracts more power from the 4 × 4 PV array system than traditional bypass diode for all interconnection schemes where it is applicable. However, SL-based DPP topology has higher cost with a complex circuitry in comparison to traditional bypass diode topology.

4. Conclusions

In this paper, a DPP converter has been used in PV modules with different static interconnection schemes including series-parallel (SP), total-cross-tied (TCT), central-cross-tied (CCT), and bridge-linked (BL). The power production from the PV modules under various interconnection schemes and mismatch conditions have been explored. More importantly, a comparison of the power production between the traditional bypass diode and the DPP-based architecture for a 4 × 4 PV array has been performed. It has been found that the two configurations—i.e., SP and CCT with the DPP converters—produce more power than traditional bypass diode-based architecture. On the other hand, TCT and BL configurations are not suitable for integrating the DPP converters due to their inherent hardware limitations. Hence, the DPP-based interconnection might be a promising solution to enhance the energy yield for PV modules with minimal mismatch power losses during partial shading conditions. It is especially suitable for the SP configuration, which is the most commonly used configuration in practice. However, the integration of DPP converters will inevitably increase the cost and complexity of the overall system, which requires further analysis.

Author Contributions: K.A.K.N., conceptualization, investigation, methodology, validation, writing—original draft preparation; Y.Y., project administration, supervision, writing—review and editing; M.N., writing—review and editing; D.S., supervision, writing—review and editing.

Funding: This research received no external funding.

Conflicts of Interest: The authors declare no conflict of interest.

References

1. Cancino-Solórzano, Y.; Paredes-Sánchez, J.P.; Gutiérrez-Trashorras, A.J.; Xiberta-Bernat, J. The development of renewable energy resources in the State of Veracruz, Mexico. *Utilities Policy* **2016**, *39*, 1–4. [CrossRef]
2. Fahrenbruch, A.; Bube, R. *Fundamentals Of Solar Cells: Photovoltaic Solar Energy Conversion*; Elsevier: Amsterdam, The Netherlands, 2012; ISBN 978-0-32-314538-1.
3. Subudhi, B.; Pradhan, R. A Comparative Study on Maximum Power Point Tracking Techniques for Photovoltaic Power Systems. *IEEE Trans. Sustain. Energy* **2013**, *4*, 89–98. [CrossRef]
4. Sahoo, S.K. Renewable and sustainable energy reviews solar photovoltaic energy progress in India: A review. *Renew. Sustain. Energy Rev.* **2016**, *59*, 927–939. [CrossRef]
5. Ahmed, J.; Salam, Z. A critical evaluation on maximum power point tracking methods for partial shading in PV systems. *Renew. Sustain. Energy Rev.* **2015**, *47*, 933–953. [CrossRef]
6. Niazi, K.A.K.; Yang, Y.; Khan, H.A.; Sera, D. Performance Benchmark of Bypassing Techniques for Photovoltaic Modules. In Proceedings of the 2019 IEEE Applied Power Electronics Conference and Exposition (APEC), Anaheim, CA, USA, 17–21 March 2019.
7. Pareek, S.; Dahiya, R. Enhanced power generation of partial shaded photovoltaic fields by forecasting the interconnection of modules. *Energy* **2016**, *95*, 561–572. [CrossRef]
8. Niazi, K.; Khan, H.A.; Amir, F. Hot-spot reduction and shade loss minimization in crystalline-silicon solar panels. *J. Renew. Sustain. Energy* **2018**, *10*, 1–8. [CrossRef]
9. Yang, Y.; Kim, K.A.; Blaabjerg, F.; Sangwongwanich, A. *Advances in Grid-Connected Photovoltaic Power Conversion Systems*; Woodhead Publishing: Cambridge, UK, 2018; ISBN 978-0-08-102342.
10. Ahsan, S.; Niazi, K.A.K.; Khan, H.A.; Yang, Y. Hotspots and performance evaluation of crystalline-silicon and thin-film photovoltaic modules. *Microelectron. Reliab.* **2018**, *88–90*, 1014–1018. [CrossRef]
11. Ishaque, K.; Salam, Z. A review of maximum power point tracking techniques of PV system for uniform insolation and partial shading condition. *Renew. Sustain. Energy Rev.* **2013**, *19*, 475–488. [CrossRef]
12. Eltawil, M.A.; Zhao, Z. MPPT techniques for photovoltaic applications. *Renew. Sustain. Energy Rev.* **2013**, *25*, 793–813. [CrossRef]
13. Nasir, M.; Zia, M.F. Global maximum power point tracking algorithm for photovoltaic systems under partial shading conditions. In Proceedings of the 16th International Power Electronics and Motion Control Conference and Exposition, Antalya, Turkey, 21–24 September 2014.
14. Alajmi, B.N.; Ahmed, K.H.; Finney, S.J.; Williams, B.W. A Maximum Power Point Tracking Technique for Partially Shaded Photovoltaic Systems in Microgrids. *IEEE Trans. Ind. Electron.* **2013**, *60*, 1596–1606. [CrossRef]
15. Niazi, K.; Akhtar, W.; Khan, H.A.; Sohaib, S.; Nasir, A.K. Binary Classification of Defective Solar PV Modules Using Thermography. In Proceedings of the 2018 IEEE 7th World Conference on Photovoltaic Energy Conversion (WCPEC), Waikoloa, HI, USA, 10–15 June 2018.
16. Liu, Y.-H.; Liu, C.-L.; Huang, J.-W.; Chen, J.-H. Neural-network-based maximum power point tracking methods for photovoltaic systems operating under fast changing environments. *Sol. Energy* **2013**, *89*, 42–53. [CrossRef]
17. Daraban, S.; Petreus, D.; Morel, C. A novel MPPT (maximum power point tracking) algorithm based on a modified genetic algorithm specialized on tracking the global maximum power point in photovoltaic systems affected by partial shading. *Energy* **2014**, *74*, 374–388. [CrossRef]
18. Salam, Z.; Ahmed, J.; Merugu, B.S. The application of soft computing methods for MPPT of PV system: A technological and status review. *Appl. Energy* **2013**, *107*, 135–148. [CrossRef]
19. Niazi, K.A.K.; Yang, Y.; Sera, D. Review of mismatch mitigation techniques for PV modules. *IET Renew. Power Gener.* **2019**. [CrossRef]
20. Kim, K.A.; Shenoy, P.S.; Krein, P.T. Converter Rating Analysis for Photovoltaic Differential Power Processing Systems. *IEEE Trans. Power Electron.* **2015**, *30*, 1987–1997. [CrossRef]
21. Jeon, Y.; Lee, H.; Kim, K.A.; Park, J. Least Power Point Tracking Method for Photovoltaic Differential Power Processing Systems. *IEEE Trans. Power Electron.* **2017**, *32*, 1941–1951. [CrossRef]
22. Khan, O.; Xiao, W. Review and qualitative analysis of submodule-level distributed power electronic solutions in PV power systems. *Renew. Sustain. Energy Rev.* **2017**, *76*, 516–528. [CrossRef]

23. Stauth, J.T.; Seeman, M.D.; Kesarwani, K. Resonant Switched-Capacitor Converters for Sub-module Distributed Photovoltaic Power Management. *IEEE Trans. Power Electron.* **2013**, *28*, 1189–1198. [CrossRef]

24. Tahmasbi-Fard, M.; Tarafdar-Hagh, M.; Pourpayam, S.; Haghrah, A. A Voltage Equalizer Circuit to Reduce Partial Shading Effect in Photovoltaic String. *IEEE J. Photovolt.* **2018**, *8*, 1102–1109. [CrossRef]

25. Jeong, H.; Lee, H.; Liu, Y.; Kim, K.A. Review of Differential Power Processing Converter Techniques for Photovoltaic Applications. *IEEE Trans. Energy Convers.* **2019**, *34*, 351–360. [CrossRef]

26. Karatepe, E.; Boztepe, M.; Çolak, M. Development of a suitable model for characterizing photovoltaic arrays with shaded solar cells. *Sol. Energy* **2007**, *81*, 977–992. [CrossRef]

27. La Manna, D.; Li Vigni, V.; Riva Sanseverino, E.; Di Dio, V.; Romano, P. Reconfigurable electrical interconnection strategies for photovoltaic arrays: A review. *Renew. Sustain. Energy Rev.* **2014**, *33*, 412–426. [CrossRef]

28. Satpathy, P.R.; Jena, S.; Sharma, R. Power enhancement from partially shaded modules of solar PV arrays through various interconnections among modules. *Energy* **2018**, *144*, 839–850. [CrossRef]

29. Kadri, R.; Andrei, H.; Gaubert, J.-P.; Ivanovici, T.; Champenois, G.; Andrei, P. Modeling of the photovoltaic cell circuit parameters for optimum connection model and real-time emulator with partial shadow conditions. *Energy* **2012**, *42*, 57–67. [CrossRef]

30. Satpathy, P.R.; Jena, S.; Jena, B.; Sharma, R. Comparative study of interconnection schemes of modules in solar PV array network. In Proceedings of the 2017 International Conference on Circuit, Power and Computing Technologies (ICCPCT), Kollam, India, 20–21 April 2017.

31. Velasco-Quesada, G.; Guinjoan-Gispert, F.; Pique-Lopez, R.; Roman-Lumbreras, M.; Conesa-Roca, A. Electrical PV Array Reconfiguration Strategy for Energy Extraction Improvement in Grid-Connected PV Systems. *IEEE Trans. Ind. Electron.* **2009**, *56*, 4319–4331. [CrossRef]

32. Lavado Villa, L.F.; Ho, T.-P.; Crebier, J.-C.; Raison, B. A Power Electronics Equalizer Application for Partially Shaded Photovoltaic Modules. *IEEE Trans. Ind. Electron.* **2013**, *60*, 1179–1190. [CrossRef]

33. Satpathy, P.R.; Sharma, R.; Jena, S. A shade dispersion interconnection scheme for partially shaded modules in a solar PV array network. *Energy* **2017**, *139*, 350–365. [CrossRef]

34. Malathy, S.; Ramaprabha, R. Reconfiguration strategies to extract maximum power from photovoltaic array under partially shaded conditions. *Renew. Sustain. Energy Rev.* **2018**, *81*, 2922–2934. [CrossRef]

35. Rana, A.S.; Nasir, M.; Khan, H.A. String level optimisation on grid-tied solar PV systems to reduce partial shading loss. *IET Renew. Power Gener.* **2017**, *12*, 143–148. [CrossRef]

Article

Ripple Vector Cancellation Modulation Strategy for Single-Phase Quasi-Z-Source Inverter

Yufeng Tang, Zhiyong Li *, Yougen Chen and Renyong Wei

School of Automation, Central South University, Changsha 410083, China
* Correspondence: lizy@csu.edu.cn; Tel.: +86-138-7312-7689

Received: 6 August 2019; Accepted: 23 August 2019; Published: 30 August 2019

Abstract: The double-frequency (2ω) power flows through the DC side of the single-phase quasi-Z-source inverter (QZSI) leads to the 2ω voltage ripple of capacitors and 2ω current ripple of inductors. This paper proposes a ripple vector cancellation modulation strategy (RVCMS) based on the thought of ripple vector cancellation. By analyzing the mechanism of ripple generation and transmission, we can obtain a variation of a shoot-through duty cycle to generate a compensated 2ω ripple used to cancel the 2ω current ripple of inductors caused by the 2ω ripple of DC link current, and the 2ω compensated variation of a shoot-through duty cycle with a specific amplitude and phase is added to the constant shoot-through duty cycle. Finally, simulation and experimental results demonstrate the correctness and effectiveness of the proposed modulation strategy for the single-phase QZSI.

Keywords: quasi-Z-source inverter; double-frequency ripple; ripple vector cancellation; shoot-through duty cycle; modulation

---

1. Introduction

The single-phase quasi-Z-source inverter (QZSI) consists of two modules, a quasi-Z-source network, and an H-bridge, and the quasi-Z-source network includes two inductors, two capacitors, and a diode [1]. Different from traditional two-stage inverters, the QZSI reduces the number of active devices, has no dead time, and can realize DC-DC and DC-AC by its unique impedance network and shoot-through modulation method [2,3]. Currently, the QZSI has been widely studied and applied in photovoltaic power generation, AC speed regulation, electric vehicles, and as a module in cascaded multilevel inverters [4–7].

For the single-phase inverter system, the double-frequency (2ω) power flows between the DC side, and the AC side causes 2ω ripples of capacitor and DC link voltage, 2ω ripples of inductor and DC link current. These 2ω ripples will cause the temperature of passive devices and the DC source to increase, which seriously affects their working life, distort the output voltage of the inverter, and the low frequency current ripple components in the DC side will affect the maximum power point tracking (MPPT) and reduce the efficiency of the photovoltaic system [8–14]. Therefore, it is necessary to suppress the 2ω ripple in the DC side. The simplest method to reduce the ripple is to increase the value of the quasi-Z-source network to buffer 2ω power in a large capacitor or inductor [8]. In [9], a parameter design method with a dynamic photovoltaic-panel and terminal capacitors for the single-phase quasi-Z-source photovoltaic inverter was proposed to reduce the 2ω ripple. However, these methods will not only lead to large volume, large weight and high cost, but also reduce reliability and efficiency due to the large value of capacitors and inductors in the quasi-Z-source network. In [10], a comprehensive model and an asymmetric quasi-Z-source network design method for a single-phase energy-stored quasi-Z-source-based photovoltaic inverter system were proposed to reduce the 2ω ripple.

The active power filter (APF) technology is usually applied to the ripple suppression for QZSI. In [11], an active-filter-integrated single-phase QZSI was proposed: The APF consists of an extra switch leg and a second-order filter; the 2ω pulsating power of the AC load is transferred to the filter and the extra leg; the capacitor voltage and the inductor current in the DC side will no longer have the 2ω ripple component with this method, but an extra circuit means a higher cost and more complicated control. In [12,13], 2ω ripple control strategies were proposed based on feed-back control. In [12], a low-pass filter was used to obtain the 2ω ripple of the inductor current in the DC side by extracting its DC component, which is used to generate a small variation of the shoot-through duty cycle, and Reference [13] regards the 2ω voltage ripple of the DC source as the feed-back signal. Reference [14] proposes a 2ω ripple suppression method based on feed-forward control: The feed-forward signal is obtained by a ripple observer, which can observe the 2ω current of the DC link by detecting the output current of the inverter. References [12–14] can realize the suppression of the 2ω ripple without an extra active filter circuit; we can call it virtual APF technology. However, they still need filters or sensors to detect the 2ω ripple components, which will increase the cost. Therefore, the relationship between the variation of the shoot-through duty cycle and the 2ω ripple in the DC side, and realizing the suppression of the 2ω ripple with less hardware, needs further study.

Reducing the modulation ratio of QZSI can effectively reduce the ripple content in the DC side; however, with the decrease of the modulation ratio and the increase of the shoot-through duty cycle, the harmonic distortion value of the inverter output voltage and the loss of switching devices will increase [15]. Some new modified modulation strategies have been addressed. In [16], a modified modulation strategy was proposed, which was different from the traditional modulation strategy; the shoot-through control lines are modified to a line with a 2ω component. Reference [17] proposes a novel dual switching frequency modulation that combines low-frequency sinusoidal pulse width modulation (SPWM) and high-frequency pulse width modulation (PWM) to remove the interdependence between the shoot-through duty cycle and the inverter modulation index. Reference [18] proposes two new space vector modulation strategies to reduce the inductor current ripple based on the principle of the volt-second balance, which can divide shoot-through times in real time. Reference [19] proposes a PWM strategy with a minimum inductor current ripple; the shoot-through time interval of three phase legs are designed according to the active state and 0 state time. In [20], a finite-control-set model-predictive control algorithm for a quasi-Z-source four-leg inverter based on the discrete time model and a predictive controller was proposed to minimize the 2ω ripple in the DC side. Reference [21] proposed a model-based current control approach based on the inherent relationship between the ripple component inductor and capacitor voltages in the DC side. This approach can reduce the DC side inductor current ripple with active damping and constant virtual time for single-phase grid-tied QZSI with an LCL filter.

This paper mainly focuses on the suppression of the 2ω current ripple of inductors and proposes a new modulation strategy based on ripple vector cancellation. In the modulation strategy, the compensated 2ω variation of the shoot-through duty cycle with a specific amplitude and phase angle is obtained to cancel the 2ω current ripple of inductors. Section 2 focuses on the operation analysis, ripple transmission, and generation mechanism of single-phase QZSI. Section 3 presents the proposed modulation strategy. Simulation and experimental studies are discussed in Section 4. Finally, conclusions are given in Section 5.

2. Single-Phase QZSI

2.1. Operation and Steady-State Analysis

The topology of single-phase QZSI is shown in Figure 1. Due to the existence of the quasi-Z-source network and the shoot-through state, both of the power switches in a leg can be turned on at the same time, which is used to step up the voltage. During the shoot-through state, the diode turns off, the inverter bridge is short circuited, the DC link voltage is 0, and the inverter does not generate the power to load. In the non-shoot-through state, the diode turns on, and the QZSI operates the same as a conventional voltage source inverter (VSI). In a switching period, the state-space model of QZSI can be obtained by using the state-space average method, as shown in (1), where i_{L_1}, i_{L_2}, v_{C_1}, and v_{C_2} are the current of inductors L_1 and L_2, the voltage of the capacitors is C_1 and C_2, v_{DC} is the voltage of the DC source, d is the shoot-through duty cycle, and i_{PN} is the DC link current, respectively.

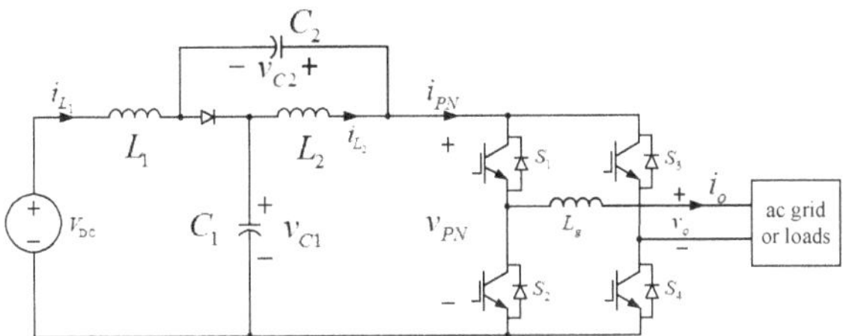

Figure 1. Single-phase quasi-Z-source inverter (QZSI).

Figure 2 shows the conventional modulation strategy (CMS) for the single-phase QZSI using the unipolar modulation strategy; its modulation waveforms are two sinusoidal waves with a difference of 180°. The CMS uses two straight lines v_p, v_n to generate the shoot-through duty cycle. When the triangular carrier is greater than v_p, the switch S_1 or S_3 turns on, and when the carrier is smaller than v_n, the switch S_2 or S_4 turns on. V_{S_1} and V_{S_2} show the switch state of the switch S_1–S_4 in the H-bridge, respectively.

$$\begin{cases} L_1 \dfrac{di_{L_1}}{dt} = v_{DC} - (1-d)v_{C_1} + dv_{C_2} \\ L_2 \dfrac{di_{L_2}}{dt} = -(1-d)v_{C_2} + dv_{C_1} \\ C_1 \dfrac{dv_{C_1}}{dt} = (1-d)(i_{L_1} - i_{PN}) - di_{L_2} \\ C_2 \dfrac{dv_{C_2}}{dt} = (1-d)(i_{L_2} - i_{PN}) - di_{L_1} \end{cases} \tag{1}$$

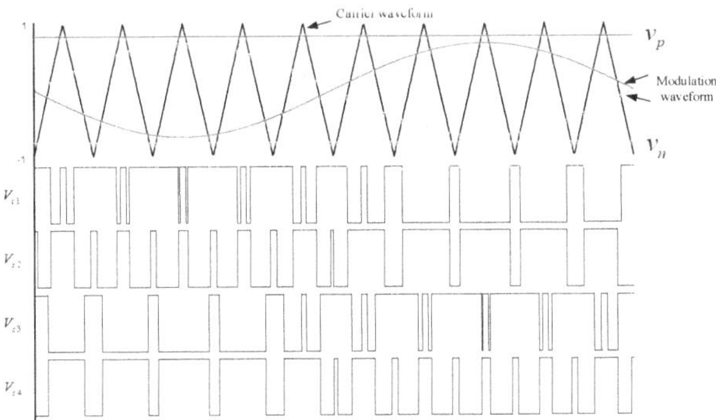

Figure 2. Conventional modulation strategy (CMS).

From (1), the DC components of the QZSI in the ideal case can be obtained as (2), where D is the average value of the shoot-through duty cycle d, $d = D + \hat{d}$, $\hat{d}$ is the variation of d, V_{C_1} and V_{C_2} are the average voltage of capacitor C_1 and C_2, I_{L_1} and I_{L_2} are the average current of inductor I_{L_1} and I_{L_2}, V_{DC} is the average voltage of dc source, and I_{PN} is the average value of the DC link current, respectively.

$$\begin{cases} V_{C_1} = \dfrac{1-D}{1-2D} V_{DC} \\ V_{C_2} = \dfrac{D}{1-2D} V_{DC} \\ I_{L_1} = I_{L_2} = \dfrac{1-D}{1-2D} I_{PN} \end{cases} \tag{2}$$

The average value of the DC link voltage during the non-shoot-through time interval is

$$V_{PN} = V_{C_1} + V_{C_2} = \frac{1}{1-2D}V_{DC} \tag{3}$$

2.2. Ripple Genertation Mechanism

The output voltage and current of the single-phase QZSI inverter can be expressed as (4), where V_o and I_o are the amplitude of the output voltage and current, ω is the fundamental frequency, and φ is the impedance angle of the load or grid, respectively.

$$v_o = V_o\sin(\omega t), \; i_o = I_o\sin(\omega t - \varphi) \tag{4}$$

The DC link voltage v_{PN} and the DC link current i_{PN} can be expressed as (5), where V_{PN}, I_{PN} are the average value and $\hat{v}_{PN}$, $\hat{i}_{PN}$ are the 2ω components of v_{PN}, i_{PN}, respectively.

$$v_{PN} = V_{PN} + \hat{v}_{PN}, \; i_{PN} = I_{PN} + \hat{i}_{PN} \tag{5}$$

From (4) and (5), the input power p_{PN} and output power p_o of the H-bridge can be calculated as (6) and (7), respectively.

$$p_{PN} = (1-D)v_{PN}i_{PN} = (1-D)(V_{PN}I_{PN} + V_{PN}\hat{i}_{PN} + I_{PN}\hat{v}_{PN} + \hat{i}_{PN}\hat{v}_{PN}) \tag{6}$$

$$p_o = \frac{1}{2}V_oI_o\cos\varphi - \frac{1}{2}V_oI_o\cos(2\omega t - \varphi) \tag{7}$$

The 2ω component of the DC link current can be given as (8), where $I_{2\omega}$ is its amplitude and α is the phase angle.

$$\hat{i}_{PN} = I_{2\omega}\cos(2\omega t - \alpha) \tag{8}$$

From $p_{PN} = p_o$ and Reference [12], the amplitude and phase angle of $\hat{i}_{PN}$ can be calculated as

$$I_{2\omega} = -\frac{V_oI_o\left[4\omega^2LC - (1-2D)^2\right]}{2(1-D)\sqrt{\left[4\omega^2LC - (1-2D)^2\right]^2V_{PN}^2 + \left[4\omega LI_{PN}(1-D)\right]^2}} \tag{9}$$

$$\alpha = \varphi + \arctan\frac{4\omega LI_{PN}(1-D)}{\left[4\omega^2LC - (1-2D)^2\right]V_{PN}} \tag{10}$$

In (9), $4\omega LI_{PN}(1-D)$ is much smaller than $\left[4\omega^2LC - (1-2D)^2\right]V_{PN}$ and can be ignored, therefore, the 2ω component of the DC link current can be expressed as

$$\hat{i}_{PN} = -\frac{V_oI_o}{2(1-D)V_{PN}}\cos(2\omega t - \alpha) \tag{11}$$

From (6) and (7), I_{PN} can be calculated as

$$I_{PN} = \frac{V_oI_o}{2(1-D)V_{PN}}\cos\varphi \tag{12}$$

2.3. Ripple Transmission Mechanism

When the shoot-through duty cycle is a variable value, from (1), the small-signal model of single-phase QZSI can be obtained as (13) using the small signal analysis method with a constant DC source voltage, where $\hat{i}_{L_1}$, $\hat{i}_{L_2}$, $\hat{v}_{C_1}$, $\hat{v}_{C_2}$ are the 2ω components of the inductor current i_{L_1}, i_{L_2}, the capacitor voltage v_{C_1}, v_{C_2}, respectively.

$$\begin{cases} L_1 \dfrac{d\hat{i}_{L_1}}{dt} = -(1-D)\hat{v}_{C_1} + D\hat{v}_{C_2} + (V_{C_1} + V_{C_2})\hat{d} \\[2mm] L_2 \dfrac{d\hat{i}_{L_2}}{dt} = -(1-D)\hat{v}_{C_2} + D\hat{v}_{C_1} + (V_{C_1} + V_{C_2})\hat{d} \\[2mm] C_1 \dfrac{d\hat{v}_{C_1}}{dt} = (1-D)\hat{i}_{L_1} - D\hat{i}_{L_2} - (1-D)\hat{i}_{PN} + (I_{PN} - I_{L_1} - I_{L_2})\hat{d} \\[2mm] C_2 \dfrac{d\hat{v}_{C_2}}{dt} = (1-D)\hat{i}_{L_2} - D\hat{i}_{L_1} - (1-D)\hat{i}_{PN} + (I_{PN} - I_{L_1} - I_{L_2})\hat{d} \end{cases} \tag{13}$$

The small signal model can be simplified as (14) under $L_1 = L_2 = L$ and $C_1 = C_2 = C$, which means $\hat{i}_L = \hat{i}_{L_1} = \hat{i}_{L_2}$ and $\hat{v}_C = \hat{v}_{C_1} = \hat{v}_{C_2}$ referring to [8].

$$\begin{cases} L \dfrac{d\hat{i}_L}{dt} = -(1-2D)\hat{v}_C + (V_{C_1} + V_{C_2})\hat{d} \\[2mm] C \dfrac{d\hat{v}_C}{dt} = (1-2D)\hat{i}_L - (1-D)\hat{i}_{PN} + (I_{PN} - 2I_{L_1})\hat{d} \end{cases} \tag{14}$$

From (2), (3) and (14), the transmission mechanism model of the inductor current ripple (15) and capacitor voltage ripple (16) can be obtained with Laplace transforms.

$$\hat{i}_L(s) = \frac{(1-2D)(1-D)}{LCs^2 + (1-2D)^2}\hat{i}_{PN}(s) + \frac{CsV_{DC} + (1-2D)I_{PN}}{(1-2D)\left[LCs^2 + (1-2D)^2\right]}\hat{d}(s) \tag{15}$$

$$\hat{v}_C(s) = \frac{-(1-D)Ls}{LCs^2 + (1-2D)^2}\hat{i}_{PN}(s) + \frac{(1-2D)V_{DC} - LsI_{PN}}{(1-2D)\left[LCs^2 + (1-2D)^2\right]}\hat{d}(s) \tag{16}$$

In (15), the transfer functions can be expressed as (17) and (18), where $G_{\hat{i}_{PN}}^{\hat{i}_L}$ represents the transfer function of $\hat{i}_{PN}$ to $\hat{i}_{L_1}$, and $G_{\hat{d}}^{\hat{i}_L}$ represents the transfer function of $\hat{d}$ to $\hat{i}_L$.

$$G_{\hat{i}_{PN}}^{\hat{i}_L}(s)\Big|_{\hat{d}=0} = \frac{(1-D)(1-2D)}{LCs^2 + (1-2D)^2} \tag{17}$$

$$G_{\hat{d}}^{\hat{i}_L}(s)\Big|_{\hat{i}_{PN}=0} = \frac{CsV_{DC} + (1-2D)I_{PN}}{(1-2D)\left[LCs^2 + (1-2D)^2\right]} \tag{18}$$

From (2), (11), (12), (15), and (16), in the steady state, assuming a constant shoot-through duty cycle, we can get

$$\begin{cases} i_{L_1} = I_{L_1} + \hat{i}_{L_1} = \dfrac{V_o I_o}{2(1-2D)V_{PN}}\cos\varphi + \dfrac{(1-2D)V_o I_o}{2\left[4\omega^2 LC - (1-2D)^2\right]V_{PN}}\cos(2\omega t - \alpha) \\[3mm] i_{L_2} = I_{L_2} + \hat{i}_{L_2} = \dfrac{V_o I_o}{2(1-2D)V_{PN}}\cos\varphi + \dfrac{(1-2D)V_o I_o}{2\left[4\omega^2 LC - (1-2D)^2\right]V_{PN}}\cos(2\omega t - \alpha) \\[3mm] v_{C_1} = V_{C_1} + \hat{v}_{C_1} = \dfrac{1-D}{1-2D}V_{DC} + \dfrac{\omega L V_o I_o}{\left[4\omega^2 LC - (1-2D)^2\right]V_{PN}}\sin(2\omega t - \alpha) \\[3mm] v_{C_2} = V_{C_2} + \hat{v}_{C_2} = \dfrac{D}{1-2D}V_{DC} + \dfrac{\omega L V_o I_o}{\left[4\omega^2 LC - (1-2D)^2\right]V_{PN}}\sin(2\omega t - \alpha) \end{cases} \tag{19}$$

From (19), the 2ω current ripple ratios of i_{L_1} and i_{L_2}, the 2ω voltage ripple ratios of v_{C_1} and v_{C_2} can be defined as (20), (21), and (22), respectively.

$$a = \frac{\left|\hat{i}_{L_1}\right|}{I_{L_1}} \times 100\% = \frac{\left|\hat{i}_{L_2}\right|}{I_{L_2}} \times 100\% = \frac{(1-2D)^2}{\left[4LC\omega^2 - (1-2D)^2\right]\cos\varphi} \times 100\% \tag{20}$$

$$b_1 = \frac{\left|\hat{v}_{C_1}\right|}{V_{C_1}} \times 100\% = \frac{(1-2D)\omega L V_o I_o}{(1-D)\left[4\omega^2 LC - (1-2D)^2\right]V_{PN}V_{DC}} \times 100\% \tag{21}$$

$$b_2 = \frac{\left|\hat{v}_{C_2}\right|}{V_{C_2}} \times 100\% = \frac{(1-2D)\omega L V_o I_o}{D\left[4\omega^2 LC - (1-2D)^2\right]V_{PN}V_{DC}} \times 100\% \tag{22}$$

3. Modulation Strategy Based on Ripple Vector Cancellation

The ripple transmission models (15) and (16) show that the 2ω inductor current ripple and the capacitor voltage ripple in the DC side are related to the variation $\hat{i}_{PN}$ and $\hat{d}$. If the shoot-through duty cycle or the DC link current has a variation, the current of the inductors will contain variation $\hat{i}_L^*$ and $\hat{v}_C^*$. This paper mainly focuses on the reducing of the 2ω current ripple in the DC side. Therefore, regarding $\hat{i}_L^*$ caused by $\hat{i}_{PN}$ as the disturbance, which caused by $\hat{d}$ as the compensation, theoretically, the 2ω current ripple will be suppressed to 0 with an appropriate $\hat{d}$. A 2ω shoot-through duty cycle $\hat{d} = \hat{d}_{2\omega}$ with specific amplitude and phase angle, according to the magnitude of the 2ω inductors current ripple, can then be calculated based on the thought of the ripple vector cancellation. Adding the 2ω compensation variation $\hat{d}_{2\omega}$ to the constant shoot-though duty cycle D, at this time, the shoot-through duty cycle can be expressed as

$$d = D + \hat{d}_{2\omega} \tag{23}$$

From (23), the shoot-through duty cycle consists of a constant value and a 2ω component, where D is determined by the output voltage of the inverter and the DC source voltage and $\hat{d}_{2\omega}$ is determined by the actual 2ω current ripple in the DC side. Different from the CMS, as shown in Figure 3, the proposed ripple vector cancellation modulation strategy (RVCMS) uses two sine waveforms with 2ω components to generate the shoot-through duty cycle, where $v_p = 1 - d$ and $v_n = -1 + d$.

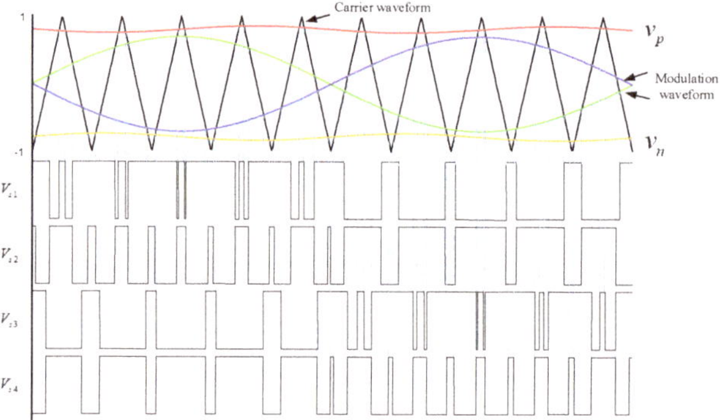

Figure 3. Proposed ripple vector cancellation modulation strategy (RVCMS).

The 2ω compensated shoot-through duty cycle can be expressed as (24).

$$\hat{d}_{2\omega} = A\sin(2\omega t + \beta) \tag{24}$$

From (17) and (18), the transfer function of $\hat{i}_{PN}$ to $\hat{d}$ used to cancel the 2ω current ripple of inductors can be calculated as

$$G(s) = \frac{G_{\hat{i}_{PN}}^{\hat{i}_L}(s)}{G_{\hat{d}}^{\hat{i}_L}(s)} = \frac{(1-D)(1-2D)^2}{CsV_{DC} + (1-2D)I_{PN}} \tag{25}$$

and $\hat{d}_{2\omega}$ can be calculated by (26), where $*$ represents the convolution.

$$\hat{d}_{2\omega} = -\hat{i}_{PN} * L^{-1}[G(s)] \tag{26}$$

From (3), (10), (11), (24)–(26), the amplitude A and phase angle β of $\hat{d}_{2\omega}$ can be calculated as

$$A = \frac{V_o I_o (1-2D)^3}{2V_{DC}\sqrt{4\omega^2 C^2 V_{DC}^2 + I_{PN}^2(1-2D)^2}} \tag{27}$$

$$\beta = \arctan\left[\frac{(1-2D)I_{PN}}{2\omega C V_{DC}}\right] - \arctan\frac{(1-2D)(1-D)4\omega L I_{PN}}{\left[4\omega^2 LC - (1-2D)^2\right]V_{DC}} - \varphi \tag{28}$$

When the proposed $\hat{d}_{2\omega}$ is applied, the entire 2ω power is buffered by the capacitors and there is no 2ω power in the inductors and DC source. It should be noted that the 2ω voltage ripple of capacitors will reduce slightly when the 2ω current ripple of the inductors are limited to 0. The phasor diagram of the 2ω power flows with a different modulation strategy are shown in Figure 4, where $P_{L-2\omega}$, $P_{C-2\omega}$, $P_{dc-2\omega}$, and $P_{o-2\omega}$ express the 2ω power of two inductors, two capacitors, and DC source and loads, respectively; the derivation and demonstration are given in Reference [12].

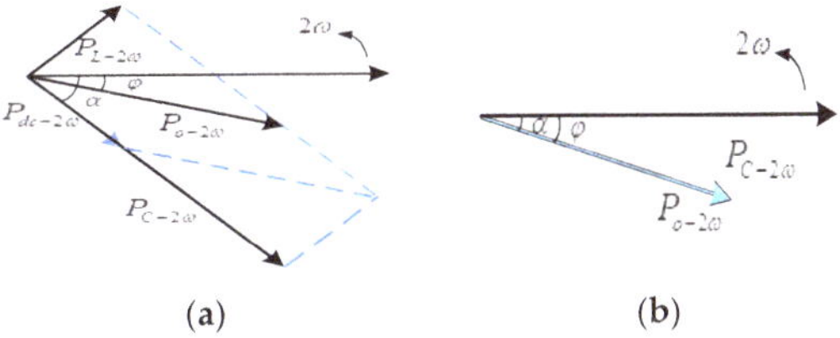

(a)   (b)

Figure 4. Phasor diagram of double-frequency (2ω) power flows with the CMS (**a**) and RVCMS (**b**).

4. Simulation and Experimental Results

In order to verify the correctness and demonstrate the effectiveness of the proposed modulation strategy for the ripple suppression effect, the 2ω inductor current ripple must be large enough with the CMS, and in Reference [8] the simulation model was built, and its parameters are shown in Table 1. Figure 5 shows the simulation waveforms of d (a), i_{L_1} (b), i_o (c), i_{PN} (d) v_{C_1} (e), and v_{C_2} (f) with the CMS. The average values of i_{L_1}, i_{PN}, v_{C_1}, and v_{C_2} are 3.014 A, 3.022 A, 90.21 V, and 30.21 V, and the amplitude of i_o is 4.154 A, respectively. From Figure 5a, the shoot-through duty cycle is a constant; Figure 5b,d–f show the 2ω ripple of the inductor current and the DC link current, the capacitor voltage are quite large with the CMS.

Table 1. Parameters of the single-phase QZSI.

Parameters	Value
Capacitors of QZS network C_1, C_2	1 mF
Inductors of QZSI network L_1, L_2	1 mH
Filter inductor L_g	4 mH
Voltage of DC source V_{DC}	60 V
Load R	20 Ω
Average shoot-through duty cycle D	0.25
Modulation ratio M	0.7
Output frequency f	50 HZ
Carrier frequency f_C	10 kHZ

Figure 6 shows the simulation results of d (a), i_{L_1} (b), i_o (c), i_{PN} (d), v_{C_1} (e), and v_{C_2} (f) with the RVCMS. With the proposed modulation strategy, the shoot-through duty cycle is a sine wave with a 2ω component; the current of inductor and the DC link have a little 2ω ripple. Figure 6e,f show the 2ω voltage ripple of capacitors reduced slightly compared to Figure 5e,f. The average values of i_{L_1}, i_{PN},

v_{C_1}, and v_{C_2} are 2.993 A, 2.995 A, 88.98 V, and 28.98 V, and the amplitude of i_o is 4.145 A, respectively. The simulation results show that the proposed RVCMS can effectively suppress the 2ω current ripple of inductors and can suppress the capacitor voltage ripple slightly compared to the CMS.

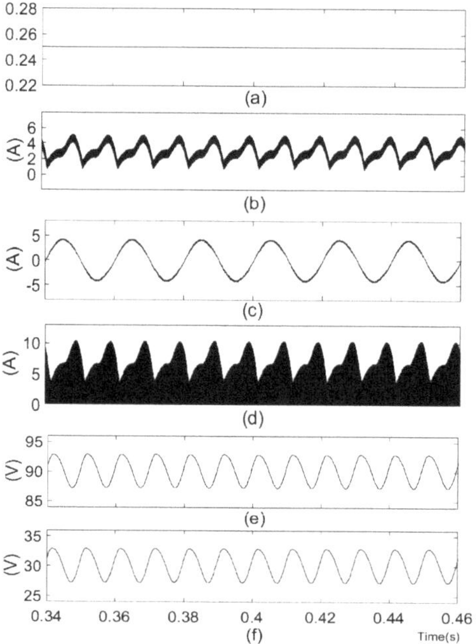

Figure 5. Simulation results of d (**a**), i_{L_1} (**b**), i_o (**c**), i_{PN} (**d**), v_{C_1} (**e**), and v_{C_2} (**f**) with the CMS.

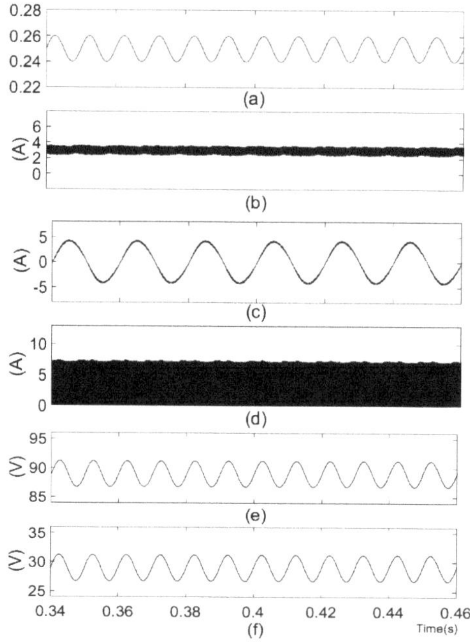

Figure 6. Simulation results of d (**a**), i_{L_1} (**b**), i_o (**c**), i_{PN} (**d**), v_{C_1} (**e**), and v_{C_2} (**f**) with the RVCMS.

Table 2 lists the 2ω ripple ratios with the CMS and RVCMS. It can be seen that the 2ω inductor current ripple ratios decreased from 40.15% to 1.69%; the 2ω voltage ripple ratios of v_{C_1} decreased from 3.14% to 2.53%, and v_{C_2} decreased from 9.40% to 7.75%. If low 2ω ripple ratios in the DC side with the CMS are required, the large inductance and capacitance are necessary. From (20) and (21), assuming that $a = 1.69\%$ and $b_1 = 2.53\%$, the inductance and capacitance of quasi-Z-source network will be $L = 36.89$ mH and $C = 1.03$ mF, respectively, and the inductance is much larger than the value with the RVCMS, achieving the same ripple suppression effect. Figure 7 shows the FFT spectrum of the AC output current with the CMS and the RVCMS, and the total harmonic distortion (THD) are 3.46% and 3.54%, respectively. It can be seen that the proposed RVCMS has little effect on the output power quality.

Table 2. The 2ω ripple ratios with a different modulation strategy.

Strategy	i_{L_1}	v_{C_1}	v_{C_2}
CMS	40.15%	3.14%	9.40%
RVCMS	1.69%	2.53%	7.75%

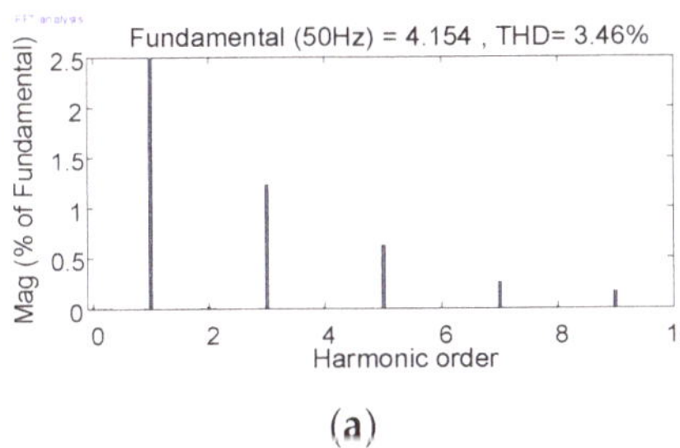
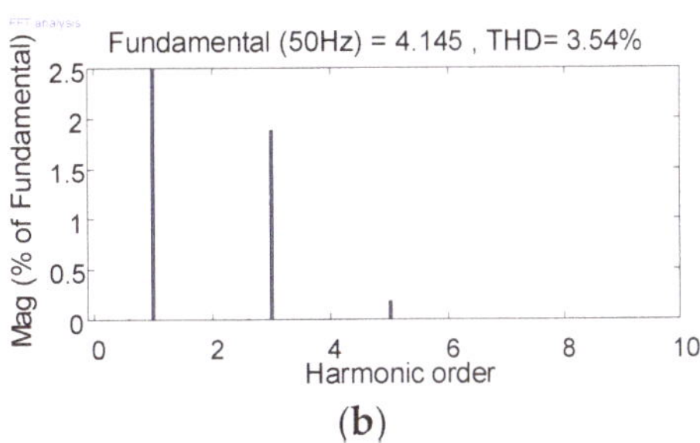

(a)   (b)

Figure 7. FFT spectrum of the AC output current with the CMS (**a**) and RVCMS (**b**).

A single-phase QZSI experimental prototype was built in the laboratory, as shown in Figure 8. The PWM control signals of the proposed RVCMS and CMS for switches were generated by a TMS320F28335 DSP. Figures 9 and 10 show the experimental results with the CMS and RVCMS, respectively. The experimental results show that the 2ω current ripple of inductors reduced greatly and the 2ω ripple of capacitors reduced slightly with the RVCMS compared to the CMS. Simulation and experimental results verify that the proposed RVCMS can realize the suppression of 2ω ripples in addition to the functions of the DC-AC and voltage boost for the single-phase QZSI.

Figure 8. Experimental prototype in the laboratory.

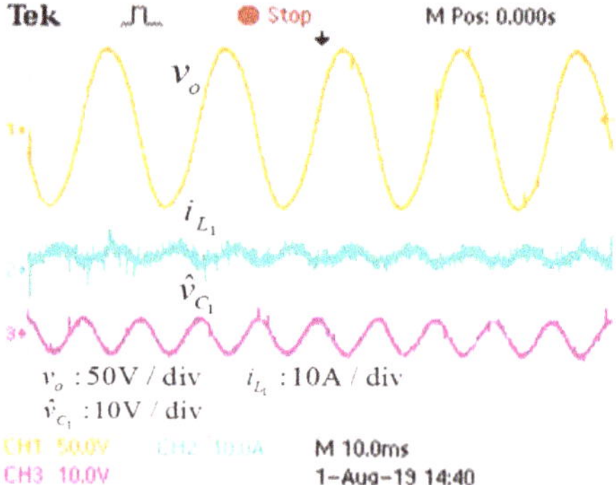

Figure 9. Experimental results with the CMS.

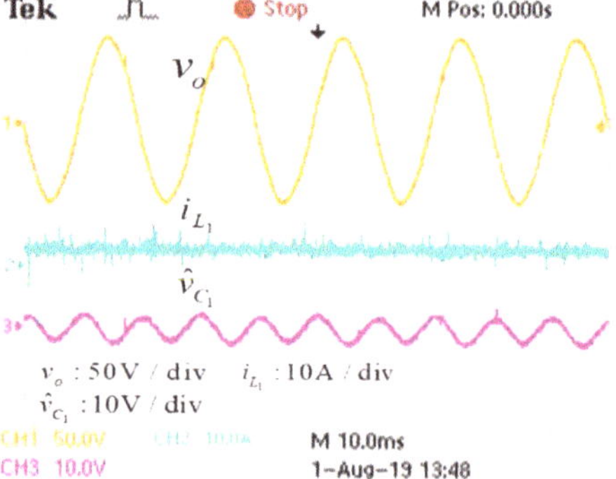

Figure 10. Experimental results with the RVCMS.

5. Conclusions

This paper proposed a minimum 2ω inductor current ripple modulation strategy based on the thought of ripple vector cancellation. The ripple generation and transmission mechanism models for single-phase QZSI are built, which reveals the causes of the 2ω ripple in the DC side. The 2ω-compensated variation of the shoot-through duty cycle used to cancel the 2ω ripple of inductors was obtained and added to the constant shoot-through duty cycle. The proposed modulation strategy in this paper can effectively reduce the 2ω current ripple of inductors and slightly reduce the 2ω voltage ripple of capacitors without adding any active or passive filter branches, and a small value of inductors and capacitors are used to receive the same effect of the ripple suppression compared to the CMS. Some comparative evaluation of simulation and experimental results demonstrated the effectiveness and validity of the proposed RVCMS for the 2ω ripple suppression of single-phase QZSI.

Author Contributions: Y.T., conceptualization, methodology, and writing—original draft; Z.L., resources and writing—review and editing; Y.C., supervision and technical analysis; R.W., formal analysis and investigation.

Funding: This research was funded by Fundamental Research Funds for the Central Universities of Central South University Central South University, grant number: 502211922.

Conflicts of Interest: The authors declare no conflict of interest.

References

1. Li, Y.; Anderson, J.; Peng, F.Z.; Liu, D. Quasi-z-source inverter for photovoltaic power generation systems. In Proceedings of the 2009 Twenty-Fourth Annual IEEE Applied Power Electronics Conference and Exposition, Washington, DC, USA, 15–19 February 2009; pp. 918–924.
2. Anderson, J.; Peng, F.Z. Four quasi-Z-source inverters. In Proceedings of the IEEE Power Electronics Specialists Conference, Rhodes, Greece, 15–19 June 2008; pp. 2743–2749.
3. Ayad, A.; Kennel, R. A comparison of quasi-Z-source inverters and conventional two-stage inverters for PV applications. *EPE J.* **2017**, *27*, 43–59. [CrossRef]
4. Battiston, A.; Martin, J.P.; Miliani, E.H.; Nahid-Mobarakeh, B.; Pierfederici, S.; Meibody-Tabar, F. Comparison criteria for electric traction system using Z-source/quasi Z-source inverter and conventional architectures. *IEEE J. Emerg. Sel. Top. Power Electron.* **2014**, *2*, 467–476. [CrossRef]
5. Battiston, A.; Miliani, E.H.; Pierfederici, S.; Meibody-Tabar, F. Efficiency improvement of a quasi-Z-source inverter-fed permanent-magnet synchronous machine-based electric vehicle. *IEEE Trans. Transp. Electrific.* **2016**, *2*, 14–23. [CrossRef]
6. Liu, Y.; Ge, B.; Abu-Rub, H.; Peng, F.Z. An effective control method for quasi-Z-source cascade multilevel inverter-based grid-tie single-phase photovoltaic power system. *IEEE Trans. Ind. Informat.* **2014**, *10*, 399–407. [CrossRef]
7. Liang, W.H.; Liu, Y.; Ge, B.; Wang, X.L. DC-Link voltage balance control strategy based on multidimensional modulation technique for quasi-Z-source cascaded multilevel inverter photovoltaic power system. *IEEE Trans. Ind. Informat.* **2018**, *14*, 4905–4915. [CrossRef]
8. Sun, D.; Ge, B.; Yan, X.; Bi, D.; Zhang, H.; Liu, Y.; Abu-Rub, H.; Ben-Brahim, L.; Peng, F. Modeling, impedance design, and efficiency analysis of quasi-Z-source module in cascade multilevel photovoltaic power system. *IEEE Trans. Ind. Electron.* **2014**, *61*, 6108–6117. [CrossRef]
9. Liu, Y.; Ge, B.; Abu-Rub, H.; Sun, D. Comprehensive modeling of single-phase quasi-Z-source photovoltaic inverter to investigate low-frequency voltage and current ripple. *IEEE Trans. Ind. Electron.* **2015**, *62*, 4194–4202. [CrossRef]
10. Liang, W.H.; Liu, Y.; Ge, B.; Abu-Rub, H.; Balog, R.S.; Xue, Y.S. Double-line-frequency ripple model, analysis, and impedance design for energy-stored single-phase quasi-Z-source photovoltaic system. *IEEE Trans. Ind. Electron.* **2018**, *65*, 3198–3209. [CrossRef]
11. Ge, B.; Abu-Rub, H.; Liu, Y.; Balog, R.S.; Peng, F.Z. An active filter method to eliminate dc-side low-frequency power for a single-phase quasi-z-source inverter. *IEEE Trans. Ind. Electron.* **2016**, *63*, 4838–4848. [CrossRef]
12. Ge, B.; Liu, Y.; Abu-Rub, H.; Balog, R.S.; Peng, F.Z.; McConnell, S.; Li, X. Current Ripple Damping Control to Minimize Impedance Network for Single-Phase Quasi-Z Source Inverter System. *IEEE Trans. Ind. Inform.* **2016**, *12*, 1043–1054. [CrossRef]
13. Zhou, Y.; Li, H.; Li, H. A single-phase PV quasi-Z-source inverter with reduced capacitance using modified modulation and double-frequency ripple suppression control. *IEEE Trans. Power Electron.* **2016**, *31*, 2166–2173. [CrossRef]
14. Wei, R.Y.; Tang, Y.F.; Wang, S.J.; Li, Z.Y. A ripple suppression strategy based on virtual self-injection APF for quasi-Z source inverter. In *IOP Conference Series: Earth and Environmental Science, Proceedings of the 2018 International Conference on New Energy and Future Energy System, Shanghai, China, 21–24 August 2018*; IOP: London, UK, 2018; Volume 188, 012065. [CrossRef]
15. Nguyen, M.K.; Choi, Y.O. Maximum boost control method for single-phase quasi-switched-boost and quasi-z-source inverters. *Energies* **2017**, *10*, 553. [CrossRef]
16. Yu, Y.; Zhang, Q.; Liang, B.; Cui, S. Single-phase Z-source inverter: Analysis and low-frequency harmonics elimination pulse width modulation. In Proceedings of the IEEE Energy Conversion Congress and Exposition (ECCE), Phoenix, AZ, USA, 17–22 September 2011; pp. 2260–2267.
17. Mohammadi, M.; Moghani, J.S.; Milimonfared, J. A Novel Dual Switching Frequency Modulation for Z-Source and Quasi Z-Source Inverters. *IEEE Trans. Ind. Electron.* **2018**, *65*, 5167–5176. [CrossRef]
18. He, Y.; Xu, Y.; Chen, J. New space vector modulation strategies to reduce inductor current ripple of z-source inverters. *IEEE Trans. Power Electron.* **2018**, *33*, 2643–2654. [CrossRef]
19. Tang, Y.; Xie, S.; Ding, J. Pulsewidth modulation of Z-source inverters with minimum inductor current ripple. *IEEE Trans. Ind. Electron.* **2014**, *61*, 98–106. [CrossRef]

20. Bayhan, S.; Trabelsi, M.; Abu-Rub, H.; Malinowski, M. Finite-control-set model-predictive control for a quasi-Z-source four-leg inverter under unbalanced load condition. *IEEE Trans. Ind. Electron.* **2016**, *64*, 2560–2569. [CrossRef]

21. Komurcugil, H.; Bayhan, S.; Bagheri, F.; Kukrer, O.; Abu-Rub, H. Model-based current control for single-phase grid-tied quasi-Z-source inverters with virtual time constant. *IEEE Trans. Power Electron.* **2018**, *65*, 8277–8286. [CrossRef]

Article

Regulation Performance of Multiple DC Electric Springs Controlled by Distributed Cooperative System

Daojun Zha [1,†], Qingsong Wang [1,†], Ming Cheng [1,*], Fujin Deng [1] and Giuseppe Buja [2]

1   School of Electrical Engineering, Southeast University, Nanjing 210096, China
2   Department of Industrial Engineering, University of Padova, 35131 Padova, Italy
*   Correspondence: mcheng@seu.edu.cn; Tel.: +86-25-8279-4152
†   These authors contributed equally to this work.

Received: 4 August 2019; Accepted: 3 September 2019; Published: 5 September 2019

Abstract: DC electric springs (DCESs) have been recently developed to improve the voltage stability of a DC microgrid. A lately proposed DCES topology is comprised of a DC/DC three port converter (TPC), a bi-directional buck-boost converter (BBC) and a battery, and is arranged as follows: The TPC input port is fed by a renewable energy source (RES) whilst the two TPC output ports supply a non-critical load (NCL) and a critical load (CL) separately; in turn, BBC together with the battery constitutes the DCES energy storage unit (ESU) and is connected in parallel to CL. In this paper, a set of DCESs with such a topology and with their CLs connected to a common DC bus is considered. The control of the DCESs is built up around a distributed cooperative system having two control levels, namely primary and secondary, each of them endowed with algorithms committed to specific tasks. The structure of the control levels is explicated and their parameters are designed. The control system is applied to a DCES set taken as a study-case and tested by simulation. The results of the tests show the excellent performance of the control system in both regulating the CL DC bus voltage and keeping the state-of-charge of the battery within predefined limits.

Keywords: DC microgrid; DC electric spring; distributed cooperative control; adaptive droop control; consensus algorithm

---

1. Introduction

As a demand-response technique, the electric spring (ES) has been proposed initially in [1] to ensure the stabilization of the supply voltage of loads when the grid power fluctuates, which today is more frequent than in the past since the grid power is mostly or entirely generated by renewable energy sources (RESs). Based on the ES concept, two types of equipment have been developed: One is named the AC electric spring (ACES) [2,3] and is intended to stabilize the supply voltage of AC appliances; the other one is named the DC electric spring (DCES) [4,5] and is intended for DC appliances. ACESs have been investigated first and actually many topologies and control strategies are available for their implementation [2,6].

Instead, there is much to explore on DCESs as the research interests on this topic date back only a few years. The overall system with the photovoltaic (PV) systems, fuel cells, and batteries that are inherently of a DC nature can be fully integrated in the DC form without any DC/AC or AC/DC converters, which contribute to a higher efficiency and reliability [7]. However, similar to the AC grids, there are also some drawbacks in the DC microgrid due to the intermittent RES, such as the power imbalance among the power sources and load, the bus voltage instability, and fluctuation. Existing solutions include the line-regulating converter [8], control for the coordinating the demands and supplies [9], and droop control strategies [10]. However, these solutions commonly suffer from additional communication system or poor regulation performances. Considering the deficiencies of the DC power system with the RES, the concept of a DC electric spring is firstly proposed in [11], for the regulation of the bus voltage and power in a DC microgrid.

According to the ES concept, the loads of a DC microgrid are sorted into critical loads (CLs) and non-critical loads (NCLs). According to the sorting terminology, CLs are critical with respect to the supply voltage in the sense that they require a well-stabilized voltage to operate correctly, while NCLs do not require it. In order to stabilize the CL voltage, the DCES transfers some of the grid power fluctuations from the DC microgrid to NCLs, and forces the remaining power fluctuations to be borne by a DCES-embedded battery, thus enabling fast and flexible energy storage with a reduced battery usage.

There are two kinds of existing versions of the topology of the DCES [4,5], which can be sequenced as DCES-1 and DCES-2. The configuration of DCES-1 is the same as the original version of ACES. Since the low-pass filter inside the ACES has a big volume, the power density of such circuit is pretty low. In contrary, DCES-2 is realized by pure DC/DC converters to avoid the disadvantage introduced by the filter with a big volume. However, there are three different power converters inside it, which leads to a more complex control and lower reliability.

To avoid these disadvantages, an improved DCES topology has been proposed lately [12], which circumvents the inconvenience of the series connection of the NCL and ES. Moreover, in the proposed DCES, the CL and NCL are isolated from each other and are both in parallel with the DC bus, which is consistent with the traditional connection type in power systems.

The topology is comprised of a DC/DC three-port converter (TPC) [13] and an energy storage unit (ESU). The TPC input port is fed by RES (or by a RES-prevalent grid), whilst the two output ports supply the NCL and CL separately. The ESU, in turn, is connected in parallel to the CL and is comprised of a bi-directional buck-boost converter (BBC) [14] and a battery; its task is to ensure the stabilization of the CL supply voltage while keeping the state-of-charge of the battery within predetermined limits. Hereafter, this DCES topology is referred to as a CL-paralleled (CLP)-ESU.

This paper focuses on DCESs and is aimed at investigating the regulation performance of a set of topology-novel DCESs when they are controlled by means of a distributed cooperative system.

Cooperative control systems of a multi-stage appliance can be classified into three categories, denoted as centralized, decentralized, and distributed [15]. A centralized system necessitates of communication between a central controller and the local controllers to control an appliance. A decentralized system dispenses from a central controller and the local controllers operate on the basis of information that they are able to find individually, like the droop control for a power system; as a counterpart, it may not be effective in the full and/or optimal utilization of the resources of the appliance [16]. A distributed system also dispenses from a central controller, but is different from a decentralized system, in which the local controllers exchange information with the other controllers through a communication network to control the appliance.

A distributed cooperative control proposed in [17], including the voltage and current controller, is based on a consensus algorithm [18], which refers to all components reaching a certain common agreement after the distributed control. In [19], the distributed cooperative control is used to establish a primary/secondary control framework for the DC microgrid. The *V-I* droop mechanism in [19] is used in the primary control for enabling a decentralized coordination of distributed energy resources. Moreover, the consensus algorithm is used in the secondary control for the DC-bus voltage regulating and battery state-of-charge (SOC) balancing. The distributed cooperative control applied in the series and shunt DCESs is proposed in [20], whilst the primary and secondary control can achieve an average DC-bus voltage consensus and SOC balance among the different DCESs.

In this paper, a distributed cooperative system is developed to control a set of multiple DCESs with a CLP-ESU topology [12]. The system has two control levels, primary and secondary, whose joint action allows the concurrent achievement of the following objectives: Local voltage stabilization, local power allocation, consensus on the DC bus voltage of the CL of each DCES, and consensus of the state-of-charge (SOC) of the battery of each DCES. The system level with the primary control is of a decentralized type and includes the phased-shift control of TPC ports, the decoupling voltage control of CLs, the adaptive droop settling of CLs, and the charging and discharging control of the batteries. In turn, the system level with the secondary control is of a distributed type and includes the voltage control of the CL DC buses and the SOC control of the batteries.

In short, the organization of the paper is as follows. Section 2 reviews the topology and operation of DCES utilized in the paper. Section 3 illustrates the distributed cooperative system arranged for the control of multiple DCESs. Section 4 shows the steady state analysis of the proposed control method. Section 5 presents the simulations carried on a study case of multiple DCESs and discusses the results. Section 6 concludes the paper.

2. DCES Topology and Operation

This paper utilizes DCESs with the CLP-ESU topology drawn within the dashed line of Figure 1. The photovoltaic (PV) array, NCL and CL are connected to the ports of the TPC, which is essentially a two-output DC/DC converter isolated by means of a high frequency transformer and equipped with a full AC/DC bridge at each port. The battery is used for voltage regulation, and is paralleled to the CL DC bus through the BBC, which can be thought of as a version of the boost DC/DC converter, modified to support the bi-directional power flow. Compared to the existing DCES ones, the CLP-ESU topology keeps the supply of the CL and NCL isolated from each other. This circuital layout is compliant with the traditional way of connecting loads with different requirements in a power system [12]. More broadly, it can be envisaged that each output port in the figure supplies a distinct DC bus with connecting as many CLs and NCLs as per the DCES sizing power.

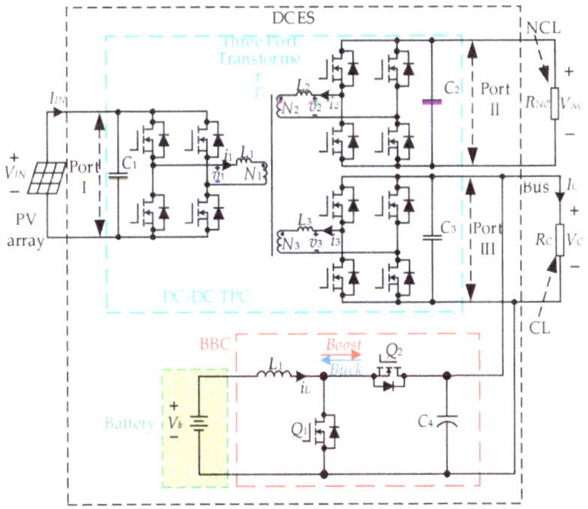

Figure 1. Critical load paralleled-energy storage unit (CLP-ESU) topology of the DC electric spring (DCES).

The typical operating modes of the DCES are three, designated as the battery-balancing mode, battery-discharging mode, and battery-charging mode.

In the battery-balancing mode (mode 1), the power delivered by the RES is enough to supply CL at the specified value. The RES power fluctuations are forced by the ESU to flow almost entirely from the RES to the NCL, whilst the remaining small part of them flows into the battery to keep the CL voltage stabilized.

In the battery-discharging mode (mode 2), the RES power is in-sufficient to supply CL at the rated value even if the NCL voltage is adjusted to consume less power than the rated one. In this mode, the battery discharges to provide the deficit of power to the CL.

At the battery-charging mode (mode 3), the RES power exceeds the power consumption of CL even if the NCL voltage is adjusted to consume more power than the rated one. In this mode, the battery is charged and stores the excess of power.

The structure of multiple DCESs in a DC microgrid is schematized in Figure 2. The figure shows that the CL DC buses (henceforth briefly referred to as the DC buses) of each DCES are connected to each other by transmission lines, whose resistances are denoted with R_{L1}, R_{L2}, R_{L3}, and R_{L4}. The link of the four DCESs together forms a typical DC microgrid.

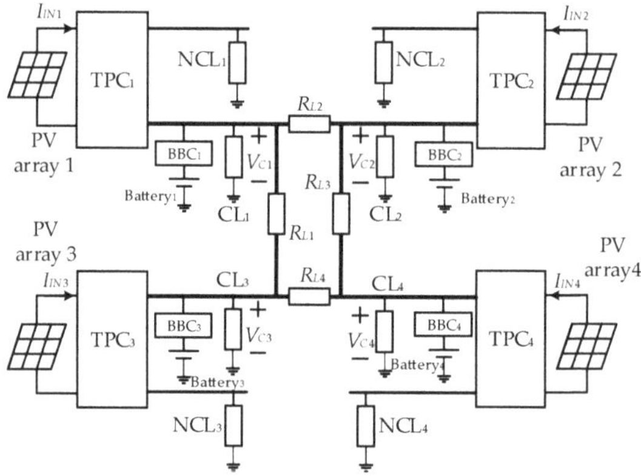

Figure 2. Structure of multiple DCESs.

3. Distributed Cooperative Control

3.1. Primary Control

The primary control, also named the local voltage control, provides the independent voltage control of each DCES. It exerts into two control actions, namely the TPC control and BBC control. As shown by the scheme in Figure 3, the TPC control regulates the power flow from the RES to the CL and NCL by changing the phase-shift angle, whilst the BBC control aims for a power regulation of the CL.

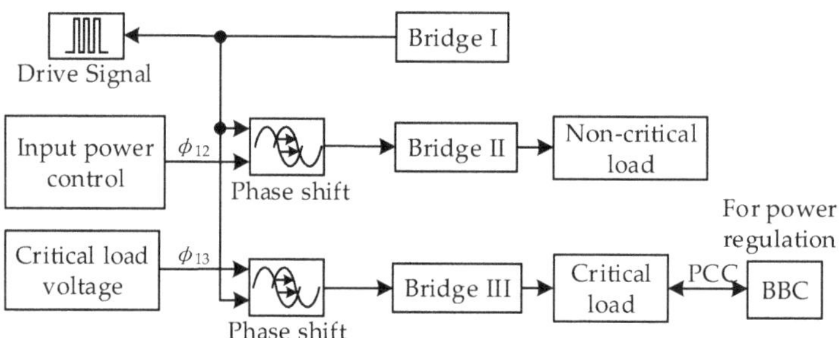

Figure 3. The proposed primary control of the DCES.

The TPC control encompasses three actions: Phase-shift control, decoupling network, and adaptive droop adjustment, as drawn in the diagram of Figure 4 for the ith DCES.

The familiar phase-shift control [21] is adopted for TPC to change the power flow from port I to port II and III, where ϕ_{12} and ϕ_{13} are the phase-shift angles of the drive signals of the H-bridges at ports I and II, and at ports I and III, respectively, so as to regulate the CL voltage to be constant.

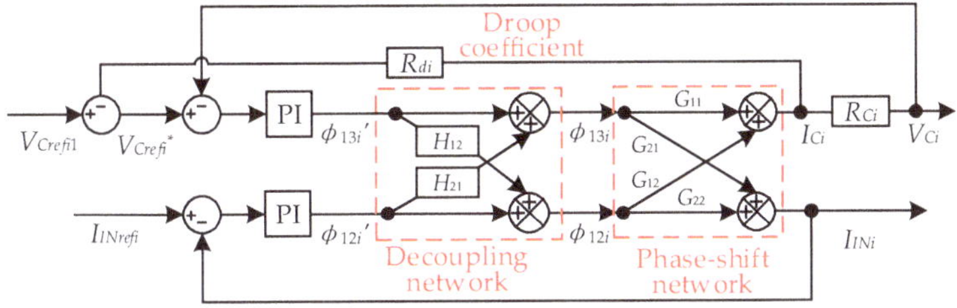

Figure 4. The control diagram of the three port converter (TPC).

The two ports of the TPC can be regarded as a DAB. As a result, the power equations can be obtained through a superposition theorem and the results are as follows [22].

$$\begin{cases} P_{12} = \dfrac{V_{IN}V_{NC}{}'}{2\pi f_s L_{12}}\phi_{12}\left(1 - \dfrac{\phi_{12}}{\pi}\right) \\ P_{13} = \dfrac{V_{IN}V_C{}'}{2\pi f_s L_{13}}\phi_{13}\left(1 - \dfrac{\phi_{13}}{\pi}\right) \\ P_{32} = \dfrac{V_{NC}{}'V_C{}'}{2\pi f_s L_{23}}\phi_{32}\left(1 - \dfrac{\phi_{32}}{\pi}\right) \end{cases} \tag{1}$$

where fs denotes the switching frequency; P12, P13, and P32 denote the power flowing from port I to port II, from port I to port III and from port III to port II, respectively. VNC′ and VC′ denote the voltages VNC and VC reflected to port I from port II and port III, respectively. It should be noticed that the power flow is determined by the phase-shift angle between the related two ports.

Based on $P_1 = P_{12} + P_{13}$ and $P_3 = P_{32} - P_{13}$, the averaged values of the currents at port I and port III can be expressed as

$$\begin{cases} I_{IN} = \dfrac{P_1}{V_{IN}} \;= \dfrac{N_1 V_{NC}}{2\pi f_s N_2 L_{12}}\phi_{12}\left(1 - \dfrac{\phi_{12}}{\pi}\right) \\ \qquad\qquad + \dfrac{N_1 V_C}{2\pi f_s N_3 L_{13}}\phi_{13}\left(1 - \dfrac{\phi_{13}}{\pi}\right) \\ I_C = \dfrac{P_3}{V_C} \;= \dfrac{N_1 V_{IN}}{2\pi f_s N_3 L_{13}}\phi_{13}\left(1 - \dfrac{\phi_{13}}{\pi}\right) \\ \qquad\qquad - \dfrac{N_1{}^2 V_{NC}}{2\pi f_s N_2 N_3 L_{23}}\phi_{32}\left(1 - \dfrac{\phi_{12}-\phi_{13}}{\pi}\right) \end{cases} \tag{2}$$

where I_{IN} and I_C denote the averaged currents at port I and port II, respectively.

It is obviously seen that both the power and current equations are nonlinear forms. To obtain the small signal model, a small-signal perturbation in the form of a small step change is applied around a quiescent operation point, which is designated as $Q(\phi_{130}, \phi_{120})$.

The elements of the relationship matrix between the current and phase angles are as follows.

$$\begin{cases} G_{11} \;= \left.\dfrac{\partial I_3}{\partial \phi_{13}}\right|_Q = \dfrac{N_1 V_1}{2\pi f_s N_3 L_{13}}\left(1 - \dfrac{2}{\pi}\phi_{130}\right) \\ \qquad\quad + \dfrac{N_1{}^2 V_2}{2\pi f_s N_2 N_3 L_{23}}\left[1 - \dfrac{2}{\pi}(\phi_{120} - \phi_{130})\right] \\ G_{12} \;= \left.\dfrac{\partial I_3}{\partial \phi_{12}}\right|_Q = -\dfrac{N_1{}^2 V_2}{2\pi f_s N_2 N_3 L_{23}}\left[1 - \dfrac{2}{\pi}(\phi_{120} - \phi_{130})\right] \\ G_{21} \;= \left.\dfrac{\partial I_1}{\partial \phi_{13}}\right|_Q = \dfrac{N_1 V_3}{2\pi f_s N_3 L_{13}}\left(1 - \dfrac{2}{\pi}\phi_{130}\right) \\ G_{22} \;= \left.\dfrac{\partial I_1}{\partial \phi_{12}}\right|_Q = \dfrac{N_1 V_2}{2\pi f_s N_2 L_{12}}\left(1 - \dfrac{2}{\pi}\phi_{120}\right) \end{cases} \tag{3}$$

The equations of the small signal modeling of the TPC could be expressed as follows.

$$\begin{bmatrix} \Delta I_{IN} \\ \Delta I_C \end{bmatrix} = \begin{bmatrix} G_{11} & G_{12} \\ G_{21} & G_{22} \end{bmatrix}\begin{bmatrix} \Delta\phi_{13} \\ \Delta\phi_{12} \end{bmatrix} = \mathbf{G}\begin{bmatrix} \Delta\phi_{13} \\ \Delta\phi_{12} \end{bmatrix} \tag{4}$$

where ΔI_{IN}, ΔI_C, $\Delta\phi_{13}$, and $\Delta\phi_{12}$ are the small-signal perturbation of I_{IN}, I_C, ϕ_{13}, and ϕ_{12}, respectively.

$$\mathbf{G} = \begin{bmatrix} G_{11} & G_{12} \\ G_{21} & G_{22} \end{bmatrix} \tag{5}$$

In Figure 4, since the CL voltage $V_C = I_C R_C$ and input port current I_{IN} are influenced by both ϕ_{12} and ϕ_{13}, a decoupling network is inserted in the TPC control [23] to eliminate the cross influence of the phase angles on the control of V_C and I_{IN}. A decoupling matrix $\mathbf{H}$ is designed to make $\mathbf{GH}$ a diagonal matrix to ensure one output is determined by one control input independently. As a result, two equations could be obtained as follows:

$$\begin{cases} \Delta\phi_{12}{}'G_{21} + \Delta\phi_{12}{}'H_{21}G_{22} = 0 \\ \Delta\phi_{13}{}'G_{12} + \Delta\phi_{13}{}'H_{12}G_{11} = 0 \end{cases} \tag{6}$$

where $\Delta\phi_{13}{}'$ and $\Delta\phi_{12}{}'$ are the small-signal perturbation of the virtual phase-shift angle derived from $\Delta\phi_{13}$ and $\Delta\phi_{12}$ in the decoupling control.

Therefore, $\mathbf{H}$ can be derived and simplified as

$$\mathbf{H} = \begin{bmatrix} 1 & H_{21} \\ H_{12} & 1 \end{bmatrix} = \begin{bmatrix} 1 & -G_{12}/G_{11} \\ -G_{21}/G_{22} & 1 \end{bmatrix} \tag{7}$$

In the control system, there are two control loops, the output voltage loop is designed for port III and the input current loop is related to port I. After the design of the decoupling network, it can be assumed that there are no interactions between these loops.

More details about the phase-shift control and decoupling control of DCES with the CLP-ESU topology can be found in [12].

The third action of the TPC control is the adaptive droop [24] adjustment based on the consensus algorithm, which is firstly utilized in the proposed primary TPC control compared to the existing literature [25]. It is utilized to establish the individual power allocation for each DCES by changing the virtual equivalent resistance R_{di} (droop coefficient) in the diagram of Figure 4. The modified CL voltage reference V^*_{Crefi} of the ith DCES is obtained by

$$V^*_{Crefi} = V_{Crefi1} - R_{di}I_{Ci} \tag{8}$$

where I_{Ci} and R_{di} is the CL current and droop coefficient of the ith DCES, respectively. V_{Crefi1} is the voltage reference generated by the secondary voltage control, as explained later on.

Thus, the voltage droop can be adjusted to comply with the different load conditions of the DCESs. The adjustment is obtained by updating the droop coefficient R_{di} as follows:

$$R_{di} = R_{d0} - \delta_{Rdi} \tag{9}$$

where, R_{d0} is the initial droop coefficient, and δ_{Rdi} is its correction term. The calculation of this term is based on the consensus algorithm.

The principle of an average consensus algorithm is explicated in [18]. Let x_i be the state variable of node i in a system, and let node i communicate with its neighboring node j under a communication weight a_{ij}, the system is in consensus only when all the state variables are equal to each other. To reach the consensus situation, the state variable x_i is updated at time t by

$$\dot{x}_i(t) = \sum_{j=1}^{n} a_{ij}(x_j(t) - x_i(t)), \quad i = 1, 2, \cdots, n \tag{10}$$

where the dot over x_i denotes the updated value of x_i. The communication weight a_{ij} is a positive quantity that indicates if there is an exchange of information from node j to i and how much valued is the information. Specifically, $a_{ij} > 0$ means that there is communication between nodes j to i, whilst $a_{ij} = 0$ means there is no communication; moreover, a large value of a_{ij} indicates that node j has a high degree of influence on the update of the state variable of node i.

The final result of a consensus algorithm is that the state variables of each nodes approach the average of the initial values of all the nodes.

$$\lim_{t \to \infty} x_i(t) = \frac{1}{n} \sum_{i=1}^{n} x_i(0), \qquad i = 1, 2, \cdots, n \tag{11}$$

Based on the consensus algorithm, the correction term δ_{Rdi} is carried in four steps: (i) Comparison of current I_{Ci} of ith CL to that of jth CL neighbor, (ii) weighting of the current difference ($I_{Cj} - I_{Ci}$) with weight a_{ij}, (iii) summation of the weighted current differences of the N neighbors, and iv) entering of the summation into a PI controller [24].

$$\delta_{Rdi} = k_{PIi} \sum_{j=1}^{N} a_{ij}(I_{Cj} - I_{Ci}) + k_{IIi} \int \sum_{j=1}^{N} a_{ij}(I_{Cj} - I_{Ci})dt \tag{12}$$

As a result, the update of R_d affects the output current and, hence, the output power of each DCES.

The BBC control for the ith DCES has the diagram in Figure 5, where the voltage reference V_{Cerfi2} is generated by the secondary SOC control. It is worthy to note right now that, in general, V_{Cerfi1} and V_{Cerfi2} are different since they are generated respectively by the secondary voltage and SOC controls. The reference V_{Crefi1} is used to control the TPC and regulate the power consumption of NCL, while the reference V_{Crefi2} is used to change the operation mode of BBC and regulate the power of the battery. The switching signals of the BBC switches Q_1 and Q_2 are delivered by two logic AND elements triggered by the output of a PI controller through the PWM generator and the outputs of two comparators.

The operation of the BBC control is as follows. When voltage V_{Ci} of ith CL DC-bus is higher than a predefined value of the reference (e.g., $V_{Ci} > 1.01V_{Crefi2}$), BBC enters into the buck mode, and DCES operates in the battery-charging mode. When V_{Ci} is lower than a predefined value of the reference (e.g., $V_{Ci} < 0.99V_{Crefi2}$), BBC enters into the boost mode, and the DCES operates in the battery-discharging mode. When V_{Ci} is in between $0.99V_{Crefi2}$ and $1.01V_{Crefi2}$, the two sides of BBC are isolated from each other, and the DCES operates in the battery-balancing mode.

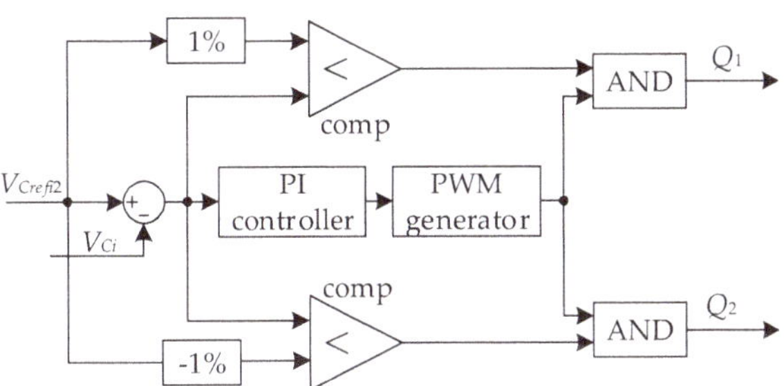

Figure 5. The charging and discharging control of BBC.

3.2. Secondary Control

The secondary control provides for generating the updated local values of references V_{Crefi1} and V_{Crefi2} for the two actions of the primary control. The diagram of the secondary control for ith DCES, drawn in Figure 6, underlines the two updating paths and the usage of two controls, one in the updating path of the CL DC-bus voltage, and the other one in that of the battery SOC. The operation of the secondary control relies on the average consensus algorithms.

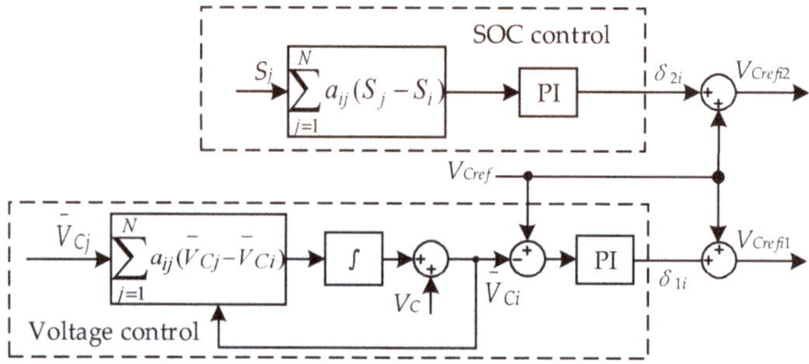

Figure 6. The proposed secondary control of the DCES.

Returning to the diagram of Figure 6, the voltage control updates the average voltage $\overline{V}_{Ci}$ of *i*th CL DC bus, by processing the local voltage measurement V_{Ci} and the neighbors' average voltage $\overline{V}_{Cj}$ [26] through a consensus algorithm. It is expressed by

$$\overline{V}_{Ci} = V_{Ci} - \int \sum_{j=1}^{n} a_{ij}(\overline{V}_{Ci} - \overline{V}_{Cj})\mathrm{d}t \tag{13}$$

After comparison of $\overline{V}_{Ci}$ to the global voltage reference V_{Cref} of the CL DC-bus voltage of the microgrid, the error term is processed by a PI controller, as formulated in (8), to generate a correction term δ_{1i} for the voltage reference V_{Cerfi1} of the TPC control of *i*th DCES.

$$\delta_{1i} = k_{PUi}(V_{Cref} - \overline{V}_{Ci}) + k_{IUi}\int (V_{Cref} - \overline{V}_{Ci})\mathrm{d}t \tag{14}$$

where the k_{PUi} and k_{IUi} are the proportional and integral coefficients of the voltage control. In agreement with (14), the correction term δ_{1i} is higher when the average voltage $\overline{V}_{Ci}$ greatly deviates from reference V_{Cref}. As shown by (15), this fact enforces reference V_{Cref1} of the TPC control so as to realize the consensus of $\overline{V}_{Ci}$ and V_{Cref}, which means that the average CL DC-bus voltage are regulated very close to the rated value [20].

$$V_{Crefi1} = V_{Cref} + \delta_{i1} \tag{15}$$

To realize the consensus of the SOC of the batteries in a multiple DCES, the SOC control of Figure 6 compels batteries with a high SOC to discharge faster and those with low SOC to charge faster. For this purpose, the following local state variable S_i is defined for the SOC of the battery of *i*th DCES [19]:

$$S_i = \frac{P_{bi}}{C_i F_{SOCi}} = \frac{V_{bi}i_{bi}}{C_i F_{SOCi}} \tag{16}$$

where P_{bi} indicates the power flow at the battery terminals (marked as negative during charging and positive during discharging), SOC_i is the SOC of the battery of *i*th DCES, and F_{SOCi} is a function of SOC_i defined as

$$F_{SOCi} = \begin{cases} SOC_i - SOC_L, & P_{bi} > 0 \\ SOC_H - SOC_i, & P_{bi} < 0 \end{cases} \tag{17}$$

where SOC_L and SOC_H are respectively the lower and upper limits of SOC_i, which are selected as 0.2 and 0.8 in this paper.

By help of a consensus algorithm, the updated value of S_i is processed by a PI controller, as formulated in (18), to generate a correction term δ_{2i} for the voltage reference V_{Cerfi2} of the BBC control of ith DCES.

$$\delta_{2i} = k_{PSi} \sum_{j=1}^{N} a_{ij}(S_j - S_i) + k_{ISi} \int \sum_{j=1}^{N} a_{ij}(S_j - S_i)\mathrm{d}t \tag{18}$$

where the k_{PSi} and k_{ISi} indicate the proportional and integral coefficient of the SOC control.

Similar to the update of the voltage reference V_{Cerfi1}, a correction term δ_{2i} becomes higher when S_i greatly deviates from the neighboring S_j. As shown by (19), this fact enforces reference V_{Crefi2} of the BBC control and, together with it, the charging or discharging process of the batteries so as to realize the consensus of SOC_i and SOC_j.

$$V_{Crefi2} = V_{Cref} + \delta_{i2} \tag{19}$$

The structure of the distributed cooperative system arranged for the control of multiple DCESs is illustrated by the block diagram in Figure 7. The diagram highlights that the system has two control levels, including the primary control and secondary control. Compared to the existing distributed cooperative control [27], the voltage references of the TPC and BBC controls that are located in the primary control level, are generated respectively by the voltage and SOC controls that are located in the secondary control level. It means that the TPC and BBC can be controlled independently, which contributes to a reduced battery usage and releases the burden of the battery.

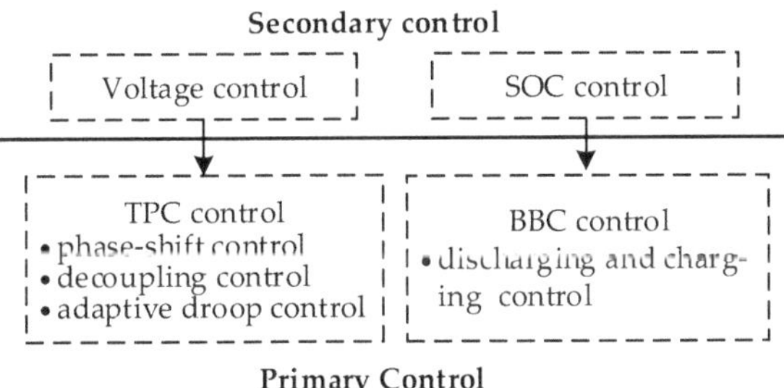

Figure 7. Structure of the proposed distributed cooperative control.

4. Steady State Analysis

In this section, the steady state analysis of the distributed cooperative control is carried out to prove the consensus of the CL average voltage and battery SOC.

4.1. Consensus of the CL Average Voltage

By differentiating (13),

$$\mathrm{d}\overline{V}_{Ci} = \mathrm{d}V_{Ci} - \sum_{j=1}^{n} a_{ij}(\overline{V}_{Ci} - \overline{V}_{Cj}) \tag{20}$$

The global observer dynamic can be formulated as

$$\mathrm{d}\overline{\mathbf{V}}_C = \mathrm{d}\mathbf{V}_C - \mathbf{L}\overline{\mathbf{V}}_C \tag{21}$$

where $\mathbf{L} = \mathbf{D} - \mathbf{A}$, in which $\mathbf{D} = diag\{\sum_{j=1}^{n} a_{ij}\}$, $\mathbf{A} = [a_{ij}] \in R^{n\times n}$, $\mathbf{V}_C = [V_{C1}, V_{C2}, \cdots, V_{Cn}]^{\mathrm{T}}$ and $\overline{\mathbf{V}}_C = [\overline{V}_{C1}, \overline{V}_{C2}, \cdots, \overline{V}_{Cn}]^{\mathrm{T}}$ is the CL DC-bus voltage vector and CL average voltage vector, respectively. Since $\overline{\mathbf{V}}_C(0) = \mathbf{V}_C(0)$, (21) can be written in the frequency domain as

$$\overline{\mathbf{V}}_C^{\mathrm{s}} = s(s\mathbf{I} + \mathbf{L})^{-1}\mathbf{V}_C^{\mathrm{s}} = \mathbf{H}\mathbf{V}_C^{\mathrm{s}} \tag{22}$$

where $\overline{\mathbf{V_C^s}}$ and $\mathbf{V_C^s}$ are the form in frequency domain of $\overline{\mathbf{V}}_C$ and $\mathbf{V}_C$.

The correction term in (14) can be written in the frequency domain as

$$\delta_1 = \mathbf{G_U}(\mathbf{V_{Cref}} - \overline{\mathbf{V_C^s}}) \tag{23}$$

where $\mathbf{V_{Cref}}$ is the Laplace transform of the reference voltage vector, and $\lim_{s\to 0} s\mathbf{V_{Cref}} = V_{Cref}\mathbf{1_n}$ in which $\mathbf{1_n}$ is a column vector with all elements equal to 1 [27]. $\mathbf{G_U} = \text{diag}\{G_{Ui}\}$, in which

$$G_{Ui} = k_{PUi} + \frac{k_{IUi}}{s} \tag{24}$$

combining (8) and (15), then write it in the frequency domain by substituting (22) and (23)

$$\mathbf{V_{Cref}^*} = \mathbf{V_{Cref1}} - \mathbf{I_C}\mathbf{R_d} = \mathbf{V_{Cref}} + \mathbf{G_U}(\mathbf{V_{Cref}} - \mathbf{HV_C^s}) - \mathbf{I_C}\mathbf{R_d} \tag{25}$$

where $\mathbf{R_d} = \text{diag}\{R_{di}\}$ is the droop coefficient matrix, and $\mathbf{I_C}$ is the Laplace transform of the CL current vector $[I_{C1}, I_{C2}, \cdots, I_{Cn}]^T$.

Thus, the dynamic behavior of the CL DC bus voltage with a closed-loop voltage regulator can be expressed as

$$\mathbf{V_C^s} = \mathbf{G_{C1}}\mathbf{V_{Cref}^*} \tag{26}$$

where $\mathbf{G_{C1}} = \text{diag}\{G_{C1i}\}$ is the transfer function matrix, in which G_{C1i} is the closed-loop transfer function of the ith TPC of DCES, and can be simplified as an inertial element

$$\lim_{s\to 0}\mathbf{G_{C1}} = \mathbf{I_n} \tag{27}$$

where $\mathbf{I_n}$ is a n-order unit matrix.

The CL current matrix $\mathbf{I_C}$ can be obtained by the admittance matrix $\mathbf{Y}$ of the DCESs

$$\mathbf{I_C} = \mathbf{YV_C^s} \tag{28}$$

By substituting (26) and (28) into (25), the dynamic behavior of the CL DC bus voltage can be written as

$$\mathbf{V_C^s} = (\mathbf{I_n} + \mathbf{G_{C1}}\mathbf{G_U}\mathbf{H} + \mathbf{G_{C1}}\mathbf{R_d}\mathbf{Y})^{-1} \times \mathbf{G_{C1}}(\mathbf{I} + \mathbf{G_U})\mathbf{V_{Cref}} \tag{29}$$

The steady state of the CL DC bus voltage can be obtained as

$$\mathbf{V_C^{ss}} = \lim_{s\to 0} s\mathbf{V_C^s} = \lim_{s\to 0}[s(s\mathbf{I_n} + s\mathbf{G_{C1}}\mathbf{G_U}\mathbf{H} + s\mathbf{G_{C1}}\mathbf{R_d}\mathbf{Y})^{-1} \times s\mathbf{G_{C1}}(\mathbf{I_n} + \mathbf{G_U})\mathbf{V_{Cref}}] \tag{30}$$

Based on the definition of $\mathbf{G_U}$, it can be obtained as

$$\lim_{s\to 0} s\mathbf{G_U} = \mathbf{K_{IU}} \tag{31}$$

Therefore,

$$\mathbf{Q}\mathbf{V_C^{ss}} = \lim_{s\to 0} s\mathbf{HV_C^s} = (\mathbf{K_{IU}})^{-1} \times \mathbf{K_{IU}}V_{Cref}\mathbf{1_n} = V_{Cref}\mathbf{1_n} \tag{32}$$

where $\mathbf{Q} = \lim_{s\to 0}\mathbf{H}$ is the $n \times n$ matrix with all the elements equal to $1/n$. So, the steady state of the CL average voltage can be written as

$$\overline{\mathbf{V_C^{ss}}} = \mathbf{Q}\mathbf{V_C^{ss}} = V_{cref}\mathbf{1_n} \tag{33}$$

Equation (33) implies that the CL average voltage will all reach consensus in the steady state at V_{Cref}.

4.2. Consensus of the Battery SOC

The correction term in (18) can be written in the frequency domain as

$$\delta_1 = \mathbf{G_S L S} \tag{34}$$

where $\mathbf{S}$ is the Laplace transform vector of the state variable in (16).

By substituting (34), (19) can be written in the frequency domain

$$\mathbf{V_{Cref2}} = \mathbf{V_{Cref}} + \mathbf{G_S L S} \tag{35}$$

Similar to (26),

$$\mathbf{V_C^s} = \mathbf{G_{C2} V_{Cref2}} \tag{36}$$

where $\mathbf{G_{C2}} = \text{diag}\{G_{C2i}\}$ is the transfer function matrix, in which G_{Ci} is the closed-loop transfer function of the ith BBC of DCES, and can be simplified as an inertial element.

$$\lim_{s \to 0} \mathbf{G_{C1}} = \mathbf{I_n} \tag{37}$$

Therefore, $\mathbf{S}$ can be written as

$$\mathbf{S} = (\mathbf{G_{C2} G_S L})^{-1} \times (\mathbf{V_C^s} - \mathbf{G_{C2} V_{Cref}}) \tag{38}$$

Based on the definition of $\mathbf{G_S}$, it can be obtained as

$$\lim_{s \to 0} s\mathbf{G_S} = \mathbf{K_{IS}} \tag{39}$$

The form of the state variable $\mathbf{S}$ in the frequency domain is $\mathbf{S^s}$, and its steady state can be obtained as

$$\mathbf{S^{ss}} = \lim_{s \to 0} s\mathbf{S^s} = \lim_{s \to 0}[s(s\mathbf{G_{C2} G_S L})^{-1} \times (s\mathbf{V_C^s} - s\mathbf{G_{C2} V_{Cref}})] \tag{40}$$

Therefore, $\mathbf{QS^{ss}}$ can be written as

$$\begin{aligned} \mathbf{QS^{ss}} &= \lim_{s \to 0} s\mathbf{HS} = \lim_{s \to 0}[s\mathbf{H}(s\mathbf{G_{C2} G_S L})^{-1} \times (s\mathbf{V_C^s} - s\mathbf{G_{C2} V_{Cref}})] \\ &= \lim_{s \to 0}[s(\mathbf{K_{IS} L})^{-1} \times (\mathbf{I_n} - \mathbf{Q}) V_{Cref} \mathbf{1_n}) \\ &= \mathbf{0_n} \end{aligned} \tag{41}$$

It can be seen that $\mathbf{S^{ss}}$ is the eigenvector of $\mathbf{Q}$ associated with the eigenvalue zero, which ensures the state variable S_i consensus. [27] According to the property of the Laplace matrix [28], the final consensus values can be obtained as

$$\mathbf{S^{ss}} = S^{ss} \mathbf{1_n} \tag{42}$$

where S^{ss} is the positive real value to which all the state variables S_i will be converged.

Equations (33) and (42) shows that the average bus voltage and battery SOC will all reach consensus in the steady state.

5. Simulation Results

Simulations are performed in the MATLAB/Simulink (2017a, MathWorks, Natick, MA, USA) to validate the performance of the proposed distributed cooperative control for the multiple DCESs. Table 1 reports the parameters of the study case utilized for simulation. To simplify the controller design, the communication weights are set to one. To verify the proposed distributed cooperative control, the resistance combinations of the CL and the NCL for the four DCESs are different from each other. Moreover, the droop coefficients of the four DCESs are also different according to the different output power.

Table 1. Parameters for simulations.

Parameter		Values
Input Voltage (V_{IN})		50 V
Global voltage reference of CL (V_{Cref})		120 V
Communication weights (a_{ij})		1
Resistance of the CL (R_C)	R_{C1}, R_{C3}	120 Ω
	R_{C2}, R_{C4}	150 Ω
Resistance of the NCL (R_{NC})	R_{NC1}, R_{NC2}	40 Ω
	R_{NC3}, R_{NC4}	60 Ω
Initial droop coefficient (R_{d0})	R_{d0}	1

The performance of the primary control and distributed cooperative control is discussed in detail as follows including the steady state, converter failure, communication weight variation, and load variation.

5.1. Steady State

This case validates the distributed cooperative control when the DCESs are at a steady state, by comparing the performance of the droop control and consensus control. The simulation results of the primary control and distributed cooperative control of this multiple DCESs are shown in Figures 8 and 9, respectively.

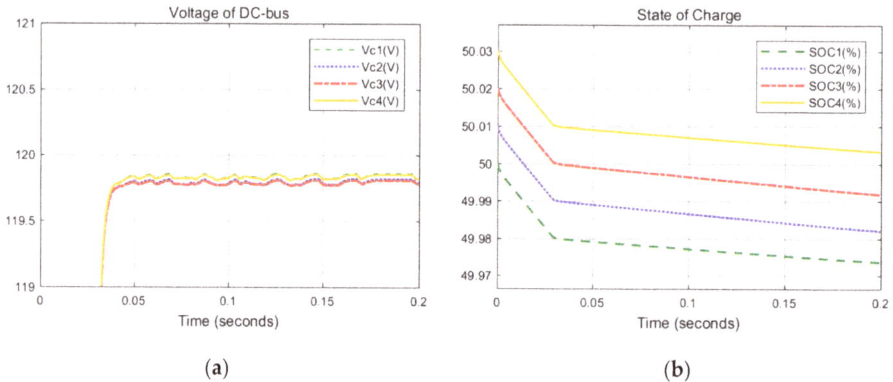

(**a**)　　　　　　(**b**)

Figure 8. Only with the primary control of multiple DCESs: (**a**) DC-bus voltage; (**b**) battery SOC.

In Figure 8, with only the primary control, the CL DC bus voltage is regulated to 119.2 V but deviates from the reference value of 120 V because of the droop gains and power allocation. The SOCs of the four batteries, whose initial value SOC_0 are different from each other, are still in difference after the discharging. In Figure 9, the distributed cooperative control is simulated. It can be seen that the CL DC bus voltages between the four DCESs are in consensus and the average value is regulated to the rated value, owing to the secondary voltage control. However, the CL DC bus voltage of the individual DCES may differ from the rated value. Moreover, the SOCs eventually converge toward consensus thanks to the fact that the secondary SOC control forces the batteries with a high SOC to be discharged faster and meanwhile the batteries with a low SOC to be charged faster.

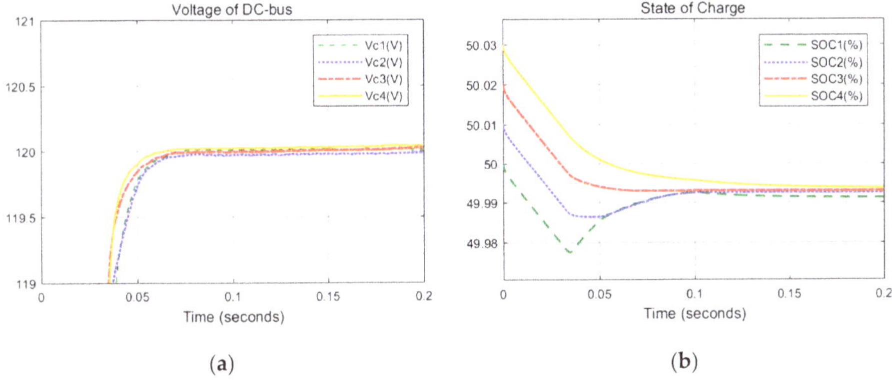

Figure 9. Distributed cooperative control of multiple DCESs: (**a**) DC-bus voltage; (**b**) battery SOC.

5.2. Converter Failure

This case illustrates the performance of the proposed control in the case of one converter failure. One DCES failure means that the output power of the DCES is zero. This happens when the DCES is shorted or broken, such as the short-circuit of the power switches and the open-circuit of the three-winding transformers. It is studied in this case to analyze the emergency performance of the proposed distributed cooperative control compared to the primary control. In this case, the output power of DCES$_1$ reduces to zero at 0.1 s. The remaining DCESs carry out the power sharing by exchanging the SOC information with neighbors. In Figure 10, the four DC-bus voltages all deviate much from the rated value after the failure. In Figure 11, although the bus voltage of the DCESs cannot be regulated to a common value, with the distributed cooperative control, the average value of the remaining three DC bus voltages are regulated close to the rated value if ignoring the small DC-bus voltage drop of DCES$_1$. Moreover, the control can drive the four battery SOCs to consensus and balance.

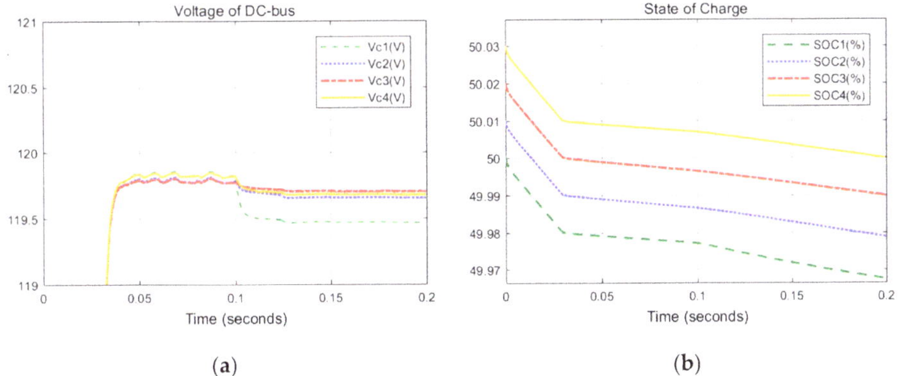

Figure 10. Only with the primary control when failure at 0.1 s of DCES1: (**a**) DC-bus voltage; (**b**) battery SOC.

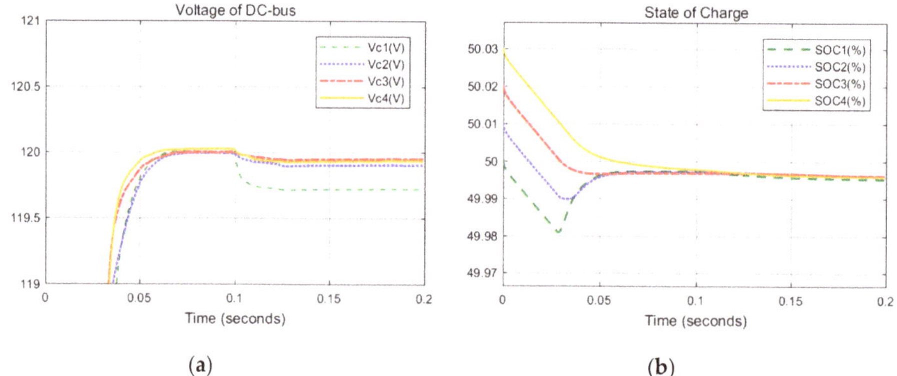

Figure 11. Distributed cooperative control when failure at 0.1 s of DCES1: (**a**) DC-bus voltage; (**b**) battery SOC.

5.3. Communication Weight Variation

In this case, the effect of the communication weight on the speed of the system convergence is investigated. It can be seen from Equations (13), (14), and (18) that the communication weight a_{ij} affects the correction term δ_{1i} and δ_{2i}, which will affect the voltage reference. To simplify the controller design, in Figures 9 and 12, the communication weights a_{ij} are set to one for normal impact levels and five for high impact levels, respectively.

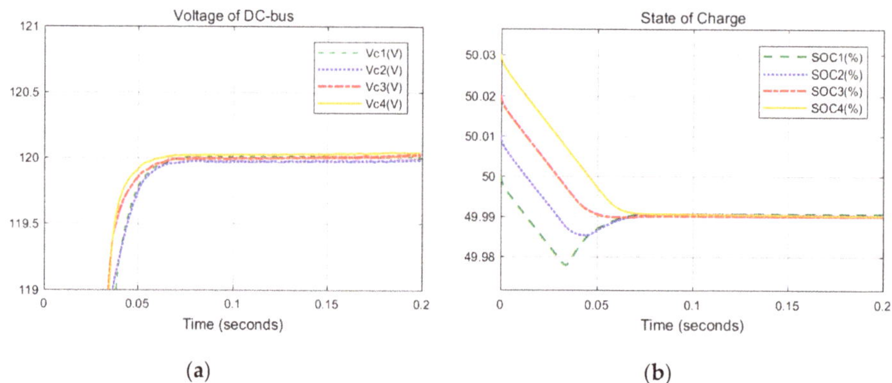

Figure 12. Distributed cooperative control when communication weights $a_{ij} = 5$: (**a**) DC-bus voltage; (**b**) battery SOC.

It can be seen in Figure 12a that the average CL DC-bus voltages reach the rated value for about 0.07 s regardless of the value of communication weight a_{ij}. Therefore, the communication weight has a weak impact on the convergence speed of the CL DC-bus voltage. However, in Figure 12b, the SOCs reach the consensus state faster with a larger a_{ij}. It can be concluded that a larger communication weight contributes to a faster convergence speed of SOC.

5.4. Load Variation

This section discusses the performance of the proposed distributed cooperative control compared to the primary control in the case of one load variation. The CL of DCES$_1$ changes from 120 Ω to 80 Ω at 0.1 s.

It can be seen in Figure 13 that the four DC-bus voltages deviated from the rated value especially after the load variation at 0.1 s. Moreover, the four SOCs are still different from each other. However, in Figure 14, with the distributed cooperative control, the four DC-bus voltages are in consensus, the average bus voltage can be regulated close to the rated value and four battery SOCs reach the consensus state although the load variation is at 0.1 s. This study indicates the good performance of the proposed control method when the CL load varies.

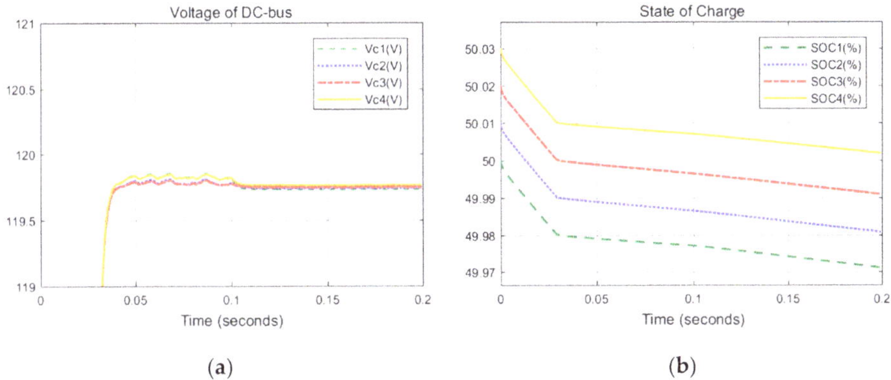

Figure 13. Primary control when the load variation is at 0.1 s of $DCES_1$: (**a**) DC-bus voltage; (**b**) battery SOC.

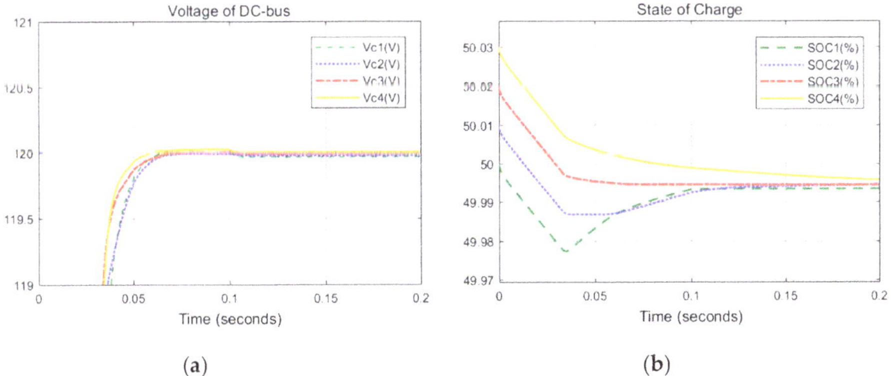

Figure 14. Distributed cooperative control when the load variation is at 0.1 s of $DCES_1$: (**a**) DC-bus voltage; (**b**) battery SOC.

6. Conclusions

A distributed cooperative control is proposed for multiple DC electric springs with the recently proposed topology in this paper. The novel DCES, which is composed of the DC/DC three port converter, bi-directional buck-boost converter, and battery system, can realize electrical isolation between the CL and the NCL and can also realize the traditional paralleled connection way of the electrical loads. The proposed distributed cooperative control composed of the primary control and the secondary control, can be used to realize the local voltage stability, power allocation of multiple DCESs and the consensus of DC-bus voltages and battery SOCs among multiple DCESs. The primary control is made up of the phased-shift control, decoupling control, adaptive droop control, and charging/discharging control. Furthermore, the secondary control consists of the voltage control and SOC control. The proposed distributed cooperative control applied in multiple DCESs with the recently proposed topologies has been validated by the simulation results.

Author Contributions: Conceptualization, D.Z., Q.W. and M.C.; methodology, Q.W.; software, D.Z.; validation, D.Z., Q.W.; formal analysis, F.D.; investigation, D.Z.; resources, Q.W.; data curation, Q.W.; writing—original draft preparation, D.Z.; writing—review and editing, D.Z., Q.W., G.B.; visualization, F.D.; supervision, M.C.; project administration, Q.W.; funding acquisition, Q.W.

Funding: This work was supported by the Natural Science Foundation of Jiangsu Province under project BK20170675, and by the National Natural Science Foundation of China under project 51877040.

Conflicts of Interest: The authors declare no conflict of interest.

References

1. Hui, S.Y.; Lee, C.K.; Wu, F.F. Electric springs—A new smart grid technology. *IEEE Trans. Smart Grid* **2012**, *3*, 1552–1561. [CrossRef]

2. Wang, Q.; Cheng, M.; Jiang, Y.; Zuo, W.; Buja, G. A Simple Active and Reactive Power Control for Applications of Single-Phase Electric Springs. *IEEE Trans. Ind. Electron.* **2018**, *65*, 6291–6300. [CrossRef]

3. Wang, Q.; Cheng, M.; Chen, Z. Steady-state analysis of electric springs with a novel δ control. *IEEE Trans. Power Electron.* **2015**, *30*, 7159–7169. [CrossRef]

4. Mok, K.; Wang, M.; Tan, S.; Hui, S.Y.R. DC electric springs—A technology for stabilizing DC power distribution systems. *IEEE Trans. Power Electron.* **2017**, *32*, 1088–1105. [CrossRef]

5. Wang, Q.; Cheng, M.; Jiang, Y.; Chen, Z.; Deng, F.; Wang, Z. DC electric springs with DC/DC converters. In Proceedings of the IEEE International Power Electronics and Motion Control Conference, Hefei, China, 28 September 2016; pp. 3268–3273.

6. Wang, M.; Yang, T.; Tan, S.; Hui, S.Y. Hybrid Electric Springs for Grid-Tied Power Control and Storage Reduction in AC Microgrids. *IEEE Trans. Power Electron.* **2019**, *34*, 3214–3225. [CrossRef]

7. Kakigano, H.; Miura, Y.; Ise, T. Distribution Voltage Control for DC Microgrids Using Fuzzy Control and Gain-Scheduling Technique. *IEEE Trans. Power Electron.* **2013**, *28*, 2246–2258. [CrossRef]

8. Ahmadi, R.; Ferdowsi, M. Improving the Performance of a Line Regulating Converter in a Converter-Dominated DC Microgrid System. *IEEE Trans. Smart Grid* **2014**, *5*, 2553–2563. [CrossRef]

9. Huang, P.; Liu, P.; Xiao, W.; El Moursi, M.S. A Novel Droop-Based Average Voltage Sharing Control Strategy for DC Microgrids. *IEEE Trans. Smart Grid* **2015**, *6*, 1096–1106. [CrossRef]

10. Rouzbehi, K.; Miranian, A.; Luna, A.; Rodriguez, P. DC Voltage Control and Power Sharing in Multiterminal DC Grids Based on Optimal DC Power Flow and Voltage-Droop Strategy. *IEEE J. Emerg. Sel. Top. Power Electron.* **2014**, *2*, 1171–1180. [CrossRef]

11. Mok, K.; Wang, M.; Tan, S.; Hui, S. DC electric springs—An emerging technology for DC grids. In Proceedings of the IEEE Applied Power Electronics Conference and Exposition, Charlotte, NC, USA, 15–19 March 2015; pp. 684–690.

12. Wang, Q.; Zha, D.; Deng, F.; Cheng, M.; Buja, G. A topology of DC electric springs for DC household applications. *IET Power Electron.* **2019**, *12*, 1241–1248. [CrossRef]

13. Phattanasak, M.; Gavagsaz-Ghoachani, R.; Martin, J.; Nahid-Mobarakeh, B.; Pierfederici, S.; Davat, B. Control of a hybrid energy source comprising a fuel cell and two storage devices using isolated three-port bidirectional DC–DC converters. *IEEE Trans. Ind. Appl.* **2015**, *51*, 491–497. [CrossRef]

14. Bahrami, H.; Farhangi, S.; Iman-Eini, H.; Adib, E. A new interleaved coupled-inductor nonisolated soft-switching bidirectional DC/DC converter with high voltage gain ratio. *IEEE Trans. Ind. Electron.* **2018**, *65*, 5529–5538. [CrossRef]

15. Xu, Y.; Zhang, W.; Hug, G.; Kar, S.; Li, Z. Cooperative control of distributed energy storage systems in a microgrid. *IEEE Trans. Smart Grid* **2015**, *6*, 238–248. [CrossRef]

16. Mokhtari, G.; Nourbakhsh, G.; Ghosh, A. Smart coordination of energy storage units (ESUs) for voltage and loading management in distribution networks. *IEEE Trans. Power Syst.* **2013**, *28*, 4812–4820. [CrossRef]

17. Zhang, X.; Dong, M.; Ou, J. A distributed cooperative control strategy based on consensus algorithm in DC microgrid. In Proceedings of the IEEE Conference on Industrial Electronics and Applications, Wuhan, China, 31 May–2 June 2018; pp. 243–248.

18. Olfati-Saber, R.; Fax, J.A.; Murray, R.M. Consensus and cooperation in networked multi-agent systems. *Proc. IEEE USA* **2007**, *95*, 215–233. [CrossRef]
19. Golsorkhi, M.S.; Shafiee, Q.; Lu, D.D.; Guerrero, J.M. A distributed control framework for integrated photovoltaic-battery-based islanded microgrids. *IEEE Trans. Smart Grid* **2017**, *8*, 2837–2848. [CrossRef]
20. Chen, X.; Shi, M.; Sun, H.; Li, Y.; He, H. Distributed cooperative control and stability analysis of multiple DC electric springs in a dc microgrid. *IEEE Trans. Ind. Electron.* **2018**, *65*, 5611–5622. [CrossRef]
21. Zhang, J.; Wu, H.; Qin, X.; Xing, Y. PWM plus secondary-side phase-shift controlled soft-switching full-bridge three-port converter for renewable power systems. *IEEE Trans. Ind. Electron.* **2015**, *62*, 7061–7072. [CrossRef]
22. Samir, H.; Subhashish, B.; Chandan, C. A novel control principle for a high frequency transformer based multiport converter for integration of renewable energy sources. In Proceedings of the IEEE Industrial Electronics Society. Annual. Conference, Vienna, Austria, 10–13 November 2013; pp. 7984–7989.
23. Wang, W.; Wang, P.; Ma, T.; Liu, H.; Wu, H. A simple decoupling control method for isolated three-port bidirectional converter in low-voltage DC microgrids. In Proceedings of the IEEE Energy Conversion Congress and Exposition, Montreal, QC, Canada, 20–24 September 2015; pp. 3192–3196.
24. Nasirian, V.; Davoudi, A.; Lewis, F.L.; Guerrero, J.M. Distributed adaptive droop control for dc distribution systems. *IEEE Trans. Energy Convers.* **2014**, *29*, 944–956. [CrossRef]
25. Kim, S.Y.; Song, H.; Nam, K. Idling port isolation control of three-port bidirectional converter for EVs. *IEEE Trans. Power Electron.* **2012**, *27*, 2495–2506. [CrossRef]
26. Nasirian, V.; Moayedi, S.; Davoudi, A.; Lewis, F.L. Distributed cooperative control of DC microgrids. *IEEE Trans. Power Electron.* **2015**, *30*, 2288–2303. [CrossRef]
27. Chen, X.; Shi, M.; Zhou, J.; Chen, Y.; Zuo, W.; Wen, J.; He, H. Distributed cooperative control of multiple hybrid energy storage systems in a DC microgrid using consensus protocol. *IEEE Trans. Ind. Electron.* **2019**, 1. [CrossRef]
28. Olfati-Saber, R.; Murray, R.M. Consensus problems in networks of agents with switching topology and time-delays. *IEEE Trans. Autom. Control* **2004**, *49*, 1520–1533. [CrossRef]

Article

Hierarchical Control with Fast Primary Control for Multiple Single-Phase Electric Springs

Qingsong Wang [1], Wujian Zuo [1,2], Ming Cheng [1,*], Fujin Deng [1] and Giuseppe Buja [3]

[1] School of Electrical Engineering, Southeast University, Nanjing 210096, China; qswang@seu.edu.cn (Q.W.); 220162254@seu.edu.cn (W.Z.); fdeng@seu.edu.cn (F.D.)
[2] State Grid Huzhou Power Supply Company, Huzhou 313000, China
[3] Department of Industrial Engineering, University of Padova, 35131 Padova, Italy; giuseppe.buja@unipd.it
* Correspondence: mcheng@seu.edu.cn; Tel.: +86-25-8279-4152

Received: 22 July 2019; Accepted: 10 September 2019; Published: 12 September 2019

Abstract: An electric spring (ES) is a new power compensation device which becomes useful for the large-scale integration of renewable energy sources into the grid. However, when the grid contains two or more ESs connected at different nodes, a voltage drop occurs between the nodes due to the line impedance. Therefore, the voltage of the ES-stabilized nodes cannot be set at the same reference (e.g., 220 V), otherwise one or more ESs break away because of the voltage windup. In this paper, a hierarchical control of multiple ESs tied to a microgrid is proposed to account for the line voltage drop. The primary control relies on the power decoupling control; it is designed for islanded operation of the microgrid, which improves the dynamics of both the voltage and frequency regulation. The secondary control relies on the droop control; it is introduced to coordinate the operation of the ESs by modifying the reference voltage of each ES dynamically. Advantages and disadvantages of the proposed hierarchical control are analyzed together with the explanations of the algorithms developed for its realization. At last, effectiveness of the arranged control system is validated by simulations on the MATLAB/Simulink platform.

Keywords: Electric spring; hierarchical control; coordinated control; power decoupling control; droop control; microgrid

---

1. Introduction

Renewable energy sources (RESs) of solar and wind type are characterized by a non-predictable intermittency of power generation. When many RESs of these types contribute to the grid power, the uncertainty on their output power affects the power quality of the power system, with the occurrence of harmonics, voltage excursions, flickers, and even with the possible collapse of the power system [1,2]. An electric spring (ES) is a new power compensation device that provides a solution to the problem of grid power uncertainty. Indeed, it changes the traditional operation mode of the power system, whereby the load consumption determines the power generation, into a new one, whereby the automatic matching between power generation and load demand is implemented. Therefore, ES represents an effective solution to the large grid power fluctuations expected when RESs of solar/ wind type penetrate the grid on a large scale [3–6].

The keystone of the ES approach is to divide the loads of a user into two groups with different requirements: one is critical load (CL), and the other one is non-critical load (NCL). CL requires highly stabilized supply voltage and/or uninterrupted supply to work correctly, and it may operate in frequent situations at nearly the rated power; examples of CL are machine rooms and medical equipment. NCL allows its supply voltage -and, hence, its absorbed power- to vary within a larger range, and can even be cut off for a certain period of time when the total power supply is not enough to ensure the required voltage for the CL supply; examples of NCL are water heater and other heating equipment. Both CL and NCL are connected at the same grid node, called the point of common coupling (PCC), but in a different way: CL in a direct way while NCL through the interposition of ES.

An ES is made of a voltage direct current (DC) source, a voltage source inverter (VSI) and an alternating current (AC) output inductor-capacitor pair; the capacitor, placed between PCC and NCL, constitutes the ES output whilst the inductor plays the role of filtering the VSI output current. By suitably controlling the capacitor voltage, the NCL voltage is adjusted such that the grid-incoming power fluctuations are transferred to the NCL so that CL is supplied at the required voltage [3]. By this reason, the branch constituted by the capacitor and NCL is called the smart load (SL).

Two basic ES versions exist. One version utilizes a capacitor as voltage DC source, and the ES exchanges only reactive power with the SL branch to stabilize the CL voltage; this version is commonly referred to as ES-1. The other version utilizes a bidirectional DC source like a battery as voltage DC source, and the ES exchanges both active and reactive power with the SL branch to stabilize the CL voltage; this version is commonly referred to as ES-2. Besides stabilizing the CL voltage, ES-2s have the capabilities of executing other tasks such as the correction of the power factor (PF) of the user [7] or the suppression of the frequency [8] and voltage [9] excursions of the power system. Due to their capabilities, hereafter only the ES-2s are considered.

Various strategies have been developed for the ES control. The δcontrol, proposed in [10], instantaneously adjusts the phase angle δbetween the PCC and line voltages to regulate the magnitude of the CL voltage and, at the same time, to keep the user PF compliant with the standards. The radial-chordal control strategy, proposed in [11], decomposes the ES output voltage into its chordal and radial components. Then it adjusts the chordal component to regulate the magnitude of the CL voltage ad the radial component to control the power angle of the SL. The active and reactive power control, proposed in [12], simplifies the ES control by using independent loops to adjust the CL voltage and the active power exchanged by the ES. Regarding the strategies for the control of multiple ESs, the droop control, presented in [13], manipulates the modulation index of the ES-embedded VSI to regulate the CL voltage at each node, which makes the solution appropriate for the ES-1 version. The consensus control, presented in [14,15], processes the voltage information coming from the adjacent ESs to calculate the reference voltage of the local ES; clearly, this control needs communication means to work.

There are other studies related to the control of multiple ESs. The simplified ES model established in [16] is intended to the simulation of a large-scale system endowed with the ESs. The modular dynamic model of the ES established in [17] is tailored for the integration of sophisticated control algorithms, reducing the order of the model by help of experimental measurements. In [18], the distributed voltage control attained with multiple ESs in a power system is compared to the single node voltage control done with the Static Synchronous Compensator (STATCOM); from the comparison executed under different voltage excursion situations, it emerges that the ESs provide a better voltage regulation than the STATCOM. The effect that the ES operation combined with an energy management strategy exerts on the suppression of the voltage and frequency excursions in microgrids is examined in [19]. In [20], hybrid ESs for grid-tied power control and storage reduction in AC microgrids is discussed. In [21], a method to stabilize a set of multiple ESs is presented, based on the small signal modeling of their behavior.

As a new paradigm of load demand management enabled by a power compensation device, the ES has outstanding advantages in adapting to the future distributed system when compared with the traditional centralized devices of reactive power compensation. However, the sizing power of a single ES is often limited. For the stabilization of the loads of a power system to take place, the joint effort of multiple ESs is requested, as exemplified in Figure 1. When multiple ESs are tied to different nodes of a power system, the node voltages are unable to reach the nominal value simultaneously (e.g., 220 V) due to the inherent voltage drop of the transmission line. Therefore, the adoption of an overall control strategy of the ESs becomes necessary to appropriately modify the local reference voltage of the ESs during the transients. The role of the strategy is to provide for a coordinate operation of the ESs while maintaining stabilized voltage across the associated CLs. Some papers have shown that the popular droop control is a useful tool on which to build up the strategy [22–26].

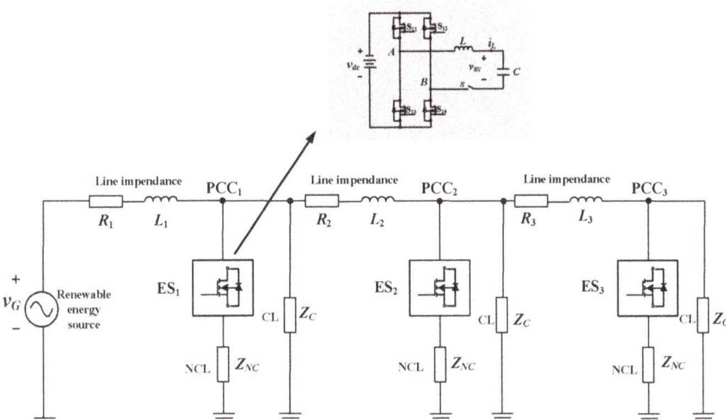

Figure 1. Microgrid with three ES-2 (electric springs (ES) exchanging both active and reactive power with the smart load (SL) branch to stabilize the critical load (CL) voltage and frequency).

The existing popular hierarchical control is well illustrated in [25] which is derived from the international standard for the integration of enterprise and control systems (ISA-95) and electrical dispatching standards to endow smartness and flexibility to Microgrids (MGs). With such control, the MGs are able to work at both islanded or stiff-source-connected modes, as well as to achieve a seamless transfer from one mode to another. Besides, multiple MG clusters can be performed to constitute a smart grid. Moreover, the system becomes more flexible and expandable, and consequently, more and more MGs could be integrated without changing the local hierarchical control system associated to each MG. Compared to the hierarchical control in [25], traditional droop control has many disadvantages which need to be improved. For instance, it is not suitable when the paralleled system must share non-linear loads due to harmonic currents, frequency deviation, accuracy sharing between active and reactive power under islanded mode. The hierarchical control proposed in [25] consists of three levels: (1) the primary control is based on the droop control; (2) the secondary control allows the restoration of the deviations obtained by the primary control; and (3) the tertiary control manages the power flow between the MG and the external electrical distribution system. Although it has so many advantages, it is complicated and hard to build prototype in a laboratory with limited resources and space. Considering the simple structure of ES system, and in order to avoid the disadvantages of the traditional droop control and to take the advantage of the hierarchical control in [25], a simplified hierarchical control only containing the droop control and the inner control loops is adopted in this paper. Another contribution is that it is proposed that the power decoupling control with fast dynamic responses is adopted at the ES level. In details, the hierarchical control is structured into two levels: primary and secondary. Compared to the existing primary control solutions, a novel scheme is developed that enhances both the dynamics and the robustness of the ES operation. After discussing the inconveniences of multiple ESs working independently, their coordinate operation is accomplished by the secondary control; besides modifying the CL reference voltage at the different nodes by using droop characteristics, it is designed to cope with the frequency excursions that often arise in the islanded operation of a microgrid. Finally, the effectiveness of the novel ES control scheme as well as of the proposed hierarchical control is verified by simulation.

The organization of the paper is as follows. Section 2 shortly reviews topology and model of an ES-2. Section 3 illustrates the proposed coordinated control for multiple ESs tied to a microgrid and explains how the control steers the ESs. Section 4 presents and discusses the results of some significant simulations carried out on a set of three ESs governed with the proposed control. Finally, Section 5 concludes the paper.

2. Electric Spring (ES-2) Operating Principles

2.1. ES-2 Topology

The circuit diagram of a single-phase ES-2 is shown in Figure 2, enclosed within the dashed line. The ES is built up around an VSI whose output voltage, designated with v_i, is proportional to V_{dc} as well as to the modulation index of VSI. The control of the ES, not shown in the figure, acts on the VSI through the delivery of the gating signals for its switches and is aimed at imposing an ES output voltage with the required values for its magnitude and phase.

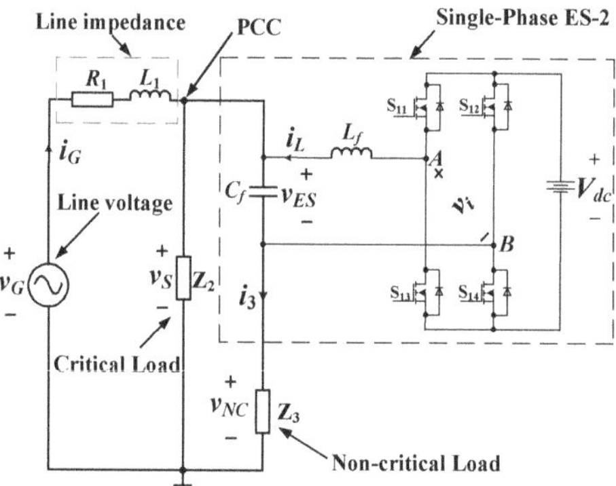

Figure 2. ES-2 circuit diagram.

In Figure 2, R_1 and L_1 denote the impedance of transmission line, and Z_2 and Z_3 denote the CL and the NCL. Moreover, v_G is the equivalent voltage of the system generators. In order to make it simple and consistent with existing definitions, it is also called line voltage here. v_{ES} represent the the ES output voltage. i_G represents the current entering into the PCC, and i_L and i_3 represent the currents at the VSI output and through the NCL, respectively. Lastly, v_S represents the PCC voltage that coincides with the voltage across the CL.

2.2. ES-2 Model

Although the CL and the NCL can be loads of resistive, capacitive or inductive type, the CL and the NCL are here taken to be of the purely resistive type to make the ES modeling simpler. The state-space equations of ES are

$$\begin{cases} \dot{x} = Ax + Bu \\ y = Cx \end{cases} \tag{1}$$

where,
$$x = \begin{bmatrix} i_L \\ v_{ES} \\ i_G \end{bmatrix}, \quad u = \begin{bmatrix} v_G \\ v_i \end{bmatrix}, \quad y = [v_s],$$

$$A = \begin{bmatrix} 0 & -\dfrac{1}{L_f} & 0 \\ \dfrac{1}{C_f} & -\dfrac{1}{C_f(R_2+R_3)} & \dfrac{R_2}{C_f(R_2+R_3)} \\ 0 & -\dfrac{R_2}{L_1(R_2+R_3)} & -\dfrac{R_1R_2+R_1R_3+R_2R_3}{L_1(R_2+R_3)} \end{bmatrix}$$

$$\begin{bmatrix} 0 & -\dfrac{1}{L_f} & 0 \\[2mm] \dfrac{1}{C_f} & -\dfrac{1}{C_f(R_2+R_3)} & \dfrac{R_2}{C_f(R_2+R_3)} \\[2mm] 0 & -\dfrac{R_2}{L_1(R_2+R_3)} & -\dfrac{R_1R_2+R_1R_3+R_2R_3}{L_1(R_2+R_3)} \end{bmatrix} A = \begin{bmatrix} 0 & -\dfrac{1}{L_f} & 0 \\[2mm] \dfrac{1}{C_f} & -\dfrac{1}{C_f(R_2+R_3)} & \dfrac{R_2}{C_f(R_2+R_3)} \\[2mm] 0 & -\dfrac{R_2}{L_1(R_2+R_3)} & -\dfrac{R_1R_2+R_1R_3+R_2R_3}{L_1(R_2+R_3)} \end{bmatrix}, B = \begin{bmatrix} 0 & \dfrac{1}{L_1} \\[2mm] 0 & 0 \\[2mm] \dfrac{1}{L_1} & 0 \end{bmatrix},$$

and $C = \begin{bmatrix} 0 & \dfrac{R_2}{R_2+R_3} & \dfrac{R_2R_3}{R_2+R_3} \end{bmatrix}$.

The block scheme of the ES, derived from the above equations, is drawn in Figure 3. The voltage v_i is equal to $d(t)V_{dc}$, where $d(t)$ is the modulation signal. In modeling the ES, it is assumed that (i) the VSI modulation frequency is much higher than the fundamental frequency, (ii) the VSI dynamics are negligible, and (iii) the pair LC at the VSI output makes almost sinusoidal the quantities in the downstream circuit. It is also assumed that the amplitude of the DC voltage source is constant.

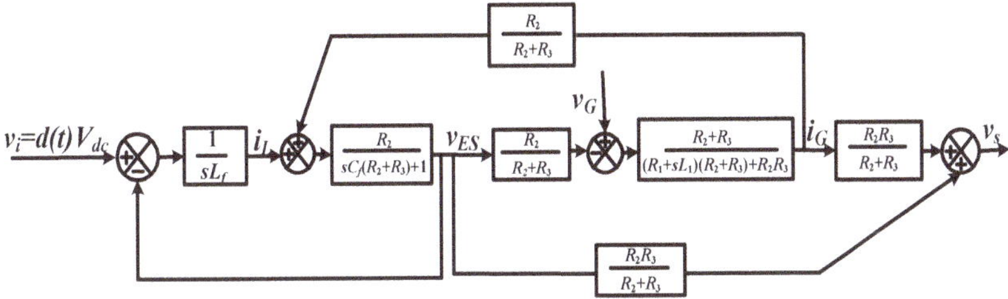

Figure 3. ES-2 block scheme.

3. The Proposed Control

3.1. ES-2 Control Scheme

Primary control, also named local voltage and frequency control, provides for the independent voltage and frequency control of each ES, which belong to the outer loops. After that, the inner loops exert upon two control actions, namely the decoupled control and filter capacitor voltage control, which can be seen by the scheme in Figure 4.

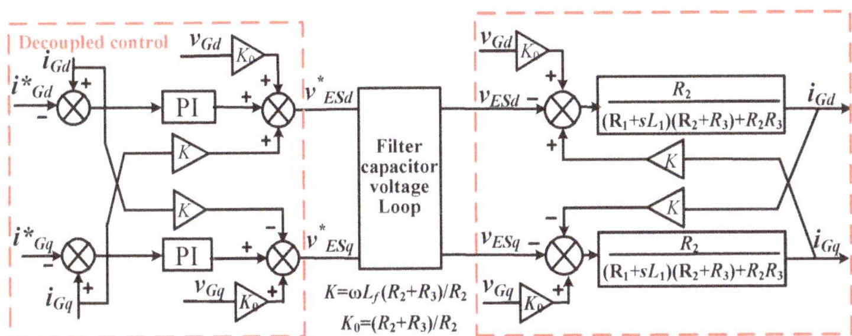

Figure 4. Current loops of primary control for ES-2.

In this subsection, a new type of decoupled power control is proposed for the ES. Compared to the existing power control of the ES [10], new functions are added. Specifically, a decoupling network as well as an inner current loop are added, which give the ES higher dynamic responses.

When the PCC voltage phasor v_s is selected as the reference phasor, the voltage and current at the PCC can be expressed as:

$$v_s(t) = \sqrt{2}V_s\cos(\omega t) \tag{2}$$

$$i_G(t) = \sqrt{2}I_G\cos(\omega t + \varnothing) \tag{3}$$

The active power (P) and reactive power (Q) at the PCC are:

$$P = Re(V_s I_G^*) = Re\big(\{v_{sd} + jv_{sq}\}\cdot\{i_{Gd} + ji_{Gq}\}\big) = v_{sd}i_{Gd} + v_{sq}i_{Gq} = v_{sd}i_{Gd} = V_s I_G \cos\varnothing \tag{4}$$

$$Q = Im(V_s I_G^*) = Im\big(\{v_{sd} + jv_{sq}\}\cdot\{i_{Gd} + ji_{Gq}\}\big) = v_{sd}i_{Gd} - v_{sq}i_{Gq} = -v_{sd}i_{Gd} = V_s I_G \sin\varnothing \tag{5}$$

where i_{Gq} and i_{Gq} represent the active and reactive components of the current i_G, and the last equalities hold in quasi-stationary conditions. Equations (4) and (5) show that, under the working hypothesis of a well stabilized PCC voltage, the decoupled power control is achieved by separately controlling the two current components. The relevant control scheme is drawn in Figure 4, where K and K_0 are the decoupling compensation term and the feedforward coefficient of the grid voltage, respectively, and the current loops are closed by means of Proportional Integral (PI) regulators.

The effectiveness of the scheme in Figure 4 confides in the accurate control of the ES output voltage rather than of the VSI output voltage. This goal is obtained by including in the scheme an auxiliary control loop, aimed at forcing the ES output voltage to accurately track the v_{ES} reference signals delivered by the decoupled control stage. The auxiliary control loop, designated with 'Filter capacitor voltage loop' in Figure 4, is explicated in Figure 5; it contains a delay block of $1.5T_s$ to account for the calculation and sampling delays in the processing of the control signals. The loop controller $G_{PR}(s)$ is a proportional resonant (PR) regulator with the following expression:

$$G_{PR}(s) = K_p + \frac{2K_r\omega_c s}{s^2 + 2\omega_c s + \omega_0^2} \tag{6}$$

where, K_p and K_r are the proportional and resonant parameters of the PR regulator. The PR regulator has the incomparable advantage of tracking closely a sinusoidally-shaped signal during its transients; furthermore, the amplitude gain of the PR regulator at resonant frequency can be set large enough to give the control loop an almost zero static error as well as a good anti-interference capability. Other quantities in Figure 5 are the VSI gain K_{PWM} and the block $G_2(s)$, whose transfer function is:

$$G_2(s) = \frac{(R_1+sL_1)(R_2+R_3)+R_2R_3}{sCf((R_1+sL_1)(R_2+R_3)+R_2R_3)+R_1+sL_1+R_2} = \frac{Z_0}{sCfZ_0+1} \tag{7}$$

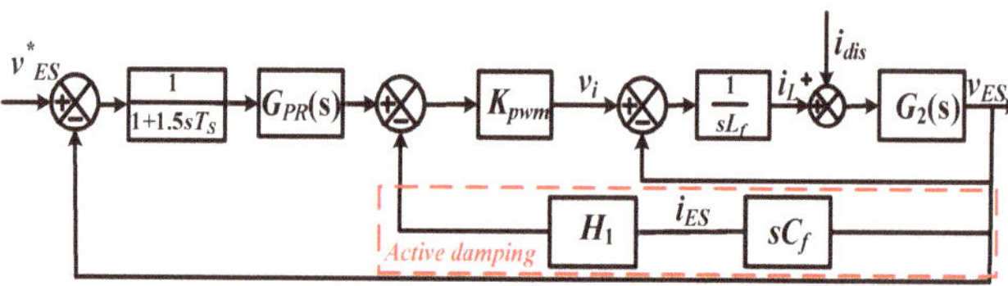

Figure 5. ES output voltage control.

Since the response of the auxiliary control loop exhibits a weakly damped behavior, an active damping is introduced in its basic scheme by means of a complementary function that increases the damping of the ES output voltage and, with it, the stability of the control. Such a function can be obtained by feeding back -into the forward path of the v_{ES} control- a signal representative of either the capacitor current or voltage or the inductance voltage, after an appropriate filtering. Irrespectively of the selected signal, its transduction has the inconvenience of necessitating one more sensor in the ES control hardware. However, the benefits obtained with the active damping function rewards for the additional hardware. In this paper, the feedback of the capacitor current is adopted as drawn in Figure 5, where sensing of the current is modeled by the time rate of the capacitor voltage.

3.2. The Proposed Hierarchical Control for Multiple ES-2

The current references in Figure 4 are delivered by PI regulators that close the control loops built up around the corresponding PCC voltage components. When multiple ESs are distributed along a microgrid as exemplified in Figure 1, it is no longer appropriate to use the same value (e.g., 220 V) for the voltage reference of each ES-tied node because the voltage drop inherent in the transmission line prevents the actual PCC voltage to match the reference. The consequence is that the integral term of the PI regulators of the voltage loops is subjected to the windup phenomenon. Let the accumulated voltage error be positive at a given node; the corresponding ES is forced to inject a useless active and reactive power into the SL, thus compromising its regulating action and affecting the safe operation of the other ESs. Therefore, different reference voltages must be set for ESs at different nodes, according to the droop characteristics of Figure 6.

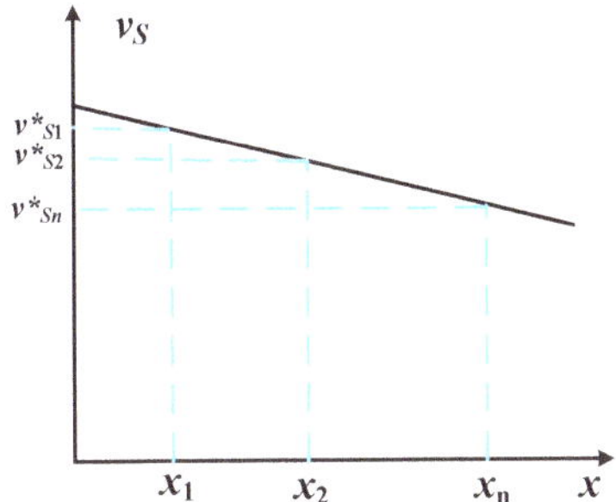

Figure 6. Droop characteristics for different node voltages.

The proposed hierarchical control determines the current components by help of a secondary control that embeds the frequency and voltage loops shown in Figure 7 within the dashed line. Both loops rely on the droop control technique. When the microgrid is running under the islanded situation, the active power flowing in the microgrid is mainly generated by the connected RESs. When an active power shortage occurs, the frequency of the microgrid declines to some extent. The tied ES compensates for the power shortage by the injection of the active power, thus improving the frequency stability of the microgrid. The stabilizing frequency action exerted by the ES allows setting of the frequency reference at the power system frequency (50Hz).

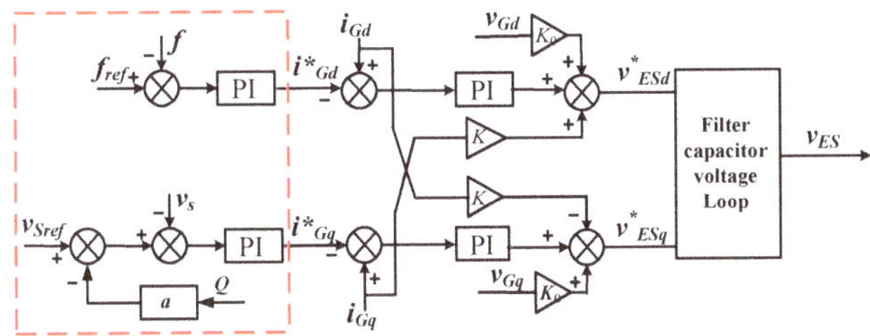

Figure 7. Proposed hierarchical control for ES-2.

In turn, the PCC voltage is subjected to a gradual downward trend, from which it has to recover. This task is supported by the voltage control loop. Since the reactive line power Q measured at PCC is directly correlated with the voltage drop of the microgrid line, the feedback value of the voltage loop is determined by multiplying Q by the droop coefficient, expressed as:

$$a = \cfrac{1}{Q_{max}\cfrac{2R_3}{V^2_S}} \quad a = \cfrac{1}{Q_{max}\cfrac{2R_3}{V^2_S}} \quad a = \cfrac{1}{Q_{max}\cfrac{2R_3}{V^2_S}} \quad a = \cfrac{1}{Q_{max}\cfrac{2R_3}{V^2_S}} \tag{8}$$

and by summing it to the PCC voltage. The difference between the reference voltage for the nodes and the feedback voltage is then processed by a PI regulator to give the reference of the reactive current component.

4. Simulations and Discussions

To validate the hierarchical control developed in the paper, the simulation code of multiple ESs tied to a microgrid has been worked out on the MATLAB/Simulink platform, as drawn in Figure 8. Parameters for the case study of an ES system is shown in Table 1. In the figure, switch S is set to activate or deactivate the load R from the system to realize the mismatch of the active power between the power source and the load, which will lead to the system frequency variation. The code includes a grid simulator and three ESs. The grid simulator is mainly used to emulate the voltage and frequency excursions during islanded situation. Each ES is packaged into a separate module so as to dispose of a code with plug-and-pull characteristics. The parameters of each module are set at the same value, which is convenient to enter the data for a microgrid with multiple ESs.

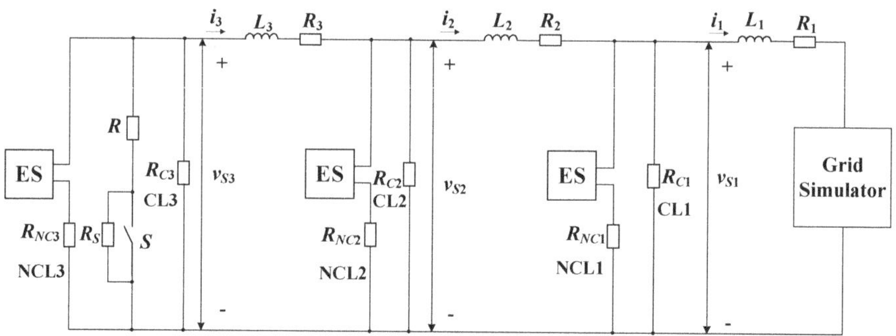

Figure 8. MATLAB/Simulink platform for microgrid with three ES-2.

Table 1. Parameters for the case study of an ES system.

Parameter	Values
Line 1 resistance (R_1)	0.3 Ω
Line 1 inductance (L_1)	5 mH
Line 2 resistance (R_2)	0.2 Ω
Line 2 inductance (L_2)	1.2 mH
Line 3 resistance (R_3)	0.2 Ω
Line 3 inductance (L_3)	1.2 mH
Critical load 1 (R_{C1})	180 Ω
Non-critical load 1 (R_{NC1})	40 Ω
Critical load 2 (R_{C2})	100 Ω
Non-critical load 2 (R_{NC2})	60 Ω
Critical load 3 (R_{C3})	100 Ω
Non-critical load 3 (R_{NC3})	60 Ω
filter Inductance (L)	2.3 mH
filter Capacitance (C)	26 μF
PCC voltage (V_S)	220 V
DC bus voltage (V_{dc})	400 V
Switching frequency (f)	20 kHz

4.1. Single ES Control

Proper operation of the microgrid with multiple ESs rests on the correct control of a single ES. In order to examine a single ES, L_2 and L_3 are cut out from the circuit in Figure 8 and only the first ES is activated. The correctness of the control is evaluated by testing the ES behavior under voltage and frequency excursions.

Figure 9 shows the simulation results of voltage control for a single ES. The figure contains three channels, namely the predefined Root Mean Square (RMS) value of the CL voltage, the measured RMS value of the CL voltage and the RMS value of the grid voltage designated as V_G. In the time interval from 0 to 0.5 s, the grid voltage is 209 V, the ES is deactivated, and the CL voltage is significantly less than the expected value of 220 V.

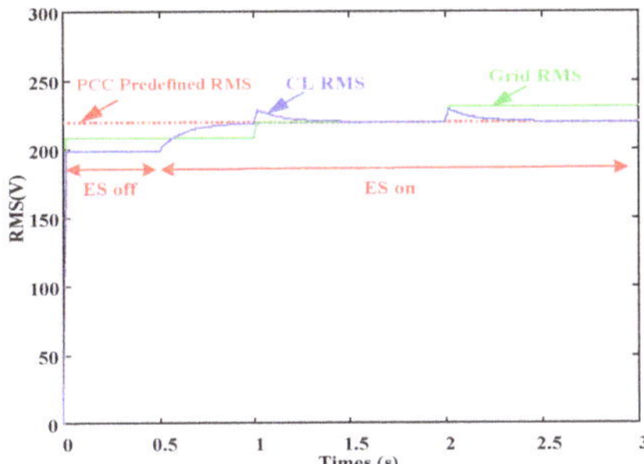

Figure 9. Simulation results of voltage control for a single ES.

From 0.5 s to 1 s, the ES starts to work, and the CL voltage reaches the predefined value of 220 V gradually. Between 1–2 s and 2–3 s, the RMS value of grid voltage is steeply increased to 220 V and then to 231 V, respectively. Under the ES voltage control, after a short and restrained transient, the CL voltage quickly stabilizes at the predefined value of 220 V, ensuring the power quality of the CL supply in agreement with the requirements.

Figure 10 shows the simulation results under frequency control of a single ES. The figure has two channels, namely the predefined and measured values of frequency. At the beginning of the simulation, ES is deactivated. Due to the lack of active power, the microgrid frequency is decreased to about 49.7 Hz, which is lower than the maximum frequency excursion of ±0.2 Hz stipulated in the national standards. At 0.5 s, the ES starts to work and quickly compensates for the active power lack of the microgrid by both reducing the active power consumption of the NCL and providing active power from the ES, thus recovering the microgrid frequency at the default value of 50 Hz. At 4 s, a switch is closed to emulate a further loading of the microgrid; its active power balance is broken again and produces a consequent fall of the microgrid frequency. The ES takes an immediate action and restores the frequency to 50 Hz at 5 s. The results above are evidence of the soundness of the designed frequency control for a single ES.

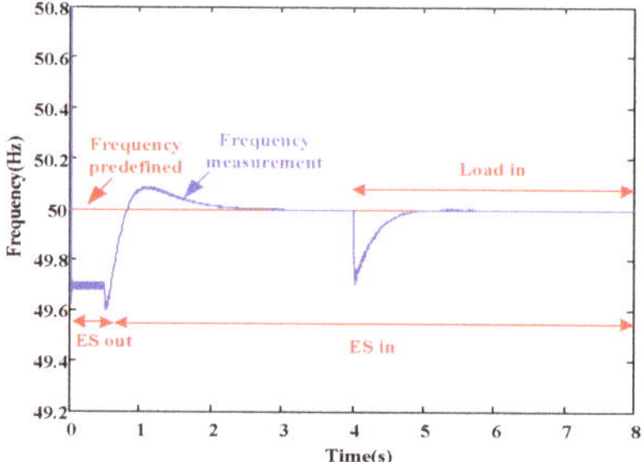

Figure 10. Simulation results of frequency control for a single ES.

4.2. Hierarchical Control of Multiple ESs

This subsection validates the effectiveness of the proposed hierarchical control by simulating the microgrid with three ESs of Figure 8 at first without any coordinated control and then by introducing the coordinated control. In the simulations, the predefined voltages at the nodes with ESs are set to 220 V.

Figure 11 reports the waveforms of the RMS value of the node voltages, and the active and reactive powers P and Q entering the nodes under the voltage control without the coordinated control. During the first 1 s-long time interval of the simulation, none of the three ESs is activated. Due to the line impedance, the CL voltages at the three PCCs greatly deviate from 220 V. At 1 s, the ES is activated. Under the action of the ESs, the CL voltages start to get closer to the predefined value of 220 V. However, as it emerges from the magnified waveforms in the central part of Figure 11(a), the three CL voltages are not able to reach the predefined value but settle on a different value. At 4 s and then at 7 s, two steeply increases of the microgrid voltage are emulated. Although the three CL voltages are close to 220 V within a certain range, they do not replicate exactly the preset value, exhibiting a steady-state error.

At 7 s, when the RMS value of the grid voltage is set to 228 V, the integral term of the voltage regulator goes saturated at 7.45 s due to the persisting voltage error, which results in a larger voltage deviation of V_{S1}. Although V_{S2} and V_{S3} are regulated at the predefined value since the integral term of their voltage regulators do not saturate, the microgrid nodes with the tied ESs are unable, on the whole, to reach the expected steady-state operation. The simulation results confirm the shortcoming arising from the combined action of the voltage drop along the microgrid line and the gradual winding up of the integral term of the voltage regulators of some ESs. The latter ESs do no longer exert the regulating action of the node voltage, as outlined in Figure 11a, while causing a rise of the ES reactive power, as outlined in Figure 11b, which is limited by either control or device protections. This behavior is against the motivations for the usage of multiple ESs, which is intended for the existing ESs to share the microgrid fluctuations in a favorable manner.

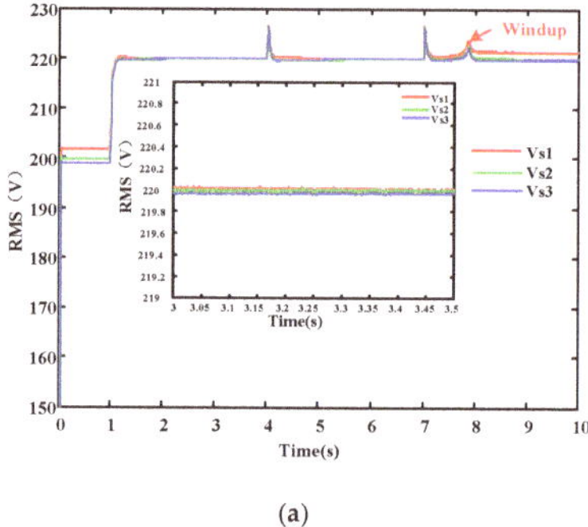

(a)

Figure 11. *Cont.*

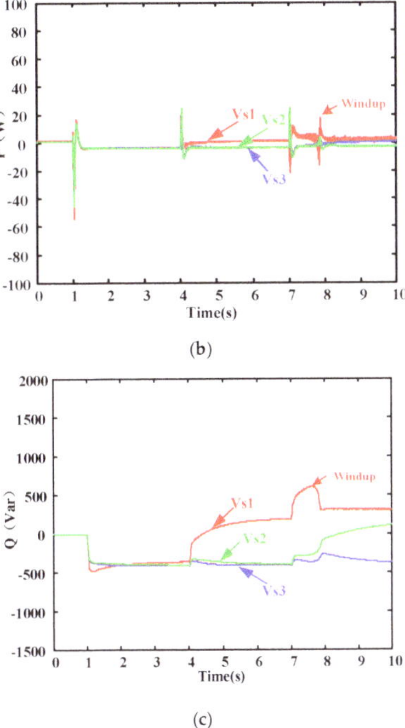

Figure 11. Simulation results of multiple ESs operating without coordinated control: (**a**) RMS value of point of common coupling (PCC) voltage; (**b**) active power; (**c**) reactive power.

As can be seen from the above section, it is far from enough for a single ES to ensure voltage deadness-free tracking, and it is necessary to have a coordinated control strategy to achieve reasonable setting of the reference values for multiple ESs. This section mainly carries out simulations under hierarchical control using droop control as the secondary control and power decoupling control with filter capacitor voltage control as the primary control for multiple ESs. Both voltage and frequency regulations are included.

Figure 12 shows the simulation results of reference voltage and measured voltage of ES when multiple ESs are distributed along the transmission line based on droop control. Within the first second during simulation, ES is deactivated at each PCC. In order to monitor the effectiveness of reference value change by each ES with the proposed droop control, all the reference values are initialized to 221 V. At 1 s, the ESs at the three PCC start to work and the reference voltage values of the ESs at three PCC are modified dynamically along with the sudden change of grid voltage. At the moments of 4 s and 7 s, when grid voltages are set to 220 V and 228 V respectively, it is seen that reference voltage value of each ES has been modified to new values and been tracked quickly.

Figure 13 shows the results of the three ESs under the action of the proposed control, showing the steady state and dynamic responses of the ESs under the case of grid voltage fluctuations. Figure 13a shows that the reference values of PCC voltages are different and tracked quickly with droop control. Figure 13b,c show the active and reactive power of each ES under the proposed control. It is seen that the ESs are three PCC operate at a dynamically stable state, which withstand the power fluctuation of the whole system in a harmony manner.

Finally, the frequency responses are verified with three sets of ESs by simulations, as shown in Figure 14. The figure contains two channels, namely the predefined value and the measured value of frequency. Within the first second, the ESs were not activated, and the system frequency was offset due to the imbalance between the active load and the active power supply, and thus the difference was about 0.05 Hz. At 1 s, the ESs start to work, and the active power sag of the whole system is compensated gradually due to the participation of multiple ESs, and the frequency of the system is kept stable at 50 Hz. At 5 s, the load demand in the system suddenly increases, leading to an increase in the power shortage of the system. At this time, the system frequency decreases to about 49.8 Hz, which violates the requirement that the maximum frequency deviation of the system should not exceed ±0.2 Hz. However, under the regulation of three ESs with the proposed control, the newly added active load of the system will achieve the purpose of sharing that partial active power are absorbed by the battery inside the ESs and meanwhile partial power fluctuations are passed to the NCLs. As a result, the system frequency is finally seen to stabilize at 50 Hz. The simulation verifies that the proposed control can enhance the stability of system frequency with the help of multiple ESs.

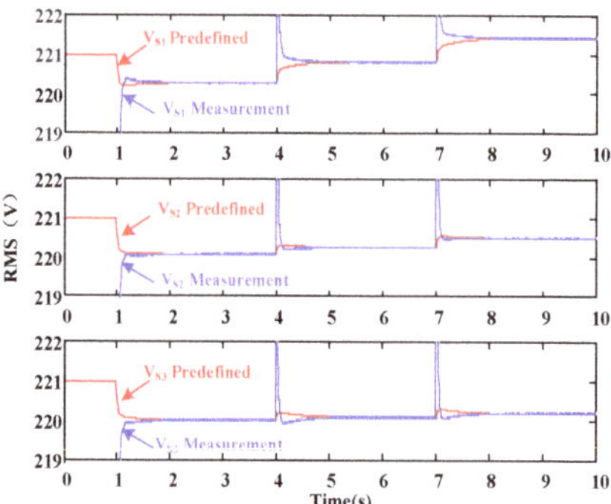

Figure 12. Reference versus measured voltages with the proposed control of multiple ESs.

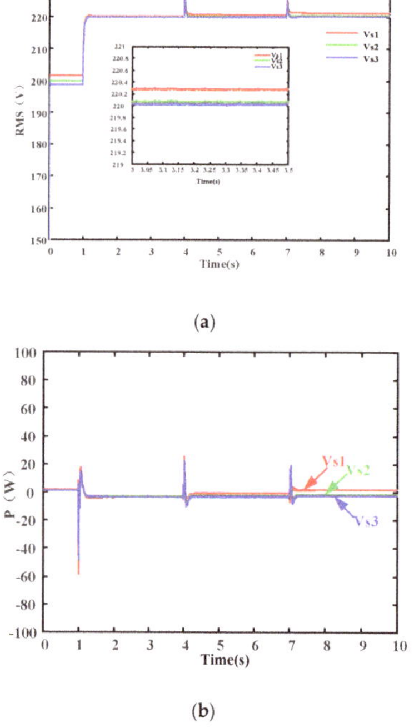

(a)

(b)

Figure 13. *Cont.*

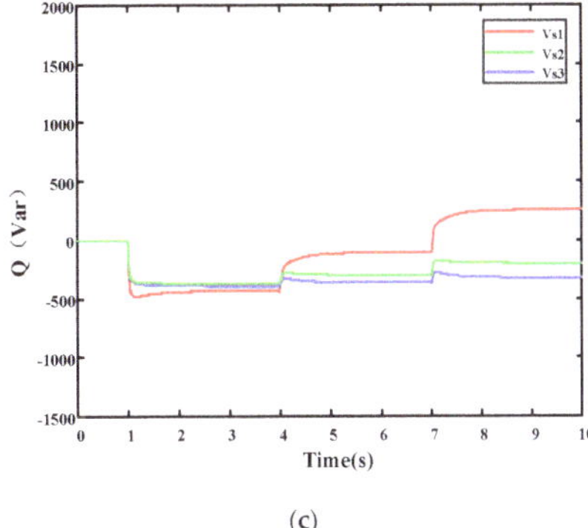

(c)

Figure 13. Simulation results of multiple ESs operating with the proposed control: (**a**) RMS value of PCC voltage; (**b**) active power; (**c**) reactive power.

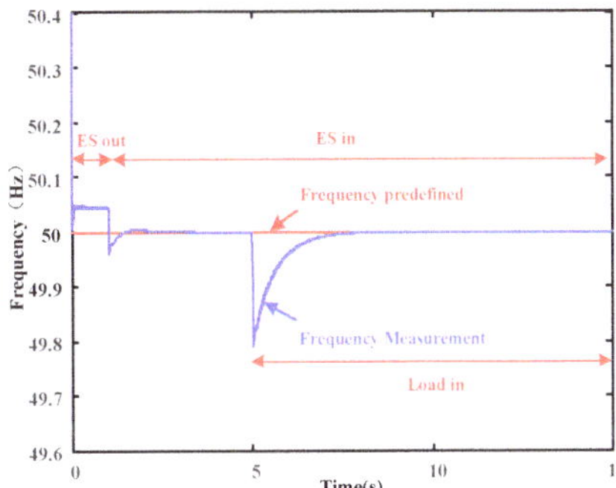

Figure 14. Simulation results of frequency responses with three ESs under the proposed control.

5. Conclusions

In this paper, a hierarchical control has been proposed for multiple ESs tied to a microgrid; it relies on the droop control as secondary control and on the power decoupling control as primary control. The issues of the existing solutions for operation of multiple ESs have been analyzed first, referring to the version-2 ES as the power compensation device. Proper setting of the reference value for the PCC voltages of the nodes with a tied ES has been done by the droop control technique, which has been instrumental in obtaining the coordinated operation of multiple ESs. In order to get fast and smooth ES dynamics, the performance of the power decoupling control has been improved by implementing an inner, closed-loop control of the ES output voltage, toughened with an active damping obtained by feedbacking the current flow in the ES output capacitor. To verify the effectiveness of the proposed hierarchical control together with the novel algorithms developed for its implementation, simulations are conducted to test the operation of single and multiple ESs. They have demonstrated that both node voltage and microgrid frequency are accurately and promptly regulated by the proposed control of multiple ESs with the proposed control. It should be noted that v_G is difficult to measure at PCC side due to the transmission line in between. Besides, the battery management system on the ES is not focused either in this paper. Both aspects will be part of future work.

Author Contributions: Conceptualization, Q.W. and W.Z.; methodology, Q.W.; software, W.Z.; validation, W.Z., Q.W.; formal analysis, F.D.; investigation, W.Z.; resources, Q.W.; data curation, Q.W.; writing—original draft preparation, W.Z., Q.W.; writing—review and editing, Q.W., G.B.; visualization, F.D.; supervision, M.C.; project administration, Q.W.; funding acquisition, Q.W.

Funding: This work was supported by the Natural Science Foundation of Jiangsu Province under project BK20170675, and by the National Natural Science Foundation of China under project 51877040.

Acknowledgments: In this section you can acknowledge any support given which is not covered by the author contribution or funding sections. This may include administrative and technical support, or donations in kind (e.g., materials used for experiments).

Conflicts of Interest: The authors declare no conflict of interest.

References

1. Cheng, M.; Zhu, Y. The state of the art of wind energy conversion systems and technologies: A review. *Energy Convers. Manag.* **2014**, *88*, 332–347. [CrossRef]
2. Hu, H.; Shao, Y.; Tang, L.; Ma, J.; He, Z.; Gao, S. Overview of harmonic and resonance in railway electrification systems. *IEEE Trans. Ind. Appl.* **2018**, *54*, 5227–5245. [CrossRef]
3. Hui, S.Y.; Lee, C.K.; Wu, F.F. Electric springs—A new smart grid technology. *IEEE Trans. Smart Grid* **2012**, *3*, 1552–1561. [CrossRef]
4. Lee, C.K.; Chaudhuri, B.; Hui, S.Y.R. Hardware and control implementation of electric springs for stabilizing future smart grid with intermittent renewable energy sources. *IEEE J. Emerg. Sel. Top. Power Electron.* **2013**, *1*, 18–27. [CrossRef]
5. Wang, Q.; Zha, D.; Deng, F.; Cheng, M.; Buja, G. A topology of DC electric springs for DC household applications. *IET Power Electron.* **2019**, *12*, 1241–1248. [CrossRef]
6. Tan, S.C.; Lee, C.K.; Hui, S.Y.R. General steady-state analysis and control principle of electric springs with active and reactive power compensations. *IEEE Trans. Power Electron.* **2013**, *28*, 3958–3969. [CrossRef]
7. Soni, J.; Panda, S.K. Electric spring for voltage and power stability and power factor correction. *IEEE Trans. Ind. Appl.* **2017**, *53*, 3871–3879. [CrossRef]
8. Chen, X.; Hou, Y.; Tan, S.C.; Lee, C.K.; Hui, S.Y.R. Mitigating voltage and frequency fluctuation in microgrids using electric springs. *IEEE Trans. Smart Grid* **2015**, *6*, 508–515. [CrossRef]
9. Liang, L.; Hou, Y.; Hill, D.J.; Hui, S.Y.R. Enhancing resilience of microgrids with electric springs. *IEEE Trans. Smart Grid* **2018**, *9*, 2235–2247. [CrossRef]
10. Wang, Q.; Cheng, M.; Chen, Z. Steady-state analysis of electric springs with a novel δ control. *IEEE Trans. Power Electron.* **2015**, *30*, 7159–7169. [CrossRef]
11. Mok, K.T.; Tan, S.C.; Hui, S.Y.R. Decoupled power angle and voltage control of electric springs. *IEEE Trans. Power Electron.* **2016**, *31*, 1216–1229. [CrossRef]
12. Wang, Q.; Cheng, M.; Jiang, Y.; Zuo, W.; Buja, G. A Simple Active and Reactive Power Control for Applications of Single-Phase Electric Springs. *IEEE Trans. Ind. Electron.* **2018**, *65*, 6291–6300. [CrossRef]
13. Lee, C.K.; Chaudhuri, N.R.; Chaudhuri, B. Droop control of distributed electric springs for stabilizing future power grid. *IEEE Trans. Smart Grid* **2013**, *4*, 1558–1566. [CrossRef]
14. Chen, X.; Hou, Y.; Hui, S.Y.R. Distributed control of multiple electric springs for voltage control in Microgrid. *IEEE Trans. Smart Grid* **2017**, *8*, 1350–1359. [CrossRef]
15. Chen, J.; Yan, S.; Yang, T.; Tan, S.C.; Hui, S.Y.R. Practical evaluation of droop and consensus control of distributed electric springs for both voltage and frequency regulation in Microgrid. *IEEE Trans. Power Electron.* **2019**, *34*, 6947–6959. [CrossRef]
16. Chaudhuri, N.R.; Lee, C.K.; Chaudhuri, B.; Hui, S.Y.R. Dynamic modeling of electric springs. *IEEE Trans. Smart Grid* **2014**, *5*, 2450–2458. [CrossRef]
17. Yang, T.; Liu, T.; Chen, J.; Yan, S.; Hui, S.Y.R. Dynamic modular modeling of smart loads associated with electric springs and control. *IEEE Trans. Power Electron.* **2018**, *33*, 10071–10085. [CrossRef]
18. Luo, X.; Akhtar, Z.; Lee, C.K.; Chaudhuri, B.; Tan, S.C.; Hui, S.Y.R. Distributed voltage control with electric springs: Comparison with STATCOM. *IEEE Trans. Smart Grid* **2015**, *6*, 209–219. [CrossRef]
19. Yang, T.; Mok, K.T.; Tan, S.C.; Lee, C.K.; Hui, S.Y.R. Electric springs with coordinated battery management for reducing voltage and frequency fluctuations in microgrids. *IEEE Trans. Smart Grid* **2018**, *9*, 1943–1952. [CrossRef]
20. Wang, M.; Yang, T.; Tan, S.C.; Hui, S.Y.R. Hybrid electric springs for grid-tied power control and storage reduction in AC microgrids. *IEEE Trans. Power Electron.* **2019**, *34*, 3214–3225. [CrossRef]

21. Yang, Y.; Ho, S.S.; Tan, S.C.; Hui, S.Y.R. Small-signal model and stability of electric springs in power grids. *IEEE Trans. Smart Grid* **2018**, *9*, 857–865. [CrossRef]
22. Bevrani, H.; Shokoohi, S. An intelligent droop control for simultaneous voltage and frequency regulation in islanded microgrids. *IEEE Trans. Smart Grid* **2013**, *4*, 1505–1513. [CrossRef]
23. Liu, J.; Miura, Y.; Ise, T. Comparison of dynamic characteristics between virtual synchronous generator and droop control in inverter-based distributed generators. *IEEE Trans. Power Electron.* **2016**, *31*, 3600–3611. [CrossRef]
24. Mahmood, H.; Michaelson, D.; Jiang, J. Reactive power sharing in islanded microgrids using adaptive voltage droop control. *IEEE Trans. Smart Grid* **2015**, *6*, 3052–3060. [CrossRef]
25. Guerrero, J.M.; Vasquez, J.C.; Matas, J.; Vicuna, L.G.; Castilla, M. Hierarchical control of droop-controlled AC and DC microgrids—A general approach toward standardization. *IEEE Trans. Ind. Electron.* **2011**, *58*, 158–172. [CrossRef]
26. Tinajero, G.A.; Aldana, N.L.D.; Luna, A.C.; Ramirez, J.S.; Cruz, N.V.; Guerrero, J.M.; Vasquez, J.C. Extended-optimal-power-flow-based hierarchical control for islanded AC microgrids. *IEEE Trans. Power Electron.* **2019**, *34*, 840–848. [CrossRef]

Article

Photovoltaic Energy Yield Improvement in Two-Stage Solar Microinverters

Andrii Chub [1,2], **Dmitri Vinnikov** [1,*], **Serhii Stepenko** [1,3], **Elizaveta Liivik** [1,4] and **Frede Blaabjerg** [4]

[1] Department of Electrical Power Engineering and Mechatronics, Tallinn University of Technology, 19086 Tallinn, Estonia; andrii.chub@taltech.ee (A.C.); serhii.stepenko@taltech.ee (S.S.); elizaveta.liivik@taltech.ee (E.L.)

[2] Department of Electronics, Federico Santa Maria Technical University, 1680 Av. España, Valparaíso, Chile

[3] Department of Information Measuring Technologies, Metrology and Physics, Chernihiv National University of Technology, 14027 Chernihiv, Ukraine

[4] Department of Energy Technology, Aalborg University, 9220 Aalborg, Denmark; fbl@et.aau.dk

[*] Correspondence: dmitri.vinnikov@taltech.ee

Received: 1 September 2019; Accepted: 30 September 2019; Published: 3 October 2019

Abstract: The focus in this paper is on the two-stage photovoltaic (PV) microinverters using a buck-boost dc/dc front-end converter. Wide input voltage range of the front-end converter enables operation under shaded conditions but results in mediocre performance in the typical voltage range. These microinverters can be controlled with either fixed or variable dc-link voltage control methods. The latter improves the converter efficiency considerably in the range of the most probable maximum power point (MPP) locations. However, the buck-boost operation of the front-end converter results in noticeable variations of the efficiency across the input voltage range. As a result, conventional weighted efficiency metrics cannot be used to predict annual energy productions by the microinverters. This paper proposes a new methodology for the estimation of annual energy production based on annual profiles of the solar irradiance and ambient temperature. Using this methodology, quantification of the annual energy production is provided for two geographical locations.

Keywords: microinverter; variable dc-link voltage; photovoltaic; solar energy; renewable energy; residential systems

---

1. Introduction

In the recent year, residential solar photovoltaic (PV) systems have been on the rise, resulting from strong governmental support either in the form of subsidies on installation or feed-in tariffs [1–4]. This trend has been supported with the rapid cost reduction of PV modules and associated hardware [5]. Deployment of PV systems relies on extensive use of power electronic converters as the critical system components [6]. Residential applications are mostly associated with either string inverters or microinverters [7]. The latter could be regarded as the class of module-level power electronic (MLPE) systems [8]. Another member of this class is PV power optimizers used for interfacing individual PV modules in series PV string and, thus, they bridge the gap between string inverters and microinverters. However, they impose limitations on system design due to voltage matching issues.

Residential PV systems are usually built with string inverters for systems over 1 kW of installed power to optimize installation costs. However, series connection of PV modules into string results in relatively high dc voltage operating on the roof. Any issues with mechanical contacts within a PV string can easily result in dc arcing and, consequently, fire hazard. In addition, a PV string inverter can be considered as a single point of failure reducing overall reliability. Moreover, the 2017 National Electrical Code was revised in section 690.12, which discards array level rapid shut-down requirements and imposes PV module-level rapid shut-down requirements [9]. Starting from 2019, this regulation requires that voltage of all conductors be dropped below 80 V within 30 s from rapid shutdown initiation. As a result, string PV systems require use of smart modules or PV power optimizers along with a PV string inverter, which makes them comparable in price to systems with a microinverter, but more labor intensive to install.

Shading of PV modules is a serious issue in residential PV systems [10,11]. Conventional string PV inverters cannot withstand harsh shading conditions due to their limited voltage regulation capability. In such cases, MLPE converters have to be used to optimize the maximum power point (MPP) tracking (MPPT) process under shading conditions [7]. PV power optimizers could be advantageous in larger residential PV systems due to the possibility of selective deployment. However, they can ensure proper maximum energy harvesting by a PV string inverter only for a limited number of PV string configurations containing a number of modules within a certain range [12]. Contrary to that, microinverters accomplish parallel grid integration of individual PV modules into the grid. Hence, rapid PV module-level shutdown is ensured. Other advantages of the microinverters, PV module health monitoring, and dust accumulation detection, are provided; flexible PV system design is possible [7].

PV microinverters can be regarded as a universal tool for deployment of small residential PV systems, which provides superior scalability and reliability [7]. Usually, they are based on either single- or two-stage energy conversion. The former features low cost of implementation at the expense of regulation range limitations, while the latter is a more complex solution with a wider input voltage regulation range. The two-stage solutions are approaching the input voltage range of the PV power optimizers. PV power optimizers usually feature a wide input voltage range of 8 to 55 V to accommodate partial shading, when the global MPP voltage can be as low as one third of the unshaded MPP voltage for conventional Si PV modules with three bypass diodes. Hence, single-stage microinverters cannot withstand opaque or significant partial shading and thus must be replaced with two-stage solutions in climatic conditions where shading causes energy yield loss.

2. DC Link Voltage Control in Two-Stage PV Microinverters

A typical two-stage PV microinverter (Figure 1) is comprised of a front-end dc/dc converter that ensures decoupling of 100/120 Hz ripple from the input port while enabling converter operation at partial and opaque shading of a PV module. The latter feature is usually realized with the application of a galvanically isolated buck-boost dc/dc converter [13] at the input. Recently, several converters based on this concept were justified as high-performance MLPE solutions [14–16]. They can include separate buck and boost switching cells [13,14], as well as a single integrated buck-boost switching cell [15,16].

Typically, a buck-boost front-end dc/dc converter features three operation modes used for the input voltage regulation: buck, boost, and pass-through. The latter mode is the most efficient among the three as a converter does not perform voltage regulation and operates as a dc transformer (DCX). In the MLPE applications, the dc-link voltage is often fixed. In such a case, the isolating high-frequency (HF) transformer has to be designed to ensure that the pass-through mode coincides with the most probable maximum power point (MPP), as shown in Figure 2a. As a result, the buck mode is utilized during the microinverter start-up or MPPT that starts from the open-circuit voltage, as well as in the operation at low ambient temperatures when a PV module features elevated operating voltages. At the same time, the boost mode comes in handy at high ambient temperatures, when the MPP voltage of a PV module is reduced, as well as at shaded conditions to optimize the MPPT.

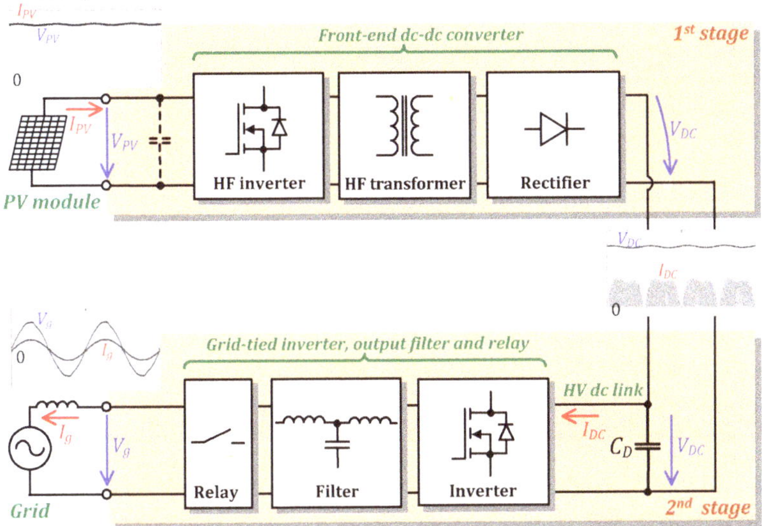

Figure 1. Generalized structure of a two-stage PV microinverter.

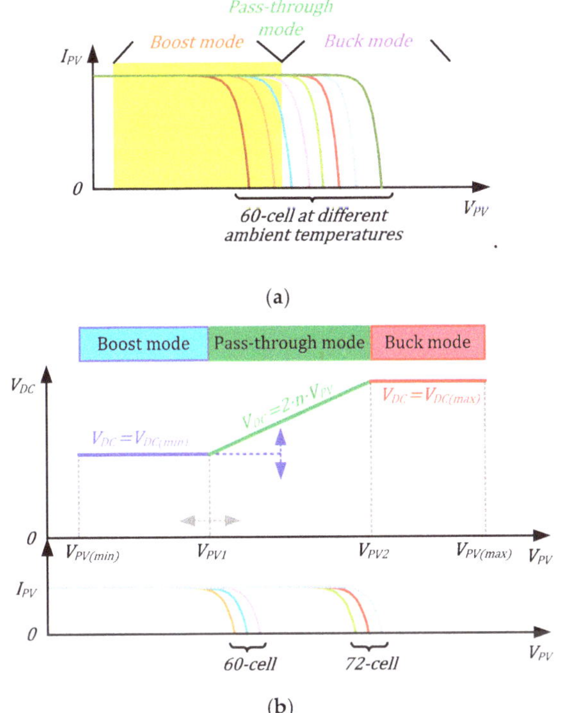

Figure 2. Utilization of operating modes of a front-end galvanically isolated buck-boost dc/dc converter at fixed (**a**) and variable and (**b**) dc-link voltage, including different PV module characteristics.

The dc-link voltage does not necessarily have to be fixed since the grid-side inverter enables this additional degree of control freedom. In this case, a microinverter operates in the boost mode at very low voltages that correspond to partial or opaque shading conditions and the dc-link voltage is limited at the minimum required value $V_{DC(min)}$, i.e., 10 V, ... , 20 V above the instantaneous grid peak voltage. The pass-through mode featuring the maximum efficiency corresponds to the most probable range of MPPs, while the grid-side inverter should perform the MPP tracking. The buck mode is used for the MPPT tracking or operating at low ambient temperatures. The dc-link voltage in the buck mode is maximum allowed by the voltage rating of the dc-link capacitor and grid-side inverter switches. Such an arrangement can result in efficient operation with the most typical Si PV modules containing 60- and 72-cell (Figure 2b). Recently, the variable dc-link voltage control was applied in several state-of-the-art two-stage microinverters [17,18] and battery chargers [19]. It resulted in higher efficiencies, extended reliability, and good compatibility with different PV modules [20,21].

The variable dc-link voltage control optimizes the efficiency of the two-stage microinverter but results in sizable variations of the efficiency around transitions between the pass-through mode and the buck or the boost mode. As a result, standardized weighted efficiencies, such as California Energy Commission (CEC) or EU metrics, would be of no use for predictions of the annual energy production by the two-stage buck-boost microinverters as their efficiency depends on the operating voltage considerably or it has to be added in the specifications. This study proposes a methodology for annual energy production estimation for the aforementioned microinverters and provides a numerical study based on experimental data to quantify the effect on converter efficiency achieved when the fixed dc-link voltage control is replaced with the variable dc-link voltage control.

3. Case Study Microinverter and Experimental Results

The microinverter for this study is described in detail in [18]. Its topology is shown in Figure 3 and the main specifications are listed in Table 1. The front-end dc/dc converter utilizes the hybrid quasi-Z-source series resonant topology. The front-end dc/dc converter implements the boost mode using the shoot-through pulse width modulation (ST-PWM) implemented by the symmetrical overlap of active states with the duty cycle D_{ST}. As a result, the input quasi-Z-source network can store energy and use it for the input voltage step-up. In the pass-through mode, it operates as the series resonant converter with very high efficiency. The grid-side inverter has to perform MPPT when the front-end dc/dc converter operates in the pass-through mode. The front-end dc/dc converter utilizes phase-shifted modulation (PSM) with an angle φ to implement voltage buck using the integrated series resonant tank. Two different dc-link voltage control principles result in a fundamentally different operation of the front-end converter, as shown in Figure 4. At the variable dc-link voltage, the region of the highest efficiency coincides with the most probable MPPs of 60- and 72-cell Si PV modules, i.e., when they operate under nominal operating cell temperature (NOCT).

The given two-stage microinverter topology [18] was selected as it has shown a wide input voltage range of over 1:6, which is wider than any other competitor and was justified as a shade-tolerant solution capable of withstanding the worst shading conditions.

Table 1. Prototype parameters and components.

Parameter or Component	Symbol or Block		Type or Value
		N	6
Transformer	TX	L_{lk}	25 µH
		L_m	1000 µH
QZS inductor	L_{qZS}	L_{lkqZS}	0.5 µH
		L_{mqZS}	11 µH
QZS capacitors	C_{qZS1}, C_{qZS2}		12 × 2.2 µF (X7R/100 V)
Resonant capacitors	C_1, C_2		43 nF (foil)
Dc-link capacitor	C_D		150 µF (Al electrolytic)
		L_{f1}	2.6 mH
LCL filter		L_{f2}	1.8 mH
		C_f	0.47 µF
Fix. dc-link voltage	V_{DC}		400 V
VAr dc-link voltage	$V_{DC(min)}$		335 V
	$V_{DC(max)}$		460 V
Grid	V_g		230 V/50 Hz
	S_{qZS}		210 kHz
Switching frequency	$S_1 \dots S_4$		105 kHz
	S_5, S_6		50 Hz
	S_7, S_8		20 kHz
LV semiconductors	$S_{qZS}, S_1 \dots S_4$		Infineon BSC035N10NS5
	D_1, D_2		Wolfspeed C3D02060E
HV semiconductors	S_5, S_6		Infineon IPB60R099C7
	S_7, S_8		ROHM SCT2120AFC

Figure 5 presents the modulation techniques used in the case study microinverter. Modulations of the front-end dc/dc converter and the grid-tied inverter are decoupled and are not synchronized in a general case. The front-end dc/dc converter operates at relatively high switching frequency in order to minimize the size of the isolation transformer, resonant tank, etc. The switch S_{qZS} is turned ON continuously in the buck and pass-through mode, as shown in Figure 5a,b. The carrier frequency of the switches $S_1, \dots, S_4$ is 105 kHz [15,18]. On the other hand, the switch S_{qZS} operates at a switching frequency twice higher than that of the switches $S_1 \dots S_4$ as it is active during active states when the isolation transformer is fed with voltage pulses (Figure 5c). Detailed description of the dc/dc converter operation is provided in [15].

The grid-tied inverter utilizes asymmetrical unipolar modulation shown in Figure 5d. In this modulation, two switches operate at high frequency (carrier frequency of the S_5, S_6 PWM signals is 20 kHz in the given case), while the other two switches S_7, S_8 operate at the grid frequency and unfold the grid voltage. This modulation was selected despite uneven losses in the switches as it provides 50 Hz rectangular common mode voltage and thus ensures low leakage current and good electromagnetic compatibility, as demonstrated in [18], where full operation of the microinverter is presented along with the description of the closed-loop control system.

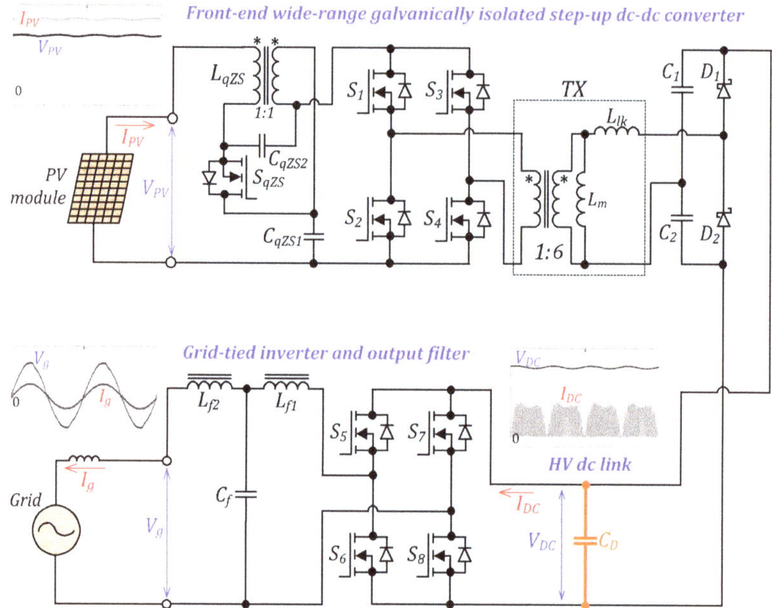

Figure 3. Topology of the case-study shade-tolerant microinverter [18].

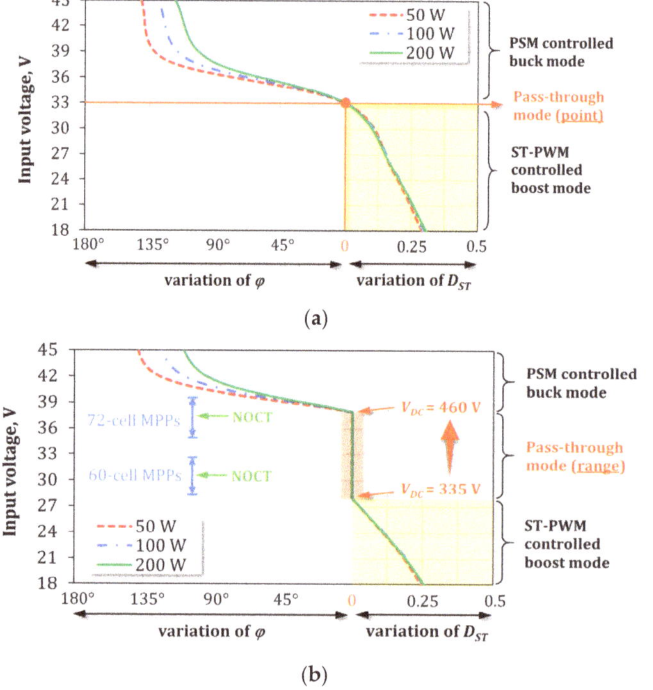

Figure 4. Control variables of the front-end converter in the microinverter topology shown in Figure 3 [18] at fixed (**a**) and variable (**b**) dc-link voltage control.

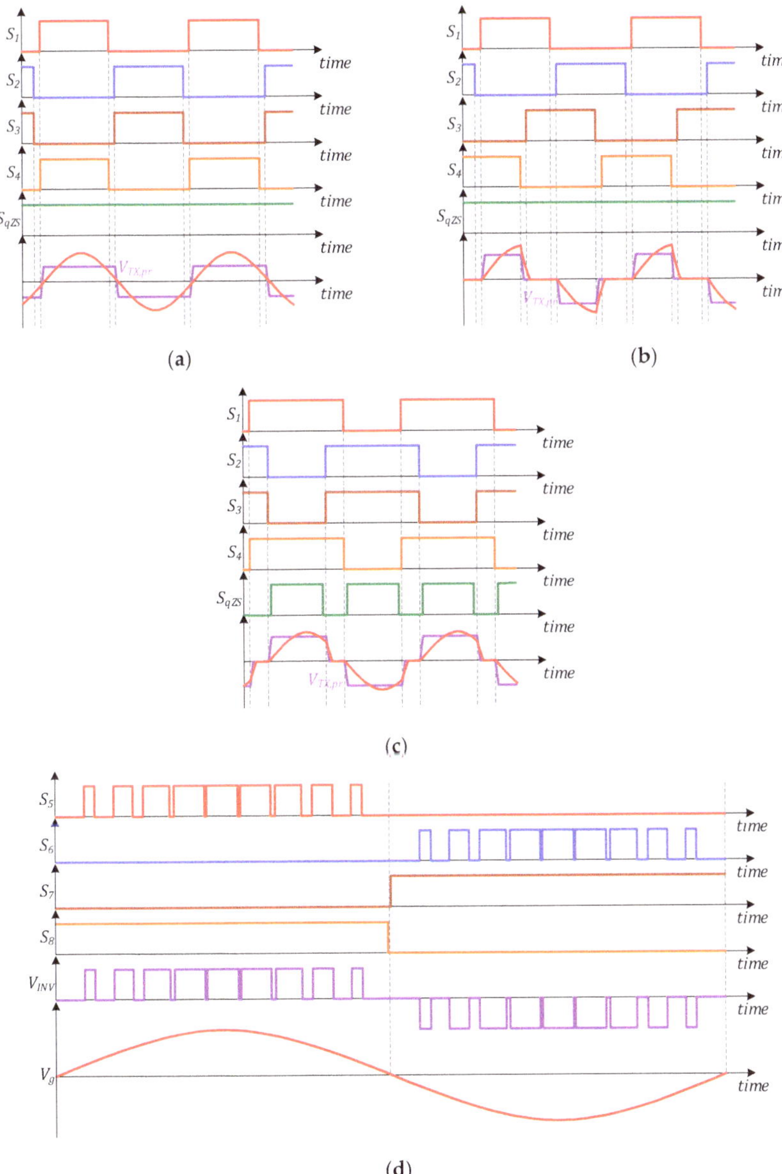

Figure 5. Modulation techniques of the front-end dc/dc converter operating in the pass-through. (**a**) buck, (**b**) boost, (**c**) modes, (**d**) asymmetrical unipolar modulation of the full bridge grid-tied inverter.

The efficiency of the experimental prototype was measured for both dc-link voltage control methods in the input voltage range of 25 to 45 V and from 10% to 100% of the rated power of 300 W. The rated power was selected during design to satisfy the requirements of the given application as will be shown in the next section. Over 50 reference points were acquired using a precision power analyzer Yokogawa WT1800. Next, the experimental efficiency surfaces were obtained in MATLAB using the surface fitting tool and thin-plate spline for interpolation. The contour plots of the experimental data interpolation shown in Figure 6a,b follow tightly the experimental data in Figure 6c,d, which results in the coefficient of determination (R^2) equal to unity. This data will be used further for annual performance prediction. It can be seen from Figure 6 that the fixed dc-link voltage control results in a limited region of high efficiency where the front-end dc/dc converter operates in or close to the pass-through mode. At the same time, the variable dc-link voltage control extends the region of the pass-through mode usage, which results in roughly 2% higher efficiency across the considered operation range.

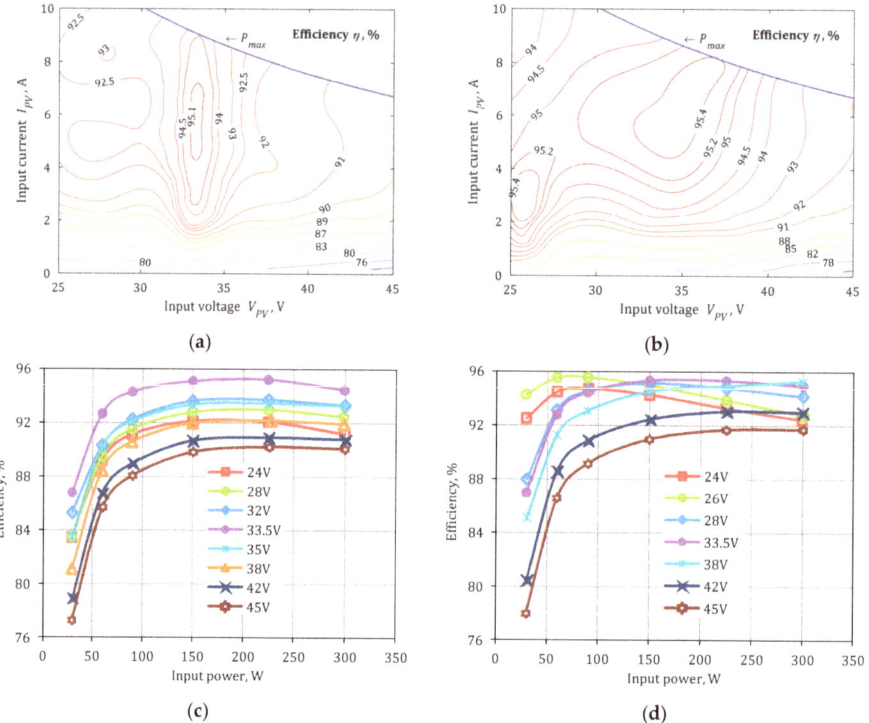

Figure 6. Interpolation results of the experimental efficiencies obtained for the prototype (Figure 3) operating with fixed (**a**) and variable (**b**) dc-link voltage control, which were derived based on experimental measurements obtained using fixed (**c**) and variable (**d**) dc-link voltage control.

To provide an experimental example of PV energy harvesting by the case study microinverter, one-day operation (moderate irradiance and no clouds) was simulated using a PV simulator Chroma 62150H-1000S. According to Figure 7, the variable dc-link voltage results in higher power injected in the grid than that for the fixed dc-link voltage control. It improves the overall efficiency of the former, thus, reducing the loss of the available PV energy from 9.2% to 7.4%, i.e., reducing the energy loss by 19.6%. This study considers the efficiency of the microinverter and leaves the MPPT efficiency out for simplicity.

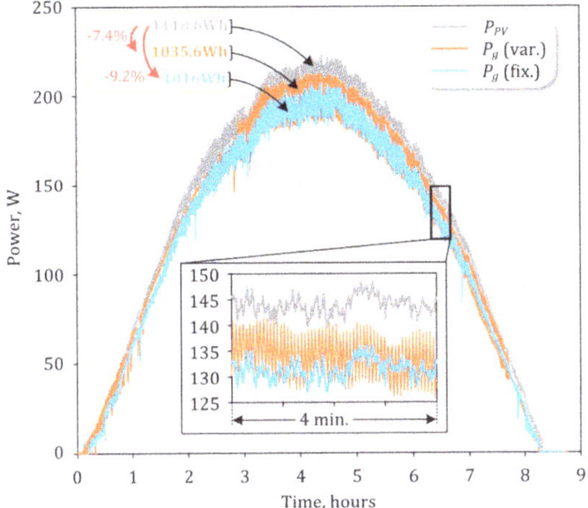

Figure 7. Experimental daily energy production for two dc-link voltage controls (var.: variable dc-link voltage control; fix.: fixed dc-link voltage control).

4. Estimation of Annual PV Energy Production

The interpolations of the experimental efficiency (Figure 6) can be used to estimate the annual energy production and energy loss for the two dc-link control methods. The efficiency includes microinverter self-powering from the input PV power. The developed methodology is explained in Figure 8. It utilizes 1-year mission profiles of the solar irradiance and the ambient temperature for Arizona (Ar.) and North Denmark (N.D.) [22] with a 1-minute resolution.

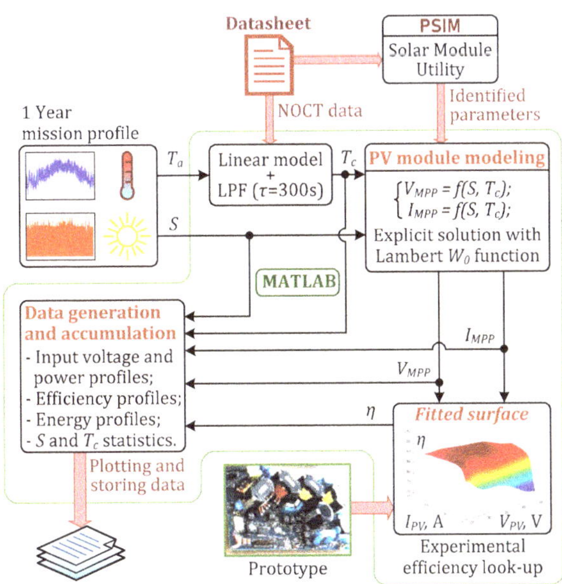

Figure 8. Flow diagram of the estimation of the annual energy production.

First, the ambient temperature (T_a) is translated into the cell temperature (T_c) using a simple linear model with a low-pass filter to account for the thermal dynamics of a PV module:

$$T_c = \left[T_a + \frac{T_{NOCT} - T_{a(NOCT)}}{S_{NOCT}} \cdot S \right] \cdot G_{LPF}(s),$$ (1)

where the nomenclature is listed Table 2. G_{LPF} is the low pass filter transfer function in the s-domain:

$$G_{LPF}(s) = \frac{1}{1 + s \cdot \tau}.$$ (2)

Table 2. Nomenclature.

Term	Explanation	Term	Explanation
k_b	Boltzmann constant	v_t	Thermal voltage
q	Elementary charge	R_s	Series resistance of PV module
a	Ideality factor	$R_{s(STC)}$	R_s in STC conditions
E_g	Band gap energy	k_{rs}	R_s thermal coefficient
T_a	Ambient temperature	R_{sh}	Shunt resistor of PV module
T_c	Cell temperature	$I_{0(STC)}$	I_0 in STC conditions
S	Solar irradiance	I_0	Saturation current
T_{STC}	T_c in STC conditions	I_{Ph}	Photocurrent
S_{STC}	S in STC conditions	N_S	Number of cells in PV module
$T_{a(NOCT)}$	T_a in NOCT conditions	η	Microinverter efficiency
S_{NOCT}	Solar irradiance in NOCT condition	P_g	Microinverter output power
T_{NOCT}	Nominal operating cell temperature (NOCT)	$I_{SC(STC)}$	Short-circuit current of PV module in STC conditions
$V_{MPP(id)}$	MPP voltage of ideal PV module	τ	Thermal time constant of PV module
$I_{MPP(id)}$	MPP current of ideal PV module	$W_0(\cdot)$	Principle branch of Lambert W function
I_{PV}	Microinverter input current	V_{MPP}	Module MPP voltage
V_{PV}	Microinverter input voltage	I_{MPP}	Module MPP current
P_{PV}	Microinverter input power	P_{MPP}	Module MPP power

In the second stage, the PV module has to be simulated by the typical single-diode model of a PV cell (Figure 9) using five parameters described in [23–25].

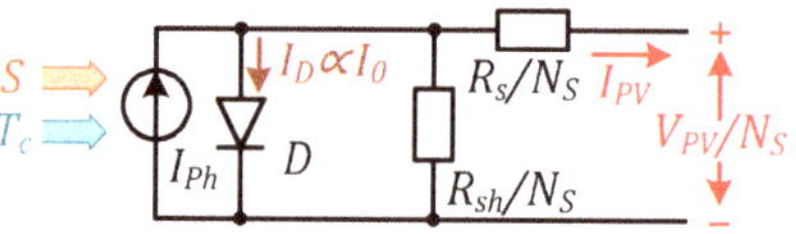

Figure 9. Single-diode model of a PV cell.

In the given model of a PV cell, diode D plays a crucial role. The fundamental value defining the operation of the p-n junction in the diode is the thermal voltage v_t:

$$\begin{cases} I_D = I_0 \cdot \left(e^{\frac{V_D}{a \cdot v_t}} - 1 \right); \\ v_t = \frac{T_c \cdot k_b}{q}. \end{cases}$$ (3)

According to Shockley's expression [26], the diode current is proportional to the reverse bias saturation current defined as:

$$I_0 = I_{0(STC)} \cdot \left(\frac{T_c}{T_{STC}} \right)^3 \cdot e^{\left(\frac{q \cdot E_g}{a \cdot k_b} \right) \cdot \left(\frac{1}{T_{STC}} - \frac{1}{T_c} \right)}.$$ (4)

The given model differs from the conventional model in the approach of simulating the series resistance R_s. Recent reports show its positive thermal dependence based on experimental results [27]. The PV module Jinko Solar JKM300M-60B used in modeling is based on the monocrystalline silicon technology and contains 60 cells. For this type of modules, the experimental thermal coefficient of R_s (k_{rs}) was reported in [27] to be used in the following equation:

$$R_s = R_{s(STC)} \cdot (1 + k_{rs}(T_c - T_{STC})). \tag{5}$$

As was mentioned before, the operation of the diode D defines the output current and the voltage of a PV cell, but its current is limited by the photocurrent generated by a PV cell:

$$I_{Ph} = I_{SC(STC)} \cdot \frac{S}{S_{STC}} \cdot \frac{R_s + R_{sh}}{R_{sh}}. \tag{6}$$

Some of the equations above contain the PV cell voltage inside and outside the exponential function, which can be resolved only using the Lambert function (in this case, its principle branch denoted as W_0). For further calculations, a helper variable must be introduced in terms of the Lambert function:

$$w = W_0\left(\frac{I_{ph} \cdot e^1}{I_0}\right). \tag{7}$$

Using this variable, the MPP voltage and current could be derived for an ideal PV cell containing only a photocurrent source and a diode:

$$V_{MPP(id)} = Ns \cdot a \cdot v_t \cdot (w - 1), \tag{8}$$

$$I_{MPP(id)} = I_0 \cdot (w - 1) \cdot e^{(w-1)}. \tag{9}$$

However, this estimation is inaccurate as the influence of the shunt and series resistors has to be taken into account [23]:

$$I_{MPP} = I_{MPP(id)} - \frac{V_{MPP(id)}}{R_{sh}}, \tag{10}$$

$$V_{MPP} = V_{MPP(id)} - I_{MPP} \cdot R_s, \tag{11}$$

$$P_{MPP} = V_{MPP} \cdot I_{MPP}. \tag{12}$$

Equations (3) to (12) constitute a short description of the methodologies described in [23–25] based on the single-diode PV cell module, but enhanced with findings from [27]. In addition, the final calculations take into account material properties of silicon from [28] and typical PV module thermal characteristics from [29], which are listed in Table 3.

Table 3. PV module parameters.

Model Parameters		Reference Data and Constants	
Parameter	Value	Parameter	Value
S_{STC}	1000 W/m^2	$V_{MPP(STC)}$	32.6 V
T_{STC}	25 °C	$I_{MPP(STC)}$	9.21 A
S_{NOCT}	800 W/m^2	$V_{OC(STC)}$	40.1 V
T_{NOCT}	45 °C	$I_{SC(STC)}$	9.72 A
$T_{a(NOCT)}$	20 °C	$V_{MPP(NOCT)}$	30.6 V
a [27]	1.1	$I_{MPP(NOCT)}$	7.32 A
E_g [28]	1.12 eV	$V_{OC(NOCT)}$	37.0 V
$I_{0(STC)}$	5.39·10^{-10} A	$I_{SC(NOCT)}$	8.01 A
R_{sh}	750 Ω	Efficiency	17.98%
$R_{s(STC)}$	0.228 Ω	N_s	60
k_{rs} [27]	0.356 %	k_b	1.38·10^{-23} J/K
τ [29]	300 s	q	1.602·10^{-19} C

Third, it is possible to define the MPP power and voltage for any solar irradiance and cell temperature. Considering the interpolated efficiency of the microinverter η, and assuming a perfect MPPT when $V_{PV} = V_{MPP}$ and $P_{PV} = P_{MPP}$, it is possible to define the microinverter output power P_g and, thus, the energy production E during an arbitrary time interval t_x:

$$P_g(t) = V_{MPP}(t) \cdot I_{MPP}(t) \cdot \eta(V_{MPP}(t); I_{MPP}(t)), \tag{13}$$

$$E = \int_0^{t_x} P_g(t)dt. \tag{14}$$

Table 3 presents the parameters of the solar module obtained from the datasheet, PSIM Solar module utility, and other literature (for fundamental and empiric constants).

The considered geographic locations differ significantly in the operating conditions. Distribution of solar irradiance has a higher expectation around 1000 W/m^2 for Arizona and around 100 W/m^2 for North Denmark, as shown in Figure 10. As a result of higher solar irradiance and ambient temperature, the PV module will experience much higher thermal stress in Arizona conditions, which was calculated using Equation (1) and plotted in Figure 11. Probability density distribution of the cell temperatures shown in Figure 12 proves that the cell temperature is mostly between 10 °C and 75 °C in the Arizona conditions, contrary to the range of −5 °C to 60 °C in the North Denmark conditions, which is calculated ignoring night time.

Next, annual mission profiles of solar irradiance and ambient temperature are translated into annual profiles of the MPP voltage and power using the model of the PV module described above. Probability distribution functions of these physical quantities are shown in Figures 13 and 14. Lower cell temperatures and solar irradiance levels resulted in higher MPP voltages but lower average MPP power in the North Denmark conditions. Contrary to that, the microinverter will process higher powers at lower MPP voltages in the Arizona conditions.

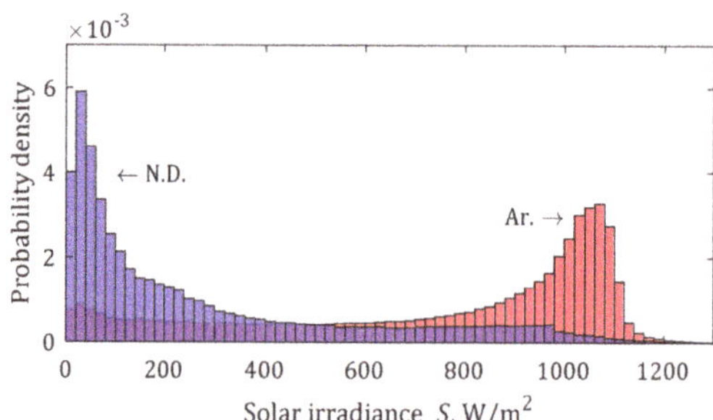

Figure 10. Density distribution of the solar irradiances.

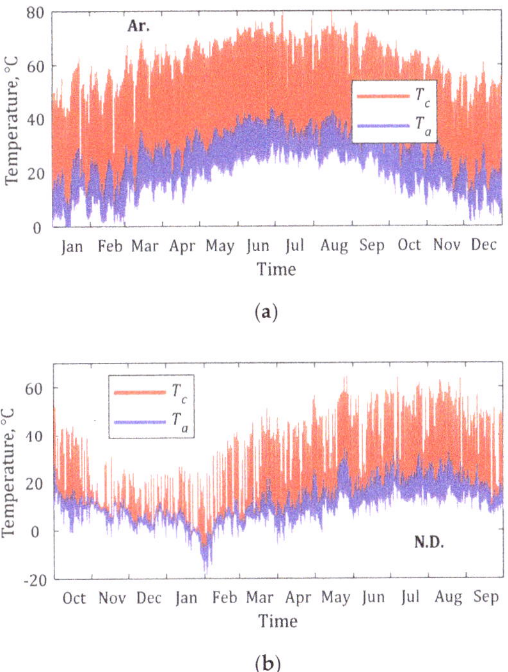

(a)

(b)

Figure 11. Annual profiles for experimental ambient temperature and estimated cell temperature for Arizona (**a**) and North Denmark (**b**) geographic locations.

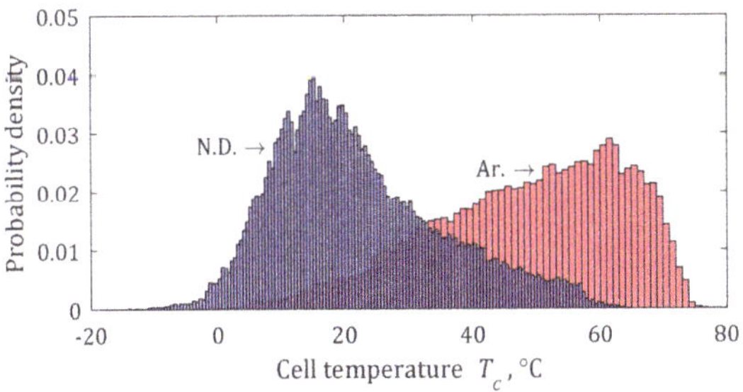

Figure 12. Density distribution of the cell temperatures.

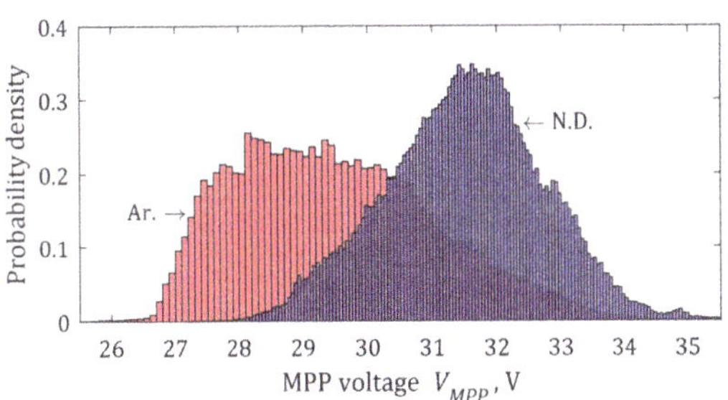

Figure 13. Density distribution of the MPP voltages.

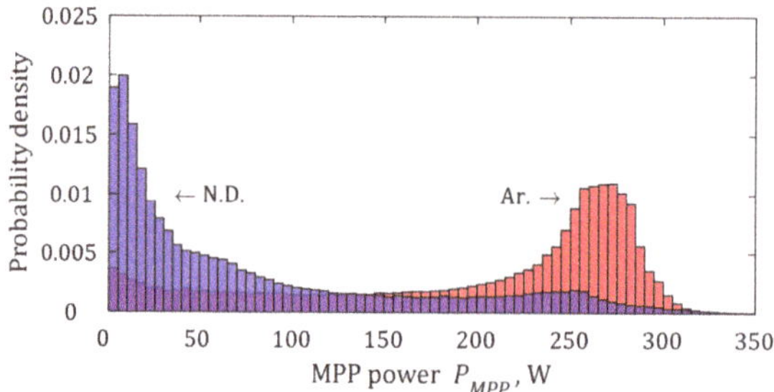

Figure 14. Density distribution of the MPP powers.

Annual mission profiles of the MPP voltage and power can be applied to the efficiency look-up block that performs interpolations for the experimental efficiency using thin-plate splines derived in MATLAB. The probability distribution of the efficiency of the microinverter strongly depends on the geographic location as well as on the dc-link voltage control. In the Arizona conditions, high average MPP power results in the narrow distribution range of 92% to 93.5% for the fixed dc-link voltage control and 94% to 95.5% for the variable dc-link voltage control, as it follows from Figure 15a. For the case of fixed dc-link voltage control, the values of efficiency can reach as low as 77% and as high as 95.2%. This wide distribution range results in a relatively low average efficiency of 91.4%. Contrary to that, in the North Denmark conditions, the microinverter will feature sizable probability values in a much wider range of efficiencies due to more intermittent solar irradiance, as shown in Figure 15b. Nevertheless, in both cases, the variable dc-link voltage control provides efficiencies higher by 2.2% on average than in the case of fixed dc-link voltage control.

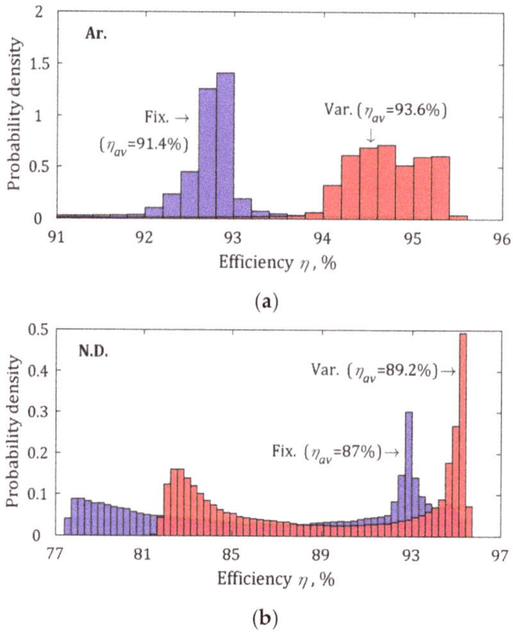

Figure 15. Density distribution of the microinverter and average efficiencies for two dc-link voltage control methods for Arizona (**a**) and North Denmark (**b**) geographic locations.

Average values of the annual microinverter efficiency are not sufficient to reach any definitive conclusion regarding annual energy production as the efficiency values have to be weighted with the input power values. Table 4 quantifies the energy production, the PV energy loss, the usable PV energy in absolute and relative values. In the North Denmark conditions, the case study PV module is capable of delivering the energy of 323.8 kW·h per year. This energy is delivered to the grid with the energy loss of 8% at the fixed dc-link voltage control and the energy loss of 6.2% at the variable dc-link voltage control. The latter provides 22.3% reduction in the annual energy loss. In the Arizona conditions, the same PV module can deliver the energy of 861.4 kW·h per year. In this case, the energy loss of 5.4% was observed for the variable dc-link voltage control, which means a reduction by 26% from the energy loss of 7.4% observed at the fixed dc-link voltage control. Hence, the variable dc-link voltage control provides over 20% reduction of the annual energy loss regardless of the climatic conditions.

Table 4. Annual energy production simulation results.

Location	North Denmark		Arizona	
Dc-link Voltage Control	**Fixed**	**Variable**	**Fixed**	**Variable**
Usable PV energy, kW·h	323.8		861.4	
Produced energy, kW·h	297.8	303.6	798	814.5
PV energy loss, kW·h	26	20.3	63.4	46.9
PV energy loss, %	8	6.2	7.4	5.4
Energy loss reduction, kW·h	-	5.7	-	16.5
Energy loss reduction, %	-	22.3	-	26

5. Conclusions

The paper has proposed a simple methodology for quantifying the annual energy production by a microinverter based on experimental mission profiles of the solar irradiance and ambient temperatures, PV module parameters identified from a datasheet using PSIM Solar module utility, and experimental efficiency interpolation based on measured values. The methodology uses a single-diode five-parameter model of a PV cell enhanced with thermally dependent model of the series resistor.

The proposed methodology was applied to a case study microinverter utilizing a buck-boost front-end converter controlled by variable and fixed dc-link voltage control methods. The experimental efficiency was found to reach over 95% in both cases, while it is increased by up to 2% on average in the tested range with the application of the variable dc-link voltage control. Experimental efficiency interpolations were used to estimate energy production by a microinverter connected to a properly sized PV module. The variable dc-link voltage control extends efficient operation from a single to several compatible PV module configurations, i.e., different number of PV cells per module.

Our one-day experimental test of PV energy harvesting by the case study microinverter showed an energy production increase by 1.8%. This value correlates well with the predicted increase of the annual energy production by 1.8% and 2% observed for a cloudy northern climate and a sunny southern climate, correspondingly. This increase could be translated in 22.3% and 26% reduction of the annual energy loss, correspondingly. Hence, the variable dc-link voltage control provides over 20% reduction of the annual energy loss regardless of the climatic conditions.

The variable dc-link voltage control is a promising solution for the two-stage microinverters with capacitive intermediate dc-link buffering 100/120 Hz power pulsations. Our future work will aim to extend the proposed methodology with a model of power loss from the MPPT, which was left out of this study.

Author Contributions: Conceptualization, A.C. and D.V.; methodology, A.C. and E.L.; software, A.C. and S.S.; validation, D.V. and S.S.; formal analysis, A.C. and D.V.; investigation, E.L. and F.B.; resources, E.L. and D.V.; data curation, F.B. and S.S.; writing—original draft preparation, A.C. and S.S.; writing—review and editing, E.L., D.V., and F.B.; visualization, A.C. and S.S.; supervision, E.L. and F.B.; project administration, D.V. and E.L.; funding acquisition, D.V., S.S., and A.C.

Funding: This research was supported in part by the Estonian Centre of Excellence in Zero Energy and Resource Efficient Smart Buildings and Districts, ZEBE, grant 2014-2020.4.01.15-0016 funded by the European Regional Development Fund, in part by the Estonian Research Council grant PSG206, in part by the European Regional Development Fund and the programme Mobilitas Pluss under the project MOBJD126 awarded by the Estonian Research Council, and in part by SERC Chile (CONICYT/FONDAP/15110019) and AC3E (CONICYT/BASAL/FB0008).

Conflicts of Interest: The authors declare no conflict of interest.

References

1. Malinowski, M.; Leon, J.I.; Abu-Rub, H. Solar Photovoltaic and Thermal Energy Systems: Current Technology and Future Trends. *Proc. IEEE* **2017**, *105*, 2132–2146. [CrossRef]
2. Poullikkas, A. A comparative assessment of net metering and feed in tariff schemes for residential PV systems. *Sustain. Energy Technol. Assess.* **2013**, *3*, 1–8. [CrossRef]
3. Lang, T.; Ammann, D.; Girod, B. Profitability in absence of subsidies: A techno-economic analysis of rooftop photovoltaic self-consumption in residential and commercial buildings. *Renew. Energy* **2016**, *87*, 77–87. [CrossRef]
4. Yamamoto, Y. Pricing electricity from residential photovoltaic systems: A comparison of feed-in tariffs, net metering, and net purchase and sale. *Sol. Energy* **2012**, *86*, 2678–2685. [CrossRef]
5. Taiyang News. Advanced Solar Module Technology, 2018 Edition. Available online: http://taiyangnews.info/TaiyangNews_Report_Advanced_Solar_Module_Technology_2018_EN_download_v_2.pdf (accessed on 11 August 2019).
6. Kjaer, S.B.; Pedersen, J.K.; Blaabjerg, F. A review of single-phase grid-connected inverters for photovoltaic modules. *IEEE Trans. Ind. Appl.* **2005**, *41*, 1292–1306. [CrossRef]
7. Kouro, S.; Leon, J.I.; Vinnikov, D.; Franquelo, L.G. Grid-Connected Photovoltaic Systems: An Overview of Recent Research and Emerging PV Converter Technology. *IEEE Ind. Electron. Mag.* **2015**, *9*, 47–61. [CrossRef]
8. Spagnuolo, G.; Kouro, S.; Vinnikov, D. Photovoltaic Module and Submodule Level Power Electronics and Control. *IEEE Trans. Ind. Electron.* **2019**, *66*, 3856–3859. [CrossRef]
9. *NFPA 70®: National Electrical Code®(NEC®)*, 2017 ed.; National Fire Protection Association: Quincy, MA, USA, 2017.
10. Sinapis, K.; Tzikas, C.; Litjens, G.; Van den Donker, M.; Folkerts, W.; Van Sark, W.G.J.H.M.; Smets, A. A comprehensive study on partial shading response of c-Si modules and yield modeling of string inverter and module level power electronics. *Sol. Energy* **2016**, *135*, 731–741. [CrossRef]
11. Sinapis, K.; Litjens, G.; Donker, M.; Folkerts, W.; Sark, W. Outdoor characterization and comparison of string and MLPE under clear and partially shaded conditions. *Energy Sci. Eng.* **2015**, *3*, 510–519. [CrossRef]
12. Kasper, M.; Bortis, D.; Kolar, J.W. Classification and Comparative Evaluation of PV Panel-Integrated DC–DC Converter Concepts. *IEEE Trans. Power Electron.* **2014**, *29*, 2511–2526. [CrossRef]
13. Yao, C.; Ruan, X.; Wang, X.; Tse, C.K. Isolated Buck–Boost DC/DC Converters Suitable for Wide Input-Voltage Range. *IEEE Trans. Power Electron.* **2011**, *26*, 2599–2613. [CrossRef]
14. LaBella, T.; Yu, W.; Lai, J.S.; Senesky, M.; Anderson, D. A Bidirectional-Switch-Based Wide-Input Range High-Efficiency Isolated Resonant Converter for Photovoltaic Applications. *IEEE Trans. Power Electron.* **2014**, *29*, 3473–3484. [CrossRef]
15. Vinnikov, D.; Chub, A.; Liivik, E.; Roasto, I. High-Performance Quasi-Z-Source Series Resonant DC–DC Converter for Photovoltaic Module-Level Power Electronics Applications. *IEEE Trans. Power Electron.* **2017**, *32*, 3634–3650. [CrossRef]
16. Chub, A.; Vinnikov, D.; Kosenko, R.; Liivik, E. Wide Input Voltage Range Photovoltaic Microconverter with Reconfigurable Buck–Boost Switching Stage. *IEEE Trans. Ind. Electron.* **2017**, *64*, 5974–5983. [CrossRef]
17. Shen, Y.; Wang, H.; Shen, Z.; Yang, Y.; Blaabjerg, F. A 1-MHz Series Resonant DC–DC Converter with a Dual-Mode Rectifier for PV Microinverters. *IEEE Trans. Power Electron.* **2019**, *34*, 6544–6564. [CrossRef]
18. Vinnikov, D.; Chub, A.; Liivik, E.; Kosenko, R.; Korkh, O. Solar Optiverter—A Novel Hybrid Approach to the Photovoltaic Module Level Power Electronics. *IEEE Trans. Ind. Electron.* **2019**, *66*, 3869–3880. [CrossRef]
19. Wang, H.; Dusmez, S.; Khaligh, A. Maximum Efficiency Point Tracking Technique for LLC-Based PEV Chargers through Variable DC Link Control. *IEEE Trans. Ind. Electron.* **2014**, *61*, 6041–6049. [CrossRef]

20. Shen, Y.; Chub, A.; Wang, H.; Vinnikov, D.; Liivik, E.; Blaabjerg, F. Wear-Out Failure Analysis of an Impedance-Source PV Microinverter Based on System-Level Electrothermal Modeling. *IEEE Trans. Ind. Electron.* **2019**, *66*, 3914–3927. [CrossRef]

21. Liivik, E.; Chub, A.; Sangwongwanich, A.; Shen, Y.; Vinnikov, D.; Blaabjerg, F. Wear-Out Failure Analysis of Solar Optiverter Operating with 60- and 72-Cell Si Crystalline PV Modules. In Proceedings of the IECON'2018, Washington, DC, USA, 21–23 October 2018; pp. 6134–6140.

22. Sangwongwanich, A.; Yang, Y.; Sera, D.; Blaabjerg, F. Lifetime Evaluation of Grid-Connected PV Inverters Considering Panel Degradation Rates and Installation Sites. *IEEE Trans. Power Electron.* **2018**, *33*, 1225–1236. [CrossRef]

23. Farivar, G.; Asaei, B.; Mehrnami, S. An Analytical Solution for Tracking Photovoltaic Module MPP. *IEEE J. Photovolt.* **2013**, *3*, 1053–1061. [CrossRef]

24. Kuperman, A. Comments on "An Analytical Solution for Tracking Photovoltaic Module MPP". *IEEE J. Photovolt.* **2014**, *4*, 734–735. [CrossRef]

25. Lineykin, S.; Averbukh, M.; Kuperman, A. An improved approach to extract the single-diode equivalent circuit parameters of a photovoltaic cell/panel. *Renew. Sustain. Energy Rev.* **2014**, *30*, 282–289. [CrossRef]

26. Sah, C.T.; Noyce, R.N.; Shockley, W. Carrier generation and recombination in P–N junctions and P–N junction characteristics. *Proc. IRE* **1957**, *45*, 1228–1243. [CrossRef]

27. Lee, K. Improving the PV Module Single-Diode Model Accuracy with Temperature Dependence of the Series Resistance. In Proceedings of the PVSC'2017, Washington, DC, USA, 25–30 June 2017; pp. 1526–1530.

28. De Soto, W.; Klein, S.A.; Beckman, W.A. Improvement and validation of a model for photovoltaic array performance. *Sol. Energy* **2006**, *80*, 78–88. [CrossRef]

29. Armstrong, S.; Hurley, W.G. A thermal model for photovoltaic panels under varying atmospheric conditions. *Appl. Therm. Eng.* **2010**, *30*, 1488–1495. [CrossRef]

Article

Active and Reactive Power Control of a PV Generator for Grid Code Compliance

Ana Cabrera-Tobar [1,*], Eduard Bullich-Massagué [2] and Mònica Aragüés-Peñalba [2] and Oriol Gomis-Bellmunt [2]

1   Department of Electrical Engineering, Universidad Técnica del Norte. Av. 17 de Julio, Ibarra 100107, Ecuador
2   Department d'Enginyeria Elèctrica, Centre d'Innovació Tecnològica en Convertidors Estàtics i Accionamients (CITCEA-UPC), Universitat Politècnica de Catalunya, UPC. Av. Diagonal 647, Pl. 2., 08028 Barcelona, Spain; eduard.bullich@citcea.upc.edu (E.B.-M.); monica.aragues@citcea.upc.edu (M.A.-P.); oriol.gomis@upc.edu (O.G.-B.)
*   Correspondence: ana.cabrera@ieee.org; Tel.: +593-972960468

Received: 4 September 2019; Accepted: 2 October 2019; Published: 12 October 2019

Abstract: As new grid codes have been created to permit the integration of large scale photovoltaic power plants into the transmission system, the enhancement of the local control of the photovoltaic (PV) generators is necessary. Thus, the objective of this paper is to present a local controller of active and reactive power to comply the new requirements asked by the transmission system operators despite the variation of ambient conditions without using extra devices. For this purpose, the control considers the instantaneous capability curves of the PV generator which vary due to the change of solar irradiance, temperature, dc voltage and modulation index. To validate the control, the PV generator is modeled in DIgSILENT PowerFactory® and tested under different ambient conditions. The results show that the control developed can modify the active and reactive power delivered to the desired value at different solar irradiance and temperature.

Keywords: PV generators; active power; reactive power; Renewable energy; grid codes; capability curves

---

1. Introduction

As more large scale photovoltaic power plants (LS-PVPPs) are being installed, the electrical system can face some challenges related to four key areas: (i) active power control, (ii) reactive power control, (iii) voltage support and (iv) frequency support [1]. Thus, many countries have updated their grid codes to permit a smooth interaction between these power plants with the transmission system. For instance, Puerto Rico requires that these PVPPs behaves similar to conventional power plants despite the intermittent conditions [2]. Considering these changes on the grid codes there are two key aspects necessary to approach: active and reactive power control.

According to the grid codes presented by Puerto Rico, Romania, South Africa and Germany, the active power management for LS-PVPPs should consider: power curtailment, ramp rate control and active power reserves (Figure 1) [2–5]. Power curtailment, also called as absolute control or limiting control, addresses the reduction of the possible active power that the power plant can generate during the day depending on the grid requirements [6]. This requirement prevents overloading at peak generation hours of PVPPs (around midday) or also when the demand is lower than the possible generated active power from the PVPP. However, due to the intermittency of the solar source, ramp rate control is also necessary to be addressed [7]. The aim of these ramp rates is to smooth the change from low to high solar irradiance and vicevesa, so the change does not affect the voltage or the frequency. As more LS-PVPPs are being introduced at the transmission system, the participation on frequency regulation is a new challenge. Thus, power reserve are already being considered in some of the grid codes. The power reserve is the reduction of the output power during some hours of the day. This reduction can oscillate between 10 to 20% of the maximum possible that the PVPP can generate [1,8].

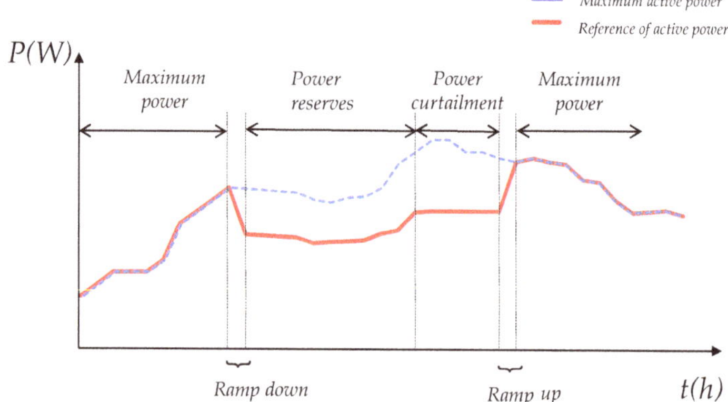

Figure 1. Active power variation applying the new control functions.

There are two main techniques used to manage the active power: (i) incorporate energy storage, and (ii) new control strategies of the PV generator [1]. In the literature, the main topology proposed is the use of the energy storage together with the PV inverter that are distributed along the PVPP [9–11]. For instance, the study developed by Muller et al, proposes the use of ultracapacitors together with a central inverter to manage the power transients due to the variability of solar irradiance. In this study, the output power fluctuations are reduced and it helps to comply the grid code limits [9]. Commonly, the management of the active power relays on the charge of the battery during high solar irradiance and the discharge in high peak demand [12]. This type of control smooths the output power of the single PV generator during the day [11,13–19]. However, the incorporation of energy storage increases the cost of installation and operation of the LS-PVPP [20]. An alternative solution is the improvement of the control by considering the characteristics of the PV generator. Commonly, the active power is managed by the maximum power point tracker (MPPT) which is part of the overall control of the PV inverter. However, the MPPT cannot withstand the power curtailment, the power reserve or the ramp rates. Therefore, this tracker should not only consider the maximum power point but also the reference of active power given by the transmission system operator (TSO). Some studies propose this method for multistring topologies (two stage inverters) used in small applications [8,21,22]. However, these studies have not been applied for central inverters that are commonly used in LS-PVPPs [23].

In the case of reactive power, the new grid codes require that the LS-PVPP injects or absorbs reactive power according to a predefined relationship between the active and the reactive power (power factor (pf)) or an specific value of reactive power. The grid codes presented by China, Germany, South Africa, Romania, and Puerto Rico requires that the LS-PVPP works under an specific capability curve (Figure 2). From this curve, it can be seen that Puerto Rico has the strictest requirement ($Q_{max} = \pm 0.623$ p.u). Meanwhile, China, Germany, Romania and South Africa require a maximum reactive power close to ± 0.33 p.u. To comply these grid codes, commonly STATCOMs or capacitors are added at the point of common coupling (PCC), as it is explained in [24]. However, limited research has been developed regarding the reactive power control of PV generators in LS-PVPPs without using extra equipment. For instance, Rakibuzzaman et al. [25] explain the control of reactive power and how the capability curve could influence in the response, but, the variation of ambient conditions is not considered on this approach. Additionally, R. Varma et al. and L. Luo are working on the control of LS-PVPPs as STATCOM to support the grid when power oscillation occurs [26,27]. However, it considers the remaining inverter capacity and depends on the solar irradiance behavior.

From a general point of view without any specific source of energy, new types of reactive power control for grid tied inverters have been presented in [28,29]. These studies do not take into account the variation of solar irradiance during the day or the corresponding capability curves of the PV generator. It is worth to point out that the control of reactive power in a LS-PVPP has not commonly been developed considering the capability curves. There are four main parameters that characterize these curves: (i) modulation index, (ii) dc voltage, (iii) solar irradiance, and (iv) ambient temperature, as it is explained in [30,31]. From this research, it can be understood that the variation of the dc voltage and the modulation index can help to have the complete curve despite the variation of ambient conditions. Although, this can reduce the active power generated by the PV generator.

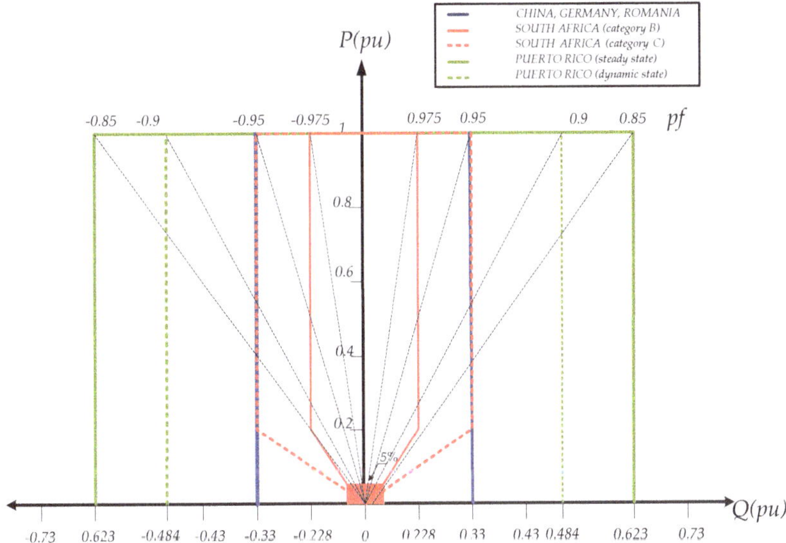

Figure 2. Reactive power requirements.

Thus, the objective of this paper is to propose a control of active and reactive power for a PV generator applied in LS-PVPPs for grid code compliance. In this paper, the PV generator has a three phase central inverter (one stage of inversion). To control the active power, two main targets are accomplished: (i) Power curtailment, and (ii) Power reserves, by using an adaptation of the Maximum Power Point Tracker (MPPT). For the reactive power control, two considerations are addressed: (i) preference of active over reactive power and (ii) preference of reactive over active power. For this control, the instant capability curves are considered by the adjustment of the dc voltage and the modulation index depending on the solar irradiance and temperature that affects to the production of active power. To validate this study, a LS-PVPP is modeled and simulated in DIgSILENT PowerFactory® under different ambient conditions. The paper is structured as follows: Section 2 explains the configuration and the control structure of a PVPP. The active power control is explained in Section 3, meanwhile the reactive power control is detailed in Section 4. Then, the simulations and the results are presented in Section 5. Finally, the discussion and the conclusions are in Sections 6 and 7 respectively.

2. Configuration and Control Structure

In a LS-PVPP, tens to hundreds of PV generators are interconnected through a collection grid in order to increase the power. ABB, SMA, Danfoss, and First Solar have described some topologies for this distribution as radial, ring and star [23]. The main difference among them is the reliability and the cost [32]. In the current paper, radial configuration is considered as it is the most used topology.

In the case of the PV generator, the configuration can be central, string or multistring. The most used configuration is the central one where the PV array is interconnected to a single stage inverter [33,34]. Then, the inverter is connected with a three winding transformer (Figure 3).

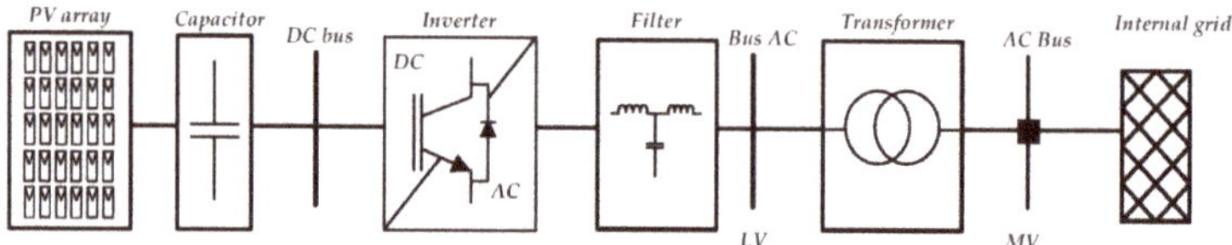

Figure 3. PV generator in central configuration.

As many PV generators are interconnected in a LS-PVPP, a central controller is necessary. Additionally, each PV generator has to perform its local control. Thus, a hierarchical control architecture is considered, as it is illustrated in (Figure 4), where the first stage is the transmission system operator (TSO) who sends the requirements, then the second stage is the power plant control (PPC) and the third stage is the PV generator's local control.

The control of the LS-PVPP is focused in two main tasks: (i) apply grid support actions, for example in case of disturbances, and (ii) coordinate the control of active and reactive power according to TSO's requirements [11,24]. For the second task, the PPC uses a Proportional-Integral (PI) controller to reduce the error between the reference given by the TSO and the power available in the grid. Then, the total active or reactive power calculated by the controller is divided by the total number of PV generators in the LS-PVPP and this is the reference value under which the PV generators should respond (Figure 5) [24,32]. After these references are calculated, the PV generator develops its corresponding control according to grid code requirements and the behavior of the internal grid to keep ac voltage and frequency constant.

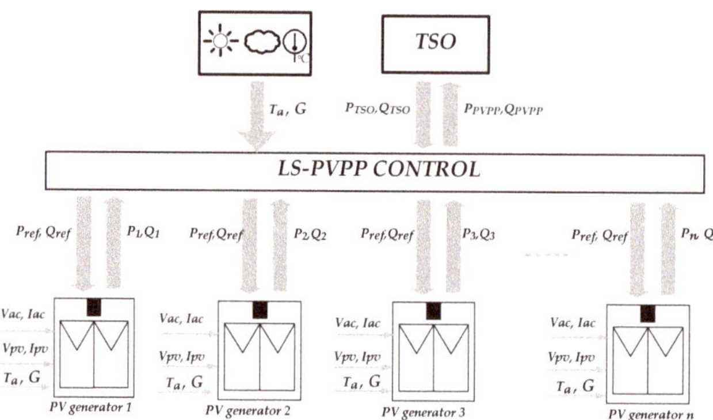

Figure 4. Proposed control architecture for a large scale photovoltaic power plant (LS-PVPP).

A general control structure of a PV generator is illustrated in Figure 6. The PV generator has three main tasks: (i) MPPT, (ii) the inverter control, and (iii) the control of active and reactive power. For the first task, the MPPT, the aim is to look for the v_{mpp} at each solar irradiance and temperature according to the P-V curves characteristics. To address this, algorithms as perturb and observe, hill climbing, incremental conductance as the more known and others as fuzzy control and swarm optimization have been developed [35,36].

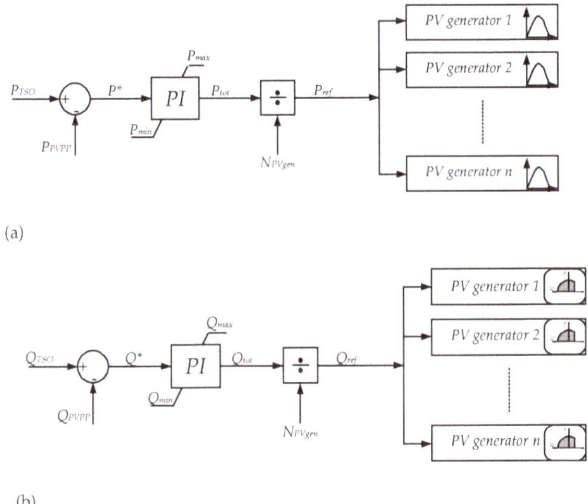

(a)

(b)

Figure 5. Power plant control (**a**) Active power and (**b**) Reactive power.

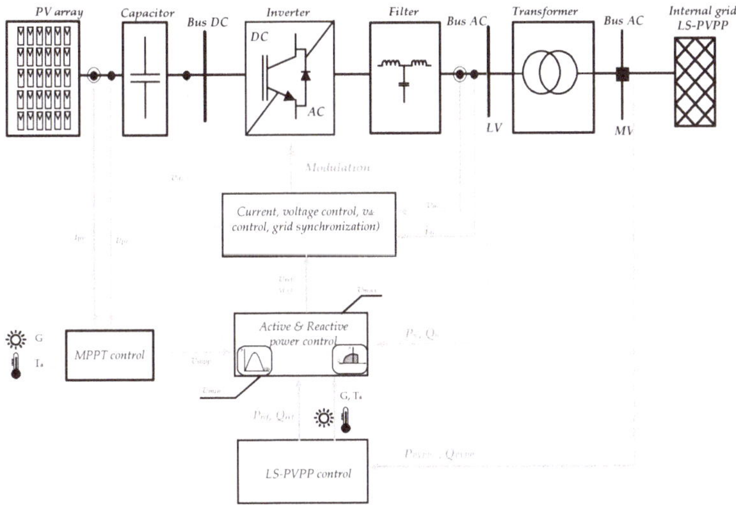

Figure 6. General control of a PV generator.

The second task, it is the one in charge of the general inverter control to interconnect the PV generator with the internal grid of the PVPP. This control performs the grid synchronization, the voltage modulation, the dc voltage regulation and the current loop. The third task, it is in charge of the delivering of the power demanded by the PPC. This control should consider the PQ capability curves of the PV generator analysed in [30] and the variation of ambient conditions as solar irradiance and ambient temperature.

3. Active Power Control

To address the active power control of a PV generator is nesessary to understand the limitations that it has as a system (PV array and the PV inverter). The active power production capability of a PV generator can be presented by a P-G curve (Figure 7). From the figure, four main regions can be identified. Region I is the starting phase, Region II is the controlling phase, Region III is the clipping phase and Region IV is the shut off phase. To go from one region to other three main points are considered: cut-in, rated and cut out solar irradiance. The solar irradiance at which the PV generator first starts to generate power is named as "cut-in solar irradiance". As the solar irradiance increases, the PV generator starts to work with the MPPT control. When the rated power is reached, the point of solar irradiance is the one named as "rated solar irradiance". Eventhough, the PV array can generate more power due to higher solar irradiance, the inverter limits the generation of active power and loses the ability to track the MPP. Thus, in this region the PV generator waste available PV power. As the solar irradiance increases, the cell temperature also does it. The active power is reduced and the system cannot track any longer the active power. When this point is reached (cut out solar irradiance), the PV generator has to shut off.

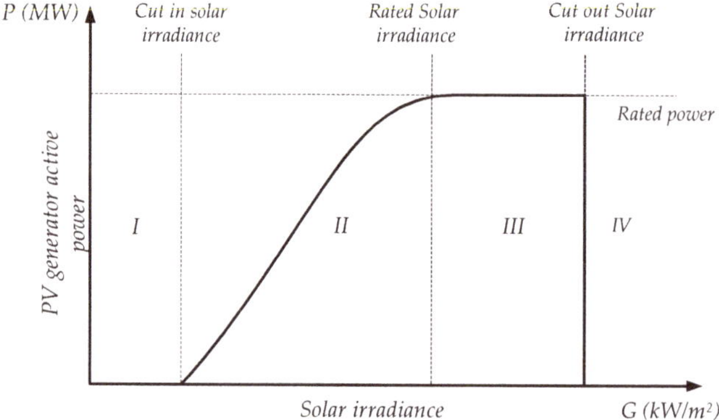

Figure 7. Control areas of a PV generator in a Power vs Solar irradiance curve (P-G curve).

Taking into consideration this curve, the present section proposes an active power control of a PV generator for power curtailment and active power reserves.

3.1. Power Curtailment

To address the power curtailment, the PV generator cannot be working at the maximum power point. Instead the control works close to the reference of active power (P_{ref}) given by the PPC. Thus, a Reference Power Point Tracker (RPPT) is used. Considering the P-G curve, it can be seen that the MPPT control is applied until a point of solar irradiance depending on the power reference. Although, the solar irradiance increases, the PV generator can only supply the reference power by using the RPPT control. In this case, the RPPT control can be applied in Region II and III of the P-G curve (Figure 8a).

The control of active power will be managed according to two variables: dc voltage variation and the active power reference (P_{ref}). For that purpose, any algorithm used for MPPT can also be used in RPPT but the target point is what changes. The most common algorithm is Perturb and Observe, that is used in the present study. In this case, the dc voltage is changed by small steps (Δv) until the active power generated by the PV array is the same as the power reference. Each time the solar irradiance changes, the algorithm should only decide if the dc voltage reference should increase or decrease its value. On this control, the dc voltage limits are also considered according to the PV array and the inverter limitations (v_{min}, v_{max}). However, as the solar irradiance changes, the reference of active power could be higher than the maximum possible power that the PV generator can supply. In this case, the algorithm starts to work as a normal MPPT (Figure 8b).

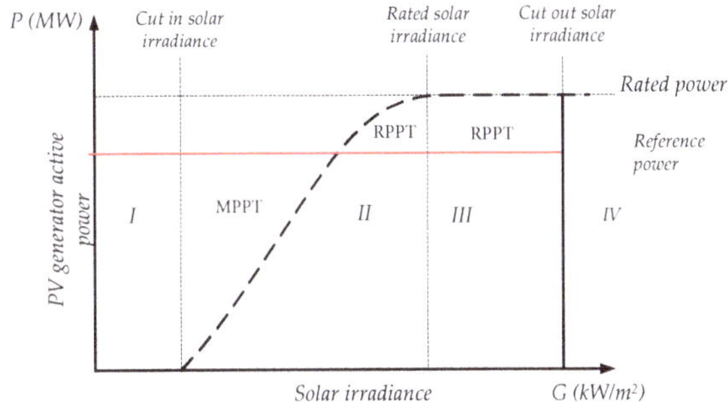

(a)

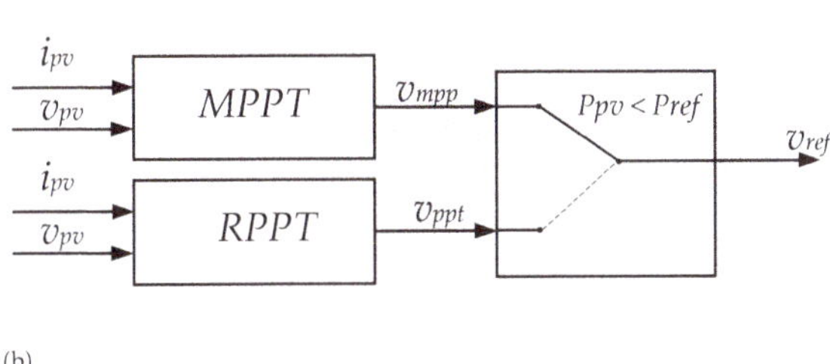

(b)

Figure 8. Active power curtailment (**a**) P-G curve when a reference is given and (**b**) logic between maximum power point tracker (MPPT) and Reference Power Point Tracker (RPPT).

An example of this control is illustrated in Figure 9, where the first point is for a given solar irradiance (blue line) and with a voltage equal to the open circuit voltage (v_{oc}). At this point, the active power that the PV generator can supply is equal to 0. Then, the dc voltage will reduce its value in small steps Δv in order to be close to the reference (2). In the case, the solar irradiance reduces, the new point of operation will be in (3). Then, again the control will change its dc voltage to get close to the reference. As the maximum power at the new solar irradiance is less than the reference, the control will change to MPPT instead of RPPT until it gets the maximum power (4). If the solar irradiance suddenly increases (red line), the PV generator operates in a new point (5). As the dc voltage is equal to v_{mpp}, this has to increase in small steps until the active power generated is the same as the reference given by the PPC (6). This algorithm is presented in the block diagram illustrated in Figure 10.

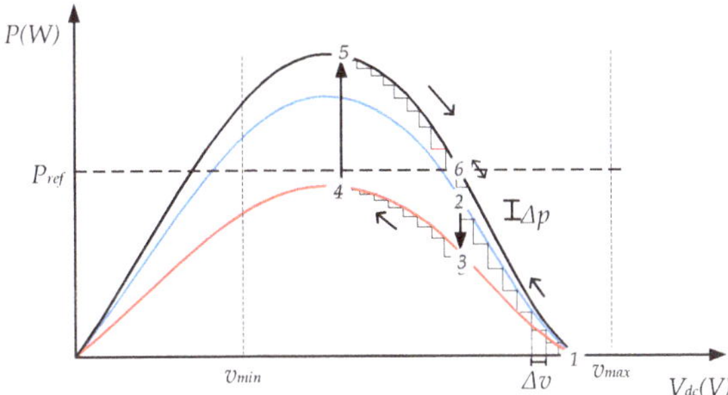

Figure 9. RPPT operation in a PV generator. (1) Initial point. (2) Point of operation by using the algorithm. (3) Change of point of operation because of solar irradiance. (4) New maximum power point. (5) Reduce of power. (6) Reference of active power.

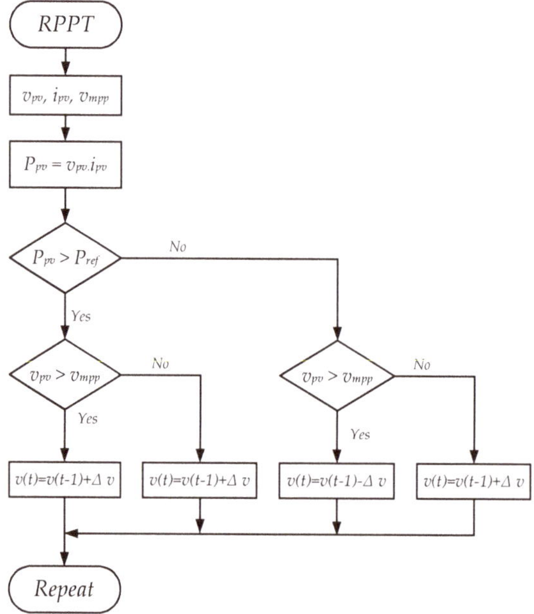

Figure 10. RPPT control algorithm.

3.2. Active Power Reserves

In this paper, any energy storage is used. So, to approach the active power reserves is necessary to work in deloaded operation. This operation consists that the PV generator supply a reduced output power instead of the maximum power. The reduction can be between 10 or 20% of the maximum power for each solar irradiance as it is illustrated in Figure 11. To obtain the new operation point, the dc voltage is different than the v_{mpp}.

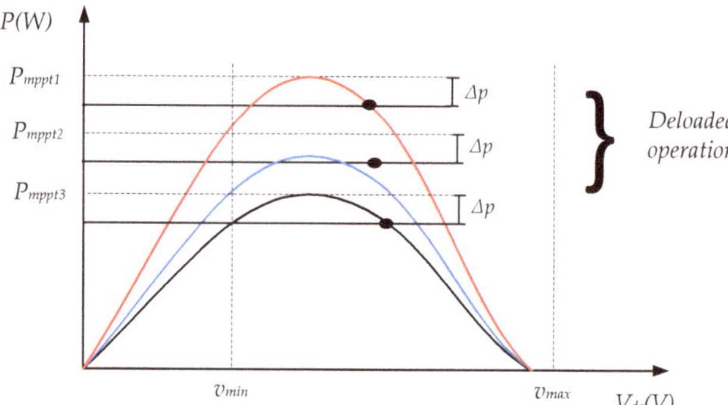

Figure 11. Deloading operation in PV generators.

To control the PV generator for active power reserves, it is necessary to calculate at each time step the possible maximum active power that the PV generator can supply (P_{mpp}). Then, the power reserve is given by the percentage required by the TSO of the maximum possible active power given by the expression:

$$P_{reserve} = \Delta P_{tso} \times P_{mpp}(G, T).\tag{1}$$

The new reference for active power can be calculated as follows:

$$P_{ref} = P_{mpp} - P_{reserve}.$$ (2)

With this new reference, the RPPT control explained in Section 3.1 can be applied. Then, the new P-G curve is presented according to the new performance of the PV generator in Figure 12. In this case, the RPPT can be used in region II and III of the P-G curve.

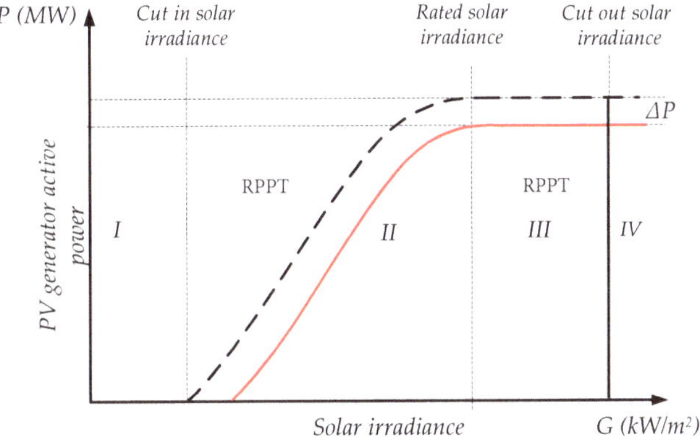

Figure 12. Control areas of a PV generator in a P-G curve when power reserve is considered.

In summary, the PV generator will be working with MPPT or RPPT depending on the ambient conditions and the PPC 's requirements. Then, the dc voltage (v_{ref}) will vary according to the control chosen until it fulfills the requirements (Figure 13).

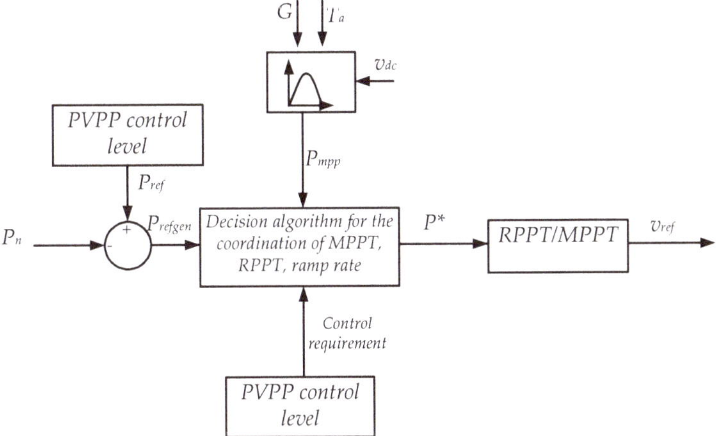

Figure 13. Control scheme for control of active power in PV generators.

After the active control has been addressed, the reactive power control is explained in the following section.

4. Reactive Power Control

The P-Q curves of a PV generator depends not only on the variation of solar irradiance or temperature but also on the dc voltage applied at the terminals of the PV arrar or the modulation index as part of the internal inverter control. In Figure 14 can be seen the variation of these parameters and how they affect to the P-Q curve.

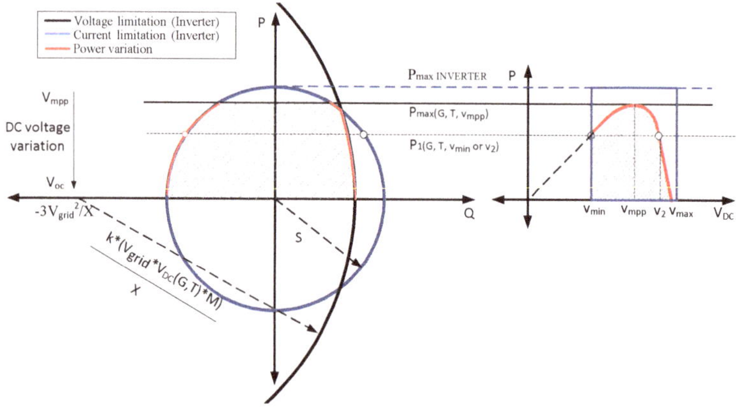

(a)

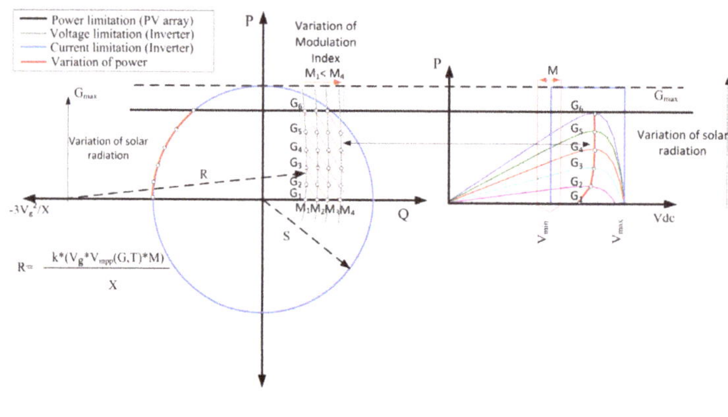

(b)

Figure 14. PQ capability curve of a PV generator (**a**) Variable dc voltage (**b**) Variable Modulation index [30] (Reproduced from Solar Energy, Vol 140, Ana Cabrera et al., "Capability curve analysis of photovoltaic generation systems", Copyright (2016), with permission from Elsevier).

These curves obeys to the following expressions:

$$P_{ref}^2 = S^2 - Q_{ref}^2,$$

(3)

$$P_{ref}^2 + \left(Q_{ref} + \frac{3V_{grid}^2}{X} \right)^2 = \left(3.\frac{V_{grid}V_{conv}}{X} \right)^2.$$

(4)

Taking into account this P-Q curve, this section presents a novel control to provide reactive power depending on the grid code requirements. In this case, the reactive power control is set as a priority

(Figure 15). This control reads the reference of reactive power given by the plant operator. If the reactive power control is not a priority, the control is developed with a conventional reactive power regulation. But if the reactive power is set as a priority, then the PV generator has to calculate the maximum possible reactive power (q_{mpp}) by taking into account the capability curves studied in [30] and the variation of ambient conditions.

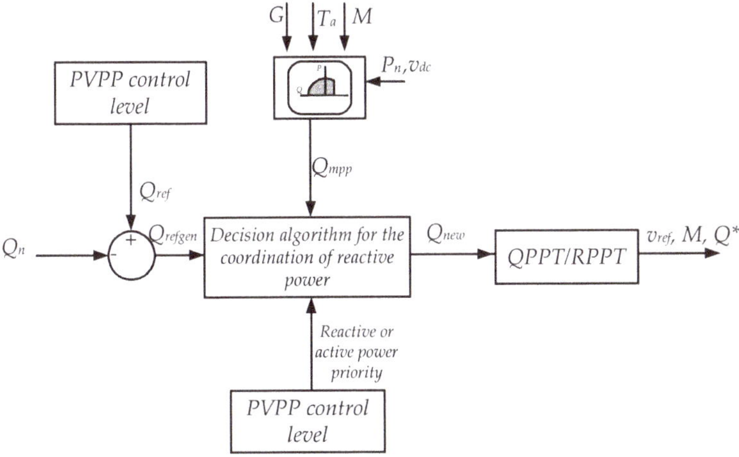

Figure 15. Control scheme for control of reactive power in PV generators.

The performance of the control will vary depending if it is absorbing or injecting reactive power.

4.1. Absorption of Reactive Power

For the absorption of reactive power, the area of operation is in the fourth quadrant of the PQ capability curve. The maximum limitation of the reactive power varies according to:

$$Q^2_{mpp}(G, T) = S^2 - P^2_{mpp}(G, T, v_{mpp}),\tag{5}$$

where, the P_{mpp} is the maximum active power that the PV generator can supply at that instant. For a given reference of absorbed reactive power q_{ref}, the control evaluates if this is higher than the q_{mpp} at each instant. If it is higher, then the PV generator has to reduce the injection of active power by the variation of dc voltage. So, the PV generator is not working any longer at MPP instead will be working in other point of operation of active power. The RPPT control should again track the reference of active power calculated due to reactive power reference. Every time the solar irradiance changes, the control has to track the reactive power point by the variation of active power. This control will be called as the reactive power point tracker (QPPT).

This behavior is illustrated in Figure 16, where the first point is for a given solar irradiance (A). At this point, the active power that the PV generator can supply is P_{mpp}. On this instant, a reference of reactive power is given to the generator's control. However, with this power the Q_{mpp} is lower than the reference. Thus, a new reference of active power is calculated (P_{ref1}). To achieve this point, the dc voltage has to change from v_{mpp} to v_{ref} so the PV generator starts to work at point B. Then, the generator can supply the value of reactive power equal to the reference (point 3). In the case the solar irradiance changes a new PV curve is generated (blue line), because of the dc voltage value, the new active power is P_2 and the PV generator starts to work in point C (PV curve) and 4 (PQ curve). As the RPPT control has to follow the reference of reactive power, then the dc voltage reduces to reach the reference (Point D and point 3).

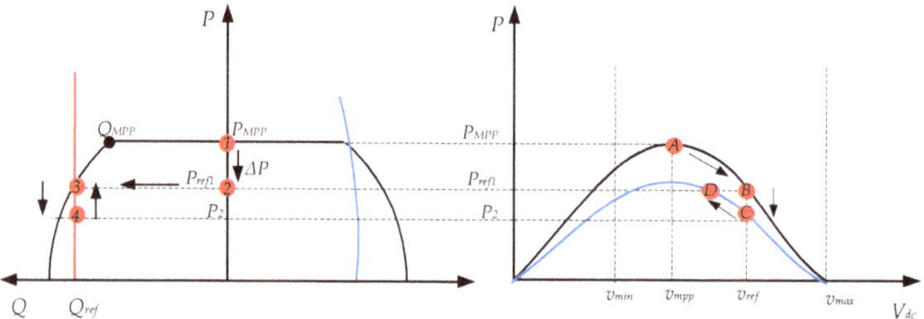

Figure 16. QPPT operation in a variable PQ and ambient conditions: Absorption of reactive power. (1) MPP at one solar irradiance, (2) Variation of active power reference, (3) Reference of reactive power, (4) New active and reactive power

4.2. Injection of Reactive Power

For the injection of reactive power, the area of operation is in the first quadrant of the PQ capability curve. The maximum limitation of the reactive power varies according to:

$$Q_{mpp} = \frac{3\sqrt{3}}{2\sqrt{2}} \cdot \frac{V_{grid} \cdot v_{mpp} \cdot M}{X},$$

(6)

where, the modulation index (M) varies between 0 and 1. The maximum possible reactive power for a given solar irradiance, temperature and v_{mpp} is when M is equal to 1. In order to increase this value of reactive power, the modulation index can be higher than 1 but it can cause the increment of harmonics [37].

In the case, the PPC asks a reference higher than Q_{mpp}, the control should manage in one hand the dc voltage to reduce the active power and on the other hand the modulation index. This behavior is illustrated in Figure 17, where the first point is for a given solar irradiance (A). At this point, the active power that the PV generator can supply is P_{mpp}. In this instant, a reference of reactive power is given to the generator's control. However, with this power the Q_{mpp} is lower than the reference. Thus, a new point of operation is calculated. First, the maximum modulation index varies to a higher value in order to increase the operation area. So, the maximum possible reactive power that the PV generator can inject increases. Then, to reach this point of operation at the specific reference of reactive power, the generated active power is reduced (P_{ref1}). To achieve this point the dc voltage has to change from v_{mpp} to v_{ref} and the PV generator starts to work at point B. The corresponding algorithm that follows the logic illustrated in Figure 18.

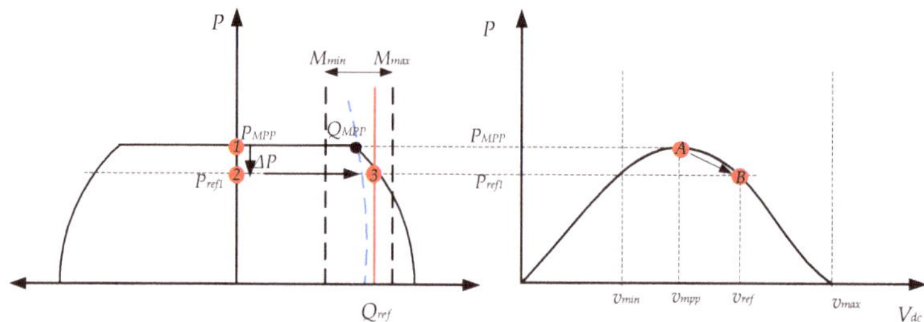

Figure 17. QPPT operation in a variable PQ and ambient conditions: Injection of reactive power. (1) MPP at one solar irradiance, (2) Variation of active power reference, (3) Reference of reactive power.

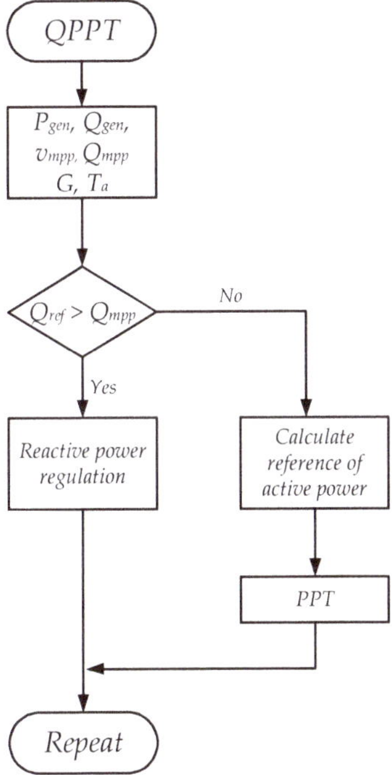

Figure 18. Logic of control for reactive power in PV generators.

The control of active and reactive power is tested for different scenarios. These tests and the performance of the control are explained in the following section.

5. Simulation and Results

A LS-PVPP of 24 MW is designed and modeled in DIgSILENT PowerFactory® and has the configuration presented in Figure 19. For the current study the results for a single PV generator are presented. The main characteristics of the PV generator are summarized in Table 1. The design of this power plant was developed according to the solar irradiance and temperature data taken from Urcuqui-Ecuador in 2014. Besides, the inverter has been oversized 20% of the maximum active power capacity of the PV array. Each PV generator has a nominal power capacity of 0.6 MVA.

Two cases studies are considered: (i) testing the active power control (case study A) and (ii) testing the reactive power control (case study B). For each type of control, the PPC is the one that sends the references of active or reactive power to the local controller. For these tests, three days are chosen with different solar irradiance and an ambient temperature around 10 °C to 25 °C (Figure 20).

Table 1. PV panel and array characteristics.

PV Panel Characteristics		PV Array Characteristics	
V_{oc}	58.8 V	P_{array}	0.5 MW
I_{sc}	5.01 A	N_{ser}	15
I_{mpp}	4.68 A	N_{par}	175
V_{mpp}	47 V	T_{min}, T_{max}	0–70 °C
k_v	0.45 1/°C	G_{max}	1100 W/m^2

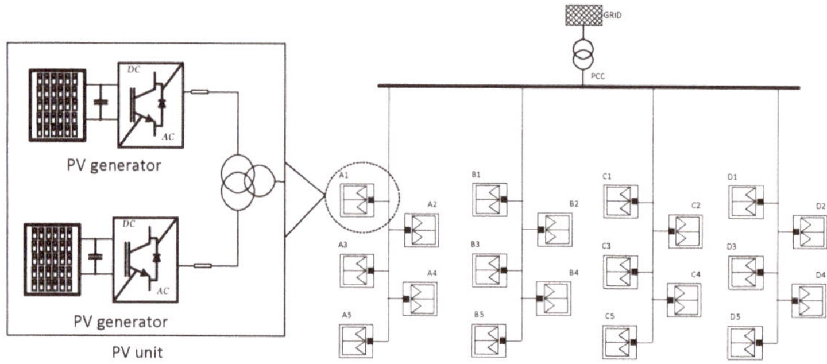

Figure 19. PVPP diagram under study.

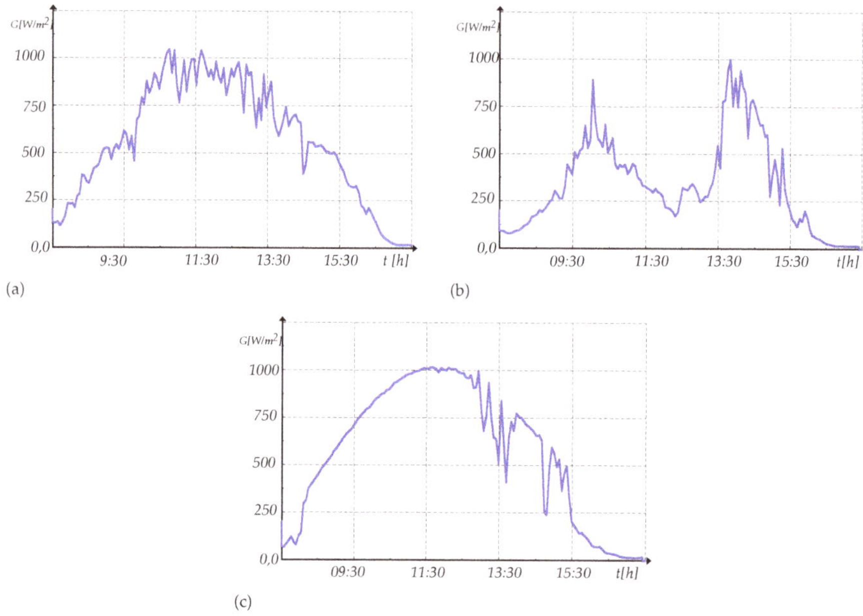

Figure 20. Solar irradiance data (**a**) Day 1 (**b**) Day 2, and (**c**) Day 3.

5.1. Case Study A

For this case study, three different active power values are set as references during the day:

- Set a power reserve of 20% of the maximum power capacity from 6:00 to 10:00 and from 15:00 to 18:00.
- Set a power curtailment of 50% of the maximum capacity from 10:00 to 15:00.
- Deactivate the power curtailment to reach the maximum power point during 10 min (11:25 to 11:35).

The test is developed for day one and two, and the results are illustrated in Figure 21. For any of the days tested, the control of active power makes possible to keep the power reserve equal to 20% of the maximum power capacity for active power higher than 0.20 p.u. When the generated active power is lower than 0.2 p.u, the power reserve is not reached and instead is equal to the maximum possible.

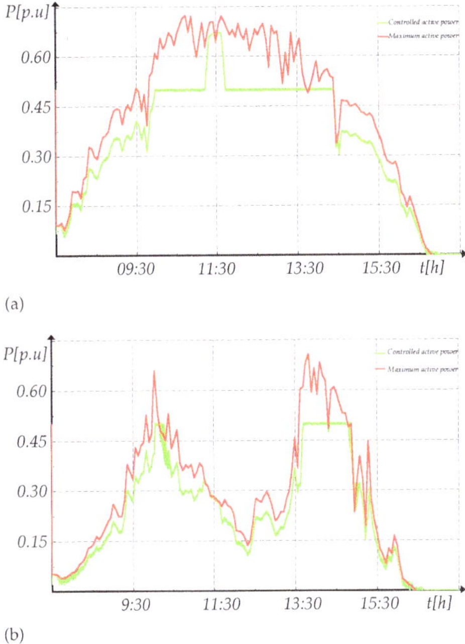

Figure 21. Control of active power for different power references. (**a**) Day one, (**b**) Day two.

However, in the case of power curtailment, there are some differences between the first and the second day. In day 1, the active power reference is reached easily with the control due to the sufficient solar irradiance. On day 2, however, the new reference of power is only reached during ten minutes in the morning and eighty minutes in the afternoon. This behavior is because of the drastic changes of solar irradiance during the day.

The deactivation of the power curtailment in order to get the maximum power during ten minutes is successful in day 1 and day 2. It is important to notice that due to the MPPT control, the ramp rate to get the maximum power is 0.05 MW/min on day 1 and 0.026 MW/min on day 2.

5.2. Case Study B

In this case study, the reactive power control is tested considering the solar irradiance of day 3. To understand the performance of this control, some tests are developed: (a) active power priority and (b) reactive power priority.

5.2.1. Active Power Priority

In this case, the normal MPPT control is used during the day as it is illustrated in Figure 22. Taking into account this control, the maximum reactive power that the PV generator can absorb is illustrated in Figure 23 together with the operational area. It can be seen, that the maximum reactive power that the PV generator can absorb is variable in the time as it is not a priority. The reactive power varies from 0.6 pu to 1 p.u depending on the time of the day.

In the case that a reference of absorbed reactive power is required ($Q_{ref} = -0.8$ p.u) by the PV generator and the MPPT is still used, the response of it is illustrated in Figure 24 together with the operational area. With these conditions, the PV generator can follow the reactive power reference between some hours (06:00 to 08:00 and 15:00 to 18:00). However, when the active power exceeds a certain value, the PV generator cannot follow the reference of reactive power.

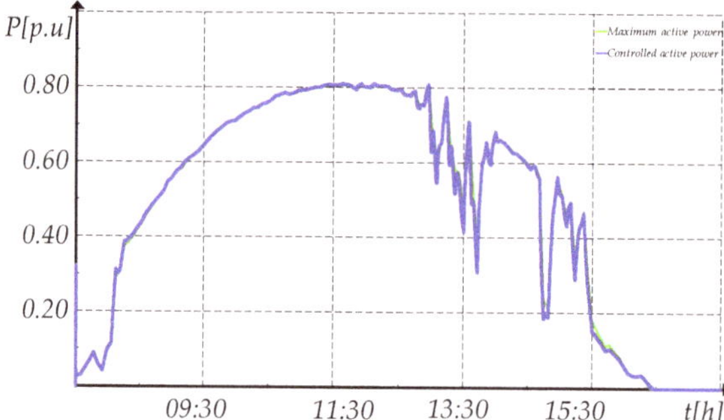

Figure 22. Active power response for day 3 considering MPPT.

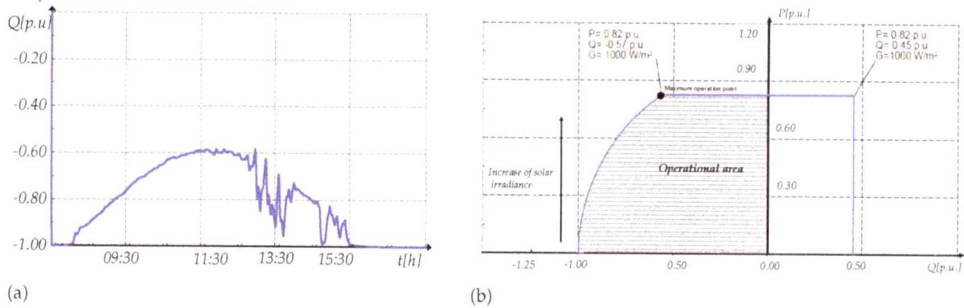

(a)   (b)

Figure 23. Absorbed reactive power when MPPT is considered (**a**) Maximum possible reactive power and (**b**) Operational area.

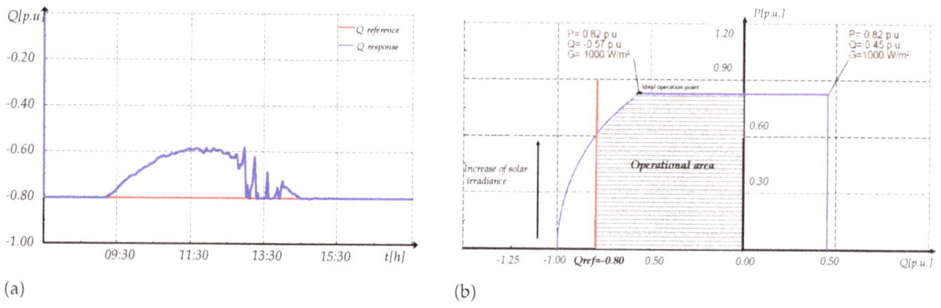

(a)   (b)

Figure 24. Absorbed reactive power when a reference of reactive power is considered (**a**) Response of reactive power (**b**) Operational area.

For a $Q_{ref} = 0.8$ p.u, the PV generator can inject the reactive power depending on the voltage limitation and the modulation index. A modulation index of 1 and 1.75 is tested and illustrated in Figures 25 and 26 respectively together with their operation area. When the modulation index is 1, the reactive power is equal to 0.45 p.u for any solar irradiance. Meanwhile, when the modulation index is 1.75, the reference of reactive power is reached from 07:00 to 10:00 and from 13:30 to 18:00. However, from 10:00 to 13:00, the reference of reactive power is not reached. Instead, the maximum possible reactive power is injected, which depends on the solar irradiance. The curve that limits its behavior from 10:00 to 13:00 is the current curve, for the rest of the day it is the voltage curve.

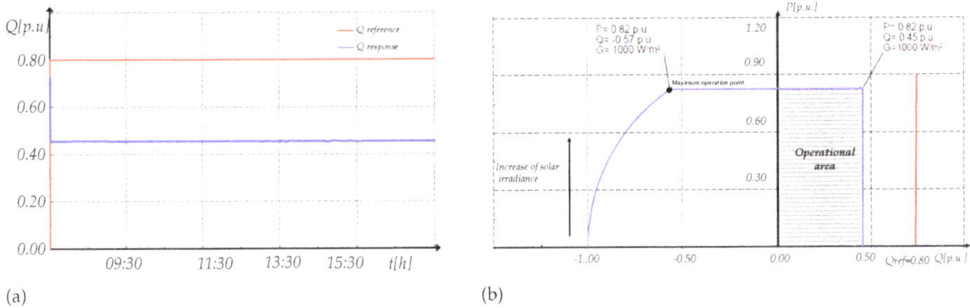

Figure 25. Injected reactive power with MPPT control and $M = 1$ (**a**) Response of reactive power and (**b**) Operational area.

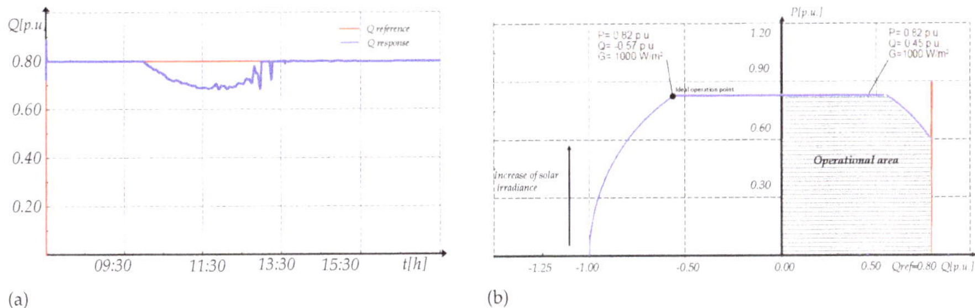

Figure 26. Injected reactive power with MPPT control and $M = 1.75$ (**a**) Response of reactive power and (**b**) Operational area.

5.2.2. Reactive Power Priority

To test this control, two references of reactive power are simulated: (i) $Q_{ref} = -0.8$ p.u and (ii) $Q_{ref} - 0.8$ p.u For the first reference (absorption of reactive power), the results of active and reactive power are illustrated in Figure 27 with the corresponding capability curve. In this case, the PV generator absorbs a value of reactive power equal to its reference almost all the time. Due to the changes of irradiance and the response time of the control, there are specific times where there is an error around 0.05 p.u. Because of the control and the ambient conditions, the active power is limited to 0.6 p.u between the hours 08:00 to 13:00.

For the second reference (injection of reactive power), two different modulation index are tested: (i) $M = 1$ and (ii) $M = 1.75$ (Figures 28 and 29). When $M = 1$, the reactive power cannot reach the reference of reactive power and stays at the maximum value (0.45 p.u). In the case $M = 1.75$, the PV generator injects a reactive power equal to the reference during all the time by the reduction of active power from 09:00 to 13:30.

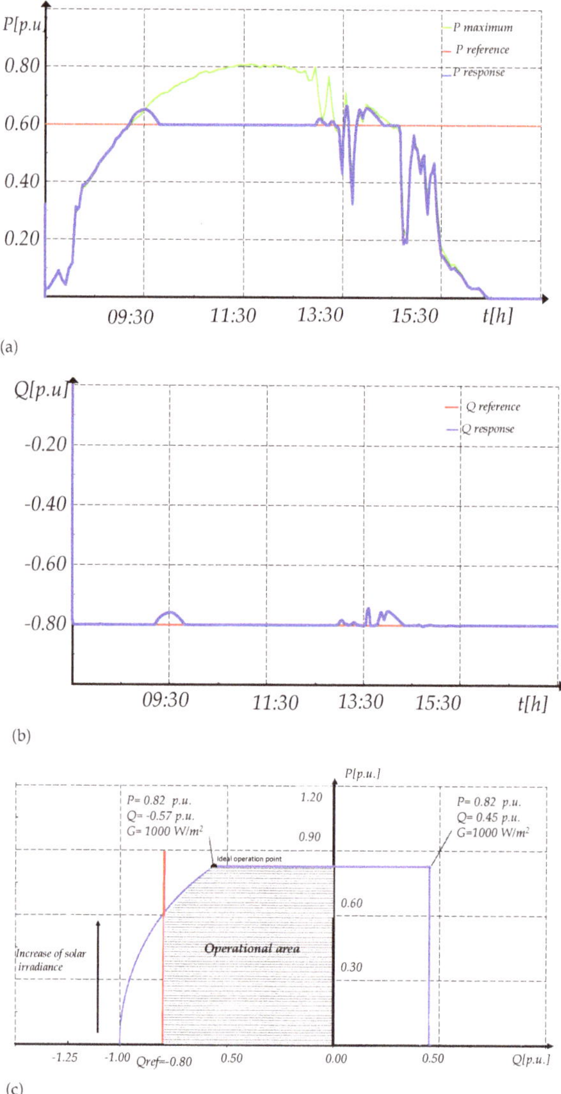

Figure 27. QPPT response when a reactive power reference is applied (**a**) Active power, (**b**) Reactive power, and (**c**) Operational area.

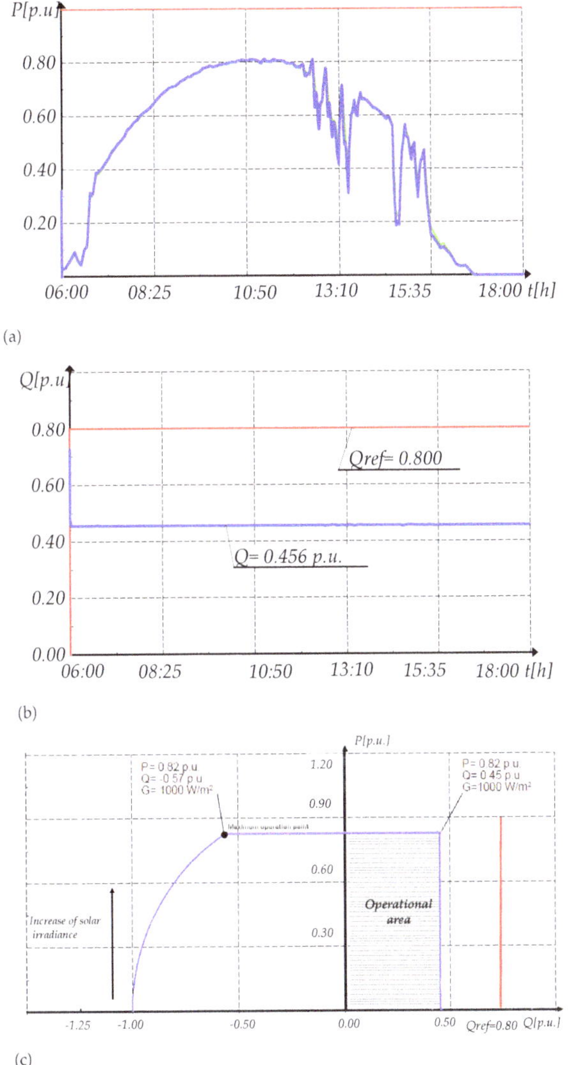

Figure 28. Power response with QPPT for $M = 1$ (**a**) Active power and (**b**) Reactive power, and (**c**) Operational area.

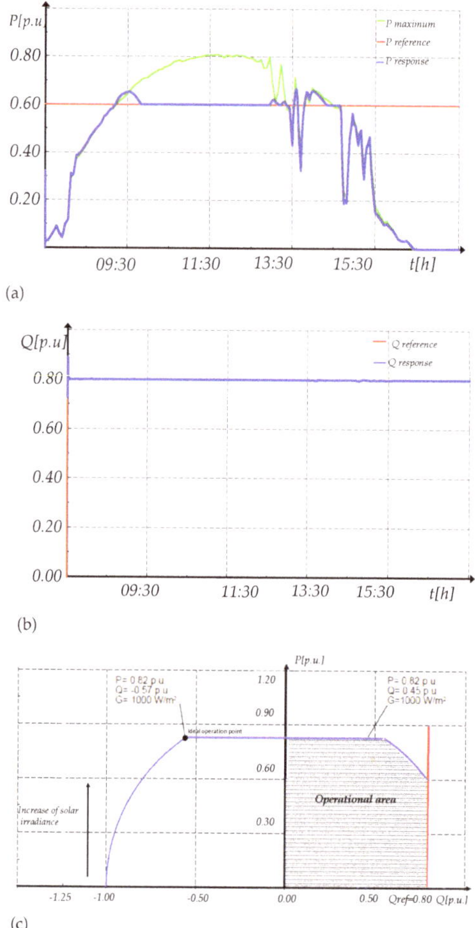

Figure 29. Power response with QPPC for $M = 1.75$ (**a**) Active power and (**b**) Reactive power, and (**c**) Operational area.

6. Discussion

From the controller presented in this paper and from the obtained results, some important issues are necessary to be addressed:

6.1. Active Power Control

The control of active power in a suboptimal point (lower than the MPP) can be developed with a RPPT control. The PV generator can supply power according to an active power reference. The response, however, depends as well on the solar irradiance fluctuations during the day. For instance, on the second day, between 14:00 to 15:00 quick solar irradiance variations are presented and the control tries to respect the 20% of power reserve but the control does not follow this reference.

6.2. Reactive Power Control

For the injection or absorption of reactive power, the response also depends on the solar irradiance when the active power generation is a priority. It can be stated that with a maximum active power, there will be a maximum reactive power that can be injected or absorbed depending on the capability curves. Because of this, if a reference of reactive power is set and at the same time the active power control is a priority, the reference will be reached only if it is lower than the maximum reactive power point possible at that instant. In the case that the reactive power is a priority and there is high solar irradiance, the reference of reactive power can be reached only if the active power point changes to other point of operation lower than the MPP.

It is important to notice that for injection of reactive power, the variation of this value does not present large fluctuation as it depends on the changes of dc voltage together with the modulation index. If the maximum modulation remains fix, then the reactive power that the PV generator can inject also remains close to a fix value. However, this value could be lower than the reference set by the control. Therefore, a change of modulation index helps to achieve the reference of reactive power asked by the PPC.

6.3. Compliance of Grid Codes

Considering the response of the PV generator for the different scenarios for reactive power, it can be analised if the requirements of the grid codes can be achieved under different scenarios. Figure 30 illustrates the capability curve given by the PV generator together with the capability curve required by Puerto Rico and Germany for steady state conditions.

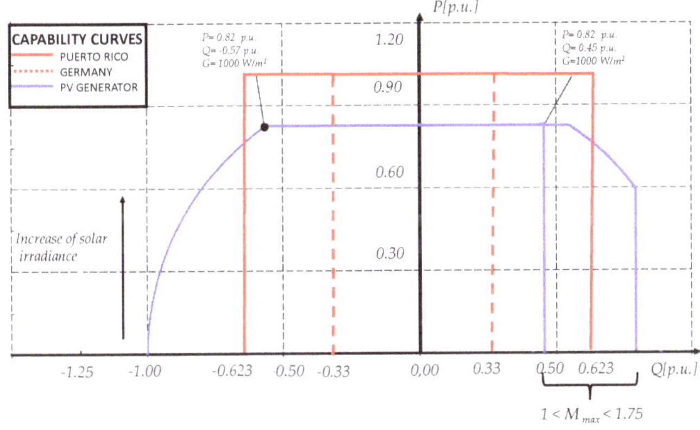

Figure 30. Capability curves comparison considering the grid codes of Puerto Rico, Germany and the capability curve extracted from the current study case.

When QPPT is utilized, the PV generator can inject or absorb reactive power according to the requirements but the active power generated could be lower than the MPP. For absorbed reactive power, if the reference is 0.623 p.u, then the new reference of active power should be 0.78 p.u. For the injection of reactive power, the modulation index has to be higher than 1 to comply this reference.

However, when the reactive power is not set as a priority then the requirements asked by the grid code of Puerto Rico cannot be accomplished for higher solar irradiance and maximum modulation index of 1. So, new equipment should be installed in order to give reactive power support as STATCOM, capacitor banks, FACTS. However, for the case of Germany, at any irradiance the PV generator can supply or inject the reference of reactive power as it is lower than 0.57 p.u without making any change on the operation of active power or the modulation index.

Additionally, it can be seen that for an active power generated lower than 0.78 p.u (corresponding to $G = 900$ W/m^2), the PV generator can absorb or inject reactive power higher than the limitations imposed by the grid codes without reducing the generated active power. Thus, it is necessary that the Grid codes will consider the effect of the PV generator performance at different solar irradiance, temperature, dc voltage and modulation index in order to set higher limitations and improve the performance of the LS-PVPP.

7. Conclusions

This paper has presented the control of active and reactive power for a PV generator considering its capability curves variation applied in a large scale photovoltaic power plant. For this purpose, the current paper has presented the general configuration and control structure used commonly in a LS-PVPP. Then the active power control for a PV generator has been presented considering active power curtailment and active power reserves. Additionally, the control of reactive power was also studied under two different considerations: active power priority or reactive power priority taking into account the corresponding capability curves. Finally, DIgSILENT PowerFactory® was used to simulate the control proposed under different conditions. From the control developed and the simulation some conclusions are presented.

The quick variation of solar irradiance affects not only to the active power response but also to the reactive power. When the solar irradiance is high, then the reactive power capability is reduced. Besides, this could disrupt the plant with quick variations of reactive power, this can be reduced with an appropriate control of the reactive power.

The modulation index and the dc voltage value play an important role on the point of operation of the PV generator when reactive power is injected. For an appropriate control the maximum modulation index can vary between 1 to 1.75 to comply grid code requirements.

The capability curves play an important role in the control of the PV generator when active and reactive power control are considered. These curves should be taken into account for each solar irradiance, ambient temperature, dc voltage and modulation index. The reactive power reference can be achieved by the consideration of these capability curves together with the control.

Considering the grid code requirements regarding the management of active power, it can be stated that working with RPPT for a given reference helps to comply the basic requirements as power reserves and power curtailment. However, a deeper study on ramp rate control must be developed considering variable solar irradiance. In the case of reactive power, grid codes should also consider the behavior of the PV generator according to ambient conditions in order to set the limitations.

Author Contributions: Research concepts were proposed by A.C.-T. and O.G.-B. Manuscript preparation and data analysis were conducted by A.C.-T., M.A.-P., E.B.-M. The edition of the manuscript was performed by all the authors.

Funding: The work was conducted in the PVTOOL project. PVTOOL is supported under the umbrella of SOLAR-ERA.NET Cofund by the Ministry of Economy and Competitiveness, the CDTI and the Swedish Energy Agency. SOLAR-ERA.NET is supported by the European Commission within the EU Framework Programme for Research and Innovation HORIZON 2020 (Cofund ERA-NET Action, n° 691664). It also was funded by the National Department of Higher Education, Science, Technology and Innovation of Ecuador (SENESCYT).

Conflicts of Interest: The authors declare no conflict of interest. The funders had no role in the design of the study; in the collection, analyses, or interpretation of data; in the writing of the manuscript, or in the decision to publish the results.

Abbreviations

The following abbreviations are used in this manuscript:

PV	Photovoltaic
LS-PVPP	Large Scale Photovoltaic Power Plant
PLL	Phase Locked Loop
PPC	Power Plant Controller
PCC	Point of Common Coupling
G	Solar Irradiance
T_a	Ambient Temperature
MPP	Maximum Power Point
MPPT	Maximum Power Point Tracker

TSO	Transmission System Operator
RPPT	Reference Power Point Tracker
P_{PVPP}, Q_{PVPP}	Active and Reactive Power of the PVPP measured at the PCC
P, Q	Active and Reactive Power
P_{ref}, Q_{ref}	Reference of Active and Reactive Power
$P_{reserve}$	Power reserve
P_{mpp}, Q_{mpp}	Active and Reactive Power at the maximum power point of operation
dc	direct current
v_{dc}	dc voltage measured at the dc bus of the PV inverter
v_{pv}, i_{pv}	values of voltage and current measured at the terminals of the PV array
v_{mpp}, i_{mpp}	values of voltage and current at the maximum power point of operation
M	Modulation Index
X	Reactance of the grid
$P\&O$	Perturb and Observe
v_{conv}	VSC ac voltage
v_{grid}	Grid voltage

References

1. Cabrera-Tobar, A.; Bullich-Massagué, E.; Aragüés-Peñalba, M.; Gomis-Bellmunt, O. Review of advanced grid requirements for the integration of large scale photovoltaic power plants in the transmission system. *Renew. Sustain. Energy Rev.* **2016**, *62*, 971–987. [CrossRef]
2. Magoro, B.; Khoza, T. *Grid Connection Code for Renewable Power Plants Connected to the Electricity Transmission System or the Distribution System in South Africa*; Technical Report; NERSA: Pretoria, South Africa, 2012.
3. BDEW. *Technical Guideline Generating Plants Connected to the Medium-Voltage Network*; BDEW: Dresden, Germany, 2008.
4. Liang, J. ; Wang, Y.; Fubao, W.; XiaoFang, W.; Chao, A.; Zhen, L.; XiaoGang, K.; Jian, Z.; Zhileir, C.; ChenHui, N.; et al. *Comparative Study of Standards for Grid Connected PV System in China, the US and European Countries*; Technical Report; National Energy Administration of China: Beijing, China, 2013.
5. Gevorgian, V.; Booth, S. *Review of PREPA Technical Requirements for Interconnecting Wind and Solar Generation*; Technical Report; NREL: Lakewood, CO, USA, 2013.
6. Bird, L.; Cochran, J.; Wang, X. *Wind and Solar Energy Curtailment: Experience and Practices in the United States*; Technical Report; NREL: Lakewood, CO, USA, 2014.
7. Cabrera-Tobar, A.; Fernandez, Y.; Huaca, J.; Pozo, M.; Bellmunt, O.G.; Massi Pavan, A. The effect of ambient temperature on the yield of a 3 MWp PV plant installed in Ecuador. In Proceedings of the IEEE International Conference on Environment and Electrical Engineering and 2019 IEEE Industrial and Commercial Power Systems Europe (EEEIC/I&CPS Europe), Genova, Italy, 11–14 June 2019; pp. 1–6. [CrossRef]
8. Sangwongwanich, A.; Yang, Y.; Blaabjerg, F. Development of flexible active power control strategies for grid-connected photovoltaic inverters by modifying MPPT algorithms. In Proceedings of the IEEE 3rd International Future Energy Electronics Conference and ECCE Asia (IFEEC 2017—ECCE Asia), Kaohsiung, Taiwan, 3–7 June 2017; pp. 87–92. [CrossRef]
9. Muller, N.; Kouro, S.; Renaudineau, H.; Wheeler, P. Energy storage system for global maximum power point tracking on central inverter PV plants. In Proceedings of the IEEE 2nd Annual Southern Power Electronics Conference (SPEC), Auckland, New Zealand, 5–8 December 2016; pp. 1–5. [CrossRef]
10. Li, X.; Hui, D.; Lai, X. Battery Energy Storage Station (BESS)-Based Smoothing Control of Photovoltaic (PV) and Wind Power Generation Fluctuations. *IEEE Trans. Sustain. Energy* **2013**, *4*, 464–473. [CrossRef]
11. Bullich-Massagué, E.; Aragüés-Peñalba, M.; Sumper, A.; Boix-Aragones, O. Active power control in a hybrid PV-storage power plant for frequency support. *Sol. Energy* **2017**, *144*, 49–62. [CrossRef]
12. Ellis, A.; Schoenwald, D.; Hawkins, J.; Willard, S.; Arellano, B. PV output smoothing with energy storage. In Proceedings of the 38th IEEE Photovoltaic Specialists Conference, Austin, TX, USA, 3–8 June 2012; pp. 001523–001528. [CrossRef]
13. Alam, M.; Muttaqi, K.; Sutanto, D. A Novel Approach for Ramp-Rate Control of Solar PV Using Energy Storage to Mitigate Output Fluctuations Caused by Cloud Passing. *IEEE Trans. Energy Convers.* **2014**, *29*, 507–518. [CrossRef]

14. Datta, M.; Senjyu, T.; Yona, A.; Funabashi, T. A Coordinated Control Method for Leveling PV Output Power Fluctuations of PV–Diesel Hybrid Systems Connected to Isolated Power Utility. *IEEE Trans. Energy Convers.* **2009**, *24*, 153–162. [CrossRef]

15. Van Haaren, R.; Morjaria, M.; Fthenakis, V. An energy storage algorithm for ramp rate control of utility scale PV (photovoltaics) plants. *Energy* **2015**, *91*, 894–902. [CrossRef]

16. Beltran, H. Energy Storage Systems Integration into PV Power Plants. Ph.D. Thesis, Universitat Politècnica de Catalunya, Barcelona, Spain, 2011.

17. De la Parra, I.; Marcos, J.; García, M.; Marroyo, L. Control strategies to use the minimum energy storage requirement for PV power ramp-rate control. *Sol. Energy* **2015**, *111*, 332–343. [CrossRef]

18. Hung, D.Q.; Mithulananthan, N.; Bansal, R. Integration of PV and BES units in commercial distribution systems considering energy loss and voltage stability. *Appl. Energy* **2014**, *113*, 1162–1170. [CrossRef]

19. Galtieri, J.; Krein, P.T. Solar Variability Reduction Using Off-Maximum Power Point Tracking and Battery Storage. In Proceedings of the IEEE 44th Photovoltaic Specialist Conference (PVSC), Washington, DC, USA, 25–30 June 2017; pp. 3214–3219. [CrossRef]

20. Hoke, A.; Maksimovic, D. Active power control of photovoltaic power systems. In Proceedings of the 1st IEEE Conference on Technologies for Sustainability (SusTech), Portland, OR, USA, 1–2 August 2013; pp. 70–77. [CrossRef]

21. Yang, Y.; Blaabjerg, F.; Wang, H. Constant power generation of photovoltaic systems considering the distributed grid capacity. In Proceedings of the IEEE Applied Power Electronics Conference and Exposition—APEC 2014, Fort Worth, TX, USA, 16–20 March 2014; pp. 379–385. [CrossRef]

22. Sangwongwanich, A.; Yang, Y.; Blaabjerg, F.; Wang, H. Benchmarking of constant power generation strategies for single-phase grid-connected Photovoltaic systems. In Proceedings of the IEEE Applied Power Electronics Conference and Exposition (APEC), Long Beach, CA, USA, 20–24 March 2016; pp. 370–377. [CrossRef]

23. Cabrera-Tobar, A.; Bullich-Massagué, E.; Aragüés-Peñalba, M.; Gomis-Bellmunt, O. Topologies for large scale photovoltaic power plants. *Renew. Sustain. Energy Rev.* **2016**, *59*, 309–319. [CrossRef]

24. Gomis-Bellmunt, O.; Serrano-Salamanca, L.; Ferrer-San-José, R.; Pacheco-Navas, C.; Aragüés-Peñalba, M.; Bullich-Massagué, E. Power plant control in large-scale photovoltaic plants: design, implementation and validation in a 9.4 MW photovoltaic plant. *IET Renew. Power Gener.* **2015**, *10*, 50–62. [CrossRef]

25. Shah, R.; Mithulananthan, N.; Bansal, R.; Ramachandaramurthy, V. A review of key power system stability challenges for large-scale PV integration. *Renew. Sustain. Energy Rev.* **2015**, *41*, 1423–1436. [CrossRef]

26. Maleki, H.; Varma, R.K. Coordinated control of PV solar system as STATCOM (PV-STATCOM) and Power System Stabilizers for power oscillation damping. In Proceedings of the IEEE Power and Energy Society General Meeting (PESGM), Boston, MA, USA, 17–21 July 2016; pp. 1–5. [CrossRef]

27. Luo, L.; Gu, W.; Zhang, X.P.; Cao, G.; Wang, W.; Zhu, G.; You, D.; Wu, Z. Optimal siting and sizing of distributed generation in distribution systems with PV solar farm utilized as STATCOM (PV-STATCOM). *Appl. Energy* **2018**, *210*, 1092–1100. [CrossRef]

28. Sadnan, R.; Khan, M.Z.R. Fast real and reactive power flow control of grid-tie Photovoltaic inverter. In Proceedings of the 9th International Conference on Electrical and Computer Engineering (ICECE), Dhaka, Bangladesh, 20–22 December 2016; pp. 570–573. [CrossRef]

29. Weckx, S.; Gonzalez, C.; Driesen, J. Combined Central and Local Active and Reactive Power Control of PV Inverters. *IEEE Trans. Sustain. Energy* **2014**, *5*, 776–784. [CrossRef]

30. Cabrera-Tobar, A.; Bullich-Massagué, E.; Aragüés-Peñalba, M.; Gomis-Bellmunt, O. Capability curve analysis of photovoltaic generation systems. *Sol. Energy* **2016**, *140*, 255–264. [CrossRef]

31. Cabrera-Tobar, A.; Aragues-Penalba, M.; Gomis-Bellmunt, O. Effect of Variable Solar Irradiance on the Reactive Power Response of Photovoltaic Generators. In Proceedings of the IEEE International Conference on Environment and Electrical Engineering and 2018 IEEE Industrial and Commercial Power Systems Europe (EEEIC/I&CPS Europe), Palermo, Italy, 12–15 June 2018; pp. 1–6. [CrossRef]

32. Van Hertem, D.; Gomis-Bellmunt, O.; Liang, J. *HVDC Grids for Offshore and Supergrid of the Future*; Wiley: Hoboken, NJ, USA, 2016.

33. Pavan, A.M.; Castellan, S.; Quaia, S.; Roitti, S.; Sulligoi, G. Power Electronic Conditioning Systems for Industrial Photovoltaic Fields: Centralized or String Inverters? In Proceedings of the International Conference on Clean Electrical Power, Capri, Italy, 21–23 May 2007; pp. 208–214. [CrossRef]

34. Sánchez Reinoso, C.R.; Milone, D.H.; Buitrago, R.H. Simulation of photovoltaic centrals with dynamic shading. *Appl. Energy* **2013**, *103*, 278–289. [CrossRef]

35. Eltawil, M.A.; Zhao, Z. MPPT techniques for photovoltaic applications. *Renew. Sustain. Energy Rev.* **2013**, *25*, 793–813. [CrossRef]

36. Belkaid, A.; Colak, I.; Isik, O. Photovoltaic maximum power point tracking under fast varying of solar radiation. *Appl. Energy* **2016**, *179*, 523–530. [CrossRef]

37. Rashid, M. *Power Electronics Handbook*; Butterworth-Heinemann: Oxford, UK, 2001.

Article

A Topology Synthetization Method for Single-Phase, Full-Bridge, Transformerless Inverter with Leakage Current Suppression Part I

Xiumei Yue [1], Hongliang Wang [1,*], Xiaonan Zhu [1], Xinwei Wei [1] and Yan-Fei Liu [2]

[1]   College of Electrical and Information Engineering, Hunan University, Changsha 410082, China; Wangyue_120109@163.com (X.Y.); zhuxn@hnu.edu.cn (X.Z.); weixinwei@hnu.edu.cn (X.W.)
[2]   Department of Electrical and Computer Engineering, Queen's University, Kingston, ON K7L 3N6, Canada; yanfei.liu@queensu.ca
*   Correspondence: liangliang-930@163.com

Received: 23 November 2019; Accepted: 10 January 2020; Published: 16 January 2020

Abstract: Single-phase full-bridge transformerless topologies, such as the H5, H6, or the highly efficient and reliable inverter concept (HERIC) topologies, are commonly used for leakage current suppression for photovoltaic (PV) applications. The main derivation methodology of full-bridge topologies has been used based on both a DC-based decoupling model and an AC-based decoupling model. However, this methodology is not suited to the search for all possible topologies, and cannot verify whether they are inclusive. Part I of this paper will propose a new topology derivation methodology based on unipolar sinusoidal pulse width modulation (USPWM) to search all possible full-bridge topologies for leakage current suppression. First of all, a unified circuit model is proposed, instead of the DC- and AC-based models. Secondly, a mathematic method called the MN principle is then proposed to search for all possible topologies, and a derivation procedure is provided. It was verified that all existing topologies could be found using the proposed method; furthermore, seven new topologies were derived. The proposed topology derivation methodology is extended to search topologies under Double-Frequency USPWM (DFUSPWM). Twenty topologies under USPWM and four topologies under DFUSPWM have been derived.

Keywords: transformerless inverter; full bridge inverter; leakage current; NPC topology

---

1. Introduction

Photovoltaic (PV) sources are among the most promising renewable energy sources, providing clean and emission free energy [1,2]. The single-phase transformerless inverter system has popularly been used, as it has high efficiency and low cost compared with the transformer inverter system. However, the leakage current is a key issue [3–5]. The leakage current generated by PV parasitic capacitors must be limited to satisfy the VDC-AR_N 4015 [6], UL1741 [7], and VDE 0126-1-1 [8] standards. In single-phase, grid-tied inverter systems, half-bridge and full-bridge inverters are typical topologies, as shown in Figure 1.

The Common Mode (CM) current path for grid-tied, transformerless, PV inverter systems is illustrated in Figure 2 [9]. The leakage current path is equivalent to an LC resonant circuit, as shown in Figure 3 [10,11]. V_{AN} and V_{BN} are the voltage difference between points A and N and points B and N, respectively, and L1 and L2 are the output filter inductors. The equivalent CM voltage V_{ecm} is defined as:

$$V_{ecm} = \frac{V_{AN} + V_{BN}}{2} + \frac{V_{AN} - V_{BN}}{2}\frac{L2 - L1}{L2 + L1} \tag{1}$$

For the half-bridge inverter in Figure 1a, only the output filter inductor L1 is employed, so L2 = 0. Thus, (1) can be simplified as follows:

$$V_{ecm} = \frac{V_{AN} + V_{BN}}{2} + \frac{V_{AN} - V_{BN}}{2} = V_{BN} \qquad (2)$$

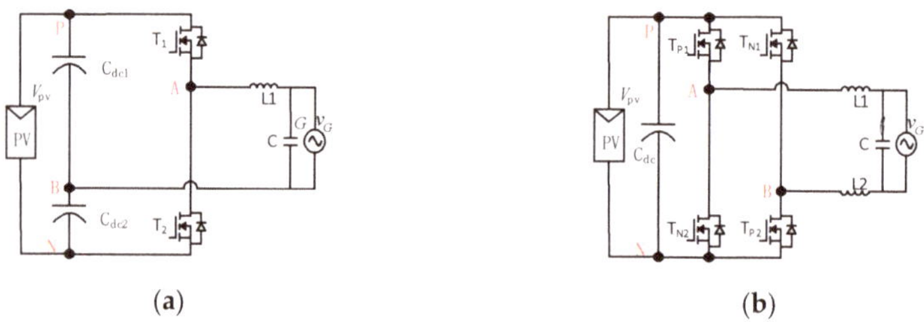

(a)   (b)

Figure 1. Topologies of (**a**) half-bridge inverter; and (**b**) full-bridge inverter (named H4).

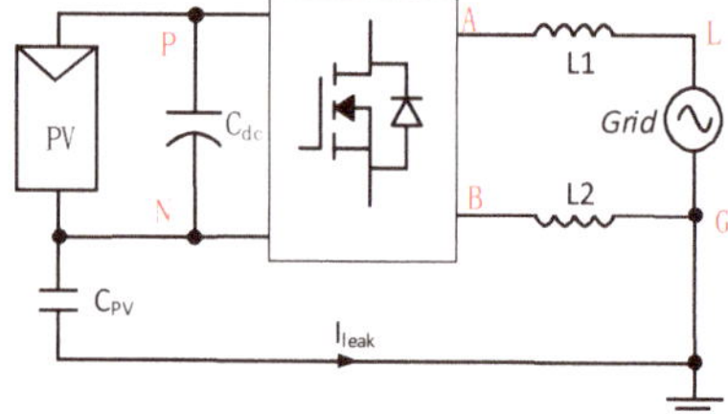

Figure 2. CM current path for transformerless PV inverter.

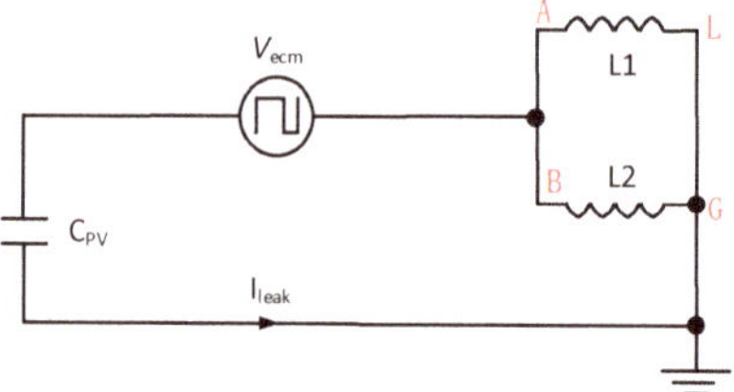

Figure 3. Equivalent circuit for the CM current path.

In Figure 1a, two capacitors, C_{dc1} and C_{dc2}, with equal capacitance values are in series. Capacitor C_{dc2} is charged or discharged by the grid current, and voltage V_{BN} equals half of the input voltage plus the voltage fluctuation of the line frequency. But the high-frequency fluctuation is so small that it can be ignored. So, V_{BN} is approximately constant. However, the DC voltage utilization of half-bridge inverters is only half that of full-bridge topologies, which means that a high-gain boost converter is needed as the first stage. As such, system efficiency and cost will be adversely affected. When two filter inductors are employed (L1 = L2), equation (1) can be simplified as:

$$V_{ecm} = \frac{V_{AN} + V_{BN}}{2} - 0 = \frac{V_{AN} - V_{BN}}{2} \qquad (3)$$

For the full-bridge topology, the leakage current can be eliminated if the common voltage is kept constant. Some state-of-the-art topologies such as H5 [12], HERIC [13,14], and H6 [10,15–43] have been developed. However, there is still a small leakage current because of the parasitic parameters. Thus, the Neutral Point Clamped (NPC) technique is introduced to achieve zero leakage current [44–48]. The full-bridge topologies are divided into DC decoupling model and AC decoupling model [49].

A few rules have been indirectly reported in the literature, as well as some topology synthetization methods such as those based on the DC- and AC-decoupling model, as well as topology derivation methods from H4, H5, and H6. None of the topology synthetization methods currently being used could answer the question of how many topologies could be derived, as there is no unified model. Part I of this paper focus on the topology derivation methodology to achieve small leakage current [50]. It proposes a unified model to replace the DC- and AC-decoupling models based on four rules, including two which have already been reported in the literature [51,52]. More importantly, a mathematic method called the "MN principle" is proposed to derive all the possible topologies. This only focuses on the number of switches in PC and NC modes. The MN principle also verifies that we only need to focus on $M \leq 4$, $N \leq 4$, because the remaining topologies can always be simplified into one of them. Thus, the method verifies that all possible topologies can be found. The derivation procedures are introduced to determine all the existing topologies and new topologies under unipolar sinusoidal pulse width modulation (USPWM) and Double-Frequency USPWM (DFUSPWM).

Part I of the paper is organized as follows. Section 2 describes the principles of the unified topology model. Section 3 introduces topology derivation under USPWM. The topology derivation under DFUSPWM is introduced in Section 4. Part I of the paper is concluded in Section 5.

2. Principle of Unified Topology Model and Symmetric Methodology

Figure 4 shows the full-bridge topology and a simplified schematic diagram.

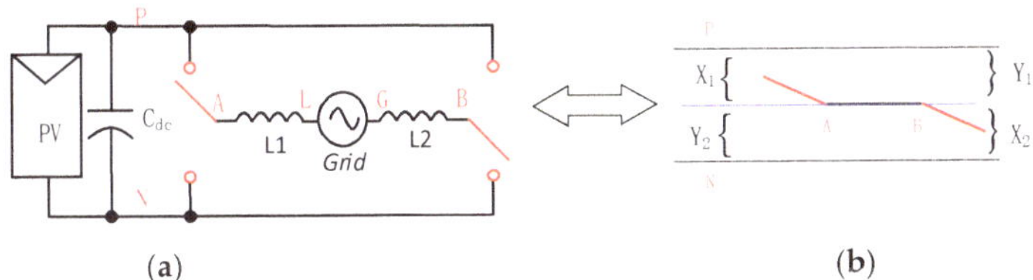

(a)   (b)

Figure 4. (**a**) Full-bridge topology; (**b**) Simplified schematic diagram of full-bridge topology.

In Figure 4a, point P and point N indicate the positive and negative DC bus terminals, respectively, and point A and point B indicate the first and second arm terminals, respectively. The semiconductor switches are always used to connect or disconnect points P and N to points A and B. Figure 4b shows a simplified schematic diagram of the full-bridge topology. Switches T_{PA}, T_{NA}, T_{PB}, and T_{NB} are the equivalent switches between points P and A, between N and A, between P and B, and between N and B, respectively. It should be noted that each equivalent switch can be a single active switch or several active switches connected in series. V_{PN} is the input voltage. The number of switches between points P and A is X_1, between points B and N is X_2, between points P and B is Y_1, and between points N and A is Y_2.

2.1. Principle of USPWM

Figure 5 shows the principle of USPWM. The differential-mode voltage V_{AB} has three levels: $+V_{PN}$, 0, and $-V_{PN}$. There are four modes in a total line-frequency period. The inverter is working in positive conduction (PC) mode and negative conduction (NC) mode when V_{AB} equals to $+V_{PN}$ and $-V_{PN}$. There are two modes if $V_{AB} = 0$: one is the positive freewheeling (PF) mode when the grid current is positive, and the other is the negative freewheeling (NF) mode when the grid current is negative.

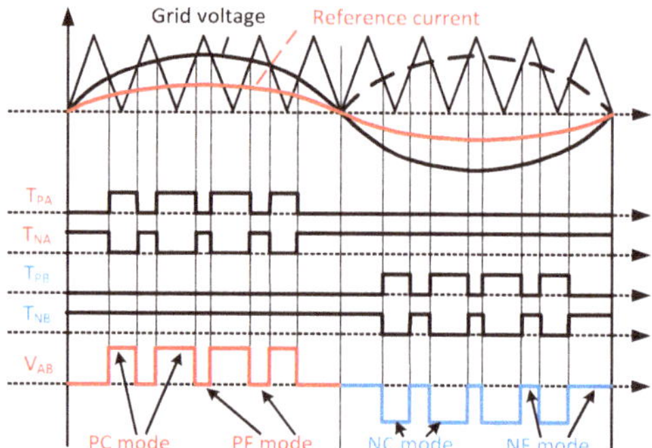

Figure 5. The principle of USPWM.

2.2. Unified Topology Model

Figure 6 shows four modes under USPWM based on the conventional H4 full-bridge topology. As shown in (4), the CM voltage, V_{cm}, is half the input voltage in PC and NC modes, equaling either input voltage V_{PN} or zero in PF mode and NF mode.

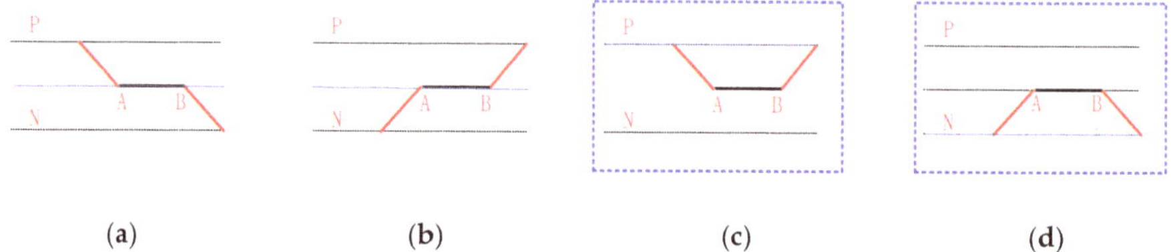

(**a**)   (**b**)   (**c**)   (**d**)

Figure 6. Four modes of H4 topology under USPWM. (**a**) PC mode; (**b**) NC mode; (**c**) PF mode; (**d**) NF mode.

The CM voltage is not constant at switching frequency, which results in high-frequency leakage current.

$$V_{cm} = \frac{V_{AN} + V_{BN}}{2} = \begin{cases} = \frac{V_{PN}}{2} & \text{PC mode or NC mode} \\ = V_{PN} \text{ or } 0 & \text{PC mode or NC mode} \end{cases} \tag{4}$$

To minimize the leakage current, the CM voltage must be kept constant. In PC and NC modes, the CM voltage is equal to half of the DC voltage. Thus, the main objective is to keep the CM voltage also being clamped to half of the input voltage in both freewheeling modes (PF and NF).

$$V_{cm} = \frac{V_{AN} + V_{BN}}{2} = \begin{cases} = \frac{V_{PN}}{2} & \text{PC mode or NC mode} \\ \cong \frac{V_{PN}}{2} & \text{PC mode or NC mode} \end{cases} \tag{5}$$

A unified topology model in PF and NF modes, as shown in Figure 7, is proposed [50]. All switches that connect points P and N are off. A controllable branch B$\hat{C}$A is added to flow positive current in PF mode, as shown in Figure 7c, and another controllable branch A$\hat{D}$B flowing negative current is added in NF mode in Figure 7d. The voltage $V_{AB} = 0$ in PF and NF modes. To regulate the leakage current, a unified topology model should be constructed according to the following four rules in PF and NF modes:

Rule #1. Turn off all the connections to points P and N.

All switches connected to the positive DC bus (point P) and the negative DC bus (point N) must be off in the PF and NF intervals.

Rule #2. Short-circuit terminals A and B to get $V_{AB} = 0$.

A, B is short-circuited through one controllable branch in PF and NF modes. One switch and one diode connected in series are used for bidirectional voltage stress and output current flow, respectively. For example, switch T_{PF} and diode D_{PF} are connected in series for positive current flowing from point B to point A. Switch T_{NF} and diode D_{NF} are connected in series for negative current flowing from point A to point B.

Rule #3. Low cost implementation to satisfy Rule #2.

For low cost, the switches which are not connected to points P and N are on to provide output current flow path in PF and NF modes.

Rule #4. Combine PF and NF modes, and cut off the redundant components.

One PF mode and one NF mode implementation are combined to form a topology. The components which are connected in parallel are merged into one as best as they can be. For example, if an extra diode is connected in parallel with the body-diode of a switch, the former is saved to reduce cost, i.e., two switches in parallel are replaced by one switch.

Based on these rules, a systematic methodology called the "MN principle" is proposed, and will be discussed in the following subsection.

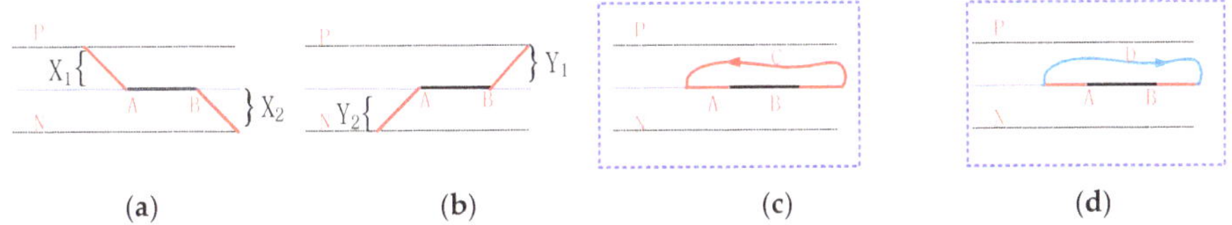

Figure 7. Four modes based on the unified topology under USPWM. (**a**) PC mode; (**b**) NC mode; (**c**) PF mode; (**d**) NF mode.

2.3. MN Principle

A systematic methodology, the MN principle, is proposed to derive all possible full-bridge topologies with small leakage currents. The MN principle can be described as follows. Let M denote the total number of switches that are turned on in PC mode. Since X_1 is the number of switches that connect point P to point A in PC mode, and X_2 is the number of switches that connect point B to point N in PC mode, then

$$M = X_1 + X_2 \qquad (6)$$

Similarly, let N denote the total number of switches that are turned on in NC mode. Since Y_1 is the number of switches that connect point P to point B and Y_2 is the number that connect point A to point N in NC mode, then

$$N = Y_1 + Y_2 \qquad (7)$$

According to rule #1, points A and B must be disconnected to points P and N, which means that at least one switch is needed for T_{PA}, T_{NB}, T_{PB}, and T_{NA}. Thus, the minimum values of X_1, X_2, Y_1, and Y_2 should be one, as shown in (8).

$$Min(X_1, X_2, Y_1, Y_2) \geq 1 \qquad (8)$$

In order to disconnect A, B to P, N in PF and NF modes, according to rule #2, one switch and an extra diode connected in series can be used. Thus, there is one possible way that two switches are in series to implement the equivalent switches T_{PA}, T_{NB}, T_{PB}, and T_{NA}, respectively. For example, two switches, T_{P1} and T_{P2}, are connected in series between points P and A to implement the equivalent switch, i.e., T_{PA}. Switch T_{P1} remains off to disconnect points P and A, and switch T_{P2} remains on to construct the freewheeling branch in PF mode. If three switches, T_{P1}, T_{P2}, and T_{P3}, are in series to implement the equivalent switch T_{PA}, switch T_{P1} remains off to disconnect points P and A, and two switches T_{P2} and T_{P3} are in series to construct the freewheeling branch in PF mode. However, the two switches, T_{P2} and T_{P3}, can be merged into a single switch. Thus, the maximum value for X_1, X_2, Y_1, and Y_2 are less than or equal to 2. Therefore,

$$Max(X_1, X_2, Y_1, Y_2) \leq 2 \tag{9}$$

From the above analysis, it may be observed that the MN principle can cover all possible topologies. Some of them can be simplified. Thus, only simplified topologies are introduced in the next section.

3. Topology Derivation under USPWM

In this section, several examples are provided to show how to derive topologies from the MN principle, such as M = 2 and N = 2, M = 2 and N = 3, or M = 3 and N = 2.

3.1. Case 1: M = 2 and N = 2

When M = 2 and N = 2, there is only one possibility to choose the combined values of X_1, X_2, Y_1, and Y_2, i.e., $X_1 = 1$, $X_2 = 1$, $Y_1 = 1$, and $Y_2 = 1$.

Figure 8 shows four modes derived from $X_1 = 1$, $X_2 = 1$, $Y_1 = 1$ and $Y_2 = 1$. The PC and NC modes are shown in Figure 8a. One switch, T_{P1}, is adopted to connect points P and A due to $X_1 = 1$; meanwhile, another switch, T_{P2}, is viewed as a connection between points B and N on $X_2 = 1$ at PC mode. Similarly, two other switches, T_{N1} and T_{N2}, are added in NC mode with $Y_1 = 1$ and $Y_2 = 1$. Figure 8b shows PF mode. According to rule #1, four switches, i.e., T_{P1}, T_{P2}, T_{N1}, and T_{N2}, are off in PF mode. There is no available switch for output current flow. Thus, an extra switch, T_{P3}, and an extra diode, D_{p3}, are added in series. This is the only choice available for PF mode. Similarly, for NF mode, an extra switch, T_{N3}, and an extra diode, D_{N3}, are added in series for output current flow, as shown in Figure 8c.

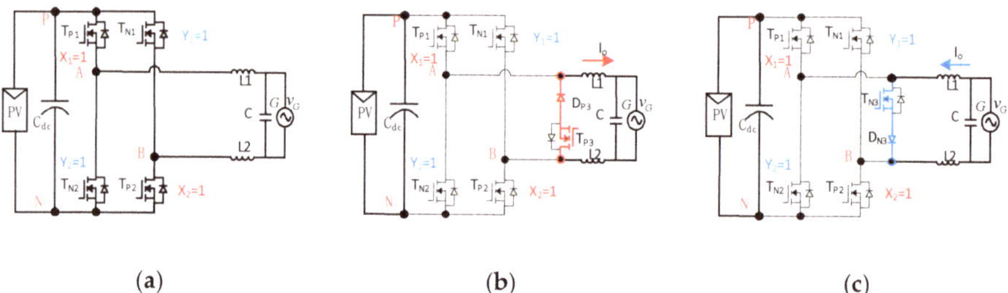

Figure 8. Four modes under $X_1 = 1$, $X_2 = 1$, $Y_1 = 1$, and $Y_2 = 1$. (**a**) PC and NC modes; (**b**) PF mode; (**c**) NF mode.

Three topologies derived from $X_1 = 1$, $X_2 = 1$, $Y_1 = 1$, and $Y_2 = 1$ are shown in Figure 9. Figure 9a can be achieved from Figure 8b,c in PF and NF modes. In Figure 9b, the body-diodes of switches T_{P3} and T_{N3} are used to replace the extra diodes in Figure 9a. Figure 9c is another topology in which four diodes plus one switch are used to replace the two switches, T_{P3} and T_{N3}. These are well-known HERIC topologies [13,30].

The topologies from the MN principle may be divided into two families. Those in the first family have extra diode for output current flow in PF and NF modes. In contrast, the topologies in the second family don't use the extra diode, and the body-diode of the switch is used to allow current to flow, as in the case in Figure 9b in PF and NF modes. Thus, two corresponding topological families under M = 2 and N = 2 are shown in Table 1.

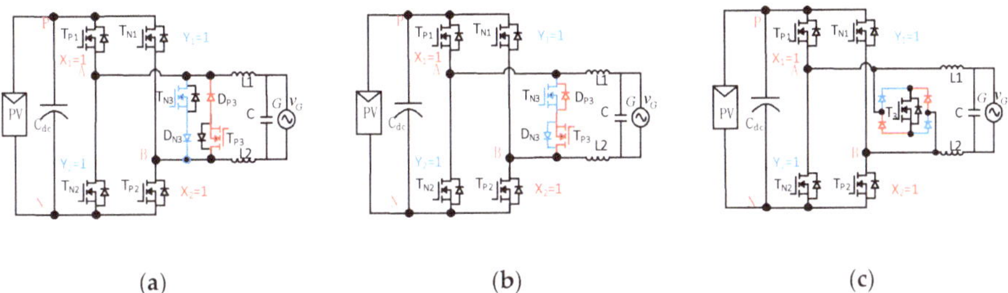

(a) **(b)** **(c)**

Figure 9. Three topologies under $X_1 = 1$, $X_2 = 1$, $Y_1 = 1$, and $Y_2 = 1$. (**a**) R1 [13]; (**b**) R2 [13]; (**c**) R3 [30].

Table 1. Topological families under M = 2 and N = 2.

(M, N)	M = X$_1$ + X$_2$	N = Y$_1$ + Y$_2$	Family with Extra Diode	Family without Extra Diode
(M = 2, N = 2)	1 + 1	1 + 1	R$_1$, R$_3$	R$_2$

3.2. Case 2: M = 3 and N = 2 or M = 2 and N = 3

For M = 3, N = 2 or M = 2, N = 3, there are same topologies between M = 3, N = 2 and M = 2, N = 3, as they are equivalent by exchanging the two bridges. Thus, M = 3 and N = 2 is made as an example to explain the derivation method. M = 3 and N = 2 means there are two possibilities to choose the combined values of X_1, X_2, Y_1, and Y_2, i.e., $X_1 = 1$, $X_2 = 2$, $Y_1 = 1$, and $Y_2 = 1$, and $X_1 = 2$, $X_2 = 1$, $Y_1 = 1$, and $Y_2 = 1$. Considering the symmetrical characteristics with respect to terminals P and N, the two cases are the same. For the sake of brevity, only the former case is analyzed below.

Figure 10 shows four modes under $X_1 = 1$, $X_2 = 2$, $Y_1 = 1$, and $Y_2 = 1$. The PC and NC modes are shown in Figure 10a. One switch, T_{P1}, is used to connect points P and A due to $X_1 = 1$; meanwhile, switches T_{P2} and T_{P3} are viewed as a connection between point B and point N as $X_2 = 2$ in PC mode. Similarly, two switches, T_{N1} and T_{N2}, are added in NC mode with $Y_1 = 1$ and $Y_2 = 1$. According to rule #1, switches T_{P1}, T_{P3}, T_{N1}, and T_{N2} are off in PF and NF modes. Switch T_{p2} is on according to rule #3, and one diode D_{p2} is added based on rule #2 in PF mode, as shown in Figure 10b. In NF mode, an extra switch, T_{N3}, plus the diode D_{N3} are added in series to flow negative output current. The diode is added or served by the body diode of switch T_{P2}. Thus, there are two circuits to realize NF mode, as shown in Figure 10c,d.

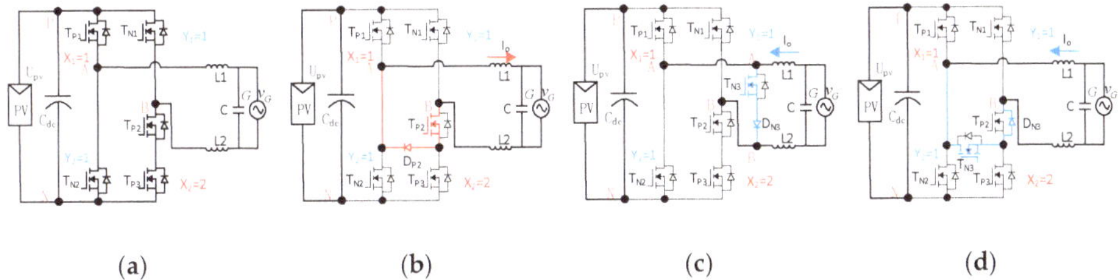

(a) **(b)** **(c)** **(d)**

Figure 10. Four modes under $X_1 = 1$, $X_2 = 2$, $Y_1 = 1$, and $Y_2 = 1$. (**a**) PC and NC modes; (**b**) PF mode; (**c**) NF mode #1; (**d**) NF mode #2.

According to Figure 10, there are one circuit in PF mode and two circuits in NF mode. There are only two possibilities to combine PF and NF modes. Correspondingly, the two topologies derived from $X_1 = 1$, $X_2 = 2$, $Y_1 = 1$, and $Y_2 = 1$ are shown in Figure 11. Figure 11a shows the topology which combines the PF mode in Figure 10b and the NF mode in Figure 10c, while Figure 11b shows the topology which combines the PF mode in Figure 10b and the NF mode in Figure 10d. Two topological families under M = 3, N = 2 or M = 2, N = 3 are shown in Table 2.

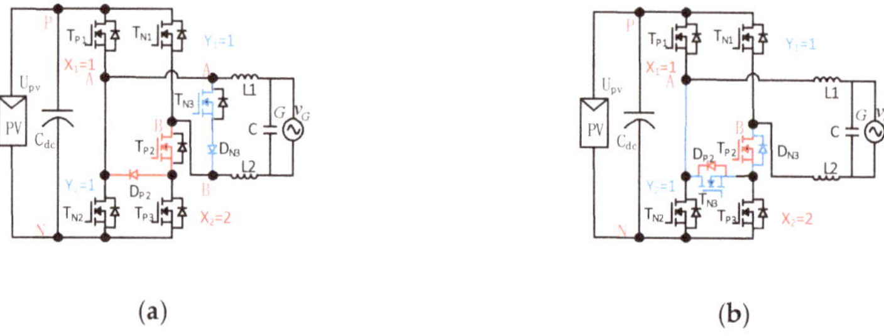

(a)   (b)

Figure 11. Two topologies derived from $X_1 = 1$, $X_2 = 2$, $Y_1 = 1$, and $Y_2 = 1$. (**a**) R4 [31]; (**b**) R5 [27].

Table 2. Topological families under M = 3 and N = 2 or M = 2 and N = 3.

(M, N)	M = $X_1 + X_2$	N = $Y_1 + Y_2$	Family with Extra Diode	Family without Extra Diode
	1 + 2	1 + 1		
(M = 2, N = 3) Or	1 + 1	1 + 2	R4	R5
(M = 3, N = 2)	2 + 1	1 + 1		
	1 + 1	2 + 1		

3.3. Case 3: M = 3 and N = 3

M = 3, which means there are two possibilities to choose the combined values of X_1 and X_2: $X_1 = 1$, $X_2 = 2$, and $X_1 = 2$, $X_2 = 1$. Similarly, N = 3 yields two possibilities to combine Y_1 and Y_2. One is $Y_1 = 1$ and $Y_2 = 2$ and the other is $Y_1 = 2$ and $X_2 = 1$. It should be noted that the same topologies exist between $X_1 = 2$, $X_2 = 1$, $Y_1 = 1$, $Y_2 = 2$, and $X_1 = 1$, $X_2 = 2$, $Y_1 = 2$, $Y_2 = 1$ when two bridges are exchanged. Thus, there are three possibilities: (1) $X_1 = 2$, $X_2 = 1$, $Y_1 = 2$, and $Y_2 = 1$; (2) $X_1 = 2$, $X_2 = 1$, $Y_1 = 1$, and $Y_2 = 2$; and (3) $X_1 = 1$, $X_2 = 2$, $Y_1 = 1$, and $Y_2 = 2$. Considering the symmetry between terminals P and N, case (1) is the same as case (3). For the sake of brevity, only cases (1) and (2) are analyzed below.

Figure 12 shows four modes under $X_1 = 2$, $X_2 = 1$, $Y_1 = 2$, and $Y_2 = 1$. The PC and NC modes are shown in Figure 12a. As shown in Figure 12a, six switches (T_{P1}, T_{P2}, T_{P3}, T_{N1}, T_{N2}, and T_{N3}) are used for $X_1 = 2$, $X_2 = 1$, $Y_1 = 2$, and $Y_2 = 1$. According to rule #1, switches T_{P1}, T_{P3}, T_{N1}, and T_{N3} are off in PF and NF modes. One rest switch, T_{P2}, is on according to rule #3, and one diode, D_{P2}, is added based on rule #2 in PF mode, as shown in Figure 12b. Diode D_{P2} can also be served by the body-diode of switch T_{N2}. The reflected PF mode is shown in Figure 12c. For NF mode, switch T_{N2} is on for negative output current flow. An extra diode, D_{N2}, is added, as shown in Figure 12d. The body-diode of switch T_{P2} is served as the diode D_{N2}, as shown in Figure 12e.

According to Figure 12, there are two circuits in PF mode and two in NF mode. There are four possibilities to combine PF and NF modes. Correspondingly, four topologies derived from $X_1 = 2$, $X_2 = 1$, $Y_1 = 2$, and $Y_2 = 1$ are shown in Figure 13. Figure 13a shows the topology combining the PF mode in Figure 12b and the NF mode in Figure 12d. Figure 13b shows the topology combining the PF mode in Figure 12b and the NF mode in Figure 12e. Figure 13c shows the topology combining the PF mode in Figure 12c and the NF mode in Figure 12d, and Figure 13d shows the topology combining the PF mode in Figure 12c and the NF mode in Figure 12e.

According to rule #4, the extra diode D_{P2} can be absent due to the presence of the body-diode of switch T_{N2} in Figure 13b. Similarly, the extra diode D_{N2} can be absent owing to the presence of the body-diode of switch T_{P2} in Figure 13c. In Figure 13b–d, the two switches, T_{P1} and T_{N1}, are combined into one switch T_1. Thus, the topologies in Figure 13b–d are the same.

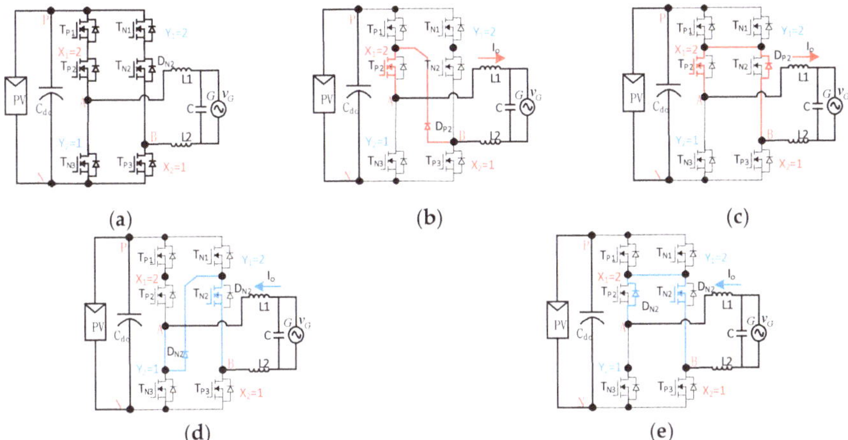

Figure 12. Four modes under $X_1 = 2$, $X_2 = 1$, $Y_1 = 2$, and $Y_2 = 1$. (**a**) PC and NC modes; (**b**) PF mode #1; (**c**) PF mode #2; (**d**) NF mode #1; (**e**) NF mode #2.

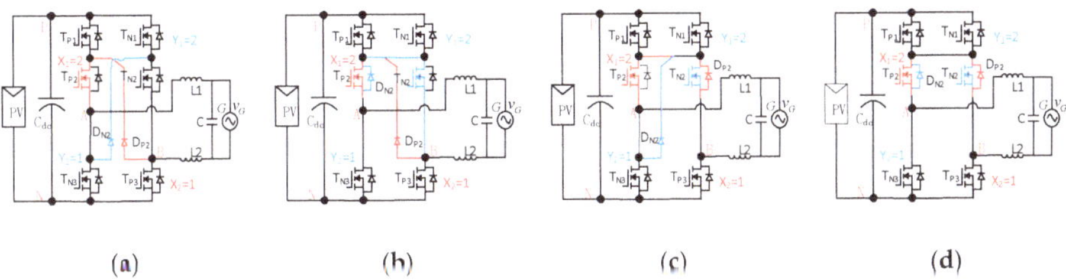

Figure 13. Four topologies derived from $X_1 = 2$, $X_2 = 1$, $Y_1 = 2$, and $Y_2 = 1$. (**a**) R6 [16,31]; (**b**) R7 (circuit one); (**c**) R7 (circuit two); (**d**) R7 (circuit three) [12].

Similarly, one topology derived from $X_1 = 2$, $X_2 = 1$, $Y_1 = 1$, and $Y_2 = 2$ is shown in Figure 14.

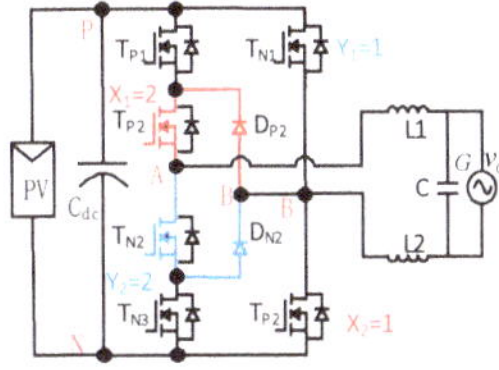

Figure 14. The topology R8 derived from $X_1 = 2$, $X_2 = 1$, $Y_1 = 1$, and $Y_2 = 2$; [38].

Correspondingly, two topological families under $M = 3$ and $N = 3$ are shown in Table 3.

Table 3. Topological families under $M = 3$ and $N = 3$.

(M, N)	$M = X_1 + X_2$	$N = Y_1 + Y_2$	Family with Extra Diode	Family without Extra Diode
(M = 3, N = 3)	2 + 1	2 + 1	R6	R7
	1 + 2	1 + 2		
	2 + 1	1 + 2	R8	None available
	1 + 2	2 + 1		

3.4. Case 4: M = 3 and N = 4 or M = 4 and N = 3

The same topologies exist between M = 3, N = 4 and M = 4, N = 3, as they are equivalent by exchanging the two bridges. For M = 3 and N = 4, M = 3 means that there are two possibilities to choose the combined values of X_1 and X_2: one is $X_1 = 2$ and $X_2 = 1$, and the other is $X_1 = 1$ and $X_2 = 2$. Similarly, N = 4 means three possibilities to combine Y_1 and Y_2. However, only one combination is available according to Equation (9), i.e., $Y_1 = 2$ and $Y_2 = 2$.

Thus, there are two possibilities to choose the combined values of X_1, X_2, Y_1, and Y_2: one is $X_1 = 1$, $X_2 = 2$, $Y_1 = 2$, and $Y_2 = 2$, and the other is $X_1 = 2$, $X_2 = 1$, $Y_1 = 2$, and $Y_2 = 2$. Considering the symmetry between terminals P and N, the two cases are the same. Figure 15 shows four modes under $X_1 = 1$, $X_2 = 2$, $Y_1 = 2$, and $Y_2 = 2$.

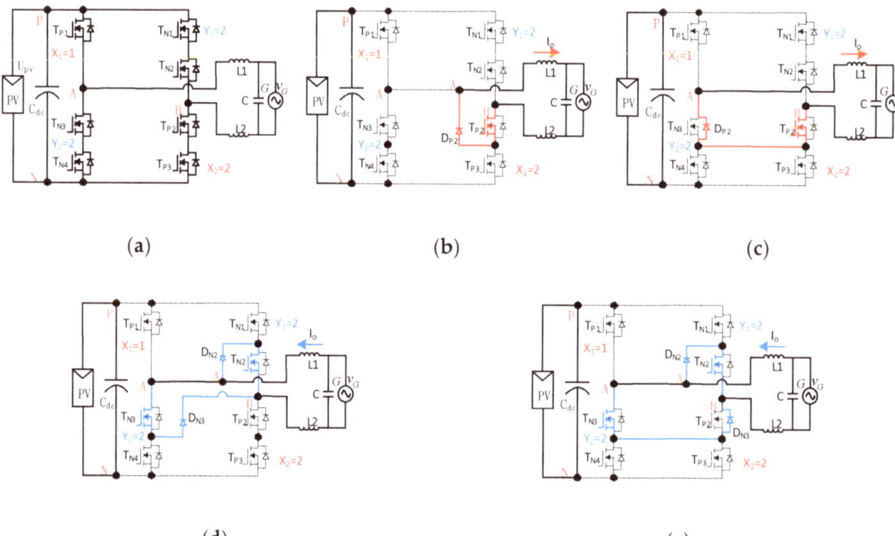

Figure 15. Four modes under $X_1 = 1$, $X_2 = 2$, $Y_1 = 2$, and $Y_2 = 2$. (**a**) PC and NC modes; (**b**) PF mode #1; (**c**) PF mode #2; (**d**) NF mode #1; (**e**) NF mode #2.

The PC and NC modes are shown in Figure 15a. Seven switches (T_{P1}, T_{P2}, T_{P3}, T_{N1}, T_{N2}, T_{N3}, and T_{N4}) are used for $X_1 = 1$, $X_2 = 2$, $Y_1 = 2$, and $Y_2 = 2$, as shown in Figure 15a. From rule #1, the switches T_{P1}, T_{P3}, T_{N1}, and T_{N4} are off in PF and NF modes. Switch T_{P2} is on according to rule #3, and one diode D_{P2} is added based on rule #2 in PF mode, as shown in Figure 15b. Diode D_{P2} is served by the body-diode of switch T_{N3}; the reflected PF mode is shown in Figure 15c. For NF mode, two switches, i.e., T_{N2} and T_{N3}, are on for negative output current flow according to rule #3, and two extra diodes, D_{N2} and D_{N3}, are added from rule #2, as shown in Figure 15d. The body-diode of switch T_{P2} serves as the diode D_{N3}, as shown in Figure 15e.

According to the above analysis, there are two circuits in PF mode and two in NF mode. Thus, four possible topologies are shown in Figure 16. Figure 16a shows the topology combining the PF mode in Figure 15b and the NF mode in Figure 15d. The other three topologies are shown in Figure 16b–d, respectively; however, they are the same, as diode D_{P2} in Figure 16b and diode D_{N3} in Figure 16c can be absent from rule #4. Furthermore, two switches, i.e., T_{P3} and T_{N4}, are merged into one switch, i.e., T_4.

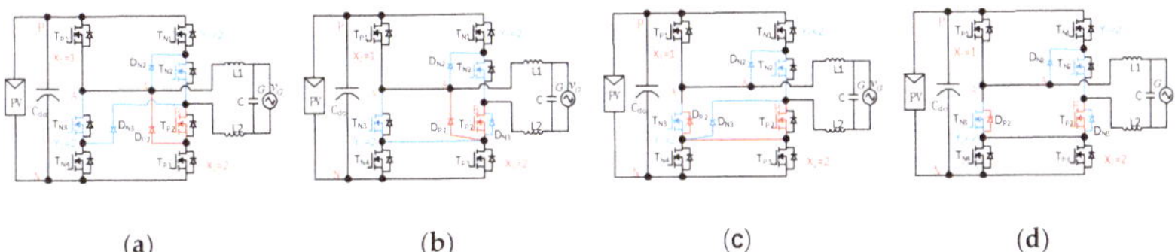

Figure 16. Two topologies derived from $X_1 = 1$, $X_2 = 2$, $Y_1 = 2$, and $Y_2 = 2$. (**a**) R9 (new); (**b**) R10 (circuit one); (**c**) R10 (circuit two); (**d**) R10 (circuit three) (new).

Correspondingly, two topological families under M = 3 and N = 4 or M = 4 and N = 3 are shown in Table 4.

Table 4. Topological families under M = 3 and N = 4 or M = 4 and N = 3.

(M, N)	$M = X_1 + X_2$	$N = Y_1 + Y_2$	Family with Extra Diode	Family without Extra Diode
(M = 3, N = 4)	1 + 2	2 + 2		
or	2 + 2	1 + 2	R9	R10
(M = 4, N = 3)	2 + 1	2 + 2		
	2 + 2	2 + 1		

3.5. Case 5: M = 4 and N = 4

In this case, M = 4 and N = 4. According to Equation (9), M = 4 and N = 4 means only one available possibility to choose the combined values of X_1, X_2, Y_1, and Y_2, i.e., $X_1 = 2$, $X_2 = 2$, $Y_1 = 2$, and $Y_2 = 2$.

Figure 17 shows four modes under $X_1 = 2$, $X_2 = 2$, $Y_1 = 2$, and $Y_2 = 2$. As shown in Figure 17a, eight switches ($T_{P1} \sim T_{P4}$ and $T_{N1} \sim T_{N4}$) are used for $X_1 = 2$, $X_2 = 2$, $Y_1 = 2$, and $Y_2 = 2$ in PC and NC modes. With the reference of the above analysis of Cases 1~4, it is easy to make a similar analysis. For the sake of brevity, this is included here.

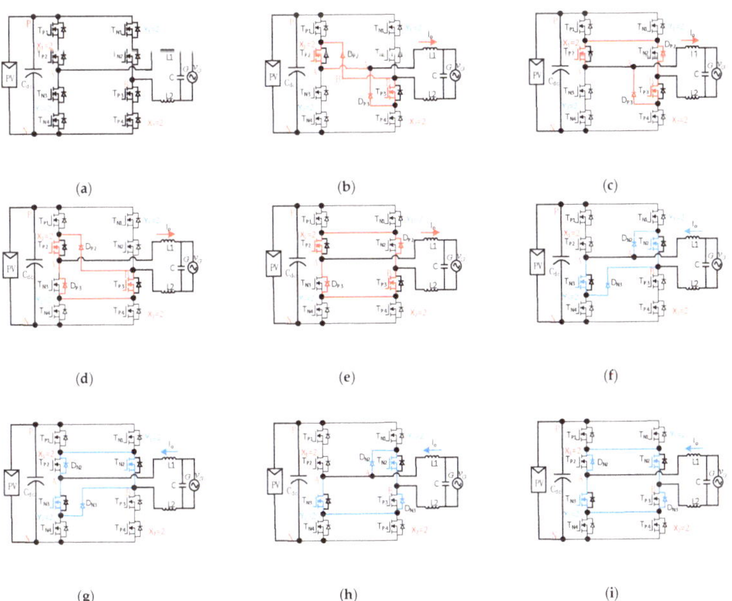

Figure 17. Four modes under $X_1 = 2$, $X_2 = 2$, $Y_1 = 2$, and $Y_2 = 2$. (**a**) PC and NC modes; (**b**) PF mode #1; (**c**) PF mode #2; (**d**) PF mode #3; (**e**) PF mode #4; (**f**) NF mode #1; (**g**) NF mode #2; (**h**) NF mode #3; (**i**) NF mode #4.

It may be observed from Figure 17 that there are four circuits in PF mode and four in NF mode. Thus, sixteen possible topologies may be derived from the MN principle. However, these topologies can be simplified based on rule #4, and the four topologies derived from $X_1 = 2$, $X_2 = 2$, $Y_1 = 2$, and $Y_2 = 2$ are shown in Figure 18.

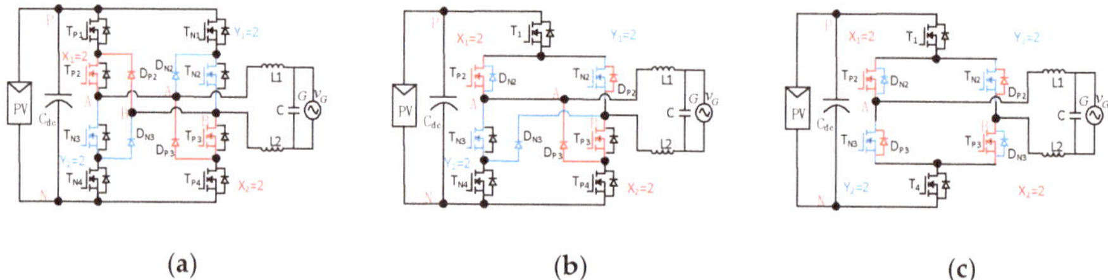

Figure 18. Four topologies derived from $X_1 = 2$, $X_2 = 2$, $Y_1 = 2$, and $Y_2 = 2$. (**a**) R11 (new); (**b**) R12 (new); (**c**) R13 [17].

Two corresponding topological families under M = 4 and N = 4 are shown in Table 5.

Table 5. Topological families under M = 4 and N = 4.

(M, N)	$M = X_1 + X_2$	$N = Y_1 + Y_2$	Family with Extra Diode	Family without Extra Diode
(4, 4)	2 + 2	2 + 2	R11, R12	R12

3.6. All Simplfied Topologies from MN Principle

From the above analysis, two corresponding topological families are summarized in Table 6.

Table 6. Two simplified topological families from MN principle.

(M, N)	$X_1 + X_2$	$Y_1 + Y_2$	Family with Extra Diode	Family without Extra Diode
(M = 2, N = 2)	1 + 1	1 + 1	R1, R3	R2
(M = 2, N = 3) or (M = 3, N = 2)	1 + 2	1 + 1	R4	R5
	1 + 1	1 + 2		
	2 + 1	1 + 1		
	1 + 1	2 + 1		
(M = 3, N = 3)	2 + 1	2 + 1	R6	R7
	1 + 2	1 + 2		
	2 + 1	1 + 2	R8	None available
	1 + 2	2 + 1		
(M = 3, N = 4) or (M = 4, N = 3)	1 + 2	2 + 2	R9	R10
	2 + 2	1 + 2		
	2 + 1	2 + 2		
	2 + 2	2 + 1		
(M = 4, N = 4)	2 + 2	2 + 2	R11, R12	R13

It may be observed from the above analysis that the MN principle can be used to derive all the possible topologies for a single-phase, full bridge, transformerless inverter. Furthermore, thirteen simplified topologies from the MN principle are provided in this paper. All existing topologies (R1–R8, R13) have been covered, and five new topologies marked from R9 to R12 have been found.

For the two topological families with the same M and N, the topologies without an extra diode are of lower cost than those with the extra diode, as the body-diode of switch in the former is used to replace the extra diode; however, the efficiency of the former will likely be a little lower, as that diode has better performance than the body-diode.

(M + N) is the number of conduction switches in PC and NC modes; the bigger (M + N), the higher the conduction loss. Thus, M = N = 2 is the best choice in terms of low conduction loss.

Although M = 4 and N = 4 means large conduction loss under USPWM, the topologies from M = 4 and N = 4 can also work in DFSPWM, where the equivalent switching frequency is double, and the size of the low pass filter is reduced. A detailed description about DFUSWPM will be given in the next section.

4. Topology Derivation under DFUSPWM

In this section, topology derivation methodology is introduced under DFUSPWM. The unified topology model and MN principle are extended to the topologies under DFUSPWM.

4.1. Principle of DFUSPWM

The principle of DFUSPWM is shown in Figure 19. Differential-mode voltage V_{AB} is a three-level waveform.

The levels are $+V_{PN}$, 0, and $-V_{PN}$. There are six modes in total line-frequency period. The inverter is working in positive conduction (PC) mode and negative conduction (NC) mode when V_{AB} equals $+V_{PN}$ and $-V_{PN}$, respectively. There are four modes if V_{AB} equals 0.

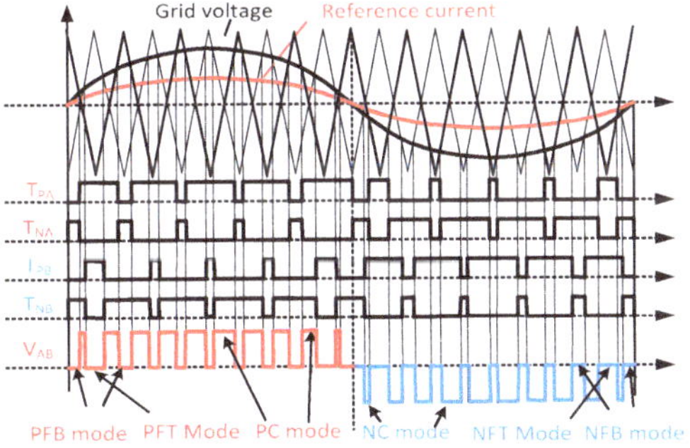

Figure 19. Principle of DFUSPWM.

To achieve double frequency of voltage V_{AB}, there are two interleaved freewheeling modes when the grid voltage is positive: one is used to refer to the freewheeling current flowing through top switch, which is defined as PFT mode. The other one, called "PFB mode" is used to flow freewheeling current through bottom switch. Similarly, two interleaved freewheeling modes, called "NFT mode" and "NFB mode", are used in NF mode. The freewheeling current flows through the top switch in NFT and through the bottom switch in NFB mode.

The modes rotate in the sequence of PFT, PC, PFB, and PC in the positive half cycle of the grid voltage. Similarly, the modes rotate in the sequence of NFT, NC, NFB, and NC in the negative half cycle. Thus, the frequency of output voltage V_{AB} is double the switch frequency.

The six modes based on H4 topology in DFUSPWM are shown in Figure 20. The PC and NC modes are shown in Figures 20a and 20b, respectively. In the positive half cycle of grid voltage, two PF modes, i.e., PFT and PFB, are shown in Figure 20c,d. Similarly, two NF modes, i.e., NFT and NFB, are shown in Figure 20e,f in the negative half cycle of the grid voltage.

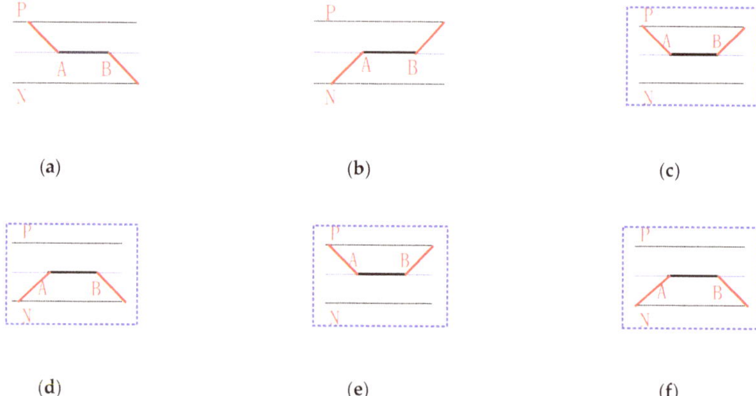

Figure 20. Six modes based on H4 topology under DFUSPWM. (**a**) PC mode; (**b**) NC mode; (**c**) PFT mode; (**d**) PFB mode; (**e**) NFT mode; (**f**) NFB mode.

4.2. Unified Topology Model of DFUSPWM

Figure 21 shows six modes from the unified topology model under DFUSPWM.

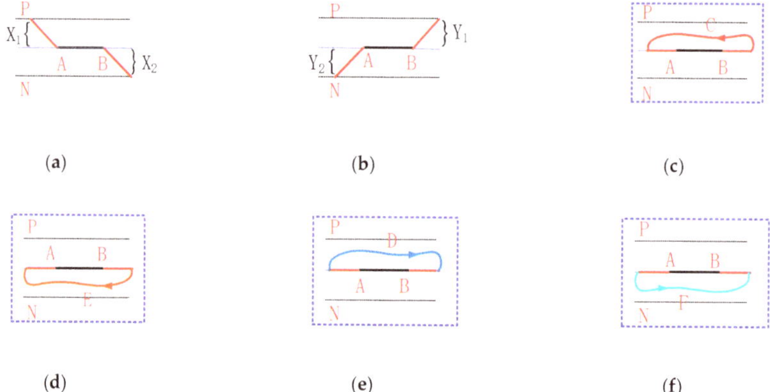

Figure 21. Six modes based on unified topology under DFUSPWM. (**a**) PC mode; (**b**) NC mode; (**c**) PFT mode; (**d**) PFB mode; (**e**) NFT mode; (**f**) NFB mode.

All rules under USPWM mentioned in Section 2 are also suitable for DFUSPWM. According to rule #1, points A and B must be disconnected from points P and N in the four freewheeling modes, i.e., the two PF modes and the two NF modes. From rule #2, the branch between points A and B is short-circuited for output current flow. Two new controlled branches, $B\hat{C}A$ and $B\hat{E}A$, are added to flow the positive current in Figure 21c,d. Two controlled branches, $A\hat{D}B$ and $A\hat{F}B$, are added to flow the negative current in Figure 21e,f.

For DFUSPWM, there are two PC modes in a switching period. Thus, according to rule #1, there are at least two couple switches to alternately turn on/off to achieve double frequency. For example, there are two switches, i.e., T_{P1} connected to point P and T_{P2} connected to point A, between point P and point A, and two, i.e., T_{P3} connected to point B and T_{P4} connected to point N, between point B and point N. Switches T_{P1} and T_{P2} are connected in series, as are T_{P3} and T_{P4}. Switches T_{P1} and T_{P3} turn on/off at the same time, and T_{P2} and T_{P4} are kept on or off at the same time. Thus, there are two possibilities to disconnect points P and A, and points B and N: one is that switches T_{P1} and T_{P3} turn off. The other is that switches T_{P2} and T_{P4} turn off. Once switches T_{P1} and T_{P3} turn off in PFT mode, switch T_{P2} is kept on and can be used to construct a freewheeling branch because it connects to point A. Similarly, switch T_{P3} is kept on and serves to construct a freewheeling branch in the PFB mode when switches T_{P2} and T_{P4} turn off. Thus, two PF modes can be achieved when $X_1 = 2$ and $X_2 = 2$. The same is true with $Y_1 = 2$, $Y_2 = 2$.

Figure 22 shows six modes under $X_1 = 2$, $X_2 = 2$, $Y_1 = 2$, and $Y_2 = 2$. The PC and NC modes are shown in Figure 22a. Eight switches ($T_{P1}{\sim}T_{P4}$ and $T_{N1}{\sim}T_{N4}$) are used. The same driving signals are provided for couples of switches T_{P1} and T_{P3}, T_{P2} and T_{P4}, T_{N1} and T_{N3}, and T_{N2} and T_{N4}.

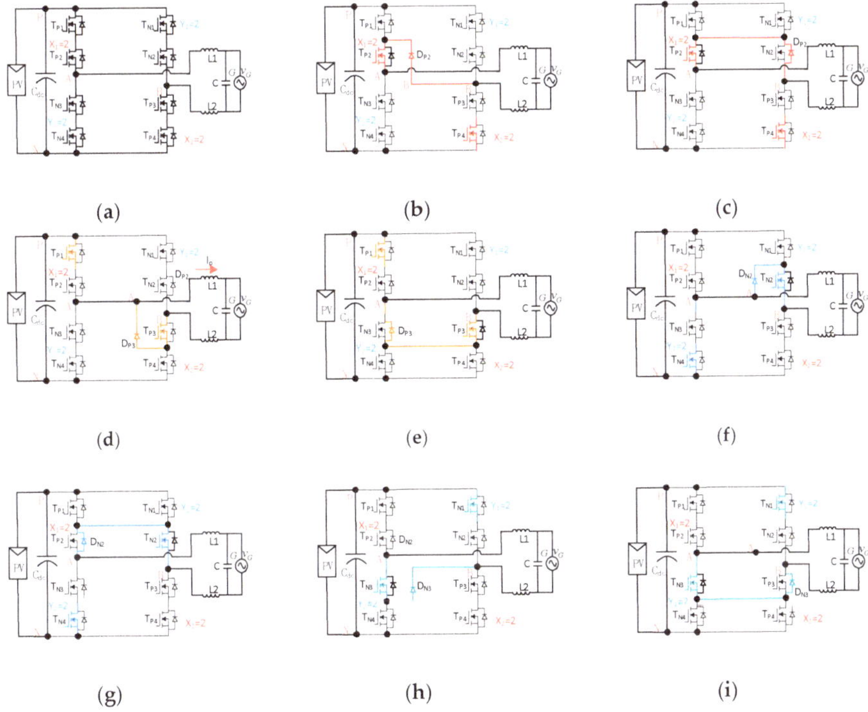

Figure 22. Six modes of M10 topology. (**a**) PC and NC modes; (**b**) PFT mode #1; (**c**) PFT mode #2; (**d**) PFB mode #1; (**e**) PFB mode #2; (**f**) NFT mode #1; (**g**) NFT mode #2; (**h**) NFB mode #1; (**i**) NFB mode #2.

In PFT mode, switches T_{P1}, T_{P3}, T_{N1}, T_{N2}, T_{N3}, and T_{N4} are off, and switches T_{P2} and T_{P4} are on. One diode, D_{P2}, is added for the positive output current flow, as shown in Figure 22b. Diode D_{P2} is served by the body-diode of switch T_{N2}. The reflected PFT mode is shown in Figure 22c.

In PFB mode, switches T_{P2}, T_{P4}, T_{N1}, T_{N2}, T_{N3}, and T_{N4} are off, and switches T_{P1} and T_{P3} are on. One diode, D_{P3}, is added for positive output current flow, as shown in Figure 22d. Diode D_{P3} is served by the body-diode of switch T_{N3}, as shown in Figure 22e. It is easy to make a similar analysis for NFT and NFB modes. Figure 22f,g show the two NFT modes, while the two NFB modes are shown in Figure 22h,i.

It should be noted from the above analysis that there are two PFT modes, two PFB modes, two NFT modes, and two NFB modes. There are eight possible topologies through choosing one PFT mode, one PFB mode, one NFT mode, and one NFB mode. According to rule #4, the four typical topologies in Figure 23 are derived from the MN principle.

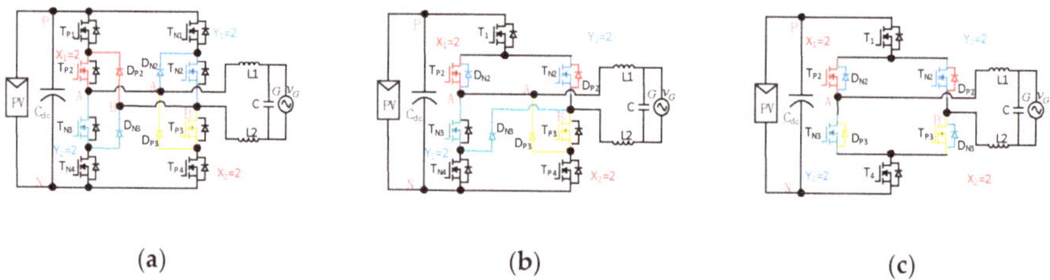

(**a**)   (**b**)   (**c**)

Figure 23. Four topologies derived from MN principle under DFUSPWM. (**a**) R11 (new); (**b**) R12 (new); (**c**) R14 [17].

Two corresponding topological families under DFUSPWM are shown in Table 7.

Table 7. Two topological families from MN principle under DFUSPWM.

(M, N)	$X_1 + X_2$	$Y_1 + Y_2$	Family with Extra Diode	Family without Extra Diode
(M = 4, N = 4)	2 + 2	2 + 2	R11, R12, R13	R14

5. Conclusions

In this paper, a topology derivation methodology under USPWM is proposed to determine all possible full-bridge topologies for small leakage current. A unified circuit model based on USPWM is provided. Secondly, a mathematic method called the "MN principle" is then proposed to determine all possible topologies. Four rules and a derivation procedure are also provided. Thirteen simplified topologies are derived using this method. All existing topologies have been covered, and four new topologies have been found. These topologies can be classified into two topology families: the first includes topologies in which extra diodes are used for current flowing, while the body-diode functions as the extra diode to flow freewheeling current in the second family topologies. Finally, the proposed method is extended to the topologies under DFUSPWM, and three corresponding topologies have been derived, and two new topologies found.

Author Contributions: Conceptualization, X.Y., X.W.; funding acquisition, X.W.; investigation, X.Y.; software, X.Z.; validation, X.Z., X.W.; writing—original draft, X.Y.; writing—review and editing, H.W. and Y.-F.L. All authors have read and agreed to the published version of the manuscript.

Funding: Research on Topology, Passive Current Sharing Mechanism and Control for Multiphase Resonant Converter with Coupled Resonant Tank: 51977069; Research on High-power and High-efficiency Electro-acoustic Transduction Mechanism and Control Method: 51837005.

Conflicts of Interest: The authors declare no conflicts of interest.

References

1. Sun, K.; Zhang, L.; Xing, Y.; Guerrero, J.M. A Distributed Control Strategy Based on DC Bus Signaling for Modular Photovoltaic Generation Systems with Battery Energy Storage. *IEEE Trans. Power Electron.* **2011**, *26*, 3032–3045. [CrossRef]

2. EPIA. *Global Market Outlook for Photovoltaics 2013–2017*; European Photovoltaic Industry Association: Brussels, Belgium, 2013.

3. Kjaer, S.B.; Pedersen, J.K.; Blaabjerg, F. A review of single-phase grid-connected inverters for photovoltaic modules. *IEEE Trans. Ind. Appl.* **2005**, *41*, 1292–1306. [CrossRef]

4. Blaabjerg, F.; Zhe, C.; Kjaer, S.B. Power electronics as efficient interface in dispersed power generation systems. *IEEE Trans. Power Electron.* **2004**, *19*, 1184–1194. [CrossRef]

5. Quan, L.; Wolfs, P. A Review of the Single Phase Photovoltaic Module Integrated Converter Topologies with Three Different DC Link Configurations. *IEEE Trans. Power Electron.* **2008**, *23*, 1320–1333. [CrossRef]

6. Power Generation Systems Connected to the Low-Voltage Distribution Network, VDE-AR-N 4105. Available online: https://www.sogou.com/link?url=hedJjaC291MdqKRdyYP6xaEfyF6gLPTmC2Ly98cd-IVBavtdX5bT4M0N-aB7Vy-zp32izvdhNQ_1-ZZN6CSHZQ (accessed on 1 August 2011).

7. Standard for Inverters, Converters, Controllers, and Interconnection System Equipment for Use with Distributed Energy Resources. UL 1741. Available online: https://standardscatalog.ul.com/standards/en/standard_1741_2 (accessed on 28 January 2010).

8. Automatic Disconnection Device between a Generator and the Public Low-Voltage Grid, DIN VDE V 0126-1-1. Available online: https://www.beuth.de/en/pre-standard/din-vde-v-0126-1-1/187485608 (accessed on 1 August 2013).

9. Lopez, O.; Teodorescu, R.; Freijedo, F.; Doval-Gandoy, J. Eliminating ground current in a transformerless photovoltaic application. *IEEE Trans. Energy Convers.* **2010**, *25*, 140–147. [CrossRef]

10. Gonzalez, R.; Gubia, E.; Lopez, J.; Marroyo, L. Transformerless Single-Phase Multilevel-Based Photovoltaic Inverter. *IEEE Trans. Ind. Electron.* **2008**, *55*, 2694–2702. [CrossRef]

11. Xiao, H.; Xie, S. Transformerless Split-Inductor Neutral Point Clamped Three-Level PV Grid-Connected Inverter. *IEEE Trans. Power Electron.* **2012**, *27*, 1799–1808. [CrossRef]

12. Victor, M.; Greizer, K.; Bremicker, A. Method of Converting a Direct Current Voltage from a Source of Direct Current Voltage, More Specifically from a Photovotatic Source of Direct Current Voltage, into a Alternating Current Voltage. U.S. Patent 7,411,802, 12 August 2008.

13. Schmidt, H.; Siedle, C.; Ketterer, J. Wechselrichter zum Unwandeln einer Elektrischen Gleichspannung in einen Wechselstrom oder eine Wechselspannung. EP Patent 2 086 102 A2, 15 May 2003.

14. Patino, D.G.; Erira, E.G.; Fuelagan, J.R.; Rosero, E.E. Implementation a HERIC inverter prototype connected to the grid controlled by SOGI-FLL. In Proceedings of the 2015 IEEE Workshop on Power Electronics and Power Quality Applications (PEPQA), Bogota, Colombia, 2–4 June 2015; pp. 1–6.

15. Zhang, L.; Sun, K.; Xing, Y.; Xing, M. H6 Transformerless Full-Bridge PV Grid-Tied Inverters. *IEEE Trans. Power Electron.* **2014**, *29*, 1229–1238. [CrossRef]

16. Yu, W.; Lai, J.S.; Qian, H.; Hutchens, C.; Zhang, J.; Lisi, G.; Djabbari, A.; Smith, G.; Hegarty, T. High-efficiency inverter with H6-type configuration for photovoltaic non-isolated ac module applications. In Proceedings of the 2010 Twenty-Fifth Annual IEEE Applied Power Electronics Conference and Exposition (APEC), Palm Springs, CA, USA, 21–25 February 2010; pp. 1056–1061.

17. Yang, B.; Li, W.; Gu, Y.; Cui, W.; He, X. Improved Transformerless Inverter with Common-Mode Leakage Current Elimination for a Photovoltaic Grid-Connected Power System. *IEEE Trans. Power Electron.* **2012**, *27*, 752–762. [CrossRef]

18. Su, X.; Sun, Y.; Lin, Y. Analysis on Leakage Current in Transformerless Single-Phase PV Inverters Connected to the Grid. In Proceedings of the Power and Energy Engineering Conference (APPEEC), 2011 Asia-Pacific, Wuhan, China, 25–28 March 2011; pp. 1–5.

19. Xiao, H.F.; Ke, L.; Li, Z. A Quasi-Unipolar SPWM Full-Bridge Transformerless PV Grid-Connected Inverter with Constant Common-Mode Voltage. *IEEE Trans. Power Electron.* **2015**, *30*, 3122–3132. [CrossRef]

20. Xiao, H.; Xie, S.; Chen, Y.; Huang, R. An Optimized Transformerless Photovoltaic Grid-Connected Inverter. *IEEE Trans. Ind. Electron.* **2011**, *58*, 1887–1895. [CrossRef]

21. Wang, J.; Ji, B.; Zhao, J.; Yu, J. From H4, H5 to H6 Standardization of full-bridge single phase photovoltaic inverter topologies without ground leakage current issue. In Proceedings of the 2012 IEEE Energy Conversion Congress and Exposition (ECCE), Raleigh, NC, USA, 15–20 September 2012; pp. 2419–2425.

22. Vazquez, G.; Martinez-Rodriguez, P.R.; Sosa, J.M.; Escobar, G.; Juarez, M.A. Transformerless single-phase multilevel inverter for grid tied photovoltaic systems. In Proceedings of the Industrial Electronics Society, IECON 2014—40th Annual Conference of the IEEE, Dallas, TX, USA, 29 October–1 November 2014; pp. 1868–1874.

23. Vazquez, G.; Martinez-Rodriguez, P.R.; Sosa, J.M.; Escobar, G.; Arau, J. A modulation strategy for single-phase HB-CMI to reduce leakage ground current in transformer-less PV applications. In Proceedings of the Industrial Electronics Society, IECON 2013—39th Annual Conference of the IEEE, Vienna, Austria, 10–13 November 2013; pp. 210–215.

24. Figueredo, R.S.; de Carvalho, K.C.M.; Ama, N.R.N.; Junior, L.M. Leakage current minimization techniques for single-phase transformerless grid-connected PV inverters—An overview. In Proceedings of the Power Electronics Conference (COBEP), Gramado, Brazil, 27–31 October 2013; pp. 517–524.

25. Hu, S.; Cui, W.; Li, W.; He, X.; Cao, F. A high-efficiency single-phase inverter for transformerless photovoltaic grid-connection. In Proceedings of the Energy Conversion Congress and Exposition (ECCE), Pittsburgh, PA, USA, 14–18 September 2014; pp. 4232–4236.

26. Salmon, J.; Knight, A.; Ewanchuk, J. Single phase multi-level PWM Inverter topologies using coupled inductors. In Proceedings of the Power Electronics Specialists Conference, Rhodes, Greece, 15–19 June 2008; pp. 802–808.

27. Ozkan, Z.; Hava, A.M. Leakage current analysis of grid connected transformerless solar inverters with zero vector isolation. In Proceedings of the Energy Conversion Congress and Exposition (ECCE), Phoenix, AZ, USA, 17–22 September 2011; pp. 2460–2466.

28. Lopez, O.; Teodorescu, R.; Freijedo, F.; DovalGandoy, J. Leakage current evaluation of a singlephase transformerless PV inverter connected to the grid. In Proceedings of the Applied Power Electronics Conference, APEC 2007—Twenty Second Annual IEEE, Anaheim, CA, USA, 25 February–1 March 2007; pp. 907–912.

29. Ma, L.; Tang, F.; Zhou, F.; Jin, X.; Tong, Y. Leakage current analysis of a single-phase transformer-less PV inverter connected to the grid. In Proceedings of the 2008 IEEE International Conference on Sustainable Energy Technologies, Singapore, 24–27 November 2008; pp. 285–289.

30. Kerekes, T.; Teodorescu, R.; Rodriguez, P.; Vazquez, G.; Aldabas, E. A New High-Efficiency Single-Phase Transformerless PV Inverter Topology. *IEEE Trans. Ind. Electron.* **2011**, *58*, 184–191. [CrossRef]

31. Ji, B.; Wang, J.; Zhao, J. High-Efficiency Single-Phase Transformerless PV H6 Inverter with Hybrid Modulation Method. *IEEE Trans. Ind. Electron.* **2013**, *60*, 2104–2115. [CrossRef]

32. Islam, M.; Mekhilef, S. A new high efficient transformerless inverter for single phase grid-tied photovoltaic system with reactive power control. In Proceedings of the Applied Power Electronics Conference and Exposition (APEC), Charlotte, NC, USA, 15–19 March 2015; pp. 1666–1671.

33. Islam, M.; Mekhilef, S. High efficiency transformerless MOSFET inverter for grid-tied photovoltaic system. In Proceedings of the 2014 Twenty-Ninth Annual IEEE Applied Power Electronics Conference and Exposition (APEC), Fort Worth, TX, USA, 16–20 March 2014; pp. 3356–3361.

34. San, G.; Qi, H.; Wu, J.; Guo, X. A new three-level six-switch topology for transformerless photovoltaic systems. In Proceedings of the 2012 7th International Power Electronics and Motion Control Conference (IPEMC), Harbin, China, 2–5 June 2012; pp. 163–166.

35. Freddy, T.K.S.; Rahim, N.A.; Wooi-Ping, H.; Seng, C.H. Comparison and Analysis of Single-Phase Transformerless Grid-Connected PV Inverters. *IEEE Trans. Power Electron.* **2014**, *29*, 5358–5369. [CrossRef]

36. Yu, W.; Lai, J.S.; Qian, H.; Hutchens, C.; Zhang, J.; Lisi, G.; Djabbari, A.; Smith, G.; Hegarty, T. A novel H6 topology and Its modulation strategy for transformerless photovoltaic grid-connected inverters. In Proceedings of the 2014 16th European Conference on Power Electronics and Applications (EPE'14-ECCE Europe), Lappeenranta, Finland, 26–28 August 2014; pp. 1–8.

37. Dong, D.; Luo, F.; Boroyevich, D.; Mattavelli, P. Leakage Current Reduction in a Single-Phase Bidirectional AC-DC Full-Bridge Inverter. *IEEE Trans. Power Electron.* **2012**, *27*, 4281–4291. [CrossRef]

38. Cui, W.; Yang, B.; Zhao, Y.; Li, W.; He, X. A novel single-phase transformerless grid-connected inverter. In Proceedings of the IECON 2011—37th Annual Conference on IEEE Industrial Electronics Society, Melbourne, Australia, 7–10 November 2011; pp. 1126–1130.

39. Gu, B.; Dominic, J.; Lai, J.-S.; Chen, C.-L.; LaBella, T.; Chen, B. High Reliability and Efficiency Single-Phase Transformerless Inverter for Grid-Connected Photovoltaic Systems. *IEEE Trans. Power Electron.* **2013**, *28*, 2235–2245. [CrossRef]

40. Gu, B.; Dominic, J.; Chen, B.; Lai, J.-S. A high-efficiency single-phase bidirectional AC-DC converter with miniminized common mode voltages for battery energy storage systems. In Proceedings of the Energy Conversion Congress and Exposition (ECCE), Denver, CO, USA, 15–19 September 2013; pp. 5145–5149.

41. Basu, K.; Mohan, N. A High-Frequency Link Single-Stage PWM Inverter with Common-Mode Voltage Suppression and Source-Based Commutation of Leakage Energy. *IEEE Trans. Power Electron.* **2014**, *29*, 3907–3918. [CrossRef]

42. Barater, D.; Buticchi, G.; Crinto, A.S.; Franceschini, G.; Lorenzani, E. Unipolar PWM Strategy for Transformerless PV Grid-Connected Converters. *IEEE Trans. Energy Convers.* **2012**, *27*, 835–843. [CrossRef]

43. Barater, D.; Buticchi, G.; Crinto, A.S.; Franceschini, G.; Lorenzani, E. A new proposal for ground leakage current reduction in transformerless grid-connected converters for photovoltaic plants. In Proceedings of the 35th Annual Conference of IEEE Industrial Electronics, 2009, IECON '09, Porto, Portugal, 3–5 November 2009; pp. 4531–4536.

44. Anandababu, C.; Fernandes, B.G. Improved full-bridge neutral point clamped transformerless inverter for photovoltaic grid-connected system. In Proceedings of the 39th Annual Conference of the IEEE Industrial Electronics Society, IECON 2013, Vienna, Austria, 10–13 November 2013; pp. 7996–8001.

45. Anandababu, C.; Fernandes, B.G. A novel neutral point clamped transformerless inverter for grid-connected photovoltaic system. In Proceedings of the IECON 2013—39th Annual Conference of the IEEE Industrial Electronics Society, Vienna, Austria, 10–13 November 2013; pp. 6962–6967.

46. Zhang, L.; Sun, K.; Feng, L.; Wu, H.; Xing, Y. A Family of Neutral Point Clamped Full-Bridge Topologies for Transformerless Photovoltaic Grid-Tied Inverters. *IEEE Trans. Power Electron.* **2013**, *28*, 730–739. [CrossRef]

47. Duan, S.; Liu, B.; Kang, Y. A Single-phase Hybrid-bridge Three-level Inverter. Chinese Patent CN101599713B, 14 September 2011.

48. Gonzalez, R.; Lopez, J.; Sanchis, P.; Marroyo, L. Transformerless Inverter for Single-Phase Photovoltaic Systems. *IEEE Trans. Power Electron.* **2007**, *22*, 693–697. [CrossRef]

49. Li, W.; Gu, Y.; Luo, H.; Cui, W.; He, X.; Xia, C. Topology Review and Derivation Methodology of Single-Phase Transformerless Photovoltaic Inverters for Leakage Current Suppression. *IEEE Trans. Ind. Electron.* **2015**, *62*, 4537–4551. [CrossRef]

50. Wang, H.; Burton, S.; Liu, Y.-F.; Sen, P.C.; Guerrero, J.M. A systematic method to synthesize new transformerless full-bridge grid-tied inverters. In Proceedings of the 2014 IEEE Energy Conversion Congress and Exposition (ECCE), Pittsburgh, PA, USA, 14–18 September 2014; pp. 2760–2766.

51. Xiao, H.; Xie, S. Leakage Current Analytical Model and Application in Single-Phase Transformerless Photovoltaic Grid-Connected Inverter. *IEEE Trans. Electromagn. Compat.* **2010**, *52*, 902–913. [CrossRef]

52. Liu, W.; Niazi, K.; Kerekes, T.; Yang, Y. A Review on Transformerless Step-Up Single-Phase Inverters with Different DC-Link Voltage for Photovoltaic Applications. *Energies* **2019**, *12*, 3626. [CrossRef]

Article

A Topology Synthetization Method for Single-Phase, Full-Bridge, Transformerless Inverter with Leakage Current Suppression—Part II

Xiumei Yue [1], Hongliang Wang [1,*], Xiaonan Zhu [1], Xinwei Wei [1] and Yan-Fei Liu [2]

[1] College of Electrical and Information Engineering, Hunan University, Changsha 410082, China; Wangyue_120109@163.com (X.Y.); zhuxn@hnu.edu.cn (X.Z.); weixinwei@hnu.edu.cn (X.W.)

[2] Department of Electrical and Computer Engineering, Queen's University, Kingston, ON K7L 3N6, Canada; yanfei.liu@queensu.ca

* Correspondence: liangliang-930@163.com

Received: 23 November 2019; Accepted: 7 January 2020; Published: 16 January 2020

Abstract: This paper proposes a topology derivation methodology to achieve small leakage current or zero leakage current for photovoltaic application. The core of the proposed method is a unified topology model and MN principle to show how to derive all possible topologies based on unipolar sinusoidal pulse width modulation (USPWM) and double-frequency USPWM (DFUSPWM). Part II of this paper discusses the topology synthetization method to achieve zero leakage current. Two types of neutral point clamped (NPC) topologies based on USPWM and DFUSPWM are elaborated. Two possible connections for the NPC cell are introduced, and detailed NPC topology derivation procedures are also provided. All existing NPC topologies are derived, and twenty-two new NPC topologies are found based on the new topology derivation methodology for a single-phase, full-bridge, transformerless inverter.

Keywords: transformerless inverter; full-bridge inverter; leakage current; NPC topology

---

1. Introduction

Photovoltaic (PV) sources have been developed as one of the most promising renewable energy sources, providing clean, reliable, and emission-free energy [1,2]. The single phase, transformerless grid-connected inverter has widely been used throughout the world. Considering the electrical connections between the PV panels and utility grid for transformerless, grid-tied systems [3–5], the leakage current generated by the PV parasitic capacitors must be limited in order to meet the safety requirement, such as standards of VDC-AR_N 4015 [6], UL1741 [7], VDE 0126-1-1 [8]. Therefore, small and even zero leakage current in transformerless PV inverter systems are a primary priority for designers.

The leakage current can be limited in full-bridge transformerless inverters if the common mode (CM) voltage is zero [9,10]. Some state-of-the art-topologies, such as the H5 inverter topology [11], as shown in Figure 1a, HERIC [12,13], and H6 inverter topologies [9,14–44], have been developed from the full-bridge inverter topology. However, there is still a small leakage current as the CM voltage no longer remains constant under parasitic parameter inflection throughout the whole line-frequency (50 Hz or 60 Hz) period. The CM voltage is indeed variable in freewheeling mode because of the potential variation induced by the charging of the switch junction capacitance. The Neutral Point Clamped (NPC) technique is introduced in these topologies to achieve zero leakage current [45–52].

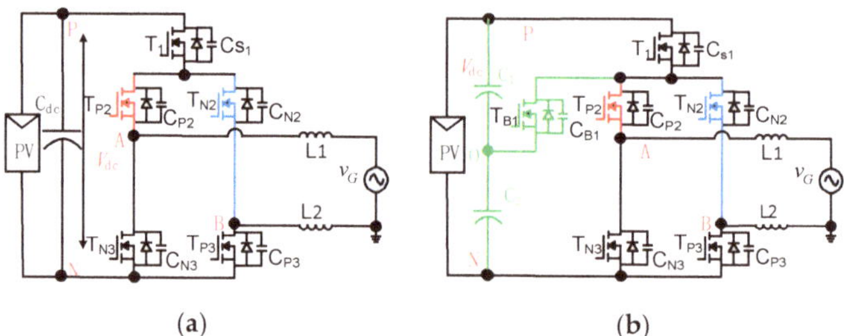

(a)   (b)

Figure 1. H5 topology without/with NPC cell. (**a**) H5 topology [11]; (**b**) Optimized H5 (OH5) [19].

The famous dc-based H5 inverter is adopted as an example to explore the inherent leakage current generation caused by the switch junction capacitors. Figure 2 shows the transient operation of H5 and optimized H5 [53].

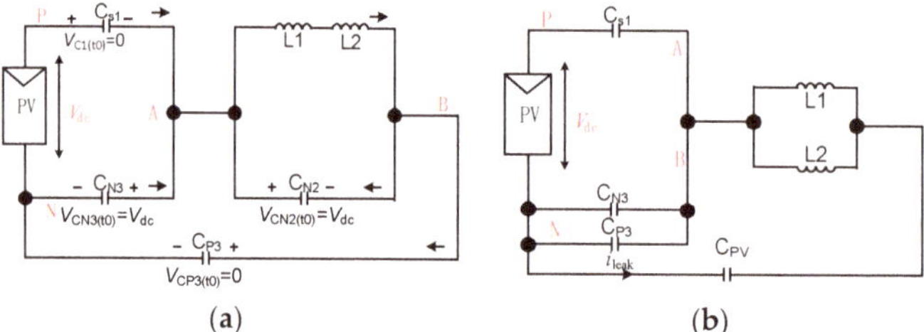

(a)   (b)

Figure 2. Transient charging and discharging operation of H5. (**a**) Transient operation [53]; (**b**) Equivalent circuit [53].

During the positive half cycle, switches T_1, T_{P2}, and T_{P3} are turned on, and switches T_{N2} and T_{N3} are turned off. Then, switches T_1 and T_{P2} are turned off. Simultaneously, the body-diode of switch T_{N2} does not conduct to go into freewheel mode. At this moment, junction capacitors C_{S1} and C_{P3} are charged while junction capacitors C_{N2} and C_{N3} are discharged. The transient charging and discharging operation from the power delivery mode to the freewheeling mode is demonstrated in Figure 2a. The voltage V_{AN} decreases and V_{BN} increases. The body-diode of switch T_{N2} is conducted to enter the freewheeling mode when the voltage C_{N2} is lowered to zero.

The equivalent circuit is illustrated in Figure 2b and the voltages V_{AN} and V_{BN} can be derived by (1)

$$V_{AN} = V_{BN} = \frac{C_{P3} + C_{N3}}{C_{S1} + C_{P3} + C_{N3}} V_{dc} \tag{1}$$

From (1), the steady-state voltage V_{AN}, V_{BN} is determined by the junction capacitors of the switches. The switch junction capacitors are approximately from several hundred picofarads to several nanofarads. The parasitic capacitor C_{pv} is around one hundred nanofarads [16,18,45]. If $C_{S1} = C_{P3} + C_{N3}$, the voltages V_{AN} and V_{BN} are both equal to $V_{dc}/2$. This indicates that the total high-frequency CM voltage is $V_{dc}/2$. Unfortunately, for most industrial applications, $C_{S1} \neq C_{P3} + C_{N3}$. This means that the total high-frequency CM voltage is not kept constant, which leads to unexpected leakage current. Furthermore, in freewheeling mode, the terminals A and B are floating, and an additional resonant path is formed, which is illustrated in Figure 2b. A high-frequency resonance occurs, which also may lead to the generation of leakage current. The resonance frequency can be calculated by

$$f_r = \frac{1}{2\pi \sqrt{L_{eq} C_{eq}}} \tag{2}$$

where

$$L_{eq} = \frac{L1 * L2}{L1 + L2}; \; C_{eq} = \frac{(C_{S1} + C_{P3} + C_{N3})C_{PV}}{C_{S1} + C_{P3} + C_{N3} + C_{PV}} \approx C_{S1} + C_{P3} + C_{N3} \tag{3}$$

To clamp the CM voltage to half of the dc-link voltage in freewheeling mode, the bus capacitors are divided into two same-series capacitors, and a clamping cell should be inserted between the midpoint of the series bus capacitors. This can provide a constant CM voltage, which can eliminate the effect of the junction capacitors of the switches. Figure 1b shows the OH5 topology with the NPC cell. Switch T_{B1} turns on to connect point A (or B) to point O in freewheeling mode. A new circuit configuration for the inverter with a single dc-link capacitor and seven insulated gate bipolar transistors (IGBTs) is proposed in [54]. The additional switch is turned on in the freewheeling modes, and two turn-OFF snubber circuits are added in parallel to the switches in order to share the input dc voltage between the snubber capacitors. As a result, the leakage current can be eliminated completely because the bridge voltages can be clamped at half dc-link voltage in the freewheeling modes. Two neutral point clamping circuits, which are common collectors and common emitters, are proposed in [55], and then a family of single-phase inverters is presented.

In theory, many topologies that could suppress leakage current. Some of them have been studied intensively and patented. However, some may not have been studied or even found. The purpose of this paper is to propose a systematic topology deduction method to obtain all the topologies with the ability to suppress leakage current to zero in theory [56]. Part II of this paper investigates the rules by which to synthesize the two type of NPC topologies to achieve zero leakage current. The first one is called "indirect connection NPC topology", and the other is called "direct connection NPC topology". Part II of this paper is organized as follows: Section 2 proposes the rules by which to synthesize the two type of NPC cells. Section 3 introduces two topology families based on indirect connection NPC cells from USPWM. Section 4 provides two topology families based on direct connection NPC cells from USPWM. The proposed topology derivation methodology is also extended to DFUSPWM in Section 5. The simulation results are provided to verify the performance of the proposed topologies in Section 6, and Part II of the paper is concluded in Section 7.

2. NPC Cells for USPWM

To substantially reduce the leakage current, the CM voltage V_{CM} should be clamped to half of the input voltage in freewheeling mode, including in PF and NF modes. Points A and B are short-circuited to achieve $V_{AB} = 0$ in freewheeling mode. The extra circuit, called "NPC cell", will be introduced to clamp the CM voltage and achieve zero leakage current. The NPC cell is made of two series capacitors and some additional switches. The DC bus capacitor is split into two identical capacitors, C_1 and C_2, in series. Due to the participation of the NPC cell, point A (or point B) is connected to midpoint O between C_1 and C_2, which guarantees that the CM voltage is half of the input voltage in PF mode and NF mode. The terms T_{PF} and T_{NF} are used to refer to the equivalent switch in the freewheeling branch in PF mode and NF mode, respectively. There are two basic rules concerning the NPC cell. Rule 1 is used to make full use of the original freewheeling switching devices in the topology and minimize the number of switches to be added. Rule 2 is used to give the locations where the switching devices of the NPC cell should be added.

Rule #1: To reduce the number of switches in NPC cell, both T_{PF} and T_{NF} are turned on in freewheeling mode, including in PF and NF modes.

Rule #2: Additional switches are added and connected to the freewheeling branch so that point A or B is clamped to point O.

Based on Rule #1, it is possible that T_{NF} works for the NPC cell in PF mode, and T_{PF} in NF mode. So, the number of switches needed in the NPC cell is reduced.

Based on Rule #2, two possible connections are available. One is that additional switches are added between point O and a point from the freewheeling branch. This type of connection is defined as an "Indirect Connection". The other is that additional switches are inserted into the freewheeling branch which is connected to point O due to the injection of additional switches. This type of connection is defined as a "Direct Connection".

Figure 3 shows the schematic diagram of the Indirect Connection of a unified NPC cell for USPWM.

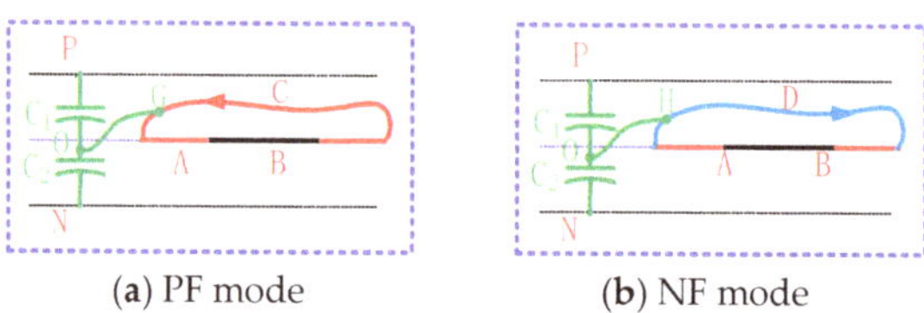

(a) PF mode   (b) NF mode

Figure 3. Indirect Connection of Unified NPC cell under USPWM. (**a**) PF mode; (**b**) NF mode.

Two capacitors with the same capacitance value are connected in series to halve the input voltage. Point O is the midpoint of these two capacitors. Additional switches are added between points O and G in Figure 3a and between points O and H in Figure 3b. It should be noted that points G and H are from the freewheeling branch. Figure 3a shows the Indirect Connection in PF mode, which means the actual current flows from point G to point O. Figure 3b shows the NF mode operation, which means that the actual current flows from point O to point H.

Figure 4 shows a schematic diagram of the Direct Connection of the unified NPC cell for USPWM. As shown in Figure 4a, additional switches marked by the green bump branch $\hat{HI}$ are inserted into the freewheeling branch $B\hat{C}A$, so that $B\hat{C}A$ is connected to the midpoint O. Points H and I are from the freewheeling branch.

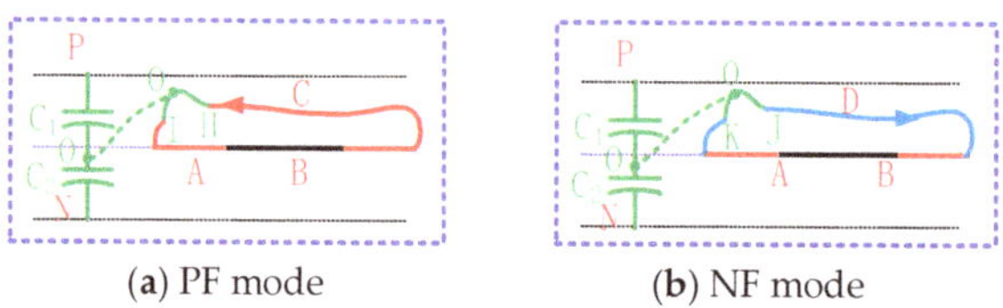

(a) PF mode   (b) NF mode

Figure 4. Direct Connection of Unified NPC cell under USPWM. (**a**) PF mode; (**b**) NF mode.

Figure 4b shows the NF mode operation. The additional switches marked by the green branch $\hat{KJ}$ are inserted into the freewheeling branch $A\hat{D}B$, and $A\hat{D}B$ is connected to the midpoint O due to the injection of $\hat{KJ}$.

3. Inverter Topologies Based on the Indirect Connection of NPC Cell from USPWM

As with the Indirect Connection of the NPC cell shown in Figure 3 the freewheeling branch is connected directly through additional switches to point O to achieve zero leakage current. Part I of this paper has described the development of all the possible topologies of inverters with low leakage current where the freewheeling branch is not connected to the capacitor midpoint O. The topologies with NPC cells can be derived based on the non-NPC topologies derived in Part I of this paper. There will be one NPC inverter topology corresponding to each non-NPC inverter topology derived in Part I.

This section focuses on how to derive the inverter topologies based on the Indirect Connection of an NPC cell under USPWM. Two topology families from the unified topology model have been described in Part I. One has an extra diode which is used to flow freewheeling current, and the other has no extra diode, but rather, body-diode switch is used to flow freewheeling current. Correspondingly, the inverter topologies with the Indirect Connection NPC cell can also be classified into two families, as shown in Table 1, R1 is a topology without an NPC cell that has been proposed in Part I. R1S1 is the topology with the NPC cell corresponding to R1, and will be examined in this paper.

Table 1. Two inverter topology families from USPWM.

(M, N)	$X_1 + X_2$	$Y_1 + Y_2$	Family With Extra Diode		Family Without Extra Diode	
			Without NPC Cell (Part I)	With NPC Cell (Part II)	Without NPC Cell (Part I)	With NPC Cell (Part II)
(M = 2, N = 2)	1 + 1	1 + 1	R1, R3	R1S1-1, R1S1-2, R3S1-1, R3S1-2	R2	R2S1
(M = 2, N = 3) or (M = 3, N = 2)	1 + 2 1 + 1 2 + 1 1 + 1	1 + 1 1 + 2 1 + 1 2 + 1	R4	R4S1-1, R4S1-2, R4S1-3	R5	R5S1
(M = 3, N = 3)	2 + 1 1 + 2	2 + 1 1 + 2	R6	R6S1	R7	R7S1
	2 + 1 1 + 2	1 + 2 2 + 1	R8	R8S1-1, R8S1-2	None available	None available
(M = 3, N = 4) or (M = 4, N = 3)	1 + 2 2 + 2 2 + 1 2 + 2	2 + 2 1 + 2 2 + 2 2 + 1	R9	R9S1-1, R9S1-2	R10	R10S1
(M = 4, N = 4)	2 + 2	2 + 2	R11, R12	R11S1-1, R11S1-2, R12S1-1, R12S1-2	R13	R13S1-1, R13S1-2

3.1. Indirect Connection NPC Cell Based on Two Topology Familes

Figure 3 shows the Indirect Connection NPC cell circuits based on the topology family with an extra diode. Two freewheeling cell circuits are introduced, and four corresponding NPC cell circuits are described.

As described in Part I of this paper, there are two freewheeling cell circuits based on the topology family with an extra diode, as shown in Figure 5a,d. For Figure 5a, the current flows from point B to point A through switch T_{PF} and diode D_{PF} in series, and from point A to point B through switch T_{NF} and diode D_{NF} in series. In Figure 5d, the current flows bidirectionally between points B and A through the switch T_F, which is kept on in both PF and NF modes.

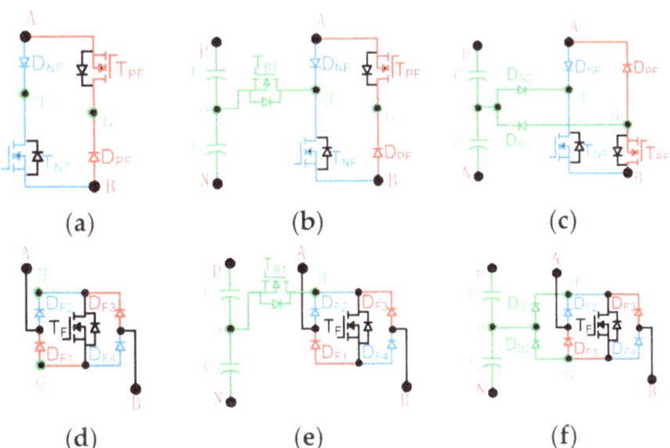

Figure 5. Indirect Connection NPC cell circuits based on topology family with an extra diode. (**a**) case #1; (**b**) NPC cell #1 under case #1; (**c**) NPC cell #2 under case #1; (**d**) case #2; (**e**) NPC cell #1 under case #2; (**f**) NPC cell #2 under case #2.

An extra switch, T_{B1}, is added between points O and H in Figure 5b,e. According to Rule #1, both switches T_{PF} and T_{NF} in Figure 5b are turned on in PF and NF modes. The midpoint O is connected to point B through the extra switch T_{B1} and switch T_{NF} based on Rule #2. Switches T_{B1} and T_{NF} are connected in series to realize bidirectional current flow between point O and point B, as shown in Figure 5b. Similarly, the switch T_F in Figure 5e is turned on in both PF and NF modes based on Rule #1. The midpoint O is connected to point B through the extra switch T_{B1} and the switch T_F based on Rule #2.

Two extra diodes, D_{B1} and D_{B2}, are added between points O and H, O and G in Figure 5c,f. Both switches T_{PF} and T_{NF} in Figure 5c are turned on in PF and NF modes based on Rule #1. Point B is connected to point O through diode D_{B1} and switch T_{PF} or through D_{B2} and T_{NF} in Figure 5c, based on Rule #2. A bidirectional current branch clamps point B (or point A) in PF mode and NF mode. It is easy to make a similar analysis of the NPC cell shown in Figure 5f. For the sake of brevity, this is not presented here.

As described in Part I of this paper, two methods can be used to construct the freewheeling cell based on the topology family without an extra diode; they are shown in Figure 6a,c. In Figure 6b, the extra switch T_{B1} is added between points O and G (H), and the midpoint O is connected to point B (or point A) through the extra switch T_{B1} based on Rule #2. Two extra diodes, D_{B1} and D_{B2}, are added between points O and G/H in Figure 6d. D_{B1} and D_{B2} are used to provide the bidirectional current path in PF and NF modes. The corresponding NPC topologies will be described in the next section.

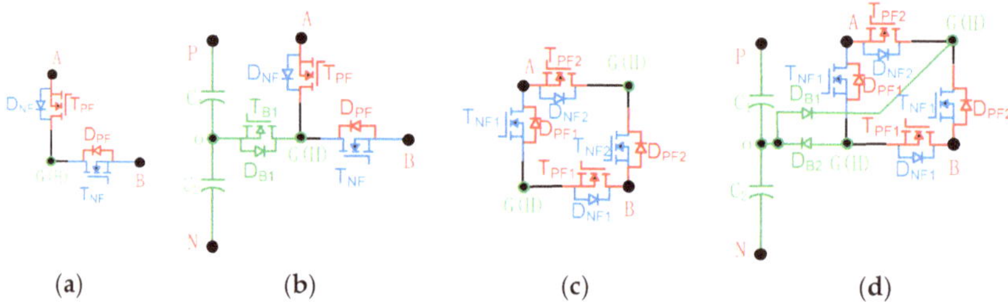

Figure 6. Indirect Connection of NPC cell circuits based on topology family without an extra diode. (**a**) case #1; (**b**) NPC cell #1 under case #1; (**c**) case #2; (**d**) NPC cell under case #2.

Figure 6 shows the Indirect Connection NPC cell circuits based on the topology family without an extra diode. Two freewheeling cell circuits are introduced, and two corresponding NPC cell circuits are described.

3.2. M = 2, N = 2

The steps of how to derive the Indirect Connection NPC topology R1S1 based on HERIC topology R1are illustrated in Figure 7.

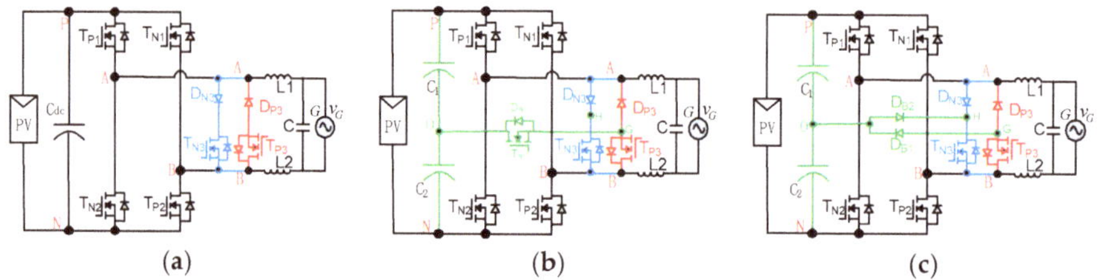

Figure 7. Indirect Connection NPC topologies R1S1 under M = 2, N = 2. (**a**) HERIC topology R1; (**b**) R1S1-1; (**c**) R1S1-2 [50,51].

The HERIC topology, named R1 in Figure 7a, is well known. Points A and B are short-circuited in PF mode (the red branch from point B to A) and NF mode (the blue branch from point A to B). The current flows to point B through T_{P3}, D_{P3} to Point A in PF mode, and flows point A through T_{N3}, D_{N3} to point B in NF mode. The DC capacitor C_{dc} is split into two series capacitors, C_1 and C_2, to provide half the input voltage shown in Figure 7b,c. According to Rule #1, switches T_{P3} and T_{N3} are on. An extra switch T_{B1} is added so that point B is clamped to point O, as shown in Figure 7b. As shown in Figure 7c, two additional diodes, D_{B1} and D_{B2}, are added to guarantee that point B is clamped to point O.

Figure 8 shows all the NPC topologies under M = 2, N = 2.

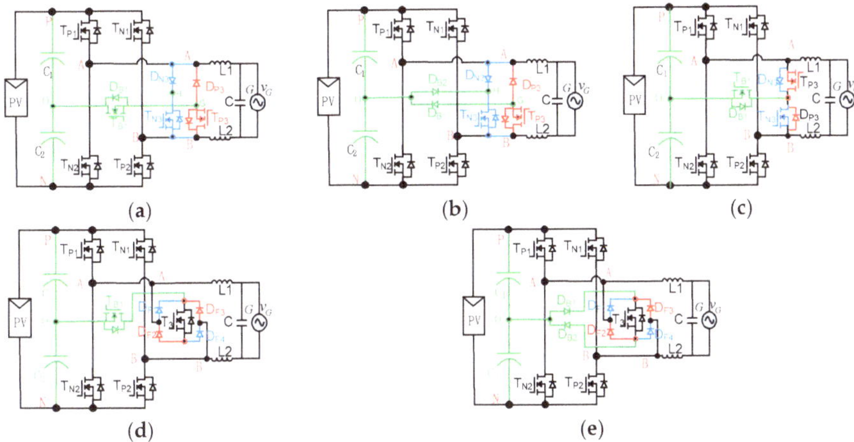

Figure 8. Indirect Connection NPC topologies from M = 2, N = 2. (**a**) R1S1-1 [50]; (**b**) R1S1-2 [18,50,51]; (**c**) R2S1 [51]; (**d**) R3S1-1 [51]; (**e**) R3S1-2 [51].

An extra switch, T_{B1}, is added to flow the bidirectional current, as shown in Figure 8a,c,d. Two extra diodes, D_{B1} and D_{B2}, are added to flow bidirectional current, as shown in Figure 8b,e.

3.3. M = 2, N = 3 or M = 3, N = 2

Figure 9 shows all the NPC topologies under M = 2, N = 3 or M = 3, N = 2. An extra switch, T_{B1}, is added to flow bidirectional current, as shown in Figures 8b and 9a,d. Two extra diodes, D_{B1} and D_{B2}, are added to flow bidirectional current, as shown in Figure 9c.

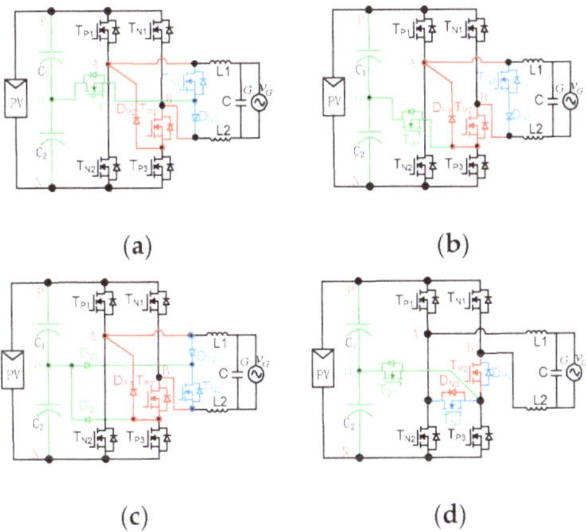

Figure 9. Indirect Connection NPC topologies from M = 2, N = 3 or M = 3, N = 2. (**a**) R4S1-1 [51]; (**b**) R4S1-2 [51]; (**c**) R4S1-3 [51]; (**d**) R5S1 [50].

3.4. M = 3, N = 3

Figure 10 shows all the NPC topologies based on the original topologies H6 and H5 under M = 3, N = 3.

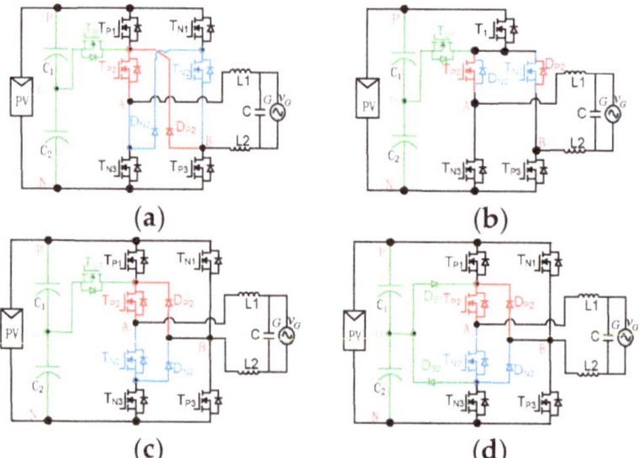

Figure 10. Indirect Connection NPC topologies from M = 3, N = 3. (**a**) R6S1 [51]; (**b**) R7S1 [19]; (**c**) R8S1-1 [51]; (**d**) R8S1-2 [48].

3.5. M = 3, N = 4 or M = 4, N = 3

Figure 11 shows all the NPC topologies under M = 3, N = 4 or M = 4, N = 3.

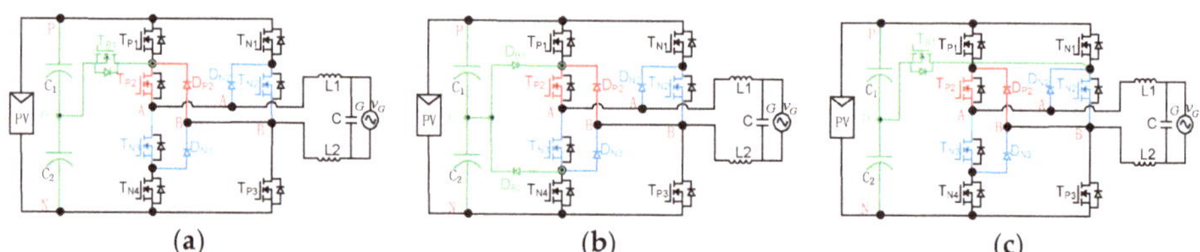

Figure 11. Indirect Connection NPC topologies from M = 3, N = 4 or M = 4, N = 3. (**a**) R9S1-1 (new); (**b**) R9S1-2 (new); (**c**) R10S1 (new).

3.6. M = 4, N = 4

Figure 12 shows all the NPC topologies under M = 4, N = 4.

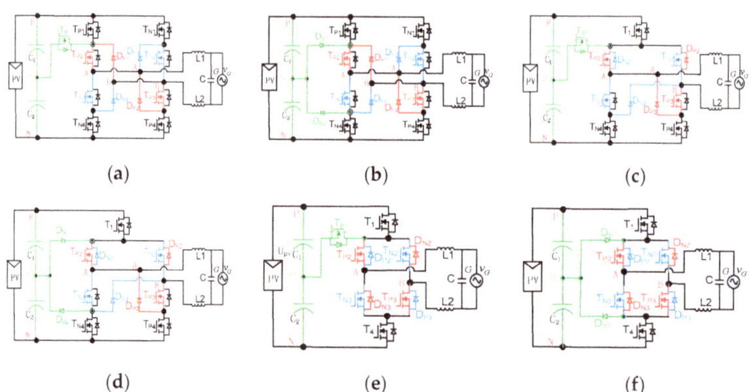

Figure 12. Indirect Connection NPC topologies from M = 4, N = 4. (**a**) R11S1-1 (new); (**b**) R11S1-2 (new); (**c**) R12S1-1 (new); (**d**) R12S1-2 (new); (**e**) R13S1-1 [49]; (**f**) R13S1-2 [49].

4. Inverter Topologies Based on the Direct Connection NPC Cell from USPWM

In this section, inverter topologies based on a Direct Connection NPC cell from USPWM are introduced. As shown in Table 2, they are also classified into two families.

Table 2. Two inverter topology families from USPWM.

(M, N)	$X_1 + X_2$	$Y_1 + Y_2$	Family With Extra Diode		Family Without Extra Diode	
			Without NPC Cell (Part I)	With NPC Cell (Part II)	Without NPC Cell (Part I)	With NPC Cell (Part II)
(M = 2, N = 2)	1 + 1	1 + 1	R1, R3	R1S2, R3S2	R2	R2S2-1, R2S2-2
(M = 2, N = 3) or (M = 3, N = 2)	1 + 2 1 + 1 2 + 1 1 + 1	1 + 1 1 + 2 1 + 1 2 + 1	R4	R4S2	R5	R5S2-1, R5S2-2
(M = 3, N = 3)	2 + 1 1 + 2	2 + 1 1 + 2	R6	R6S2	R7	R7S2-1, R7S2-2
	2 + 1 1 + 2	1 + 2 2 + 1	R8	R8S2	None available	R8S2
(M = 3, N = 4) or (M = 4, N = 3)	1 + 2 2 + 2 2 + 1 2 + 2	2 + 2 1 + 2 2 + 2 2 + 1	R9	R9S2	R10	R10S2
(M = 4, N = 4)	2 + 2	2 + 2	R11, R12	R11S2, R12S2	R13	R13S2

4.1. Direct Connection NPC Cell Based on Two Topology Families

Figure 13 shows the Direct Connection NPC cell circuits based on the topology family with an extra diode. Two freewheeling cell circuits are introduced in Figure 13a,c, respectively. The switches T_{PF} and T_{NF} in the NPC cell are on in both PF and NF mode, based on Rule #1. According to Rule #2, the extra switches, T_{B1}, T_{B2}, and extra diodes, D_{B3}, D_{B4}, are inserted into the freewheeling cell in Figure 13b. As shown in Figure 13b, the midpoint O is connected to point B (or point A) through the extra switch T_{B1} (T_{B2}) and extra diode D_{B4} (D_{B3}). The midpoint O is connected to point B through the extra switch T_{B1} shown in Figure 13d. It is observed that the freewheeling cell is changed due to the injection of extra switches and diodes.

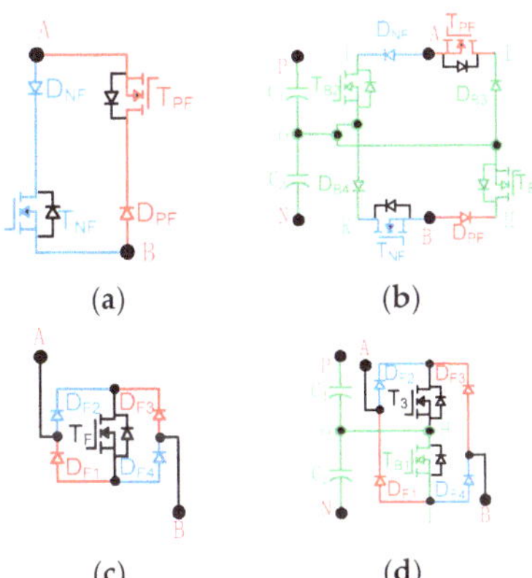

Figure 13. Direct Connection NPC cell circuits based on topology family with an extra diode. (**a**) Case #1; (**b**) NPC cell under case #1; (**c**) Case #2; (**d**) NPC cell under case #2 (new).

Figure 14 shows the Direct Connection NPC cell circuits based on the topology family without an extra diode. There is one circuit which may be used to construct the freewheeling cell, as shown in Figure 14a. Figure 14b,c show the NPC cell circuits. The original freewheeling branch is cut and two extra switches, T_{B1} and T_{B2}, are inserted to form the new freewheeling branch Figure 14b. The midpoint O is connected to point B (or point A) through the extra switches T_{B1} (or T_{B2}) based on Rule #2. One extra switch, T_{B1}, and two extra diodes, D_{B2} and D_{B3}, are inserted to construct the new freewheeling branch in Figure 14c. The current flows from point B to point A through D_{PF}, T_{B1}, D_{B2}, T_{PF} in PF mode and point B is connected to point O. The current flows from point A to point B through D_{NF}, D_{B3}, and T_{NF}.

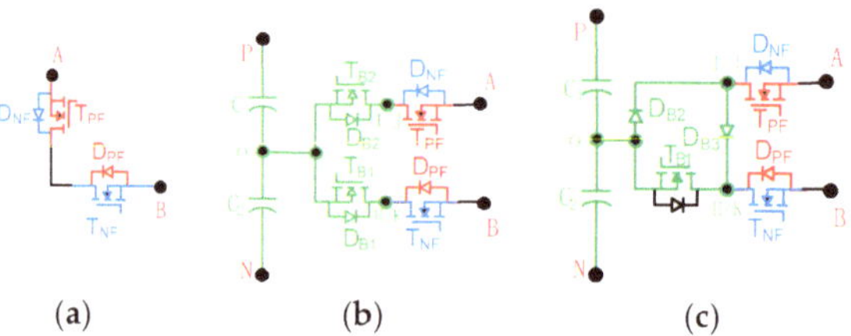

Figure 14. Direct Connection NPC cell circuits based on topology family without an extra diode. (**a**) freewheeling circuit; (**b**) NPC cell #1; (**c**) NPC cell #2.

4.2. M = 2, N = 2

In order to show how the Direct Connection NPC topology R2S2 is derived based on HERIC topology R2, it is illustrated in Figure 15. According to Rule #1, switches T_{P3} and T_{N3} are on in PF and NF modes. As shown in Figure 15b, two extra switches, T_{B1} and T_{B2}, are inserted so that point B or point A is clamped to point O based on Rule #2. In Figure 15c, two extra diodes, D_{B2} and D_{B3}, and one extra switch, T_{B1}, are inserted to guarantee that point B is clamped to point O.

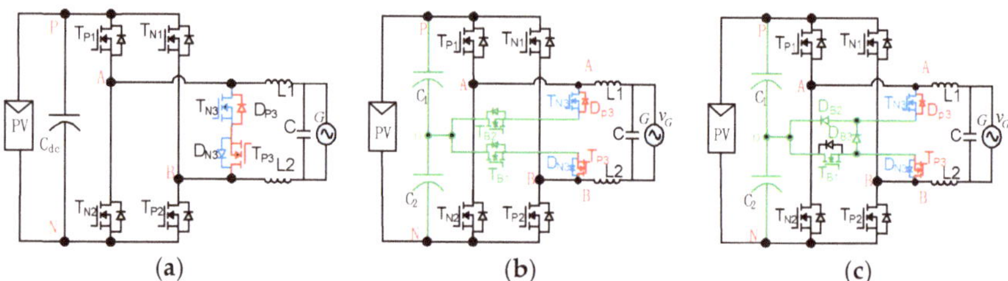

Figure 15. Direct Connection NPC topologies R1S2 under M = 2, N = 2. (**a**) HERIC topology R2; (**b**) R2S2-1 from R2; (**c**) R2S2-2 from R2 (new).

Figure 16 shows the other Direct Connection NPC topologies based on HERIC topology under M = 2, N = 2. Two extra switches, T_{B1} and T_{B2}, are inserted as shown in Figure 16a, and T_{B1} is inserted in Figure 16b.

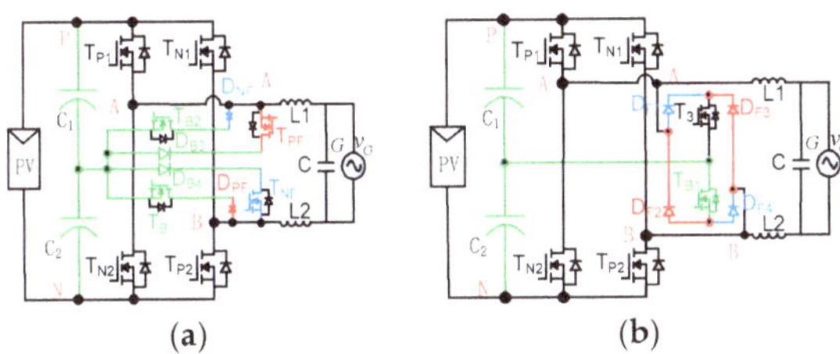

Figure 16. Direct Connection NPC topologies from M = 2, N = 2. (**a**) R1S2 (new); (**b**) R3S2 (new).

4.3. M = 2, N = 3 or M = 3, N = 2

Figure 17 shows all the NPC topologies under M = 2, N = 3 or M = 3, N = 2. Two extra switches, T_{B1}, T_{B2}, and two extra diodes, D_{B3} and D_{B4}, are inserted so that point B or A is clamped to point O, as shown in Figure 17a. Two extra switches, T_{B1}, T_{B2}, are inserted in Figure 17b. One switch, T_{B1}, and two diodes, D_{B2} and D_{B3}, are inserted in Figure 17c.

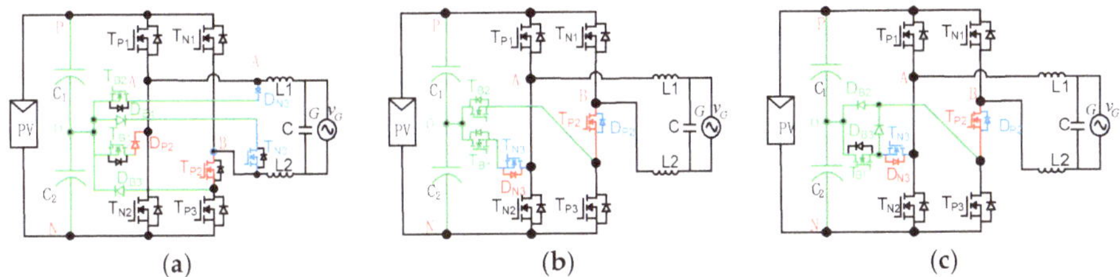

Figure 17. Direct Connection NPC topologies from M = 2, N = 2. (**a**) R4S2 (new); (**b**) R5S2-1 [47]; (**c**) R5S2-2 (new).

4.4. M = 3, N = 3

Figure 18 shows all the NPC topologies under M = 3, N = 3. Two extra switches, T_{B1} and T_{B2}, and two extra diodes, D_{B3} and D_{B4}, are inserted in Figure 18a,d. Two extra switches, T_{B1}, T_{B2}, are added in Figure 18b. One switch, T_{B1}, and two diodes, D_{B2}, D_{B3}, are added in Figure 18c.

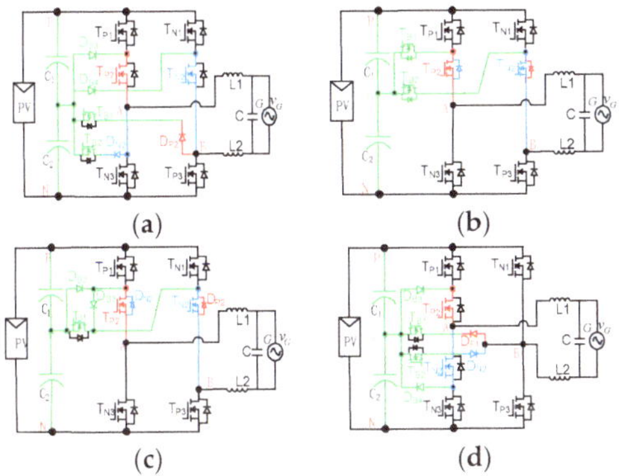

Figure 18. Direct Connection NPC topologies from M = 3, N = 3. (**a**) R6S2 (new); (**b**) R7S2-1 [47]; (**c**) R7S2-2 (new); (**d**) R8S2 (new).

4.5. M = 3, N = 4 or M = 4, N = 3

Figure 19 shows the NPC topologies under M = 3, N = 4 or M = 4, N = 3.

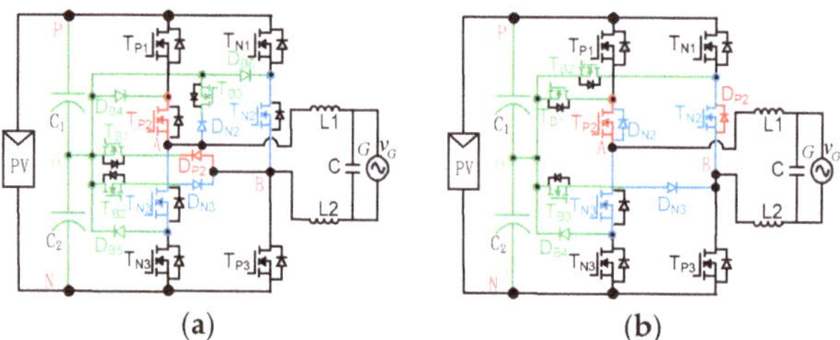

Figure 19. Direct Connection NPC topologies from M = 3, N = 3. (**a**) R9S2 (new); (**b**) R10S2 (new).

4.6. M = 4, N = 4

Figure 20 shows all the NPC topologies under M = 4, N = 4. Figure 20a is given as an example to describe the current flowing. In PF or NF mode, there are two branches for the flow of positive current. One is as follows: Point B→D_{P2}→T_{B1}→D_{B5}→T_{P2}→Point A, and the other is from Point B→T_{P3}→D_{B6}→T_{B2}→D_{P3}→Point A. Similarly, there are two branches for the flow of negative current. The first one is as follows: Point A→D_{N2}→T_{B3}→D_{B7}→T_{N2}→Point B. The other is from Point A→T_{N3}→D_{B8}→T_{B4}→D_{N3}→Point B.

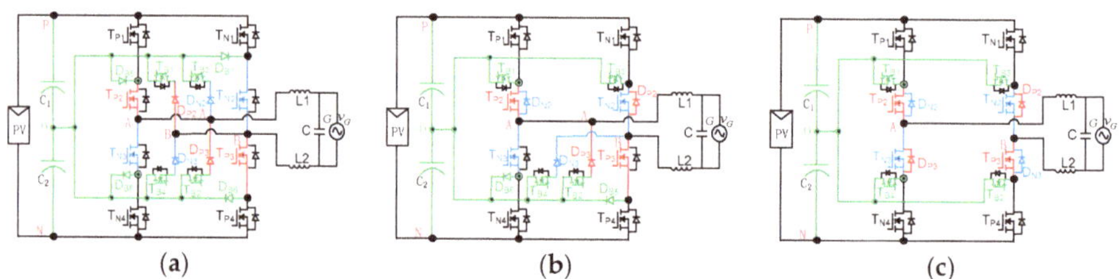

Figure 20. Direct Connection NPC topologies from M = 4, N = 4. (**a**) R11S2 (new); (**b**) R12S2 (new); (**c**) R13S2 (new).

5. Two Types of NPC Cells and Reflected Topologies under DFUSPWM

Similar to USPWM, there are two type of NPC cells under DFUSPWM. Part A introduces the principle of the Indirect Connection NPC cell and reflected topologies under DFUSPWM. Part B introduces the principle of the Direct Connection NPC cell and reflected topologies under DFUSPWM.

5.1. Principle of Indirect Connection NPC Cell and Reflected Topologies under DFUSPWM

Figure 21 shows a schematic diagram of the first type of unified NPC cell. Two capacitors with the same capacitance are connected in series to achieve half of the input voltage for each one. The Indirect Connection NPC cell under DFUSPWM is connected to the freewheeling cell through extra branches in PFT mode, PFB mode, NFT mode, and NFB mode to keep the CM voltage constant. The extra branches are $\hat{O}G$ and $\hat{O}H$ flowing positive current in Figure 21a,b, and $\hat{O}I$ and $\hat{O}J$ flowing negative current in Figure 21c,d. It can be seen that the internal connections of the freewheeling cell remain the same in freewheeling mode.

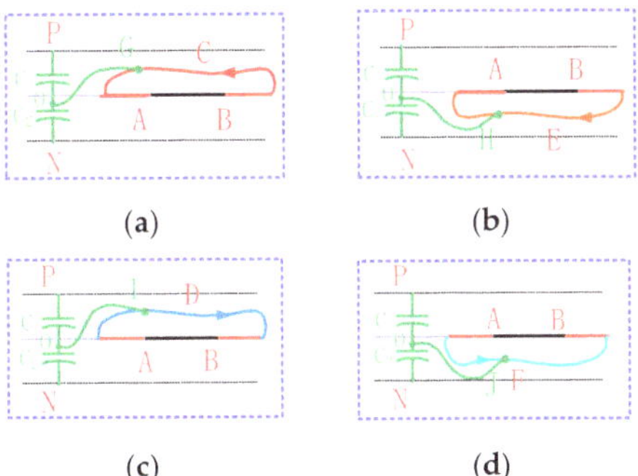

Figure 21. Indirect Connection NPC cell under DFUSPWM. (**a**) PFT mode; (**b**) PFB mode; (**c**) NFT mode; (**d**) NFB mode.

The inverter topologies with the Indirect Connection NPC cell under DFUSPWM can also be classified into two types, as shown in Table 3.

Table 3. Two NPC topology families under DFUSPWM.

(M, N)	$X_1 + X_2$	$Y_1 + Y_2$	Family With Extra Diode		Family Without Extra Diode	
			Without NPC Cell	With NPC Cell	Without NPC Cell	With NPC Cell
(M = 4, N = 4)	2 + 2	2 + 2	R11, R12	R11S1, R12S1	R13	R13S1

Figure 22 shows the Indirect Connection NPC cell circuits based on freewheeling cell circuits. There are three methods by which to construct the freewheeling cell based on the topology family with an extra diode. They are shown respectively in Figures 5a and 22c,e.

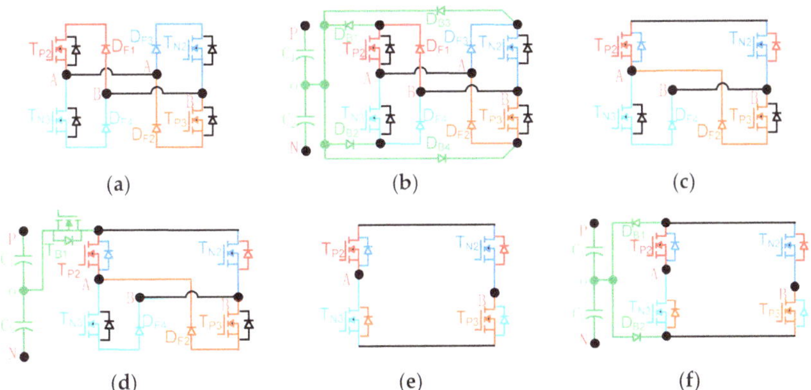

Figure 22. Indirect Connection NPC cell circuits under DFUSPWM. (**a**) case #1; (**b**) NPC cell under case #1; (**c**) case #2; (**d**) NPC cell under case #2; (**e**) case #3; (**f**) NPC cell under case #3.

For the NPC cells in Figure 22b,d,f, T_{P2} and T_{N3} have the same signals to remain on or off, and T_{P3} and T_{N2} have the same driving signals under DFUSPWM. Thus, additional diodes or switches can be used to provide a bidirectional current branch to clamp point A or point B to point O. The corresponding inverter topologies are shown in Figure 23.

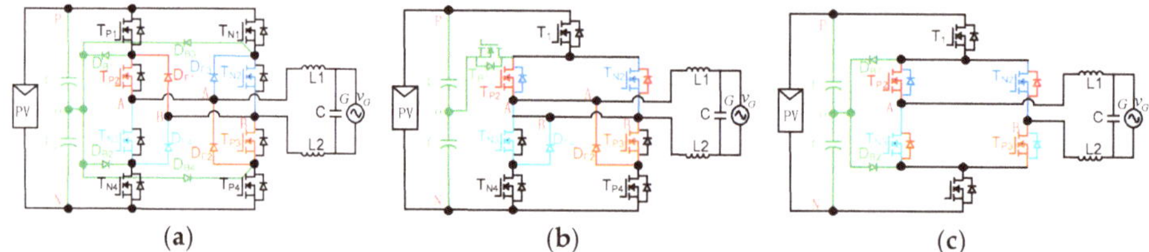

Figure 23. Inverter topologies with the Direct Connection NPC cell under DFUSPWM. (**a**) R11S1 (new); (**b**) R12S1 (new); (**c**) R13S1 (new).

5.2. Principle of the Direct Connection NPC Cell and Reflected Topology under DFUSPWM

Figure 24 shows a schematic diagram of the second type of unified NPC cell under DFUSPWM. The extra branch $\hat{GH}$ is injected into the freewheeling path $B\hat{C}A$ in Figure 24a, and $\hat{IJ}$ is injected into the freewheeling path $B\hat{E}A$ in Figure 24b. Similarly, the extra branch $\hat{KL}$ is injected into the freewheeling path $A\hat{D}B$ in Figure 24c and $\hat{MN}$ is injected into the freewheeling path $A\hat{F}B$ in Figure 24d. Clearly, the internal connections of the freewheeling cell have been changed due to the injection of the Direct Connection NPC cell.

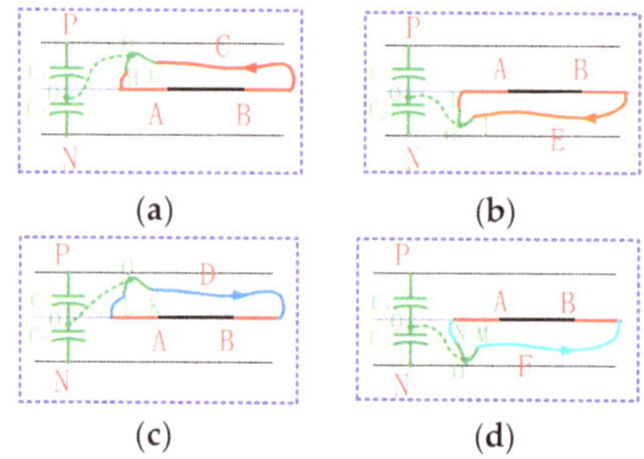

Figure 24. Direct Connection NPC cell under DFUSPWM. (**a**) PFT mode; (**b**) PFB mode; (**c**) NFT mode; (**d**) NFB mode.

The inverter topologies with the Direct Connection NPC cell under DFUSPWM can also be classified into two types, as shown in Table 4.

Table 4. Direction connection NPC cell under DFUSPWM.

(M, N)	$X_1 + X_2$	$Y_1 + Y_2$	Family with Extra Diode		Family Without Extra Diode	
			Without NPC Cell	With NPC Cell	Without NPC Cell	With NPC Cell
(M = 4, N = 4)	2 + 2	2 + 2	R11, R12	R11S2, R12S2	R13	R13S2

Figure 25 shows the Direct Connection NPC cell circuits based on the freewheeling cell circuits under DFUSPWM. Three freewheeling cell circuits are introduced in Figure 25a,c,e.

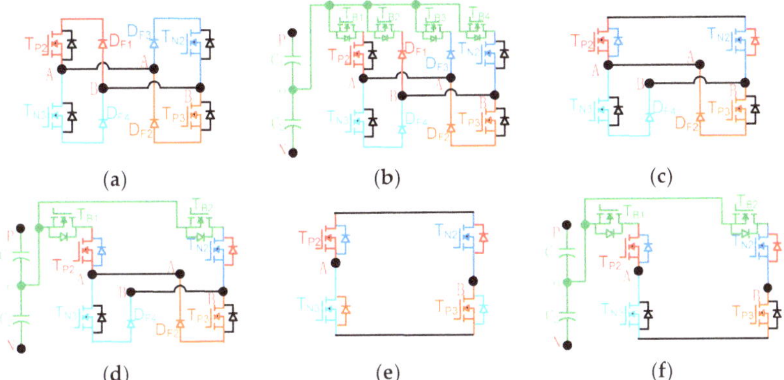

Figure 25. Direct Connection NPC cell circuits under DFUSPWM. (**a**) case #1; (**b**) NPC cell under case #1; (**c**) case #2; (**d**) NPC cell under case #2; (**e**) case #3; (**f**) NPC cell under case #3.

There are two current branches which allow the current to flow from point B to A, or from point A to B. Point A is of the same voltage as point B so that point A or point B is connected to NPC cell.

For NPC cells, as shown in Figure 25b,d,f, additional diodes and/or switches are used to provide a bidirectional current branch to clamp point A or B to point O. The corresponding inverter topologies are shown in Figure 26.

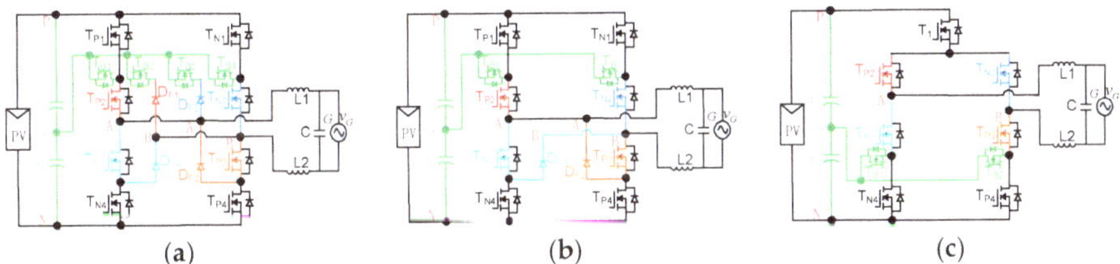

Figure 26. Inverter topologies with the Direct Connection NPC cell under DFUSPWM. (**a**) R11S2 (new); (**b**) R12S2 (new); (**c**) R13S2 (new).

In order to make an overall comparison of all the derived topologies, Tables 5 and 6 are presented, wherein the number of switches, the number of diodes and the economic cost of all the derived topologies are taken into consideration. Among them, the economic cost part is obtained based on the formula that each switch costs 1, while each diode costs 0.3. Based on the two tables, we may select a circuit topology which is suitable for our situation.

Table 5. Comparison of the Inverter topologies under USPWM.

Topology Name	Number of Switches	Number of Diodes	Economic Cost	Topology Name	Number of Switches	Number of Diodes	Economic Cost
R1S1-1	7	2	7.6	R10S1	8	3	8.9
R1S1-2	6	4	7.2	R13S1-1	7	0	7
R3S1-1	6	4	7.2	R13S1-2	6	2	6.6
R3S1-2	5	6	6.8	R1S2	8	4	9.2
R4S1-1	7	2	7.6	R3S2	6	4	7.2
R4S1-2	7	2	7.6	R4S2	8	4	9.2
R4S1-3	6	4	7.2	R6S2	8	4	9.2
R6S1	7	2	7.6	R8S2	8	4	9.2
R8S1-1	7	2	7.6	R9S2	10	6	11.8
R8S1-2	6	4	7.2	R11S2	12	8	14.4
R9S1-1	8	3	8.9	R12S2	12	4	13.2
R9S1-2	7	5	8.5	R2S2-1	8	0	8
R11S1-1	9	4	10.2	R2S2-2	7	2	7.6
R11S1-2	8	6	9.8	R5S2-1	8	0	8
R12S1-1	8	2	8.6	R5S2-2	8	2	8.6
R12S1-2	7	4	8.2	R7S2-1	8	0	8
R2S1	7	0	7	R7S2-2	7	2	7.6
R5S1	7	0	7	R10S2	10	2	10.6
R7S1	6	0	6	R13S2	12	0	12

Table 6. Comparison of the Inverter topologies under DFUSPWM.

Connection Mode of NPC Cell	Topologies Name	Number of Switches	Number of Diodes	Economic Cost
Indirect Connection	R11S1	8	8	10.4
	R12S1	8	2	8.6
	R13S1	6	2	6.6
Direct Connection	R11S2	12	4	13.2
	R12S2	10	2	10.6
	R13S2	9	0	9

6. Simulation Results

Two types NPC topologies with a Direct Connection NPC cell or an Indirect Connection NPC cell are provided based on the topologies provided in Part I of this paper. To further verify the theoretical analysis in the coming sections, simulations based on two proposed topologies are made, and the simulation results are given. One is the H6 topology (R5) and the proposed H6 topology with a Direct Connection NPC cell (R5S2-2) under USPWM. The other is the proposed H8 topology (R13) and H8 topology with an Indirect Connection NPC cell (R11S1) under DFUSPWM. 0shows the simulation parameters.

6.1. H6 Topology Without/With Direct Connection NPC Cell under USPWM

Figure 27 shows the H6 topology without/with a Direct Connection NPC cell. The H6 topology R5 is illustrated in Figure 27a. Switches T_{P2} and T_{N3} are used to allow the freewheeling current to flow. The freewheeling branch is cut off and the Direct Connection NPC cell is injected into the topology R5 to form R5S2-2, as shown in Figure 27b.

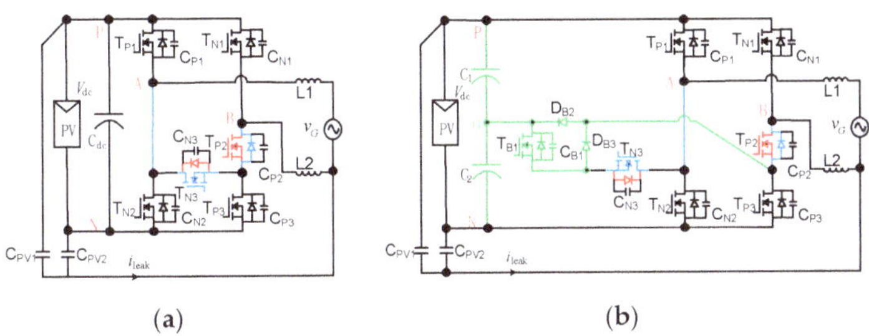

(a)   (b)

Figure 27. Topology H6 without /with NPC cell under USPWM. (**a**) R5; (**b**) R5S2-2 (new).

Figure 28 shows the waveforms of CM voltage and leakage current. In t_0–t_1, the topology H6 without the NPC cell works, and the CM voltage $(V_{AN} + V_{BN})/2$ does not remain constant. The CM voltage includes high frequency resonant voltage in freewheeling mode, as terminals A and B are floating $(V_{AN} = V_{BN})$. The leakage current is 100 mA. In t_1–t_2, the topology H6 with the NPC cell works and the CM voltage $(V_{AN} + V_{BN})/2$ always remains constant. The leakage current is reduced from 100 mA to 2 mA.

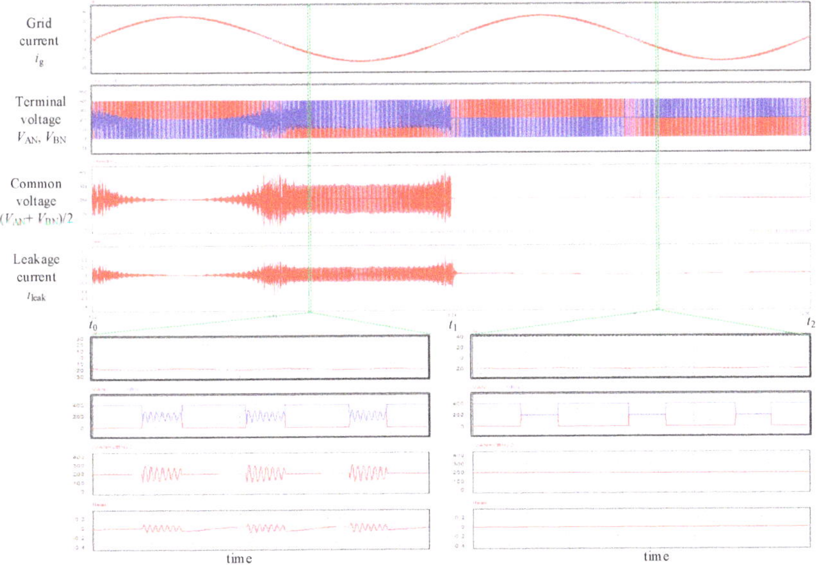

Figure 28. Waveforms of grid current, terminal voltage, CM voltage, and leakage current.

Figure 29 shows the waveforms of the grid current, terminal voltage, CM voltage, and leakage current with the parameters listed in Table 7. The THD of the grid current is about 1.12%. As we can see, the CM voltage is maintained at 200 V with small fluctuations over the whole power line period; thus, the RMS value of the leakage current is only 3 mA. Figure 30 shows the waveforms when the load steps at $t = 0.06$ s. The THD is still 1.12%. The CM voltage and leakage current have the same values as those in Figure 29.

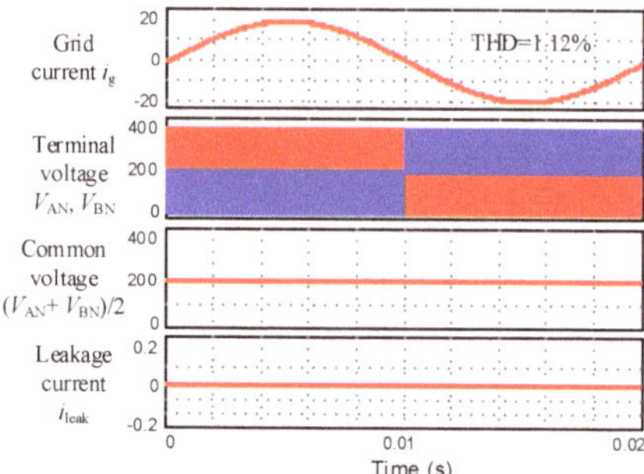

Figure 29. Waveforms of grid current, terminal voltage, CM voltage, and leakage current.

Table 7. Simulation Parameters.

Parameter	Value
Rated power	3000 W
Input voltage	400 V
Grid voltage/frequency	220 V/50 Hz
Filter inductor L1, L2	1 mH
Switching frequency	20 kHz
DC-bus Capacitor C_{dc1}, C_{dc2}	470 μF
Junction capacitor of each switch	100 pF
PV parasitic capacitor C_{PV1}, C_{PV2}	0.1 μH

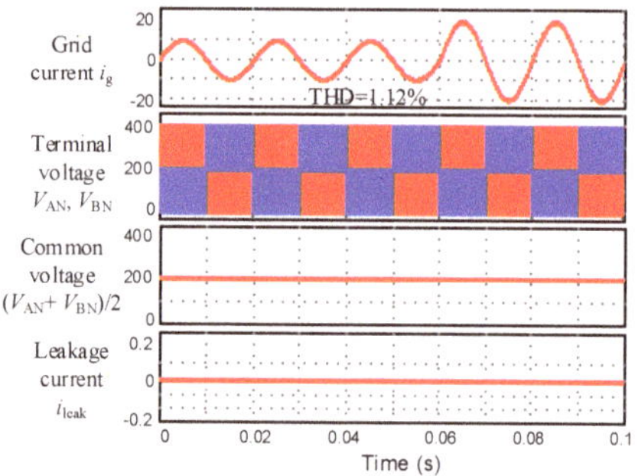

Figure 30. Waveforms of grid current, terminal voltage, CM voltage, and leakage current in the case.

6.2. H8 Topology Without/With Indirect Connection NPC Cell under DFUSPWM

Figure 31 shows the topology H8 without/with "Indirect Connection" NPC cell under DFUSPWM.

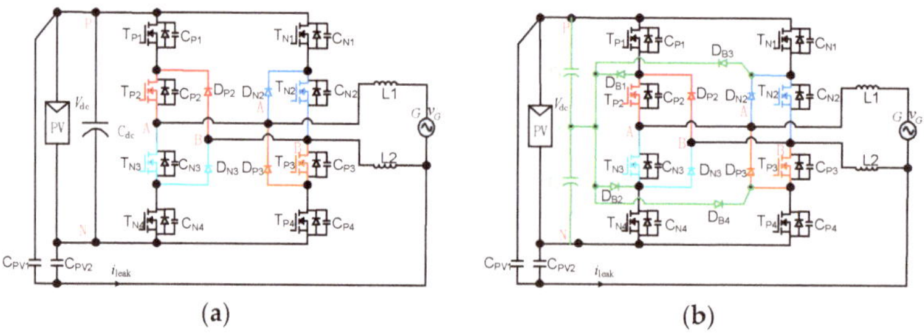

Figure 31. Topology H8 without /with NPC cell under DFUSPWM. (**a**) R13 (new); (**b**) R11S1 (new).

Figure 32 shows the waveforms of CM voltage and leakage current. It may be observed that the topology H8 without the NPC cell works, and the CM voltage is not constant between t_0 and t_1. The CM voltage contains high frequency resonant voltage in freewheeling mode due to the floating terminals A and B. The leakage current is 100 mA. In t_1–t_2, the topology H8 with NPC cell works, and the CM voltage always remains constant. The leakage current is reduced from 100 mA to 3 mA.

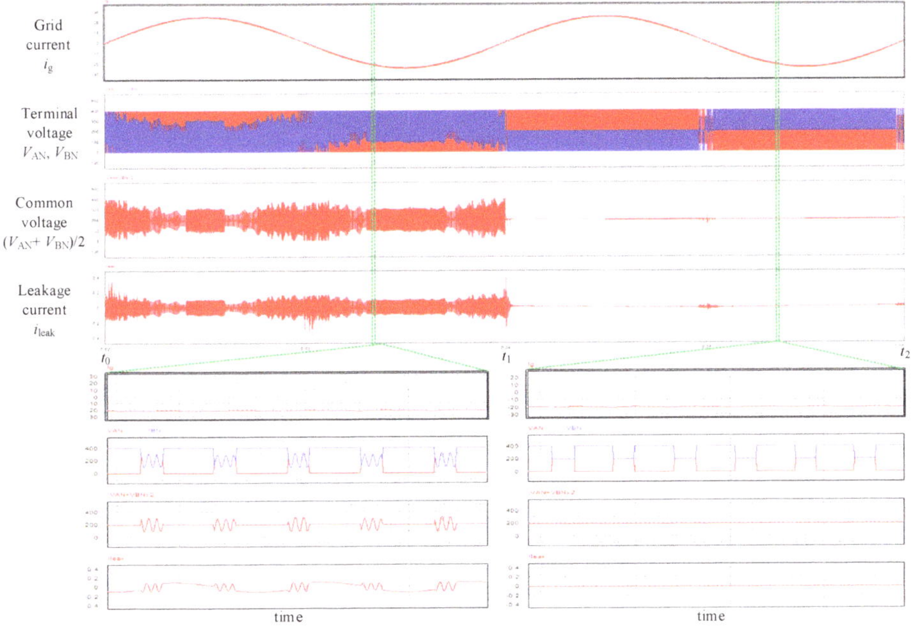

Figure 32. Waveforms of grid current, terminal voltage, CM voltage and leakage current.

Figure 33 shows the waveforms of the leakage current, voltage V_{AB}, and PWM signal. The same PWM signals are provided to couple switches T_{P1} and T_{P3}, T_{P2} and T_{P4}, T_{N1} and T_{N3}, as well as T_{N2} and T_{N4}. To achieve good clamping performance, complementary PWM signals are given to couple switches T_{P1} and T_{N2}, as well as T_{P2} and T_{N1}. The frequency of voltage V_{AB} is double the switching frequency.

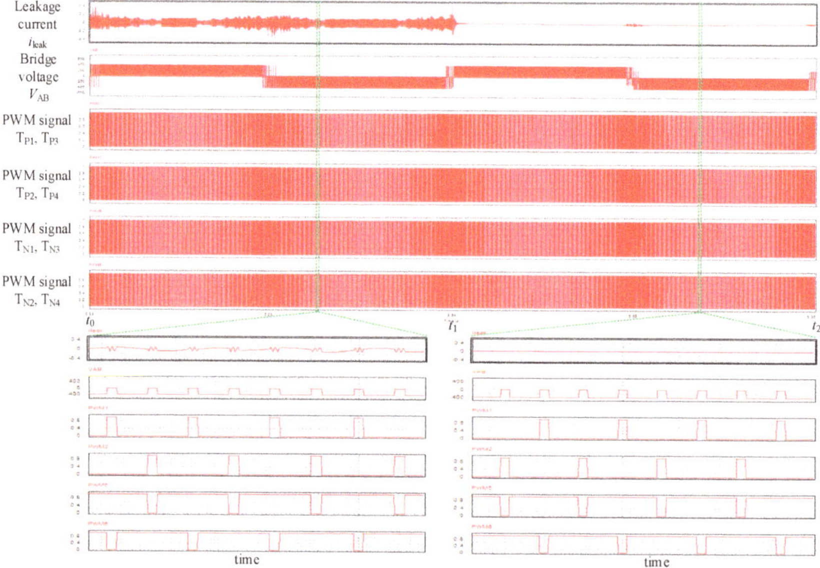

Figure 33. Waveforms of leakage current, voltage V_{AB}, and PWM signal.

7. Conclusions

In this paper, a topology derivation methodology, named the "MN principle", is proposed to cover all possible full-bridge topologies with leakage current suppression under USPWM and DFUSPWM. Two types of NPC cells have been developed: one is called "indirect connection NPC topology", and the other "direct connection NPC topology". Under USPWM, twenty-three newly-found indirect connection NPC topologies have been derived along with the existing ones, and twenty-two newly-found direct connection NPC topologies have been derived along with the existing ones. The proposed method is also extended to the topologies under DFUSPWM. As a result, four corresponding topologies have been derived and one existing topology was also covered. Finally, simulations are given to verify the performance of the leakage current suppression of the proposed topologies. And it can be inferred that the contribution of this paper has important guiding significance for topological derivation.

Author Contributions: Conceptualization, X.Y., X.W.; funding acquisition, X.W.; investigation, X.Y.; software, X.Z.; validation, X.Z., X.W.; writing—original draft, X.Y.; writing—review and editing, H.W. and Y.-F.L. All authors have read and agreed to the published version of the manuscript.

Funding: Research on Topology, Passive Current Sharing Mechanism and Control for Multiphase Resonant Converter with Coupled Resonant Tank: 51977069; Research on High-power and High-efficiency Electro-acoustic Transduction Mechanism and Control Method: 51837005.

Conflicts of Interest: The authors declare no conflict of interest.

References

1. Sun, K.; Zhang, L.; Xing, Y.; Guerrero, J.M. A Distributed Control Strategy Based on DC Bus Signaling for Modular Photovoltaic Generation Systems With Battery Energy Storage. *IEEE Trans. Power Electron.* **2011**, *26*, 3032–3045. [CrossRef]
2. EPIA. *Global Market Outlook for Photovoltaics 2013–2017*; European Photovoltaic Industry Association: Brussels, Belgium, 2013.
3. Kjaer, S.B.; Pedersen, J.K.; Blaabjerg, F. A review of single-phase grid-connected inverters for photovoltaic modules. *IEEE Trans. Ind. Appl.* **2005**, *41*, 1292–1306. [CrossRef]

4. Blaabjerg, F.; Zhe, C.; Kjaer, S.B. Power electronics as efficient interface in dispersed power generation systems. *IEEE Trans. Power Electron.* **2004**, *19*, 1184–1194. [CrossRef]

5. Quan, L.; Wolfs, P. A Review of the Single Phase Photovoltaic Module Integrated Converter Topologies With Three Different DC Link Configurations. *IEEE Trans. Power Electron.* **2008**, *23*, 1320–1333. [CrossRef]

6. Power Generation Systems Connected to the Low-Voltage Distribution Network, VDE-AR-N 4105. Available online: https://www.sogou.com/link?url=hedJjaC291MdqKRdyYP6xaEfyF6gLPTmC2Ly98cd-1VBavtdX5b14M0N-aB7Vy-zp32izvdhNQ_1-ZZN6CSHZQ (accessed on 1 August 2011).

7. Standard for Inverters, Converters, Controllers, and Interconnection System Equipment for Use With Distributed Energy Resources. Available online: https://standardscatalog.ul.com/standards/en/standard_1741_2 (accessed on 28 January 2010).

8. Automatic Disconnection Device Between a Generator and the Public Low-Voltage Grid, DIN VDE V 0126-1-1. Available online: https://www.beuth.de/en/pre-standard/din-vde-v-0126-1-1/187485608 (accessed on 1 August 2013).

9. Gonzalez, R.; Gubia, E.; Lopez, J.; Marroyo, L. Transformerless Single-Phase Multilevel-Based Photovoltaic Inverter. *IEEE Trans. Ind. Electron.* **2008**, *55*, 2694–2702. [CrossRef]

10. Xiao, H.; Xie, S. Transformerless Split-Inductor Neutral Point Clamped Three-Level PV Grid-Connected Inverter. *IEEE Trans. Power Electron.* **2012**, *27*, 1799–1808. [CrossRef]

11. Victor, M.; Greizer, K.; Bremicker, A. Method of Converting a Direct Current Voltage from a Source of direct Current Voltage, More Specifically from a Photovotatic Source of Direct Current Voltage, into a Alternating Current Voltage. U.S. Patent 7411802B2, 12 August 2008.

12. Schmidt, H.; Siedle, C.; Ketterer, J. Inverter for Converting an Electric Direct Current into an Alternating Current or an Alternating Voltage. European Patent EP2 086 102 A2, 5 August 2009.

13. Patino, D.G.; Erira, E.G.; Fuelagan, J.R.; Rosero, E.E. Implementation a HERIC inverter prototype connected to the grid controlled by SOGI-FLL. In Proceedings of the 2015 IEEE Workshop on Power Electronics and Power Quality Applications (PEPQA), Bogota, Colombia, 2–4 June 2015; pp. 1–6.

14. Zhang, L.; Sun, K.; Xing, Y.; Xing, M. H6 Transformerless Full-Bridge PV Grid-Tied Inverters. *IEEE Trans. Power Electron.* **2014**, *29*, 1229–1238. [CrossRef]

15. Yu, W.; Jai, J.-S.; Qian, H.; Hutchens, C.; Zhang, J.; Lisi, G.; Djabbari, A.; Smith, G.; Hegarty, T. High-efficiency inverter with H6-type configuration for photovoltaic non-isolated ac module applications. In Proceedings of the 2010 Twenty-Fifth Annual IEEE Applied Power Electronics Conference and Exposition (APEC), Palm Springs, CA, USA, 21–25 February 2010; pp. 1056–1061.

16. Yang, B.; Li, W.; Gu, Y.; Cui, W.; He, X. Improved Transformerless Inverter With Common-Mode Leakage Current Elimination for a Photovoltaic Grid-Connected Power System. *IEEE Trans. Power Electron.* **2012**, *27*, 752–762. [CrossRef]

17. Su, X.; Sun, Y.; Lin, Y. Analysis on Leakage Current in Transformerless Single-Phase PV Inverters Connected to the Grid. In Proceedings of the 2011 Asia-Pacific Power and Energy Engineering Conference (APPEEC), Wuhan, China, 25–28 March 2011; pp. 1–5.

18. Xiao, H.F.; Ke, L.; Li, Z. A Quasi-Unipolar SPWM Full-Bridge Transformerless PV Grid-Connected Inverter with Constant Common-Mode Voltage. *IEEE Trans. Power Electron.* **2015**, *30*, 3122–3132. [CrossRef]

19. Xiao, H.; Xie, S.; Chen, Y.; Huang, R. An Optimized Transformerless Photovoltaic Grid-Connected Inverter. *IEEE Trans. Ind. Electron.* **2011**, *58*, 1887–1895. [CrossRef]

20. Wang, J.; Ji, B.; Zhao, J.; Yu, J. From H4, H5 to H6 Standardization of full-bridge single phase photovoltaic inverter topologies without ground leakage current issue. In Proceedings of the 2012 IEEE Energy Conversion Congress and Exposition (ECCE), Raleigh, NC, USA, 15–20 September 2012; pp. 2419–2425.

21. Vazquez, G.; Martinez-Rodriguez, P.R.; Sosa, J.M.; Escobar, G.; Juarez, M.A. Transformerless single-phase multilevel inverter for grid tied photovoltaic systems. In Proceedings of the IECON 2014—40th Annual Conference of the IEEE Industrial Electronics Society, Dallas, TX, USA, 29 October–1 November 2014; pp. 1868–1874.

22. Vazquez, G.; Martinez-Rodriguez, P.R.; Sosa, J.M.; Escobar, G.; Arau, J. A modulation strategy for single-phase HB-CMI to reduce leakage ground current in transformer-less PV applications. In Proceedings of the IECON 2013—39th Annual Conference of the IEEE Industrial Electronics Society, Vienna, Austria, 10–13 November 2013; pp. 210–215.

23. Figueredo, R.S.; de Carvalho, K.C.M.; Ama, N.R.N.; Matakas, L. Leakage current minimization techniques for single-phase transformerless grid-connected PV inverters—An overview. In Proceedings of the 2013 Brazilian Power Electronics Conference (COBEP), Gramado, Brazil, 27–31 October 2013; pp. 517–524.

24. Hu, S.; Cui, W.; Li, W.; He, X.; Cao, F. A high-efficiency single-phase inverter for transformerless photovoltaic grid-connection. In Proceedings of the 2014 IEEE Energy Conversion Congress and Exposition (ECCE), Pittsburgh, PA, USA, 14–18 September 2014; pp. 4232–4236.

25. Salmon, J.; Knight, A.; Ewanchuk, J. Single phase multi-level PWM Inverter topologies using coupled inductors. In Proceedings of the 2008 IEEE Power Electronics Specialists Conference (PESC), Rhodes, Greece, 15–19 June 2008; pp. 802–808.

26. Ozkan, Z.; Hava, A.M. Leakage current analysis of grid connected transformerless solar inverters with zero vector isolation. In Proceedings of the 2011 IEEE Energy Conversion Congress and Exposition (ECCE), Phoenix, AZ, USA, 17–22 September 2011; pp. 2460–2466.

27. Lopez, O.; Teodorescu, R.; Freijedo, F.; DovalGandoy, J. Leakage current evaluation of a singlephase transformerless PV inverter connected to the grid. In Proceedings of the APEC 2007—Twenty Second Annual IEEE Applied Power Electronics Conference and Exposition, Anaheim, CA, USA, 25 February–1 March 2007; pp. 907–912.

28. Ma, L.; Tang, F.; Zhou, F.; Jin, X.; Tong, Y. Leakage current analysis of a single-phase transformer-less PV inverter connected to the grid. In Proceedings of the 2008 IEEE International Conference on Sustainable Energy Technologies (ICSET), Singapore, 24–27 November 2008; pp. 285–289.

29. Kerekes, T.; Teodorescu, R.; Rodriguez, P.; Vazquez, G.; Aldabas, E. A New High-Efficiency Single-Phase Transformerless PV Inverter Topology. *IEEE Trans. Ind. Electron.* **2011**, *58*, 184–191. [CrossRef]

30. Ji, B.; Wang, J.; Zhao, J. High-Efficiency Single-Phase Transformerless PV H6 Inverter With Hybrid Modulation Method. *IEEE Trans. Ind. Electron.* **2013**, *60*, 2104–2115. [CrossRef]

31. Islam, M.; Mekhilef, S. A new high efficient transformerless inverter for single phase grid-tied photovoltaic system with reactive power control. In Proceedings of the 2015 IEEE Applied Power Electronics Conference and Exposition (APEC), Charlotte, NC, USA, 15–19 March 2015; pp. 1666–1671.

32. Islam, M.; Mekhilef, S. High efficiency transformerless MOSFET inverter for grid-tied photovoltaic system. In Proceedings of the 2014 Twenty-Ninth Annual IEEE Applied Power Electronics Conference and Exposition (APEC), Fort Worth, TX, USA, 16–20 March 2014; pp. 3356–3361.

33. San, G.; Qi, H.; Wu, J.; Guo, X. A new three-level six-switch topology for transformerless photovoltaic systems. In Proceedings of the 2012 7th International Power Electronics and Motion Control Conference (IPEMC), Harbin, China, 2–5 June 2012; pp. 163–166.

34. Freddy, T.K.S.; Rahim, N.A.; Wooi-Ping, H.; Che, H.S. Comparison and Analysis of Single-Phase Transformerless Grid-Connected PV Inverters. *IEEE Trans. Power Electron.* **2014**, *29*, 5358–5369. [CrossRef]

35. Du, D.; Hao, R.; Li, H.; Zheng, T.Q. A novel H6 topology and Its modulation strategy for transformerless photovoltaic grid-connected inverters. In Proceedings of the 2014 16th European Conference on Power Electronics and Applications (EPE'14-ECCE Europe), Lappeenranta, Finland, 26–28 August 2014; pp. 1–8.

36. Dong, D.; Luo, F.; Boroyevich, D.; Mattavelli, P. Leakage Current Reduction in a Single-Phase Bidirectional AC-DC Full-Bridge Inverter. *IEEE Trans. Power Electron.* **2012**, *27*, 4281–4291. [CrossRef]

37. Cui, W.; Yang, B.; Zhao, Y.; Li, W.; He, X. A novel single-phase transformerless grid-connected inverter. In Proceedings of the IECON 2011—37th Annual Conference on IEEE Industrial Electronics Society, Melbourne, VIC, Australia, 7–10 November 2011; pp. 1126–1130.

38. Gu, B.; Dominic, J.; Lai, J.-S.; Chen, C.-L.; LaBella, T.; Chen, B. High Reliability and Efficiency Single-Phase Transformerless Inverter for Grid-Connected Photovoltaic Systems. *IEEE Trans. Power Electron.* **2013**, *28*, 2235–2245. [CrossRef]

39. Gu, B.; Dominic, J.; Chen, B.; Lai, J.-S. A high-efficiency single-phase bidirectional AC-DC converter with miniminized common mode voltages for battery energy storage systems. In Proceedings of the 2013 IEEE Energy Conversion Congress and Exposition (ECCE), Denver, CO, USA, 15–19 September 2013; pp. 5145–5149.

40. Basu, K.; Mohan, N. A High-Frequency Link Single-Stage PWM Inverter With Common-Mode Voltage Suppression and Source-Based Commutation of Leakage Energy. *IEEE Trans. Power Electron.* **2014**, *29*, 3907–3918. [CrossRef]

41. Barater, D.; Buticchi, G.; Crinto, A.S.; Franceschini, G.; Lorenzani, E. Unipolar PWM Strategy for Transformerless PV Grid-Connected Converters. *IEEE Trans. Energy Convers.* **2012**, *27*, 835–843. [CrossRef]

42. Barater, D.; Buticchi, G.; Crinto, A.S.; Franceschini, G.; Lorenzani, E. A new proposal for ground leakage current reduction in transformerless grid-connected converters for photovoltaic plants. In Proceedings of the 2009 35th Annual Conference of IEEE Industrial Electronics (IECON'09), Porto, Portugal, 3–5 November 2009; pp. 4531–4536.

43. Islam, M.; Afrin, N.; Mekhilef, S. Efficient single phase transformerless inverter for grid-tied PVG system with reactive power control. *IEEE Trans. Sustain. Energy* **2016**, *7*, 1205–1215. [CrossRef]

44. Islam, M.; Mekhilef, S. H6-type transformerless single-phase inverter for grid-tied photovoltaic system. *IET Power Electron.* **2015**, *8*, 636–644. [CrossRef]

45. Anandababu, C.; Fernandes, B.G. Improved full-bridge neutral point clamped transformerless inverter for photovoltaic grid-connected system. In Proceedings of the IECON 2013—39th Annual Conference of the IEEE Industrial Electronics Society, Vienna, Austria, 10–13 November 2013; pp. 7996–8001.

46. Anandababu, C.; Fernandes, B.G. A novel neutral point clamped transformerless inverter for grid-connected photovoltaic system. In Proceedings of the IECON 2013—39th Annual Conference of the IEEE Industrial Electronics Society, Vienna, Austria, 10–13 November 2013; pp. 6962–6967.

47. Zhang, L.; Sun, K.; Feng, L.; Wu, H.; Xing, Y. A Family of Neutral Point Clamped Full-Bridge Topologies for Transformerless Photovoltaic Grid-Tied Inverters. *IEEE Trans. Power Electron.* **2013**, *28*, 730–739. [CrossRef]

48. Duan, S.; Liu, B.; Kang, Y. A Single-phase Hybrid-Bridge Three-Level Inverter. Chinese Patent ZL200910063079.1, 14 September 2011.

49. Gonzalez, R.; Lopez, J.; Sanchis, P.; Marroyo, L. Transformerless Inverter for Single-Phase Photovoltaic Systems. *IEEE Trans. Power Electron.* **2007**, *22*, 693–697. [CrossRef]

50. Li, W.; Gu, Y.; Luo, H.; Cui, W.; He, X.; Xia, C. Topology Review and Derivation Methodology of Single-Phase Transformerless Photovoltaic Inverters for Leakage Current Suppression. *IEEE Trans. Ind. Electron.* **2015**, *62*, 4537–4551. [CrossRef]

51. Zhou, L.; Gao, F.; Yang, T. Neutral-point-clamped circuits of single-phase PV inverters: Generalized principle and implementation. In Proceedings of the 2015 IEEE Energy Conversion Congress and Exposition (ECCE), Montreal, QC, Canada, 20–24 September 2015; pp. 442–449.

52. Islam, M.; Mekhilef, S. Efficient transformerless MOSFET inverter for a grid-tied photovoltaic system. *IEEE Trans. Power Electron.* **2016**, *31*, 6305–6316. [CrossRef]

53. Mei, Y.; Hu, S.; Lin, L.; Li, W.; He, X.; Cao, F. Highly efficient and reliable inverter concept-based transformerless photovoltaic inverters with tri-direction clamping cell for leakage current elimination. *IET Power Electron.* **2016**, *9*, 1675–1683. [CrossRef]

54. Akpınar, E.; Balıkcı, A.; Durbaba, E.; Azizoğlu, B.T. Single-phase transformerless photovoltaic inverter with suppressing resonance in improved H6. *IEEE Trans. Power Electron.* **2019**, *34*, 8304–8316. [CrossRef]

55. Zhou, L.; Gao, F. Low leakage current single-phase PV inverters with universal neutral-point-clamping method. In Proceedings of the 2016 IEEE Applied Power Electronics Conference and Exposition (APEC), Long Beach, CA, USA, 20–24 March 2016; pp. 410–416.

56. Wang, H.; Burton, S.; Liu, Y.-F.; Sen, P.C.; Guerrero, J.M. A systematic method to synthesize new transformerless full-bridge grid-tied inverters. In Proceedings of the 2014 IEEE Energy Conversion Congress and Exposition (ECCE), Pittsburgh, PA, USA, 14–18 September 2014; pp. 2760–2766.

 *energies*

Article

A Family of Single-Stage, Buck-Boost Inverters for Photovoltaic Applications

Ben Zhao [1,*], Alexander Abramovitz [2], Chang Liu [1], Yongheng Yang [3] and Yigeng Huangfu [1]

[1] School of Automation, Northwestern Polytechnical University, Xi'an 710072, China; lcy@mail.nwpu.edu.cn (C.L.); yigeng@nwpu.edu.cn (Y.H.)

[2] Department of Electrical Engineering, Holon Institute of Technology, Holon 58102, Israel; alabr@hotmail.com

[3] Department of Energy Technology, Aalborg University, 9220 Aalborg, Denmark; yoy@et.aau.dk

* Correspondence: benz@nwpu.edu.cn; Tel.: +86-29-8843-1329

Received: 5 February 2020; Accepted: 21 March 2020; Published: 3 April 2020

Abstract: This paper introduces a family of single-stage buck-boost DC/AC inverters for photovoltaic (PV) applications. The high-gain feature was attained by applying a multi-winding tapped inductor, and thus, the proposed topologies can generate a grid-level AC output voltage without using additional high step-up stages. The proposed topologies had a low component count and consisted of a single magnetic device and three or four power switches. Moreover, the switches were assembled in a push-pull or half/full-bridge arrangement, which allowed using commercial low-cost driver-integrated circuits. In this paper, the operation principle and comparison of the proposed topologies are presented. The feasibility of the proposed topologies was verified by simulations and experimental tests.

Keywords: PV microinverters; converter topologies; single-stage; buck-boost; tapped inductor

---

1. Introduction

The continuous development of distributed photovoltaic (PV) power generation systems arouses much interest in MIEs/MICs, also known as microinverters. Unlike the string inverters using series-connected PV panels to achieve a high voltage, microinverters are designed to directly connect a single PV panel with a low voltage to the grid while providing an individual MPPT and, in turn, avoiding mismatch losses within the PV array. The "plug-and-play" feature of the microinverter allows the incorporation of PV modules of different types into a single array, also facilitating its future expansion and maintenance. To some extent, the labor cost can also be reduced.

In practice, the low DC voltage produced by the PV module (e.g., 20–30 V) and the relatively high AC voltage of the utility (e.g., 230 V RMS) imply that a high step-up DC-DC stage followed by a regular inverter is required. Such a straightforward scheme is referred to as the two-stage approach and is quite popular due to its ease of implementation and control. Yet, the two-stage solution is costly and the efficiency is reduced. The single-stage microinverter that combines both the voltage step-up and inversion functions in one power stage can possibly lead to a lower component count and a reduced cost. Thus, the single-stage inverters have been the focus of recent research activities. Numerous single-stage boost-derived topologies have been proposed in the literature due to the inherent voltage step-up capability [1]. The limited voltage gain of the boost-type converter can be improved by means of integrating tapped inductors, as discussed in [2,3].

Additionally, due to the voltage step-up/down capability, the buck-boost derived topologies can also be a viable solution for single-stage inverter applications. Thus, a number of buck-boost type single-stage inverters with low component counts were reported. For instance, single-stage buck-boost inverters with only three switches were proposed in [4,5], as shown in Figure 1a, where a tapped inductor was used as a regular inductor in one half-line cycle and as a fly-back transformer in the subsequent half-line cycle. Unfortunately, this type of inverter cannot attain the required voltage step-up. As shown in Figure 1b, a four-switch, single-stage, buck-boost inverter was then presented in [6], which employed a tapped inductor and the SEPIC converter to increase the voltage gain. However, according to the operational principles, the turns ratio of the tapped inductor has to be equal to unity, and consequently, the voltage gain is still limited. Topologies in [7,8] also have only four switches to realize the single-stage conversion and have the merit of a common terminal between input and output ports. Figure 1c shows the circuit diagram of the converter in [7]. Another single-stage, buck-boost inverter has the advantage of reduced magnetic volume and low leakage currents [9]. The topologies in [10–12] were conceived to also eliminate the leakage currents, but the number of active switches is increased, as observed in Figure 1d. Furthermore, a differential buck-boost inverter with active power decoupling capability was proposed in [13,14], where no extra components are required. It has only four switches; on the contrary, a rather complicated control method is needed. An active buck-boost inverter using an "AC/AC unit" to realize the buck-boost conversion was introduced in [15,16], as presented in Figure 1e. Yet, each unit consisted of four switches, and, thus, in total, eight switches are needed for the microinverter. The authors of [17] expanded this idea to cascaded multilevel buck-boost inverters using H-bridges for each PV panel and a central AC/AC unit. To improve the efficiency and system reliability, a solution for the current shoot-through issue was discussed in [18,19] to eliminate the dead-time effect. Moreover, ref. [18] presented a converter with eight switches and four inductors, while [19] has four switches, four diodes, and six inductors, which make the topologies quite complicated. The topology in [20] has merits of a wide input voltage range, low leakage currents, small grid current ripples, and low common-mode voltages. However, as seen in Figure 1f, it has four high-frequency switches and two bidirectional switches, which are realized by connecting back-to-back MOSFETs in series. Doing so significantly increases the total number of switches (i.e., eight). Although the ideas of [4–20] are very interesting, their attained voltage gain is comparable to the traditional buck-boost converter.

Additional attempts to increase the gain of the buck-boost derived topologies were reported. For example, in [21] a series connection between a buck-boost converter and the PV array was introduced to have a higher gain, but the gain improvement was limited. The topology in [22], see Figure 2a, employed a switched inductor, which can improve the gain by the factor of $\sqrt{2}$ over that of the traditional buck-boost converter. However, in total, the topology in [22] had four switches, eight diodes, and four inductors. The tapped-inductor buck-boost inverter topologies presented in [23,24], as shown in Figure 2b,c, respectively, can achieve a much higher voltage gain than the traditional ones, but the switch counts were up to eight, whereas [25,26] had five switches, as presented in Figure 2d. The advantage of the topologies in [25,26] is that only one high-frequency switch was used, and thus, the switching losses were lower. For the topologies in Figure 2, the main characteristics are further compared in Table 1. According to Table 1, most of the topologies had a high semiconductor count, from 7 up to 12. The experimental efficiency of more than 96% was reported in [23]. However, the test was with an input of 100–200 V and a 110-V output, which cannot support the performance with a high-voltage step-up. An efficiency of 86% was achieved in [25] with a 60-V input, a 230-V output, and 100-W output power, which is reasonable for a tapped-inductor buck-boost inverter. Yet, the experimental efficiency of the other two proposals was not reported clearly in the literature.

Table 1. Comparison of the main topologies of the existing single-stage, buck-boost inverters.

Ref.	Switches Count	Diodes Count	Inductors Count	Input Voltage	Output Voltage	Output Power	Efficiency
[22]	4	8	4	20 V	314 V	100 W	/
[23]	8	0	1 Tapped	100–200 V	110 V	500 W	>96%
Figure 2b [24]	8	0	1 Tapped	40 V	230 V	/	/
[25]	5	2	1 Tapped	60 V	230 V	100 W	86%

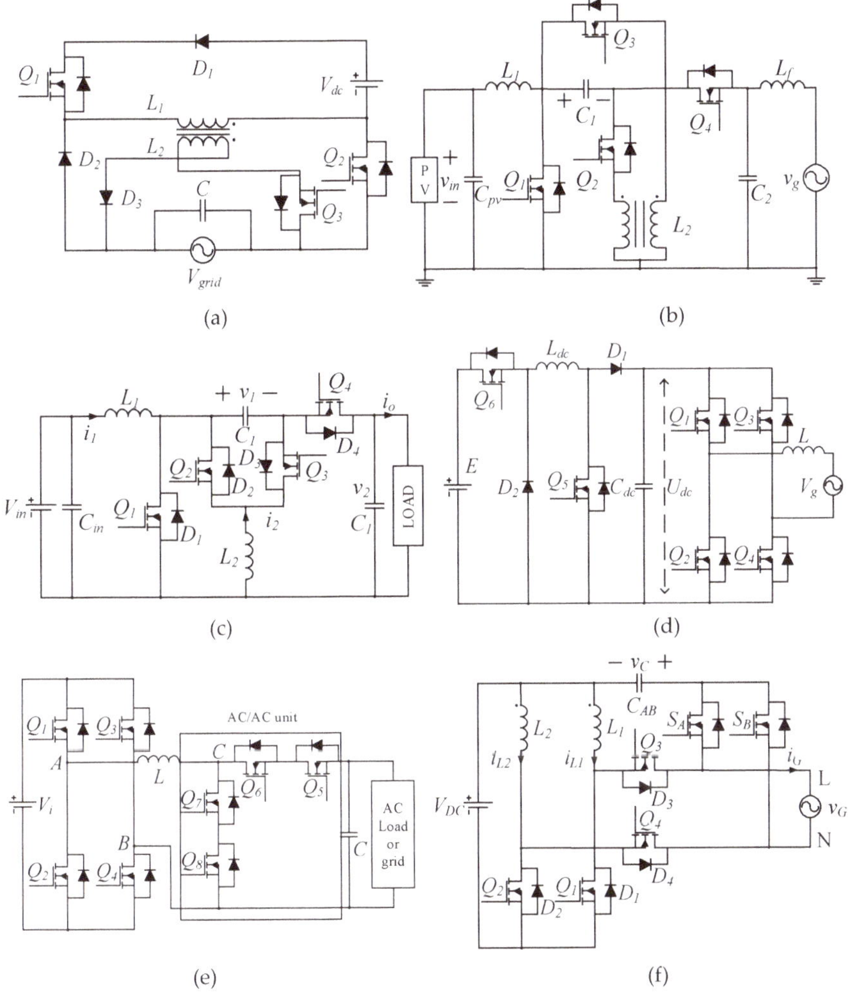

Figure 1. Prior-art, single-stage, buck-boost inverters: (**a**) [5], (**b**) [6], (**c**) [7], (**d**) [12], (**e**) [15], and (**f**) [20].

The high switch count of the reviewed converters, the resulting circuit complexity, higher cost, and lower efficiency, counter the main design goal of producing a simple and low-cost single-stage inverter. Therefore, more efforts have been made to develop more single-stage, buck-boost inverter topologies with a high gain and a low switch count. Recently, a family of single-stage, buck-boost rectifiers with high power factor were proposed in [27], analyzed, and verified in [28]. With the same principles, a family of tapped-inductor, buck-boost microinverters can be derived by reversing the power flow. This calls for the application of bidirectional switches. The proposed tapped-inductor, buck-boost type inverter family is illustrated in Figure 3. The basic operation and the preliminary simulation study of the two topologies in the family were reported in [29,30], while the converters have not been experimentally verified, and the design considerations are not fully addressed.

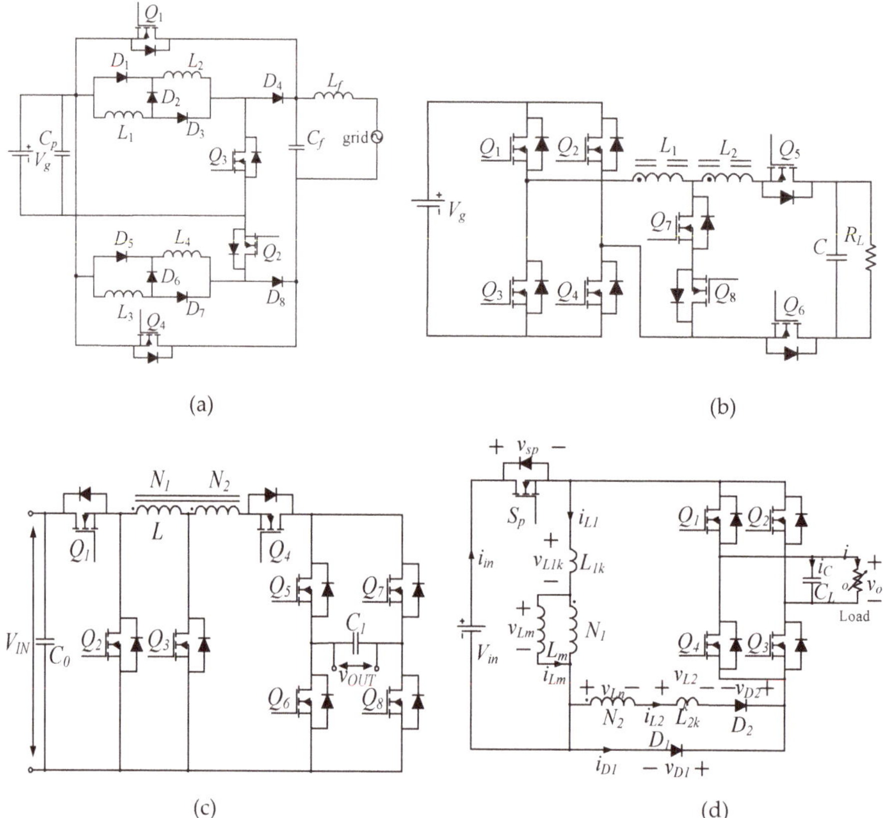

Figure 2. Prior-art, single-stage, buck-boost inverters with high gains: (**a**) [22], (**b**) [23], (**c**) [24], and (**d**) [25].

Accordingly, in addition to the topologies in [29,30], this paper further introduces two more practical topologies and all four topologies in the family are presented in detail. More importantly, a comparison of the proposed family was done thoroughly in terms of the component count, the voltage conversion ratio, the voltage stress, the peak current stress, and the RMS current stress, which can be used in the design phase. What is more, more detailed simulation studies for all the topologies in the family were presented. A prototype of the SSBBI of the proposed family was built and experimental results are illustrated in this paper. The rest of the paper is organized as follows. Section 2 introduces the proposed family, and the operation principles of the proposed family are demonstrated on a topology (i.e., the SSBBI) in Section 3. Circuit characteristics are discussed in Section 4, including the analysis of the conversion ratio, turns ratio, and duty cycle constraints together with voltage and current stresses, as design considerations. Simulation results are given in Section 5, where the comparison of the family is provided. Experimental tests are presented in Section 6 to validate the discussion. Finally, concluding remarks are provided in Section 7.

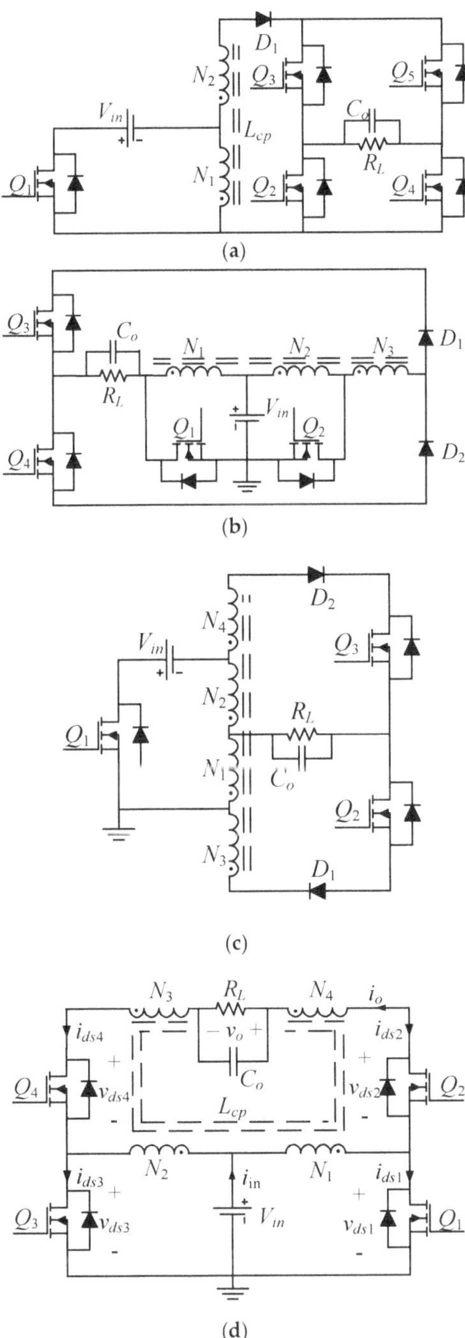

Figure 3. Proposed family of single-stage, buck-boost inverters: (**a**) Variant 1, (**b**) Variant 2, (**c**) Variant 3, (**d**) Variant 4 (SSBBI).

2. Single-Stage, Buck-Boost Inverter Family

As shown in Figure 3, the proposed inverter family makes use of a tapped inductor to attain a high step-up voltage conversion ratio. This helps to generate a grid-compatible voltage from a low DC voltage source. Two, three, and four winding, tapped-inductor structures are needed. The turns ratio, n, of the tapped inductor is defined as follows. For the two-windings inverter topology in Figure 3a, $n = N_2/N_1$. The three-windings topology in Figure 3b has an equal number of primary turns, $N_1 = N_2$, and the turns ratio is defined as $n = N_3/N_1 = N_3/N_2$. The topologies in Figure 3c,d rely on a symmetrical tapped-inductor structure with an equal turns ratio, defined as $n = N_3/N_1 = N_4/N_2$.

The topology in Figure 3a includes a floating source, a single ground-referenced PWM switch, Q_1, and a ground-referenced line frequency unfolding bridge, Q_2–Q_5. The topology in Figure 3b includes a grounded source, a ground-referenced push-pull pair of PWM switches, Q_1–Q_2, and a floating line frequency unfolding totem pole, Q_3–Q_4. The topology in Figure 3c includes a floating source, a single ground-referenced PWM switch, Q_1, and a floating line frequency unfolding totem pole, Q_2–Q_3. The topology in Figure 3d includes a grounded source and a ground-referenced full bridge. Here, the lower switches, Q_1–Q_3, are PWM devices, whereas the high switch pair can perform either a simple line frequency unfolding function or be operated as synchronous rectifiers. Since the body diodes of the high switches are exploited as rectifiers, the reverse recovery capability should be considered. This can be an issue for silicon-based devices, while the emerging GaN MOSFETs can deliver the required performance.

To summarize, the proposed inverters have the merits of:

(1) Generating a grid-level AC output voltage from a relatively low DC input voltage without extra high gain DC-DC converters.
(2) Having a low component count as single-stage topologies consisting of a single magnetic device and three or four switches.
(3) A push-pull or half/full-bridge arrangement of the switches, where the commercial low-cost driver-integrated circuits can be easily used.

The proposed tapped-inductor, buck-boost inverter family in Figure 3 was then studied through simulations. The exploration indicated that the topology in Figure 3d can also help to avoid much of the practical grounding, driving, and controller interface issues. Additionally, considering the lowest semiconductor count (see Table 2), the topology in Figure 3d appears as the most attractive candidate in the family. Hereafter, this topology (i.e., the SSBBI in Figure 3d) is considered in the following detailed analysis to exemplify the converter operation.

Table 2. Comparison of the component count of the tapped-inductor, buck-boost inverter family.

Topologies	Switches	Diodes	Windings	Filter Cap.
Figure 3a	5	1	2	1
Figure 3b	4	2	3	1
Figure 3c	3	2	4	1
Figure 3d	4	0	4	1

3. Operation Principles of the Proposed SSBBI

As shown in Figure 3d, the power stage of the proposed SSBBI included four switches, Q_1–Q_4, in a full-bridge arrangement. A tapped inductor, L_{cp}, with four windings was employed. The output filter capacitor here was C_o and the load was an equivalent resistance, R_L, for stand-alone applications. The voltage across them was the AC output, v_o. As mentioned previously, two symmetrical pairs of windings were used for the tapped inductor. The turns of the primary windings must be the same, i.e., $N_1 = N_2$. Similarly, equal secondary windings were used, i.e., $N_3 = N_4$. The turns ratio of the tapped inductor was then obtained as $n = N_3/N_1 = N_4/N_2$. The SSBBI can generate a bipolar output voltage with the help of the symmetrical structure, and thus, it can achieve the DC-AC inversion. The desired output voltage can be obtained using any common control strategy of a constant frequency duty cycle. The operation principle is detailed in the following.

Supposing the converter was operating in the CCM, the SSBBI had two switching states in each half-line cycle, denoted as states A and B in the positive half-line cycle and A′ and B′ in the negative half-line cycle. The switching states of the four switches are listed in Table 3, and further illustrated in Figure 4.

Table 3. Switching states of semiconductor devices.

Switches	Positive Output Voltage		Negative Output Voltage	
	State A	State B	State A′	State B′
Q_1	On	Off	Off	Off
Q_2	Off	On	On	On
Q_3	Off	Off	On	Off
Q_4	On	On	Off	On

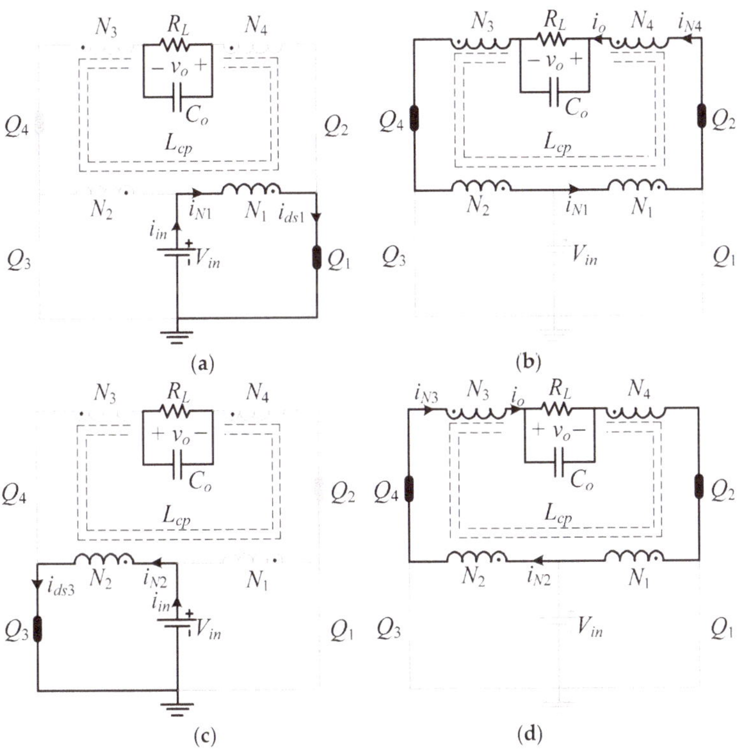

Figure 4. Equivalent circuits (switching states) of the proposed SSBBI: (**a**) State A, (**b**) State B, (**c**) State A′, (**d**) State B′.

According to the equivalent circuit of state A shown in Figure 4a, the state started at the beginning of each switching cycle in the positive half-line cycle. Here, the switch Q_1 was turned on and the state lasted for the duration of DT_s. In this state, the tapped inductor was charged by the input source, V_{in}, through the primary winding N_1. The output capacitor, C_o, can sustain the output voltage on the load. As shown in Figure 4b, state B began when the switch Q_1 was turned off and lasted for the duration of

$(1-D)T_s$. In this state, the energy stored in the tapped inductor was discharged and released to the output side through all the four windings of the tapped inductor. During states A and B, when the output voltage was positive, Q_1 and Q_2 were switched, while the switch Q_3 was maintained off and Q_4 remained on. In comparison, the states A and B were replaced by the states A' and B' during the negative output half-line cycle due to the symmetrical operation principle. The equivalent circuits of state A' and B' are shown in Figure 4c,d, respectively.

The key waveforms of the SSBBI are described in Figure 5, where S_{Q1}–S_{Q4} are the gating signals for Q_1–Q_4 switches, respectively. Due to the symmetry of the SSBBI, it was sufficient to consider its operation during the positive half cycle. When Q_1 was turned on and Q_2 was turned off, the primary winding of the tapped inductor was energized. This caused the magnetizing current of the tapped inductor to ramp up. When Q_1 was turned off and Q_2 was turned on, the tapped inductor was discharged to support the output through all the windings. Thus, the magnetizing current of the tapped inductor ramped down. Notably, in terms of control of the converter, in grid-tied applications, the task of the control circuit is to shape the average output current, I_o, into a sinusoidal waveform (see i_{N4} in Figure 5), while the controller should regulate the output voltage in stand-alone applications.

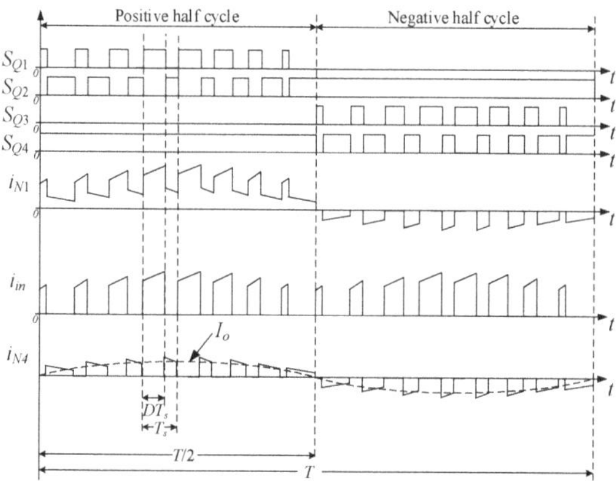

Figure 5. Illustration of key waveforms of the proposed SSBBI.

4. Analysis and Design Considerations of the Proposed SSBBI

4.1. CCM Voltage Gain

In the CCM, the tapped inductor, L_{cp}, was charged by the input voltage source, V_{in}, only through the primary winding N_1 or N_2 during the time of DT_s (state A or A'). However, the output voltage, v_o, was stressed on all the four windings of the tapped inductor during the time of $(1-D)T_s$ (state B or B'). Thus, according to the volt-sec balance, it gives

$$\int_0^{DT_s} V_{in}dt + \int_{DT_s}^{T_s} \frac{-v_o}{2n+2}dt = 0 \tag{1}$$

which led to that the quasi-steady-state voltage gain of the SSBBI to be calculated as

$$M = \frac{v_o}{V_{in}} = 2(n+1)\frac{D}{1-D}. \tag{2}$$

It can be recognized from Equation (2) that the SSBBI was a buck-boost type topology and had the function of voltage step-up/down. A higher gain can be achieved by choosing a proper turns ratio, n.

4.2. Turns Ratio and Duty Cycle Constraints

It should be noticed that when the tapped inductor is discharged to the output side (see states B and B′), the voltage across the primary winding must be always less than the DC input voltage, V_{in}. Accordingly,

$$\frac{v_o}{2(n+1)} < V_{in}.\tag{3}$$

In this way, it prevented the discharging current of the tapped inductor to go back to the DC input source through the body diode of the switch at the lower side. Such a condition should be avoided since the output voltage would be clamped and the circulating current will lower the efficiency as well. With this concern, the turns ratio should be designed sufficiently large to make the SSBBI work properly. Thus,

$$n > \frac{V_{omax}}{2V_{in}} - 1.\tag{4}$$

Moreover, it can be obtained by combining (2) and (3) that

$$\frac{D}{1-D} < 1.\tag{5}$$

Subsequently, the maximum duty ratio, D_{max}, should be limited to

$$D_{max} < 0.5.\tag{6}$$

4.3. Voltage and Current Stress

4.3.1. Voltage Stress of Switches

During state A, the input voltage, V_{in}, was imposed on the primary winding N_1 of the tapped inductor when the switch Q_1 was on. Therefore, the voltage stress on the switch Q_3 was the sum of the input voltage and the induced voltage across the primary winding N_2, which was twice the input voltage, V_{in} as

$$V_{Q3max} - 2V_{in}.\tag{7}$$

Meanwhile, since the switch Q_4 was in on-state, the voltage across the four windings of the tapped inductor as well as the output voltage, v_o, was stressed on the off-state switch Q_2. Thus, the maximum stress of the Q_2 will lead to:

$$V_{Q2max} = 2(n+1)V_{in} + V_{omax}.\tag{8}$$

The same results can be obtained for the switches Q_1 and Q_4 in state A′ because of the symmetrical operation of the SSBBI. The voltage stresses for all the switches are summarized in Table 4.

Table 4. SSBBI switch voltage and current stresses.

Switches	Voltage Stress	Current Stress	
		Peak	RMS
Q_1, Q_3	$2V_{in}$	$2(n+1)I_m + \dfrac{I_m V_m}{V_{in}}$	$I_{acrms}\sqrt{\dfrac{3}{8}\dfrac{V_m^2}{V_{in}^2} + \dfrac{8}{3\pi}\dfrac{(n+1)V_m}{V_{in}}}$
Q_2, Q_4	$2(n+1)V_{in}+V_{omax}$	$I_m + \dfrac{I_m V_m}{2(n+1)V_{in}}$	$I_{acrms}\sqrt{1 + \dfrac{4}{3\pi}\dfrac{V_m}{(n+1)V_{in}}}$

4.3.2. Analysis of Current Stress

It was assumed that the output voltage and current of the SSBBI were ideally in phase without harmonics as

$$\begin{cases} v_o(t) = V_m \sin\omega t \\ i_o(t) = I_m \sin\omega t \end{cases}\tag{9}$$

Furthermore, by applying Equations (2) and (9), and replacing the steady-state duty ratio D with the time-varying duty ratio $d(t)$, it can be obtained that

$$\frac{v_o(t)}{V_{in}} = 2(n+1)\frac{d(t)}{1-d(t)} = \frac{V_m \sin \omega t}{V_{in}} \tag{10}$$

from which the duty ratio, $d(t)$, can be derived as

$$d(t) = \frac{V_m \sin \omega t}{2(n+1)V_{in} + V_m \sin \omega t}. \tag{11}$$

For the proposed SSBBI, the average output current equaled to the average current of the upper switch, $\langle i_o(t)\rangle = i_{Q2}(t)[1 - d(t)]$, as shown in Figure 6. Therefore, assuming that the current ripples are negligible, the current amplitude of the switch Q_2 can be obtained by combining Equations (9) and (11) as

$$i_{Q2}(t) = \frac{\langle i_o(t)\rangle}{1-d(t)} = I_m \sin \omega t + \frac{I_m V_m \sin^2 \omega t}{2(n+1)V_{in}}. \tag{12}$$

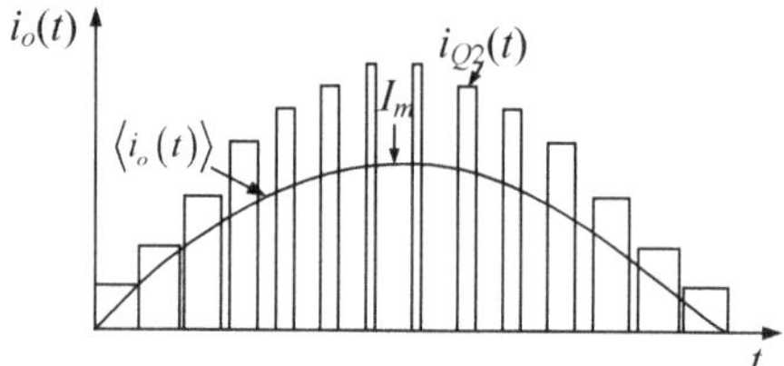

Figure 6. Illustration of the switch current, $i_Q(t)$, and the average output current, $<i_o(t)>$, throughout the half-line cycle.

Thus, the maximum current of the switch Q_2 at the peak output voltage can be obtained as

$$I_{Q2max} = I_m + \frac{I_m V_m}{2(n+1)V_{in}}. \tag{13}$$

The squared RMS current of the switch Q_2 within a switching period is:

$$i^2_{Q2rmsTs} = \frac{1}{T_s}\int_t^{t+T_s} i^2_{Q2}(t)dt = [1 - d(t)]i^2_{Q2}(t). \tag{14}$$

Subsequently, the squared value of the switch RMS current is:

$$I^2_{Q2rms} = \frac{1}{T/2}\int_0^{T/2} i^2_{Q2rmsTs}dt \tag{15}$$

with T being the generated output voltage period. Substituting Equations (11), (12), and (14) into (15) yields

$$I^2_{Q2rms} = \frac{1}{T/2}\int_0^{T/2} I_m^2 \sin^2 \omega t + \frac{I_m^2 V_m \sin^3 \omega t}{2(n+1)V_{in}}dt = I^2_{acrms}\left(1 + \frac{4}{3\pi}\frac{V_m}{(n+1)V_{in}}\right). \tag{16}$$

Thus, the RMS current of the switch Q_2 is obtained as

$$I_{Q2rms} = I_{acrms}\sqrt{1 + \frac{4}{3\pi}\frac{V_m}{(n+1)V_{in}}}. \tag{17}$$

The current amplitude of the lower switch Q_1 is $2(n + 1)$ times higher than the upper switch current due to the function of the tapped-inductor turns ratio, n. Thus,

$$i_{Q1}(t) = 2(n + 1)i_{Q2}(t) = 2(n + 1)I_m \sin \omega t + \frac{I_m V_m \sin^2 \omega t}{V_{in}}. \tag{18}$$

Therefore, the peak current through the lower switch, Q_1, is:

$$i_{Q1max} = 2(n + 1)I_m + \frac{I_m V_m}{V_{in}}. \tag{19}$$

The squared value of the lower switch RMS current through the switching period, T_s, is:

$$i^2_{Q1rmsTs} = \frac{1}{T_s} \int_t^{t+T_s} i^2_{Q1}(t)dt = d(t)i^2_{Q1}(t). \tag{20}$$

Since the low switch conducts for half the line period, the squared value of its RMS current on the line period scale can be calculated as:

$$I^2_{Q1rms} = \frac{1}{T} \int_0^T i^2_{Q1rmsTs}dt. \tag{21}$$

Substituting Equations (11), (18), and (20) into (21), gives

$$I_{Q1rms} = I_{acrms} \sqrt{\frac{3}{8} \frac{V_m^2}{V_g^2} + \frac{8}{3\pi} \frac{(n+1)V_m}{V_{in}}}. \tag{22}$$

With the above analysis, the voltage and current stresses of the SSBBI are summarized in Table 4.

5. Simulation Results and Comparison

5.1. Basic System Operation

Referring to Figure 3d, simulations were carried out to verify the feasibility of the proposed SSBBI in PSIM software. The key simulation parameters were: Output power P_o = 200 W, input voltage V_{in} = 48 V, output voltage v_o = 110 V/60 Hz, switching frequency f_s = 20 kHz, tapped-inductor magnetizing inductance L_m = 150 µH, turns ratio n = 1.5, and output capacitance C_o = 2 µF. Several control strategies can be applied to control the proposed SSBBI. Initially, to validate the basic operational principle, the simple open-loop SPWM was used. Simulation results are shown in Figure 7, which demonstrates that the SSBBI can generate the desired output voltage. This provides proof of concept of the proposed circuit family for single-stage microinverter applications.

Furthermore, as can be observed in Figure 7a, the circuit simulation results (key waveforms) were in a close agreement with the analytical results in Figure 5. The gate-driving signals are further shown in Figure 7b to demonstrate the controllability of the converter. Moreover, the output voltage of the proposed inverter is given in Figure 7c, as well as the voltage across the switches. It can be observed in Figure 7c that the SSBI can produce high-quality sinusoidal outputs, and the voltage stresses on the switches were also in consistency with the analysis. Additionally, the currents flowing through the power devices under the 200-W output power are presented in Figure 7d, which again agrees with the theoretical analysis presented in Section 4.

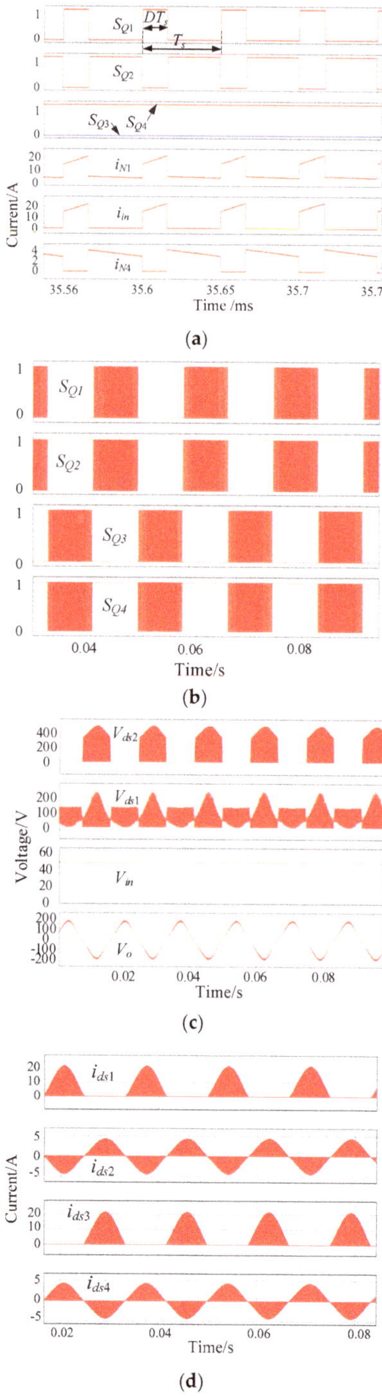

Figure 7. Key simulation waveforms of the proposed SSBBI: (**a**) Driving signal and currents on the switching period scale; (**b**) driving signals for switches; (**c**) V_{ds} of the switches in one leg, input, and output voltage; (**d**) switch currents on the output period scale.

The analytical results were further verified by simulations. Key simulated waveforms of the proposed topologies in Figure 3a–c are shown in Figure 8. It is observed in Figure 8 that all the topologies of the proposed family can generate a good-quality sinusoidal output voltage. Simulations also support the theoretically predicted results of the current stress analysis. When comparing the performance of the topologies in Figure 3a–c with the SSBBI, it can be seen that the four topologies had similar high-quality output voltage waveforms and the comparable current stress at the same output power. However, the SSBBI had the lowest semiconductor count and the easier driver implementation, which proved again the competitiveness of the SSBBI in the family.

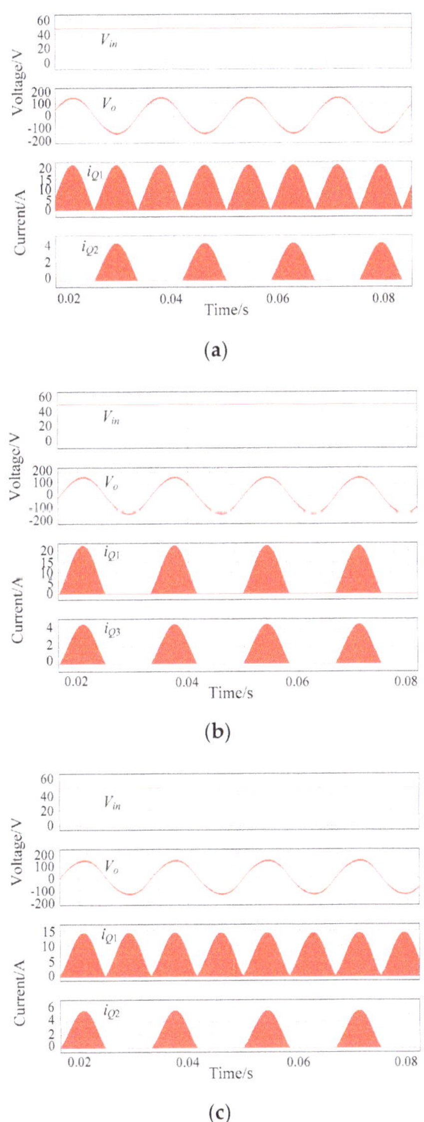

Figure 8. Simulation waveforms of the input voltage, output voltage, and switches' current of the variant topologies: (**a**) Figure 3a, (**b**) Figure 3b, (**c**) Figure 3c.

5.2. Comparison of the Proposed Single-Stage, Buck-Boost Inverter Family

To better appreciate the merits of the proposed single-stage inverter family, a detailed comparison of the proposed topologies is conducted in this section. The voltage conversion ratio of the proposed family and its derivation under the assumption of the CCM operation is summarized in Table 5. The benchmarking of the proposed topologies' voltage conversion ratio with the same turns ratio $n = 2$ is further shown in Figure 9a and with the same duty ratio $D = 0.5$ in Figure 9b. According to Table 5 and Figure 9, the SSBBI had the largest voltage gain in the family. The peak voltage stress analysis was performed and is summarized in Table 6. Lastly, Tables 7 and 8 present the results of the peak current and the RMS current stress analysis of semiconductor devices. As can be seen from Tables 6–8, the voltage and current stresses of the SSBBI were comparable to other topologies in the family. Moreover, as mentioned previously, the SSBBI component count was lower by one or two diodes. Thus, the SSBBI had the optimum circuit composition and characteristics in the family.

Table 5. Comparison of the voltage conversion ratio of the proposed topologies.

Topology	Voltage Gain $M = v_o/V_{in}$
Figure 3a	$M_a = (n+1)\frac{D}{1-D}$
Figure 3b	$M_b = (n+2)\frac{D}{1-D}$
Figure 3c	$M_c = \frac{(n+1)}{2}\frac{D}{1-D}$
SSBBI	$M = 2(n+1)\frac{D}{1-D}$

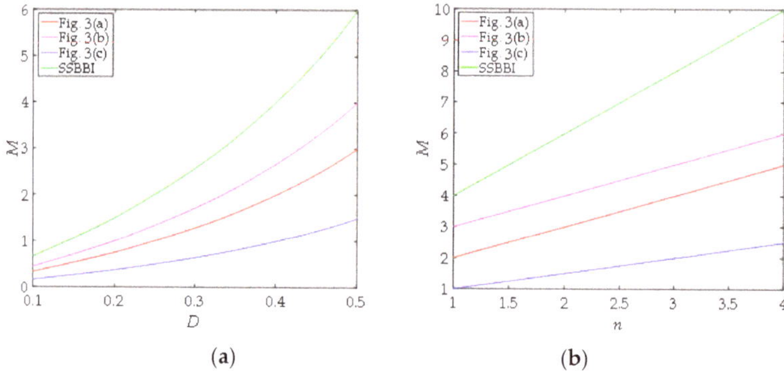

(a)   (b)

Figure 9. Comparison of the voltage conversion ratio, M, of the proposed single-stage inverter family: (**a**) As function of the duty ratio D (for $n = 2$), (**b**) as function of the turn ratio n (for $D = 0.5$).

Table 6. Comparison of the voltage stress.

Topology	Voltage Stress		
	Low Side Switches	**High Side Switches**	**Diodes**
Figure 3a	$V_{in} + \frac{V_{o\max}}{n+1}$	$V_{o\max}$	$(n+1)V_{in}+V_{o\max}$
Figure 3b	$2V_{in}$	$(n+2)V_{in}+V_{o\max}$	$(n+2)V_{in}+V_{o\max}$
Figure 3c	$V_{in} + \frac{2V_{o\max}}{n+1}$	$2V_{o\max}$	$\frac{(n+1)V_{in}}{2} + V_{o\max}$
SSBBI	$2V_{in}$	$2(n+1)V_{in}+V_{o\max}$	/

Table 7. Comparison of the peak current stress.

Topology	Peak Current Stress		
	Low Side Switches	**High Side Switches**	**Diodes**
Figure 3a	$(n+1)I_m + \frac{I_m V_m}{V_{in}}$	$I_m + \frac{I_m V_m}{(n+1)V_{in}}$	$I_m + \frac{I_m V_m}{(n+1)V_{in}}$
Figure 3b	$(n+2)I_m + \frac{I_m V_m}{V_{in}}$	$I_m + \frac{I_m V_m}{(n+2)V_{in}}$	$I_m + \frac{I_m V_m}{(n+2)V_{in}}$
Figure 3c	$\frac{(n+1)}{2}I_m + \frac{I_m V_m}{V_{in}}$	$I_m + \frac{2I_m V_m}{(n+1)V_{in}}$	$I_m + \frac{2I_m V_m}{(n+1)V_{in}}$
SSBBI	$2(n+1)I_m + \frac{I_m V_m}{V_{in}}$	$I_m + \frac{I_m V_m}{2(n+1)V_{in}}$	/

Table 8. Comparison of the RMS current stress.

Topology	RMS Current Stress		
	Low Side Switches	**High Side Switches**	**Diodes**
Figure 3a	$I_{acrms}\sqrt{\frac{3}{4}\frac{V_m^2}{V_{in}^2} + \frac{8}{3\pi}\frac{(n+1)V_m}{V_{in}}}$	$I_{acrms}\sqrt{\frac{1}{2} + \frac{4}{3\pi}\frac{V_m}{(n+1)V_{in}}}$	$I_{acrms}\sqrt{1 + \frac{8}{3\pi}\frac{V_m}{(n+1)V_{in}}}$
Figure 3b	$I_{acrms}\sqrt{\frac{3}{8}\frac{V_m^2}{V_{in}^2} + \frac{4}{3\pi}\frac{(n+2)V_m}{V_{in}}}$	$I_{acrms}\sqrt{\frac{1}{2} + \frac{4}{3\pi}\frac{V_m}{(n+2)V_{in}}}$	$I_{acrms}\sqrt{\frac{1}{2} + \frac{4}{3\pi}\frac{V_m}{(n+2)V_{in}}}$
Figure 3c	$I_{acrms}\sqrt{\frac{3}{4}\frac{V_m^2}{V_{in}^2} + \frac{4}{3\pi}\frac{(n+1)V_m}{V_{in}}}$	$I_{acrms}\sqrt{\frac{1}{2} + \frac{8}{3\pi}\frac{V_m}{(n+1)V_{in}}}$	$I_{acrms}\sqrt{\frac{1}{2} + \frac{8}{3\pi}\frac{V_m}{(n+1)V_{in}}}$
SSBBI	$I_{acrms}\sqrt{\frac{3}{8}\frac{V_m^2}{V_{in}^2} + \frac{8}{3\pi}\frac{(n+1)V_m}{V_{in}}}$	$I_{acrms}\sqrt{1 + \frac{4}{3\pi}\frac{V_m}{(n+1)V_{in}}}$	/

6. Experimental Results and Discussion

6.1. Experimental Results of SSBBI

A 100-W laboratory prototype of the proposed SSBBI was built and tested. The key operation parameters were: Input voltage, V_{in} = 48 V; output voltage, v_o = 110 V/60 Hz; and switching frequency, f_s = 20 kHz. The prototype's view and the components arrangement are shown in Figure 10. The board was designed larger to reserve additional space needed for experimenting with various snubbers and control schemes. The main components of the prototype are summarized in Table 9. The tapped inductor was designed according to the design guide provided by Magnetics-Inc [31], including the magnetic core, the turns, and the wire. A dSPACE system was used to implement the control for the quick experimental study of the SSBBI.

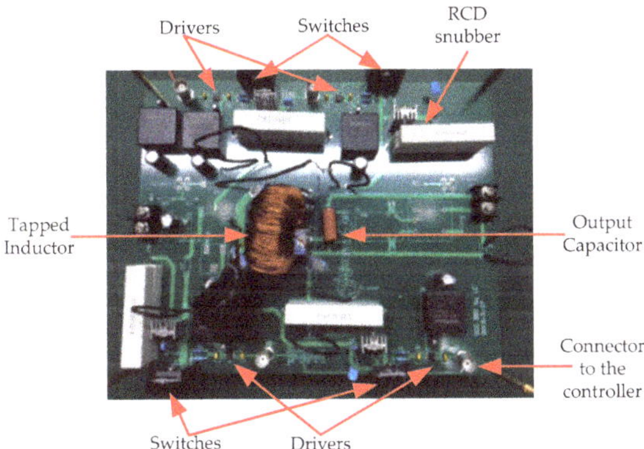

Figure 10. Photo of the experimental prototype of the proposed SSBBI.

Table 9. Main components of the prototype of the proposed SSBBI.

Components	Value/Model
High side switches	IPW90R340C3
Low side switches	IPW65R125C
Driver ICs	1EDI20N12AF
Primary magnetizing inductance	100 µH
Inductor core	55439A2
Inductor Turns	30/45
Output capacitor	2.2 µF

Experimental results are shown in Figures 11 and 12. Figure 11 presents the gate-driving signals for switches at the line period scale and at the switching period scale, respectively. The output voltage and the switch voltage are shown in Figure 12. Observations in Figure 12 clearly indicate that the output voltage was sinusoidal. The THD of the experimental output voltage was around 5% with the open-loop control. This verified that the experimental SSBBI prototype operated according to the theoretical expectations. That is, the proposed SSBBI can achieve the inversion and produce a high-quality sinusoidal output.

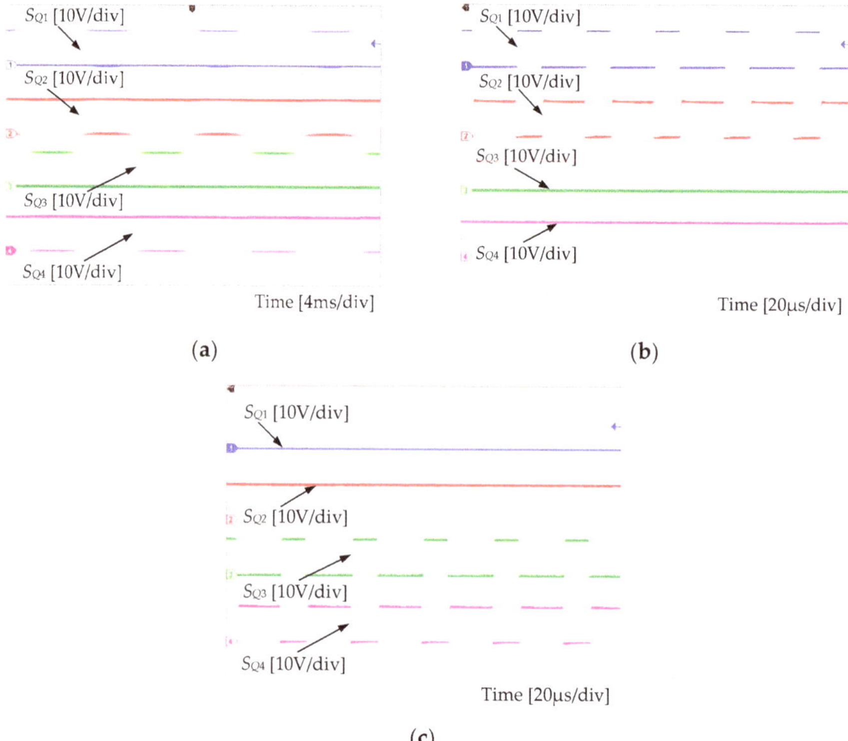

Figure 11. SSBBI's driving signals: (**a**) At the line period scale, (**b**) during positive half-line cycle (at switching period scale), (**c**) negative half-line cycle (at switching period scale).

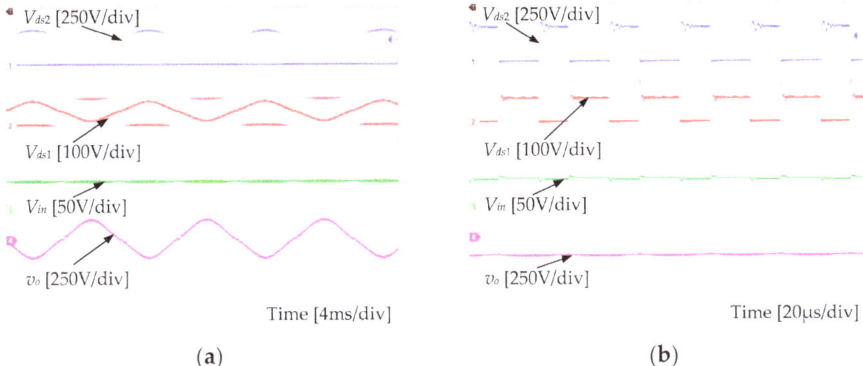

Figure 12. Experimental waveforms of V_{ds2}, V_{ds1}, V_{in}, and v_o: (**a**) At the line period scale, (**b**) at the switching period scale.

In addition, as shown in Figure 12, when zooming into the switch voltage waveform, it was revealed that a voltage spike appeared at the instant of the switch turning off. This is typical for converters with coupled inductors [32]. For the first version of the prototype, a simple RCD snubber was used to verify the basic operation principle of the proposed topologies. The efficiency of 75% was achieved with 100-W output power, where the RCD snubber accounted for a large portion of the total power losses. Moreover, the voltage spike can be suppressed with an appropriate snubber arrangement and design to capture and recycle the leakage energy to achieve much higher efficiency according to the analysis. Snubber details and verification are the subjects of the follow-up research work. What is more, the voltage gain was slightly lower than the theoretical one due to the power losses. With the planned regenerative snubber, the power losses will be less and, thus, the practical voltage gain should be closer to the theoretical one. Overall, the simulation and experimental results were in agreement with the theoretical analysis. Thus, the effectiveness of the proposed inverter family was verified, which had the merits of single-stage conversion, low component count, and easy implementation. These advantages are significant from PV applications, while the efficiency should be further enhanced.

6.2. Comparison of the SSBBI and the State of the Art

After the preliminary experimental test of the SSBBI prototype, the non-optimized performance of the SSBBI could be compared with its counterparts. The comparison results are shown in Table 10. According to Table 10, it is known that the SSBBI had the lowest semiconductor count, almost half of its counterparts. The lower component count makes the SSBBI a simple structure, requiring simpler driving and auxiliary power supplies. These advantages will lead to lower cost, which is a practical concern for the microinverters.

Table 10. Comparison of the SSBBI with the state of the art.

Topologies	Switches Count	Diodes Count	Inductors Count	Input Voltage	Output Voltage	Output Power	Efficiency
[22]	4	8	4	20 V	314 V	100 W	/
[23]	8	0	1 Tapped	100–200 V	110 V	500 W	>96%
Figure 2b [24]	8	0	1 Tapped	40 V	230 V	/	/
[25]	5	2	1 Tapped	60 V	230 V	100 W	86%
SSBBI	4	0	1 Tapped	48 V	110 V	100 W	75%

The efficiency performance of the SSBBI was not outperforming, as mentioned previously. With the theoretical analysis and simulations, the power losses on the RCD snubber were around 15%. Thus, with a proper regenerative snubber, the efficiency will be more than 85% as predicted, where component optimization can further be applied to improve the efficiency. Nevertheless, the efficiency of 85% will be reasonable for a 100-W, single-stage, buck-boost inverter and comparable with the experimental efficiency in [25].

7. Conclusions

This paper introduced a family of single-stage, buck-boost inverter topologies. Compared to the counterparts, the proposed topologies had a lower component count. The key feature of the proposed family was the application of a multi-winding tapped inductor that helped to attain a higher voltage gain required in PV applications, as microinverters. The operational principle was discussed in this paper, which was supported by simulation and experimental results. A stand-alone experimental SSBBI prototype was designed, built, and tested. Experimental results showed that the proposed topology is capable of delivering a well-shaped sinusoidal output. However, the practical voltage gain was slightly lower than theoretical prediction and the efficiency was not at a very satisfactory level due to the RCD snubber losses and the un-optimized components of the converter, which will be the future work. Overall, the proposed family can present a viable solution to single-stage microinverter applications.

Author Contributions: Conceptualization, A.A.; methodology, A.A. and B.Z.; software, B.Z.; validation, C.L., B.Z., and Y.H.; formal analysis, B.Z.; investigation, C.L.; resources, Y.H.; data curation, C.L.; writing—original draft preparation, B.Z.; writing—review and editing, A.A and Y.Y.; visualization, Y.Y.; supervision, Y.H.; project administration, B.Z.; funding acquisition, B.Z. All authors have read and agreed to the published version of the manuscript.

Funding: This research was funded by the National Natural Science Foundation of China under Grant No. 51807164 and by Shaanxi Key R&D Program under Grant No. 2018GY-073 and by Xi'an Science and Technology Association Youth Talent Support Project.

Conflicts of Interest: The authors declare no conflict of interest.

Nomenclature

n	Turns ratio of the tapped inductor
N_1, N_2, N_3, N_4	Windings of the tapped inductor
Q_1, Q_2, Q_3, Q_4, Q_5	Switches (MOSFETs)
D_1, D_2	Diodes
L_{cp}	Tapped inductor
R_L	Equivalent load resistance
C_o	Output capacitor
V_{in}	Input voltage
i_{in}	Input current
v_o	Output voltage
i_o	Output current
$v_{ds1}, v_{ds2}, v_{ds3}, v_{ds4}$	Drain-source voltage of the switches Q_1–Q_4
$i_{ds1}, i_{ds2}, i_{ds3}, i_{ds4}$	Currents through the switches Q_1–Q_4
D	Duty cycle
T_s	Switching period
$i_{N1}, i_{N2}, i_{N3}, i_{N4}$	Currents through the windings
$S_{Q1}, S_{Q2}, S_{Q3}, S_{Q4}$	Gating signals the switches Q_1–Q_4
I_o	Average output current
M	Voltage gain
$V_{o\max}$	Maximum output voltage
$D_{\max}$	Maximum duty ratio
$V_{Q1\max}, V_{Q2\max}, V_{Q3\max}, V_{Q4\max}$	Voltage stress on the switches Q_1–Q_4

$v_o(t)$	Time-varying output voltage
$i_o(t)$	Time-varying output current
V_m	Peak output voltage
I_m	Peak output current
ω	Angular frequency
$d(t)$	Time-varying duty ratio
I_{Q1max}, I_{Q2max}	Maximum current of the switch Q_1, Q_2
$i^2_{Q1rmsTs}, i^2_{Q2rmsTs}$	Squared RMS current of the switch Q_1, Q_2 within a switching period
I^2_{Q1rms}, I^2_{Q2rms}	Squared RMS current of the switch Q_1, Q_2
I_{Q1rms}, I_{Q2rms}	RMS current of the switch Q_1, Q_2
f_s	Switching frequency
L_m	Tapped-inductor magnetizing inductance

Abbreviations

DC	Direct current
AC	Alternating current
PV	Photovoltaic
MIE/MIC	Module-integrated electronic/converter
MPPT	Maximum power point tracking
SEPIC	Single ended primary inductor converter
PWM	Pulse width modulation
MOSFET	Metal oxide semiconductor field-effect transistor
GaN	Gallium nitride
SSBBI	Single-stage, buck-boost inverter
CCM	Continuous conduction mode
SPWM	Sinusoidal pulse width modulation
THD	Total harmonic distortion
RMS	Root mean square
RCD	Resistor-capacitor-diode

References

1. Meneses, D.; Blaabjerg, F.; Garcia, O.; Cobos, J.A. Review and comparison of step-up transformerless topologies for photovoltaic AC-module application. *IEEE Trans. Power Electron.* **2013**, *28*, 2649–2663. [CrossRef]
2. Fang, Y.; Ma, X. A novel PV microinverter with coupled inductors and double-boost topology. *IEEE Trans. Power Electron.* **2010**, *25*, 3139–3147. [CrossRef]
3. Abramovitz, A.; Zhao, B.; Smedley, K.M. High-gain single-stage boosting inverter for photovoltaic applications. *IEEE Trans. Power Electron.* **2016**, *31*, 3550–3558. [CrossRef]
4. Zhang, Y.; Woldegiorgis, A.T.; Chang, L. Design and test of a novel buck-boost inverter with three switching devices. In Proceedings of the 2012 Twenty-Seventh Annual IEEE Applied Power Electronics Conference and Exposition (APEC), Orlando, FL, USA, 5–9 February 2012.
5. Chowdhury, A.S.K.; Chakraborty, S.; Salam, K.M.A.; Razzak, M.A. Design of a single-stage grid-connected buck-boost photovoltaic inverter for residential application. In Proceedings of the 2014 Power and Energy Systems: Towards Sustainable Energy, Bangalore, India, 13–15 March 2014.
6. Kumar, A.; Gautam, V.; Sensarma, P. A SEPIC derived single-stage buck-boost inverter for photovoltaic applications. In Proceedings of the 2014 IEEE International Conference on Industrial Technology (ICIT), Busan, Korea, 26 February–1 March 2014.
7. Kumar, A.; Sensarma, P. A four-switch single-stage single-phase buck-boost inverter. *IEEE Trans. Power Electron.* **2017**, *32*, 5282–5292. [CrossRef]
8. Kumar, A.; Sensarma, P. New switching strategy for single-mode operation of a single-stage buck-boost inverter. *IEEE Trans. Power Electron.* **2018**, *33*, 5927–5936. [CrossRef]

9. Akbar, F.; Cha, H.; Lee, S.; Nguyen, T. A single-phase transformerless buck-boost inverter for standalone and grid-tied applications with reduced magnetic volume. In Proceedings of the 2019 10th International Conference on Power Electronics and ECCE Asia (ICPE 2019-ECCE Asia), Busan, Korea, 27–30 May 2019.

10. Mostaan, A.; Abdelhakim, A.; Soltani, M.; Blaabjerg, F. Single-phase transformer-less buck-boost inverter with zero leakage current for PV systems. In Proceedings of the 2017 43rd Annual Conference of the IEEE Industrial Electronics Society (IECON), Beijing, China, 29 October–1 November 2017.

11. Muhammedali Shafeeque, K.; Subadhra, P.R. A novel single-stage DC-AC boost inverter for solar power extraction. In Proceedings of the 2013 Annual International Conference on Emerging Research Areas and 2013 International Conference on Microelectronics, Communications and Renewable Energy, Kanjirapally, India, 4–6 June 2013.

12. Wu, W.; Ji, J.; Blaabjerg, F. Aalborg inverter -a new type of "buck in buck, boost in boost" grid-tied inverter. *IEEE Trans. Power Electron.* **2015**, *30*, 4784–4793. [CrossRef]

13. Xu, S.; Chang, L.; Shao, R.; Mohomad, A.R.H. Power decoupling method for single-phase buck-boost inverter with energy-based control. In Proceedings of the 2017 IEEE Applied Power Electronics Conference and Exposition (APEC), Tampa, FL, USA, 26–30 March 2017.

14. Xu, S.; Shao, R.; Chang, L.; Mao, M. Single-phase differential buck-boost inverter with pulse energy modulation and power decoupling control. *IEEE J. Emerg. Sel. Top. Power Electron.* **2018**, *6*, 2060–2072. [CrossRef]

15. Tang, Y.; Dong, X.; He, Y. Active buck-boost inverter. *IEEE Trans. Ind. Electron.* **2014**, *61*, 4691–4697. [CrossRef]

16. Tang, Y.; Xu, F.; Bai, Y.; He, Y. Comparative analysis of two modulation strategies for an active buck-boost inverter. *IEEE Trans. Power Electron.* **2016**, *31*, 7963–7971. [CrossRef]

17. Huang, Q.; Huang, A.Q.; Yu, R.; Liu, P.; Yu, W. High-efficiency and high-density single-phase dual-mode cascaded buck-boost multilevel transformerless PV inverter with GaN AC switches. *IEEE Trans. Power Electron.* **2019**, *34*, 7474–7488. [CrossRef]

18. Khan, A.A.; Cha, H.; Ahmed, H.F.; Kim, J.; Cho, J. A highly reliable and high-efficiency quasi single-stage buck-boost inverter. *IEEE Trans. Power Electron.* **2017**, *32*, 4185–4198. [CrossRef]

19. Khan, A.A.; Cha, H. Dual-buck-structured high-reliability and high-efficiency single-stage buck-boost inverters. *IEEE Trans. Ind. Electron.* **2018**, *65*, 3176–3187. [CrossRef]

20. Ho, C.N.M.; Siu, K.K.M. Manitoba inverter—single-phase single-stage buck-boost VSI topology. *IEEE Trans. Power Electron.* **2019**, *34*, 3445–3456. [CrossRef]

21. Testa, A.; Caro, S.D.; Scimone, T.; Panarello, S. A buck-boost based DC/AC converter for residential PV applications. In Proceedings of the International Symposium on Power Electronics, Electrical Drives, Automation and Motion, Sorrento, Italy, 20–22 June 2012.

22. Abdel-Rahim, O.; Orabi, M.; Ahmed, M.E. Buck-boost interleaved inverter for a grid-connected photovoltaic system. In Proceedings of the 2010 IEEE International Conference on Power and Energy, Kuala Lumpur, Malaysia, 29 November–1 December 2010.

23. Tang, Y.; He, Y.; Dong, X. Active buck-boost inverter with coupled inductors. In Proceedings of the 2014 IEEE Energy Conversion Congress and Exposition (ECCE), Pittsburgh, PA, USA, 14–18 September 2014.

24. Husev, O.; Matiushkin, O.; Roncero-Clemente, C.; Blaabjerg, F.; Vinnikov, D. Novel family of single-stage buck-boost inverters on the unfolding circuit. *IEEE Trans. Power Electron.* **2019**, *34*, 7662–7676. [CrossRef]

25. Sreekanth, T.; Lakshminarasamma, N.; Mishra, M.K. Coupled inductor-based single-stage high gain DC-AC buck-boost inverter. *IET Power Electron.* **2016**, *9*, 1590–1599. [CrossRef]

26. Sreekanth, T.; Lakshminarasamma, N.; Mishra, M.K. A single-stage grid-connected high gain buck-boost inverter with maximum power point tracking. *IEEE Trans. Energy Convers.* **2017**, *32*, 330–339. [CrossRef]

27. Zhao, B.; Abramovitz, A.; Smedley, K. Family of bridgeless buck-boost PFC rectifiers. *IEEE Trans. Power Electron.* **2015**, *30*, 6524–6527. [CrossRef]

28. Zhao, B.; Ma, R.; Abramovitz, A.; Smedley, K. Bridgeless buck-boost PFC rectifier with a bi-directional switch. In Proceedings of the 2016 IEEE 8th International Power Electronics and Motion Control Conference (IPEMC-ECCE Asia), Hefei, China, 22–26 May 2016.

29. Zhao, B.; Abramovitz, A.; Ma, R.; Huangfu, Y. High gain single stage buck-boost inverter. In Proceedings of the 2017 20th International Conference on Electrical Machines and Systems (ICEMS), Sydney, Australia, 11–14 August 2017.

30. Zhao, B.; Abramovitz, A.; Ma, R.; Liang, B. One-cycle-controlled high gain single stage buck-boost inverter for photovoltaic application. In Proceedings of the 43rd Annual Conference of the IEEE Industrial Electronics Society (IECON), Beijing, China, 29 October–1 November 2017.
31. Design Guide of Inductor. Available online: www.mag-inc.com (accessed on 10 January 2020).
32. Abramovitz, A.; Smedley, M.K. Analysis and design of a tapped-inductor buck-boost PFC rectifier with low bus voltage. *IEEE Trans. Power Electron.* **2011**, *26*, 2637–2649. [CrossRef]

Article

Modular Multilevel Converter for Photovoltaic Application with High Energy Yield under Uneven Irradiance

Anirudh Budnar Acharya [1,*], Dezso Sera [2], Remus Teodorescu [1] and Lars Einar Norum [3]

[1] Department of Energy Technology, Aalborg University, 9220 Aalborg, Denmark; ret@et.aau.dk
[2] School of Electrical Engineering and Robotics, Queensland University of Technology, Brisbane, QLD 4000, Australia; dezso.sera@qut.edu.au
[3] Department of Electric Power Engineering, Norwegian University of Science and Technology, 7491 Trondheim, Norway; norum@ntnu.no
* Correspondence: abac@et.aau.dk

Received: 3 April 2020; Accepted: 18 May 2020; Published: 21 May 2020

Abstract: The direct integration of Photovoltaic (PV) to the three-phase Modular Multilevel Converter (MMC) without *dc–dc* converters results in high-efficiency PV power plant with increased energy yield. The arm power control method for the MMC further improves the extraction of available power under uneven irradiance across different phases of the MMC. However, the uneven irradiance between the sub-modules results in residual voltage that results in harmonics and unbalance components. In this paper, the effect of uneven irradiance across the sub-module of the MMC is investigated with arm power control method. A modified balancing algorithm for the arm power control of the MMC is proposed which enables balanced power to be injected into *ac* grid despite uneven irradiance across the sub-modules in the MMC. The modified balancing algorithm enables to keep the unbalance in the phase currents below 10% and the Total Harmonic Distortion (THD) is confined as per IEEE 519 standard.

Keywords: modular multilevel converter; photovoltaic power system; grid integration; control system; distributed renewable energy source

---

1. Introduction

The aim of decreasing the emission of greenhouse gas to minimize the impact on the environment has given a tremendous push to power plants based on renewable energy sources. Solar power is abundantly available and many countries have pledged to use 100% renewable energy by 2050 [1]. The large share of energy consumption from renewable sources will be contributed by solar power in the near future. As the extraction of solar power is highly weather-dependent, efficient power converters are necessary that can harvest the available power at all weather conditions.

Modular PV power plants are preferred in locations where energy yield is impacted due to varying weather conditions. Furthermore, modular PV power plants are preferred for commercial installation where partial shading of the panel is a concern. The modular power converters decrease the effects of PV panel mismatch as compared to central power converters where the PV panels are connected to form an array. The modularity of such converters can be a panel, string, or array level. However, in most of the cases the modularity is achieved at the cost of additional *dc–dc* converters [2].

The PV power plant using Cascaded H-Bridge (CHB) converter is studied in [3,4], it operates at high efficiency and increases energy yield due to increase in number of Maximum Power Point Tracking (MPPT). The Modular Multilevel Converter (MMC) proposed in [5] increases the number of MPPT for the same number of switches compared to the CHB converter. The MMC topology, its variants, and applications are discussed in [6,7]. In [3,8,9], MMC is proposed as an inverter for PV plant. The detailed discussion on topology and control methods for the MMC are presented in [10]. The Figure 1a shows a three-phase double star MMC with Half-Bridge (HB) sub-modules. Each phase of the MMC can be divided into sub-units referred to as upper and lower arm, respectively. Each arm of the MMC has series-connected power electronic blocks referred to as "sub-modules" and an inductor referred to as "arm inductor". The sub-modules can be identical or a combination of different power converter topologies [11]. Typically used sub-modules are half-bridge or full-bridge converters.

Three distinct variants of the topology for connecting the PV panels to the sub-modules are shown in Figure 1b–d. In [12,13], the PV panels are directly connected to the sub-module of the MMC as shown in Figure 1b. The overall efficiency of the PV plant is considerably high as the MMC efficiency is in the range of 99% [14]. Such a system is comparable to the central PV power plant with the additional benefit of an increased number of independent MPPT algorithms, which in this case is equal to the number of sub-modules. This results in higher energy yield and better efficiency than the central PV power plant. In [15], the authors show that the Levelized Cost of Energy (LCOE) for the MMC-based PV plant can be brought lower than that of the central PV plant. The PV panels connected to the sub-modules using the *dc–dc* converters is shown in Figure 1c,d. The use of a *dc-dc* converter allows the decoupling of the PV control and the MMC control. The advantage is that the sub-module capacitor voltages across the MMC are equal; therefore, no modification is necessary in the MMC control. However, in this configuration the overall efficiency is lower compared to PV plant without *dc-dc* converters. In cases where the isolated converters are used, the *dc*-link voltage can be scaled to Medium Voltage (MV) facilitating direct connection of the MMC PV plant to the distribution grid. Thereby, avoiding the need for a step-up transformer typically used for connection to the MV grid.

In [16], the control method for the MMC uses individual Pulse Width Modulators (PWM) for each of the sub-modules. In [17], the method presented in [16] is further extended to control the MMC with the energy sources connected to the sub-modules. The energy in each of the sub-module is locally controlled, which effectively provides the possibility of distributing the control between the main and local controllers. It uses phase-shifted PWM and additional sub-modules are necessary as energy buffers to avoid: (1) large variation of capacitor voltage in the sub-module with an energy source and (2) to avoid very high switching frequency of the sub-module. In [18,19] the non-carrier based approach is used for controlling the MMC when the energy sources are connected to the sub-modules. The non-carrier based control method relies on calculating the fundamental positive and negative sequence circulating current references required to balance the energy between the upper and lower arms of the MMC. In [12], a cost function is presented to optimize the calculation of fundamental circulating current references for extracting the maximum power from the PV and injection of balanced power to the grid. Calculating weights for the cost function is not straight forward and is usually obtained from trial-and-error or extensive simulation cases. In [13], arm power control of MMC is presented, the control system is distributed such that each arm of the MMC is controlled independently. This method also avoids the mathematical computation of the fundamental circulating current references.

Using arm power control the MMC is controlled such that maximum power is extracted from the PV panels and a balanced power is injected to the *ac* grid. The sum of sub-module capacitor voltages in an arm of the MMC is allowed to be different across the upper and lower arms of the MMC. However, within the arm of the MMC, all capacitor voltages are maintained to be equal. This is achieved with the help of sorting and tracking algorithm. The variation of the irradiance is assumed at arm-level for the three-phase MMC leading to six independent MPPT. Such an assumption is viable in large power plants were the effects of shading is minimal. In the case of residential and commercial PV plants, the consequence of shading between the sub-modules cannot be neglected. The shading of PV panels will result in a decrease of power extracted as the MMC is only capable of MPPT at arm-level. This will reduce the yield ratio and LCOE compared to the module-level power converters.

The MMC with arm power control can enable MPPT at the sub-module level. This is achieved by providing individual sub-module capacitor voltage references obtained from the MPPT algorithm. As a result, the sub-module capacitor voltages within the arm of the MMC will not be equal to its average value. Therefore, the voltage inserted by each arm of the MMC will not be equal. As the output voltage in a phase of the MMC is the difference of the upper and the lower arm voltages, the unequal arm voltages will result in a residual voltage at the output terminal. A high deviation in the magnitude of the sub-module capacitor voltages in the arm of the MMC might result in higher residual voltage. This will result in undesired current harmonics and unbalance current components.

In this paper, the effect of unequal sub-module capacitor voltages in the arm of the MMC using arm power control is investigated. A modified sorting and balancing algorithm is proposed that allows the MMC-based PV plant with arm power control to track the MPPT at the sub-module level and inject balanced power to the *ac* grid. The effect of phase current THD is analyzed in the case of uneven irradiance on the sub-modules. The modified sorting and tracking algorithm mitigates the residual voltage between the converter and grid voltages thereby reducing the THD in the phase currents. As a consequence of lower residual voltage the unbalance in the phase current is mitigated. The modified algorithm ensures balanced power injection to the *ac* grid despite extreme unbalance in power generation. The proposed solution makes the arm power control for the MMC suitable for PV applications which are prone to uneven irradiance.

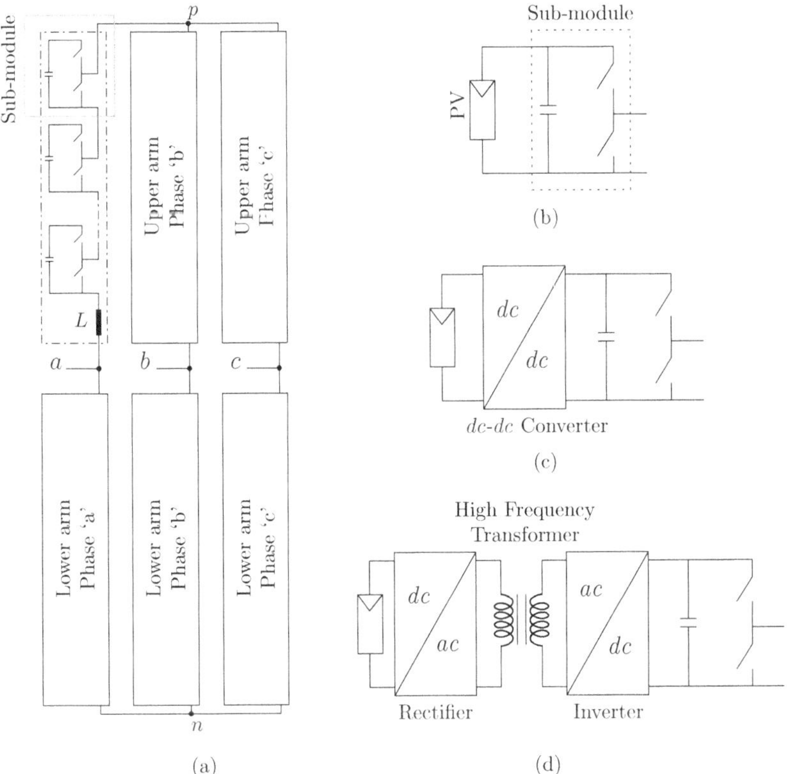

Figure 1. The Modular Multilevel Converter (MMC) and sub-modules with photovoltaic (PV) panels. (**a**) Three-phase MMC indicating the upper and lower arms and the sub-module, (**b**) the HB sub-module with direct connection of the PV panel, (**c**) the PV is connected to the sub-module using a non-isolated *dc-dc* converter, and (**d**) the PV is connected to the sub-module using a isolated converter.

2. Direct Connection of PV Panels to the MMC

To utilize the modularity, increase efficiency, and reliability of the MMC, either a group or individual PV panels is connected directly to the sub-modules of the MMC. Such a configuration inherits the advantages of the MMC such as redundancy, fault-tolerant operation, improved harmonic performance, and hot-swap.

The topology of the MMC with the direct connection of PV panels to the sub-module is shown in Figure 2. Two PV panels are connected in series to form a string which is connected to the sub-module with a series diode to avoid power flow into the PV string. The number of PV panels connected in series or parallel depends on the sizing of the PV plant. Such a configuration is versatile and can have "6N" independent MPPT algorithms. The MPPT granularity is defined as the number of independent MPPT. The MMC can be controlled such that MPPT is performed either at sub-module, arm, or MMC level depending on the irradiance pattern. This will ensure high energy yield under different operating conditions.

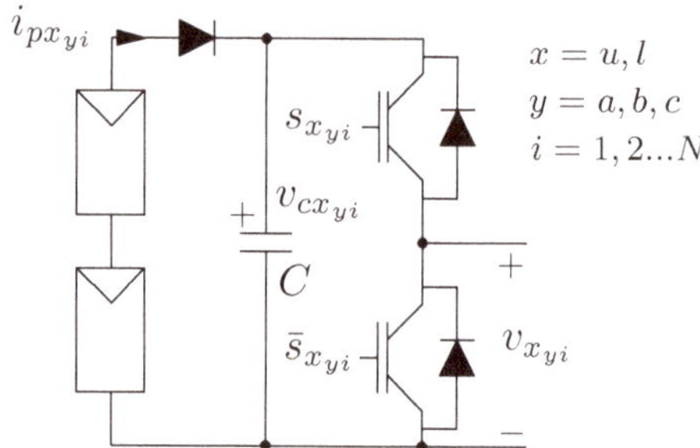

Figure 2. The PV string, two PV panels in series, is connected to the *i*th sub-module. A diode is included to avoid the power flow into PV string.

The sub-module is said to be inserted when the capacitor is included in the arm of the MMC, i.e., when insertion index $n_{x_{yi}} = 1$. When the capacitor is not included in the arm of the MMC, the corresponding sub-module is said to be bypassed, i.e, when insertion index is $n_{x_{yi}} = 0$. When the sub-module is inserted the output voltage of the sub-module is $v_{x_{yi}} = n_{x_{yi}} \cdot v_{cx_{yi}}$. Therefore, the voltage across the arm of the MMC is sum of the individual sub-module output voltages expressed as

$$v_x = \sum_{i=1}^{N} n_{x_{yi}} \cdot v_{cx_{yi}} \tag{1}$$

The current through the sub-module capacitor voltage is expressed as

$$C\frac{d}{dt}\left(v_{cx_{yi}}\right) = i_{px_{yi}} + n_{x_{yi}} \cdot i_{xy} \tag{2}$$

The current from the PV string, $i_{px_{yi}}$, depends on the irradiance level, temperature, and the capacitor sub-module voltage. To track the maximum power on each sub-module, the capacitor voltage is varied and retained at an operating point where the maximum power is extracted from the PV string. When the sub-module is inserted the magnitude of the capacitor voltage changes based on the net current through the capacitor. In this configuration, the PV string current always has a positive average value, however, the arm current alternates sinusoidally. Therefore, the sub-modules in the arm of the MMC have to be selectively inserted or bypassed to reduce the error between the capacitor voltage and the Maximum Power Point (MPP) voltage.

The fundamental sub-module capacitor ripple voltage also influences the power extracted from the PV string. In [20], the effective power loss per panel is studied concerning the sub-module capacitor voltage ripple. For a fixed switching frequency, irradiance of 1000 W/m², and at a constant temperature, it is shown that the decrease of sub-module capacitor voltage ripple from 10% to 5% results in a decrease of effective loss of power extracted from PV panel, i.e., from 2.47% to 0.56%, respectively.

In Figure 3a, the voltage across the sub-module capacitor is shown for capacitance between 20 mF to 100 mF incremented in steps of 10 mF. The data from the Canadian Solar CS6K-285M-FG PV panel is used for the analysis. The maximum allowed sub-module capacitor voltage is 75 V. The switching frequency is selected to be 10 kHz, the irradiance is maintained at 1000 W/m². Figure 3 shows the capacitance of the sub-module against the capacitor voltage ripple, to keep the fundamental ripple voltage within 5% of the rated sub-module capacitor voltage the capacitance has to be greater than 50 mF. This capacitance is easily attainable as the sub-module operates at low voltage in the order of few tens of volts.

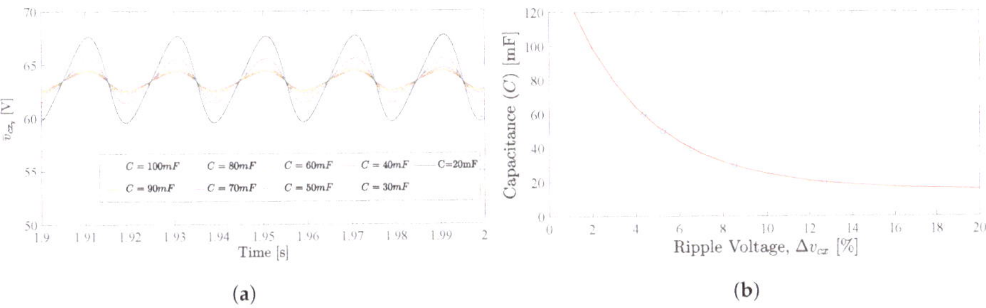

(a)   (b)

Figure 3. (**a**) Sub-module capacitor voltages for different value of capacitance ranging from 20 mF to 100 mF in steps of 10 mF. (**b**) Sub-module capacitance as a function of ripple voltage at maximum rated capacity of the plant operating with 10 kHz switching frequency.

3. Arm Power Control of MMC Based PV Plant

The block diagram of arm power control proposed in [13] is shown in Figure 4. The power in each arm of the MMC is independently controlled such that (1) each phase of the MMC delivers the same balanced power to the grid, and the (2) maximum power from the PV is extracted in each arm of the MMC. Such a control method leads to MPPT granularity of six.

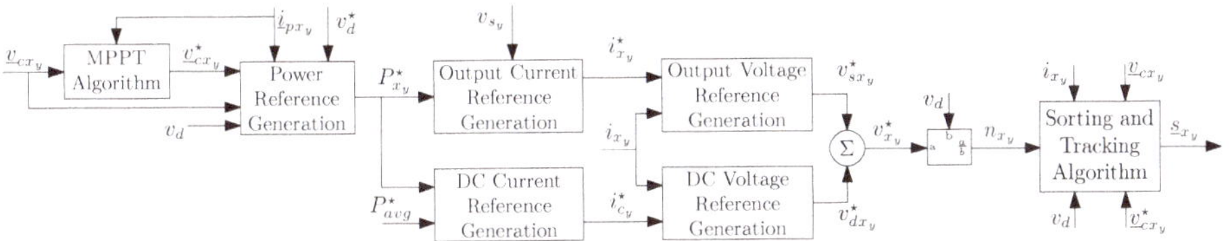

Figure 4. The block diagram of arm power control method of the MMC for PV application proposed in [13].

The MPPT Algorithm provides the individual voltage references for the sub-modules in arm of the MMC as a vector, $\underline{v}_{cx_y}^\star$ [1 × N]. These voltage references are added to obtain the desired sum-capacitor voltage reference for individual arm of the MMC, i.e., $v_{cx_y}^{\Sigma\star} = \sum_{i=1}^{N} v_{cx_{yi}}^\star$. A Proportional-Integral (PI) controller is used to generate the power reference such that the voltage error between $v_{cx_y}^{\Sigma\star}$ and v_{dc} is driven to zero as

$$P_{x_y}^\star(s) = \left[v_{cx_y}^{\Sigma\star}(s) - v_{dc}(s) \right] \cdot \left(k_{p_{dc}} + \frac{k_{i_{dc}}}{s} \right) \tag{3}$$

The *ac* current reference for the arm of the MMC is calculated using the power reference ($P_{x_y}^\star$) and the grid voltage at the point of common coupling. The *dc* current reference for the arm of the MMC is obtained with a PI controller to drive the error between the arm power reference and the average arm power to zero. The average arm power is the mean of power extracted from the PV in each arm of the MMC, defined as in (4).

$$P_{avg} = \frac{1}{6}\left(\sum_{y=a,b,c}\left[\sum_{x=u,l}\left\{\sum_{i=1}^{N} v_{cx_{yi}} \cdot i_{px_{yi}}\right\}\right]\right) \tag{4}$$

The desired voltage reference for each arm of the MMC is obtained as sum of outputs from "output voltage reference generation" and the "*dc* voltage reference generation" blocks, respectively. In the *dc* voltage reference generation, a separate Proportional Resonant (PR) controller is used to suppress the second harmonic circulating current. The insertion index for the arm is calculated as (5) using the arm voltage reference, $v_{x_y}^\star$.

$$n_{x_y} = \frac{v_{x_y}^\star}{v_{dc}} = \frac{\displaystyle\sum_{i=1}^{N} n_{x_{yi}} \cdot v_{cx_{yi}}}{v_{dc}} \tag{5}$$

The number of sub-module inserted in a switching period is positive integer value of N_{x_y}, i.e., $N_{x_y} = \left\lceil n_{x_y} \cdot N \right\rceil$. The "Sorting and Tracking Algorithm" is shown in Figure 5, the sub-modules are referred as SM in the algorithm. It enables the insertion and bypass of the sub-module in a switching period such that the sub-module voltages in an arm of the MMC are maintained to their desired values. However, in [13], all the sub-module capacitor voltages in an arm of the MMC are maintained equal. The scenario of the uneven irradiance within the arm of the MMC has not been considered. Such an uneven irradiance within the arm of the MMC will result in different sub-module capacitor voltage references from the MPPT algorithm. The algorithm provides the provision to address unequal irradiance between the sub-modules in an arm of the MMC. The list L_1 contains all the sub-modules with voltage less than their MPPT references, and L_2 contains all the sub-modules with voltage greater than their MPPT references. Based on the polarity of the arm current and the magnitude of the sub-module capacitor voltages, the sub-modules are either inserted or bypassed to maintain the voltage within a threshold ϵ. The only limitation is that all the sub-module capacitor voltage references are identical for an arm of the MMC, i.e., $\forall\, i = 1$ to N, $v_{cx_{yi}} = v_{cx}^\star$.

For this study, the parameters of the MMC are identical to the case considered in [13], as tabulated in Table A1. The PI- and PR-controller parameters are shown in Table A2.

Scenario 1

In this scenario, all the PV panels connected to the sub-module of the MMC receive equal irradiance. At the Standard Test Condition (STC), the irradiance is 1000 W/m^2, cell temperature is 25 °C, and airmass is 1.5. The operation of the MMC under STC, where all the sub-module receive equal irradiance of 1000 W/m^2 is shown in Appendix B.

All the sub-module capacitor voltages (v_{cx_y} [V] for x = u, l and y = a, b, c, respectively) are maintained at the desired MPP voltage references, as shown in Appendix B Figure A1a. Active power (P [kW]) is injected to the grid by maintaining zero reactive power (Q [$KVAr$]). During the entire operation of the MMC the *dc* and *ac* circulating currents are zero, as shown in Appendix B Figure A1b.

The frequency spectrum of the phase currents injected to the grid for scenario 1 is analyzed in this paper and are shown in Figure 6a–c. The THD for each phases are 1.01%, 1.1%, and 1.04% for phase "a", "b", and "c" currents, respectively. The THD of currents in each phase do not vary significantly. The control of the MMC makes sure that the distortion in all the three phases are minimized by maintaining the desired *ac* voltage reference. The THD is well below the 5% limit as required by IEEE 519 [21] for the scenario 1.

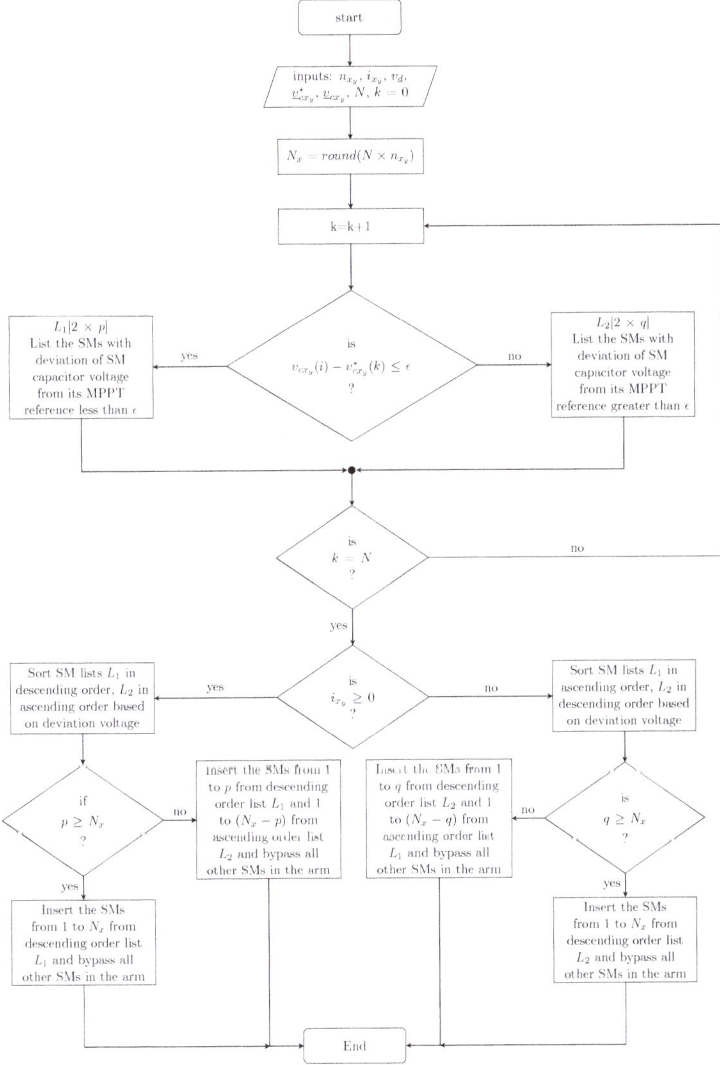

Figure 5. The sorting and tracking algorithm used in the arm power control of the MMC for PV application [13].

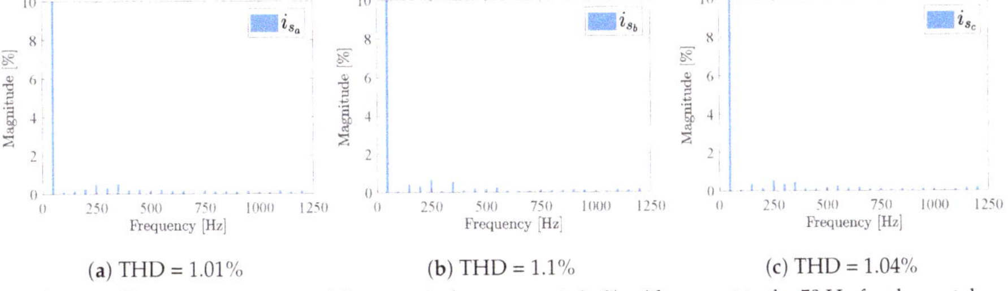

(**a**) THD = 1.01%　　　　(**b**) THD = 1.1%　　　　(**c**) THD = 1.04%

Figure 6. Frequency spectrum of the output phase currents in % with respect to the 50 Hz fundamental current (100%).

The positive, negative and zero sequence components of the three phase currents are shown in Figure 7. The negative sequence component under steady state is less that 1 A. The amount of unbalance in the currents is 0.3% for scenario 1.

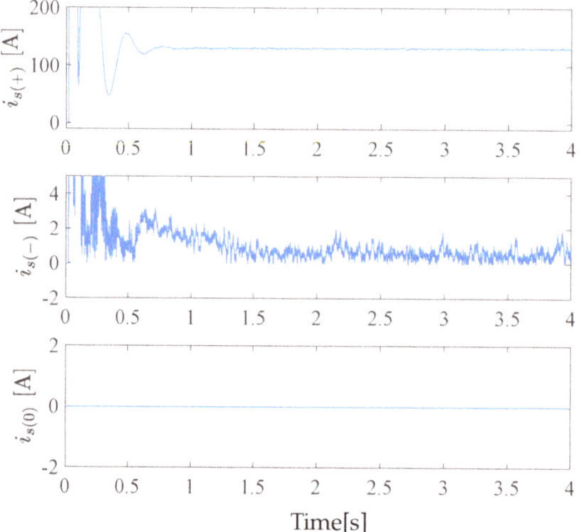

Figure 7. The positive sequence current ($i_{s(+)}$), negative sequence current ($i_{s(-)}$), and zero sequence current ($i_{s(0)}$) for the currents injected to the grid.

4. Uneven Irradiance and Its Consequence

The distribution of irradiance pattern within an arm of the MMC is highly dependent on weather conditions and shading. In [13], the irradiance across the sub-modules in an arm of the MMC are assumed to be identical, and the MPPT is allowed only at arm-level. Such a restriction decreased the harvested power when the irradiance is uneven across the sub-modules in an arm of the MMC. Therefore, the sub-module level MPPT is investigated in this section as scenario 2.

Scenario 2

In this section, a scenario is considered where the irradiance across the PV panels connected to sub-modules in an arm of the MMC is uneven. The sub-modules in an arm of the MMC are allowed to track MPPT by providing individual MPP references from the MPPT algorithm to the power reference generation block in the controller.

If the sub-module capacitor voltage is allowed to follow the MPPT reference within the arm of the MMC, then each sub-module in the arm will deviate from the average value i.e., $v_{cx_{yi}} \neq v_{cx_y}^{\Sigma} / N$.

The current controllers will increase or decrease the inserted arm voltage reference to compensate for the voltage difference due to unequal sub-module voltages in the arm of the MMC. However, the sorting and tracking algorithm does not account for the voltage error between the desired arm voltage and the arm voltage to be inserted. This voltage error varies based on choice of sub-modules to be inserted. This leads to a voltage error in each switching period per arm of the MMC, resulting in a residual voltage. This residual voltage per phase (sub-script 'y' is dropped for simplicity) can be expressed as

$$v_{x,\epsilon} = N \left(\frac{v_x^{\star}}{v_{dc}} \right) - \sum_{j=1}^{N_x} v_{cx_{K(j)}} \tag{6}$$

where the 'K' is a row matrix $[1 \times N_{x_y}]$ with the sub-module indexes to be inserted. Therefore, $v_{cx_{yK(j)}}$ will yield the value of the sub-module capacitor whose index is stored in the jth location of the row matrix 'K'.

If the residual error is large then it will lead to increased harmonics in the output current. Such a variation is acceptable until the THD is well below 5% as required by IEEE 519 [21] and that no *dc* current greater than 0.5% of the rated current is injected to the grid [22].

The simulation results are shown where the irradiance is linearly distributed across all of the upper and lower arms of the MMC from 10 W/m^2 to 1000 W/m^2. For the scenario considered, the sub-module capacitor voltages are shown in Figure 8a for each of the six arms of the MMC. Figure 8b shows the upper and lower arm currents (i_{u_y} [A], i_{l_y} [A]), output currents (i_{s_y} [A]), circulating currents (i_{c_y} [A] $\forall$ y = a, b, c), the active and reactive power injected to the grid (P [kW], Q [kVAr]), and the last plot shown the voltages (v_{s_y} [V] $\forall$ y = a, b, c) at PCC along with the phase currents (i_{s_y} [A]) for 100 ms duration between 4.9 s to 5 s, $\forall$ y = a, b, c.

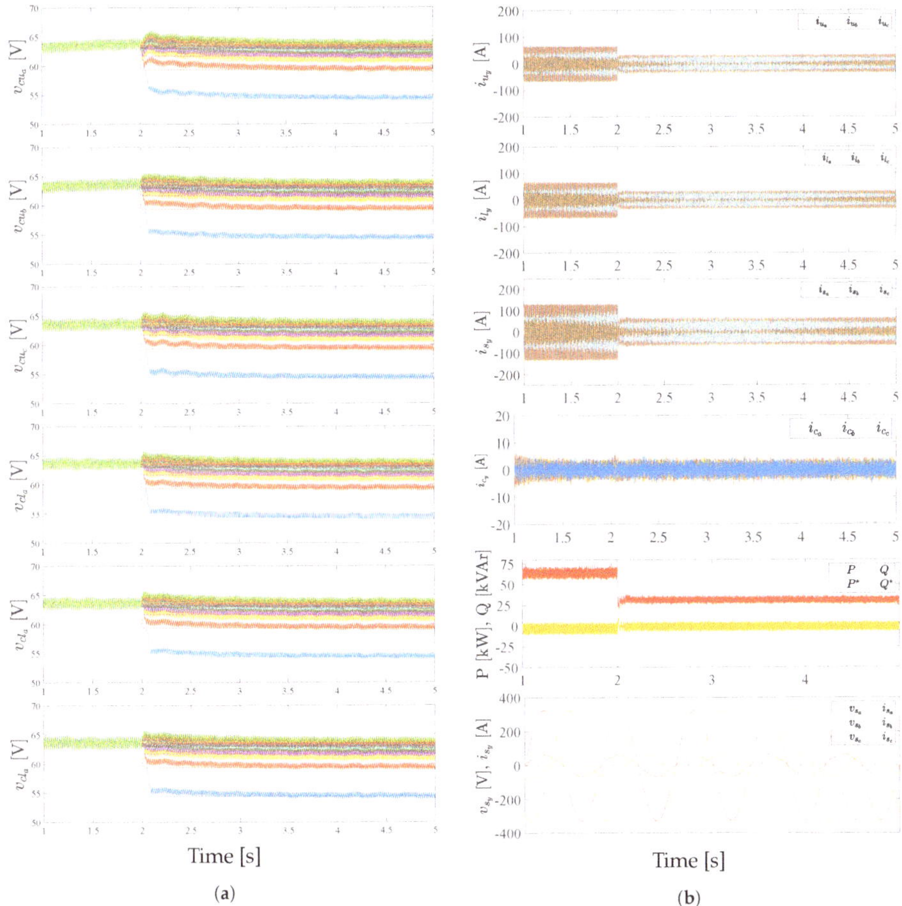

Figure 8. Simulation results for scenario 2: (**a**) Capacitor voltages for all the sub-modules in an arm of the MMC for all three phases. From the top, upper arm phase "a", upper arm phase "b", upper arm phase "c", lower arm phase "a", lower arm phase "b", and lower arm phase "c", respectively. (**b**) The upper and lower arm currents (i_{u_y} [A], i_{l_y} [A]), output currents (i_{s_y} [A]), circulating currents (i_{c_y} [A] $\forall$ y = a, b, c), the active and reactive power injected to the grid (P [kW], Q [kVAr]), and the plot shown the voltages (v_{s_y} [V] $\forall$ y = a, b, c) at PCC and the phase currents (i_{s_y} [A]) for 100 ms duration between 4.9 s to 5 s, $\forall$ y = a, b, c.

It is seen that the sorting and tracking algorithm [13] can be used for tracking the MPP voltages for the respective sub-modules by providing the individual references from the MPPT algorithm instead of a average voltage. Moreover, balanced active power is injected to the grid at unity power factor. Since each arm of the MMC produces equal power there is no need to transfer power between the phases of the MMC. Hence the circulating current is zero. The Figure 9 shows the residual voltage defined as in (6). The Figure 10a–c shows the frequency spectrum of the phase currents; the THDs are 5.11%, 5.28%, and 5.36% for phase a, b, and c currents, respectively. It is seen that the THD is higher that the permitted level as per IEEE 519 standard.

The positive, negative, and zero sequence components of the three-phase currents are shown in Figure 11 for the scenario 2. It is seen that the unbalance current injected to the grid is well within 0.5% of the rated magnitude of phase current for the scenario 2.

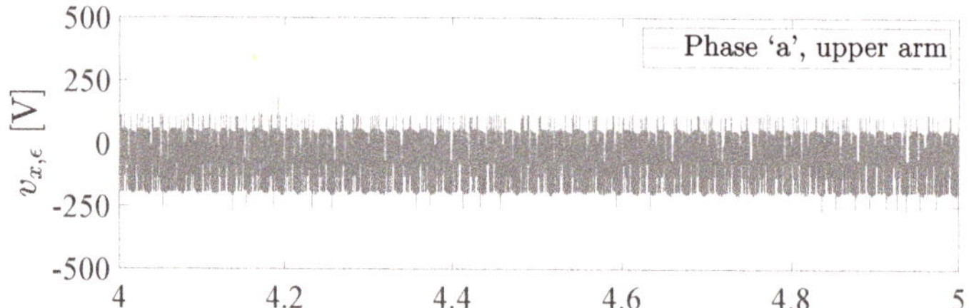

Figure 9. The residual voltage as defined in (6) for the phase a upper arm of the MMC.

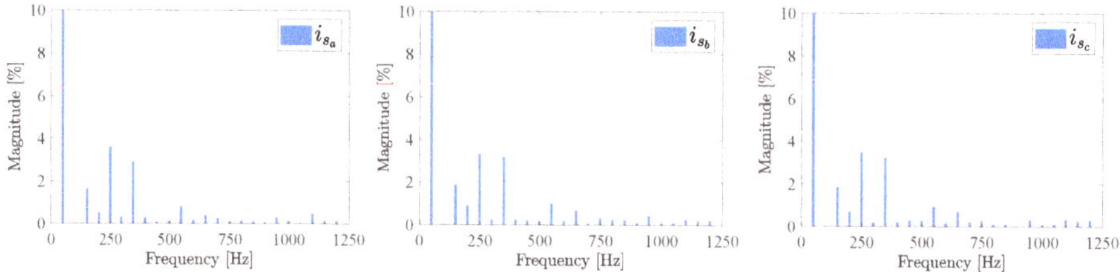

(**a**) THD = 5.11%, $\|i_{s_{a_1}}\| = 43.06\text{A}$　　(**b**) THD = 5.11%, $\|i_{s_{b_1}}\| = 43.11\text{A}$　　(**c**) THD = 5.11%, $\|i_{s_{c_1}}\| = 43.06\text{A}$

Figure 10. Frequency spectrum of the output phase currents in % with respect to the 50 Hz fundamental current, $\|i_{s_{y_1}}\| \ \forall \ y = a, b, c$.

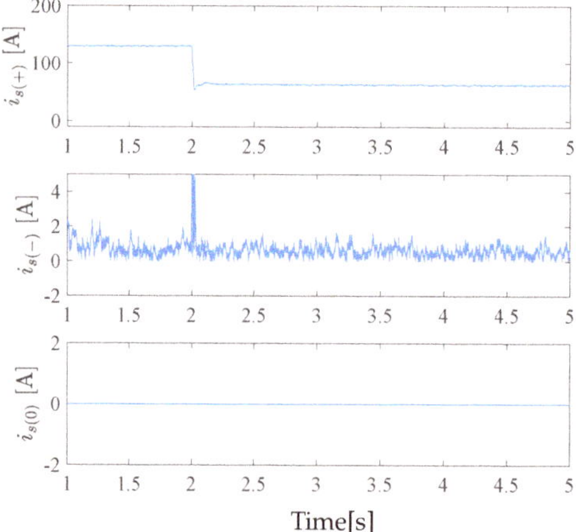

Figure 11. The positive sequence current ($i_{s(+)}$), negative sequence current ($i_{s(-)}$), and zero sequence current ($i_{s(0)}$) for the currents injected to the grid for scenario 2.

5. Modified Sorting and Tracking Algorithm

The sorting and tracking algorithm enables the MMC to have individual MPPT for each sub-module, as seen in scenario 2. This increases the MPPT granularity of the MMC-based PV plant to 6N. For the plant considered in this paper, the MPPT granularity will be 114. The drawback is that the residual error leads to harmonic distortion at the output current. Based on the operating condition, the value of the harmonic distortion might not adhere to the value permitted by the IEEE standard 519 [21]. Therefore, to ensure that for all operating steady-state conditions the harmonic distortion is within the limits, the residual voltage has to be alleviated. The voltage error as per (6) has to be mitigated to reduces the harmonic distortion and any unbalance in current injected to the grid.

In this section, a modified sorting and tracking algorithm is proposed that takes into account the voltage error and increases or decreases the insertion indexes. Further, during a switching period one of the inserted sub-modules is pulse-width modulated such that the average value of the inserted arm voltage inserted matches the desired arm voltage in a switching period. Sub-modules with minimum or maximum voltage deviation from their MPP voltage value are selected, based on the arm current polarity, for PWM in every switching period. Therefore, the duty ratio and the sub-module index for the PWM changes every switching period. By doing so, the loss of power extraction from the PV panel due to the PWM of the sub-module is minimized.

The sorting and tracking algorithm selects the N_x sub-modules to be inserted per arm of the MMC in a given phase, with this the residual voltage is computed as per (6). If the error is negative, then the insertion index is increased to minimize the error. If the error is positive, then the insertion index is decreased to mitigate the residual voltage. The insertion index is either increased or decreased until the magnitude of the ratio as per (7) is less than one, this will be the modified number of sub-modules to be inserted "$N_x^\star$".

$$w = \frac{|v_{x,\epsilon}|}{\sum_{j=1}^{N_x} v_{cx_{K(j)}}} \tag{7}$$

If the arm current is positive (or negative) then the sub-module with the lowest (or highest) voltage in the set of sub-modules to be inserted is selected for modulation. The duty ratio is the calculated as

$$d = \frac{\left| N\left(\frac{v_x^\star}{v_{dc}}\right) - \sum_{j=1}^{N_x^\star} v_{cx_{K(j)}} \right|}{\sum_{j=1}^{N_x^\star} v_{cx_{K(j)}}} < 1 \tag{8}$$

Scenario 3

This scenario is identical to scenario 2; however, the modified sorting and tracking algorithm is used to mitigate the THD which is observed in scenario 2. The index of the sub-module to be modulated and the duty ratio "d" is shown in Figure 12 for 10 ms duration. The index and the duty ratio is modified every switching period so that the average value of the arm voltage is equal to the desired arm voltage.

Figure 13 shows the residual voltage as a result of using modified sorting and a tracking algorithm. The average value of the residual voltage is now zero, and the instantaneous magnitude of the residual voltage over a switching period is lower than the value seen in scenario 2.

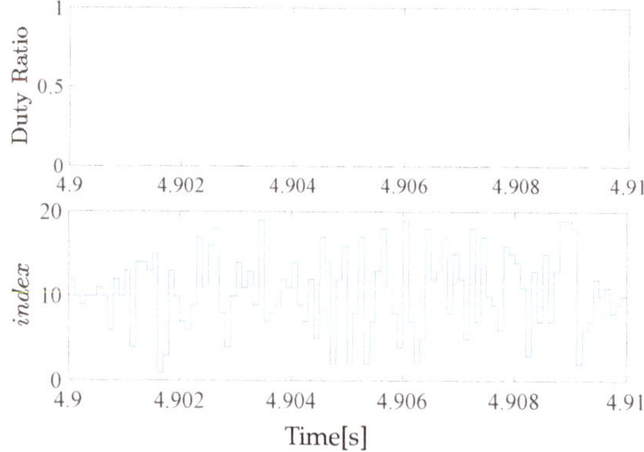

Figure 12. The duty ratio for Pulse Width Modulators (PWM) and the index of the sub-module to be modulated.

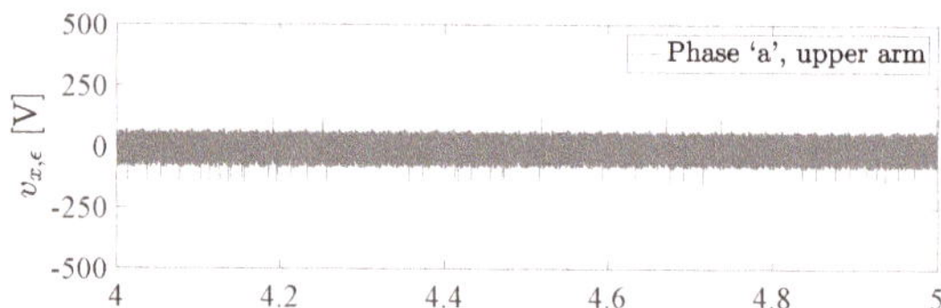

Figure 13. The residual voltage as defined in (6) for the phase "a" upper arm of the MMC with modified ST algorithm.

The frequency spectrum of the output phase currents is shown in Figure 14. the THD is calculated to be 3.68%, 3.75%, and 3.56% for phase a, b, and c currents, respectively. The THD is decreased by 30% bringing it well within the permitted level as per IEEE 519 standard.

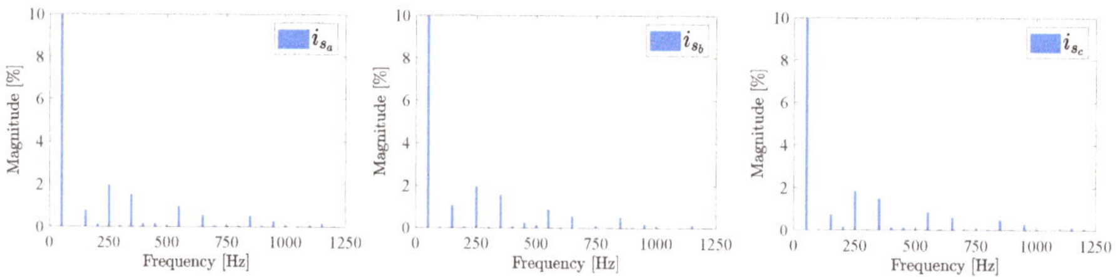

(**a**) THD = 3.68%, $\|i_{s_{a_1}}\|$ = 44.71A   (**b**) THD = 3.75%, $\|i_{s_{b_1}}\|$ = 44.88A   (**c**) THD = 3.56%, $\|i_{s_{c_1}}\|$ = 44.74A

Figure 14. Frequency spectrum of the output phase currents in % with respect to the 50 Hz fundamental current, $\|i_{s_{y_1}}\| \; \forall \; y = a, b, c$.

6. Conclusions

The simulation results of the MMC-based PV plant with arm power control are presented specifically when the irradiance is uneven within the arms of the MMC. The consequence of uneven irradiance on each sub-modules of the arm of the MMC is discussed in terms of harmonic distortion and unbalance in the phase currents. It is seen that the MMC-based PV plant is capable of tracking the maximum power at individual sub-module level brining the MPPT granularity to "6· N". This is achieved without any additional *dc-dc* converters.

It is further noticed that, based on the operating conditions, the harmonic distortion in the output currents increases due to residual voltage between the actual inserted arm voltage and the desired arm voltage. The effect of this residual voltage is the increase in THD and the amount of unbalance in the output current. Though there is no strict requirement on the unbalance, usually a balanced operation is desired for stable operating conditions. Furthermore, there is a strict requirement on the THD of the currents injected into the *ac* grid. It is seen that based on the operating load and irradiance pattern on the PV panels connected to the MMC the THD values can be higher than 5%, which is the allowed limit.

In this paper a modified sorting and tracking algorithm is proposed to the arm power control of the MMC. It enables the effective operation of the MMC-based PV plant even under unequal irradiance patterns across the sub-modules. For the case considered, it is shown that the THD is reduced by 30%, bringing it well within the permitted limit.

Author Contributions: Conceptualization, methodology, and validation: A.B.A., formal analysis: R.T. and D.S. Article is written by A.B.A. Article is reviewed by R.T., L.E.N., and D.S. All authors have read and agreed to the published version of the manuscript.

Funding: This project has received funding from the European Union's Horizon 2020 research and innovation programme under the Marie Skłodowska-Curie grant agreement No 845122.

Conflicts of Interest: The authors declare no conflicts of interest

Nomenclature

$x = u\ or\ l$	Upper (u) or Lower (l) arm
$y = a,\ b\ or\ c$	Phase a, b or c
$i = 1, 2, 3...$	Sub-Module index
N	Number of Sub-Modules
n_{xy_i}	Insertion index of ith Sub-Module in upper or lower arm per phase
n_{xy}	Insertion index of upper or lower arm per phase
i_{xy}	Upper or lower arm current per phase
i_{sy}	Output current per phase
i_{cy}	Circulating current per phase
i_{pxyi}	PV string current in ith Sub-Module per phase
v_{cxyi}	Capacitor voltage of ith Sub-Module in upper or lower arm per phase
v_{cxy}^{Σ}	Sum capacitor voltage of upper or lower arm per phase
v_{xy}	Inserted upper or lower arm voltage per phase
v_{sy}	Output voltage in each phase
v_{cxy}	Average arm capacitor voltage per phase
v_{dc}	Effective DC link voltage
P	Three phase active power
Q	Three phase reactive power

Appendix A. Parameters of Modular Multilevel Converter

Table A1. Parameters of MMC Converters.

Parameters	Symbol	Value
Rated Apparent Power	S_s	65 kVA
Rated Output Voltage	v_s	400 V
Rated Output Current	i_s	141 A
Output Frequency	f_s	50 Hz
Maximum DC Voltage	v_{dc}	1.4 kV
SM Capacitance	C	20 mF
Arm Inductance	L_a	1.2 mH
Rated SM Voltage	v_{cx}	63.4 V
Maximum SM Voltage	$v_{cx(max)}$	75 V
Switching Frequency	f_{sw}	10,000 Hz
Number of SMs	N	19

Table A2. Controller parameter.

Power Reference Generation	$k_{p_{dc}} = 40$ W/V $\tau_{i_{dc}} = 10$ s $k_{p_x} = 100$ W/V $\tau_{i_x} = 333$ ms
Output Current Controller	$k_{p_\alpha} = 2$ V/A $k_{i_\alpha} = 209$ V/A $k_{p_\beta} = 2$ V/A $k_{i_\beta} = 209$ V/A
Circulating Current Controller	$k_{p_c} = 0.04$ V/A $k_{i_c} = 400$ V/A $k_{p_d} = 40$ V/A $\tau_{i_d} = 4$ s

Appendix B. Simulation Results of MMC Based PV Plant at STC

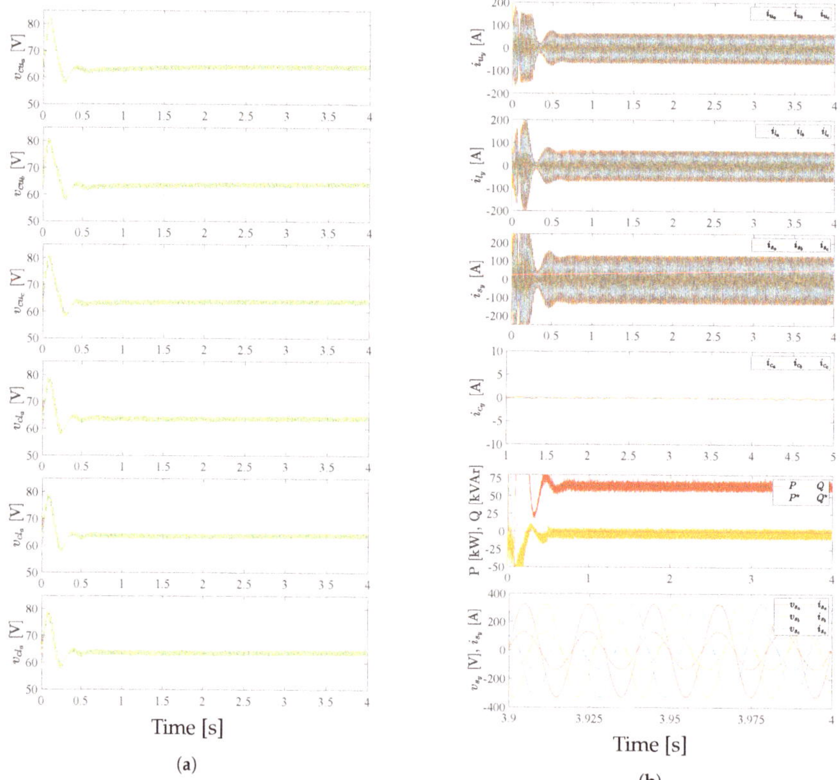

Figure A1. Simulation results for scenario 1 (**a**) Capacitor voltages for all the sub-modules in an arm of the MMC for all three phases. From the top, upper arm phase "a", upper arm phase "b", upper arm phase "c", lower arm phase "a", lower arm phase "b", and lower arm phase "c", respectively. (**b**) The upper and lower arm currents (i_{u_y} [A], i_{l_y} [A]), output currents (i_{s_y} [A]), circulating currents (i_{c_y} [A] $\forall$ y = a, b, c), the active and reactive power injected to the grid (P [kW], Q [$kVAr$]) and the plot shown the voltages (v_{s_y} [V] $\forall$ y = a, b, c) at PCC and the phase currents (i_{s_y} [A]) for 100 ms duration between 4.9 s to 5 s, $\forall$ y = a, b, c.

The simulation results under STC is presented. Figure A1 shows the sub-module capacitor voltages in each arm of the MMC. The voltage is retained at their respective MPP voltage at STC.

Figure A1 shows the arm currents (i_{u_y} [A], i_{l_y} [A]), the output phase currents (i_{s_y} [A]), the insertion indexes (n_{u_y}) for the upper arm in all the phases, the active and reactive power (P [kW], Q [$kVAr$]), along with the reference values and the voltages at PCC (v_{s_y} [V]) and output currents (i_{s_y} [A]) for a 100 ms duration between 3.9 s to 4 s.

References

1. Bogdanov, D.; Farfan, J.; Sadovskaia, K.; Aghahosseini, A.; Child, M.; Gulagi, A.; Oyewo, A.S.; de Souza Noel Simas Barbosa, L.; Breyer, C. Radical Transformation Pathway towards Sustainable Electricity via Evolutionary Steps. *Nat. Commun.* **2019**, *10*, 1–16. [CrossRef]
2. Carrasco, J.M.; Franquelo, L.G.; Bialasiewicz, J.T.; Galván, E.; Portillo Guisado, R.C.; Prats, M.N.M.; León, J.I.; Moreno-Alfonso, N. Power-Electronic Systems for the Grid Integration of Renewable Energy Sources: A Survey. *IEEE Trans. Ind. Electron.* **2006**, *53*, 1002–1016. [CrossRef]
3. Rivera, S.; Wu, B.; Lizana, R.; Kouro, S.; Perez, M.; Rodriguez, J. Modular Multilevel Converter for Large-Scale Multistring Photovoltaic Energy Conversion System. In Proceedings of the 2013 IEEE Energy Conversion Congress and Exposition, Denver, CO, USA, 15–19 September 2013; IEEE: Piscataway, NJ, USA, 2013; pp. 1941–1946. [CrossRef]
4. Zhao, W.; Choi, H.; Konstantinou, G.; Ciobotaru, M.; Agelidis, V.G. Cascaded H-Bridge Multilevel Converter for Large-Scale PV Grid-Integration with Isolated DC-DC Stage. In Proceedings of the 2012 3rd IEEE International Symposium on Power Electronics for Distributed Generation Systems (PEDG), Aalborg, Denmark, 25–28 June 2012; pp. 849–856. [CrossRef]
5. Lesnicar, A.; Marquardt, R. An Innovative Modular Multilevel Converter Topology Suitable for a Wide Power Range. In Proceedings of the 2003 IEEE Bologna Power Tech Conference Proceedings, Bologna, Italy, 23–26 June 2003; Volume 3, p. 6. [CrossRef]
6. Akagi, H. Classification, Terminology, and Application of the Modular Multilevel Cascade Converter (MMCC). In Proceedings of the 2010 International Power Electronics Conference-ECCE ASIA, Jeju, Korea, 29 May–2 June 2011; pp. 508–515. [CrossRef]
7. Dekka, A.; Wu, B.; Fuentes, R.L.; Perez, M.; Zargari, N.R. Evolution of Topologies, Modeling, Control Schemes, and Applications of Modular Multilevel Converters. *IEEE J. Emerg. Sel. Top. Power Electron.* **2017**, *5*, 1631–1656. [CrossRef]
8. Hakimi, S.M.; Hajizadeh, A. Integration of Photovoltaic Power Units to Power Distribution System through Modular Multilevel Converter. *Energies* **2018**, *11*, 2753. [CrossRef]
9. Nademi, H.; Das, A.; Burgos, R.; Norum, L. A New Circuit Performance of Modular Multilevel Inverter Suitable for Photovoltaic Conversion Plants. *IEEE J. Emerg. Sel. Top. Power Electron.* **2015**, *4*, 393–404. [CrossRef]
10. Sharifabadi, K.; Harnefors, L.; Nee, H.P.; Norrga, S.; Teodorescu, R. *Design, Control and Application of Modular Multilevel Converters for HVDC Transmission Systems*; Wiley-IEEE Press: Hoboken, NJ, USA, 2016.
11. Nami, A.; Liang, J.; Dijkhuizen, F.; Demetriades, G.D. Modular Multilevel Converters for HVDC Applications: Review on Converter Cells and Functionalities. *IEEE Trans. Power Electron.* **2014**, *30*, 18–36. [CrossRef]
12. Stringfellow, J.D.; Summers, T.J.; Betz, R.E. Control of the Modular Multilevel Converter as a Photovoltaic Interface under Unbalanced Irradiance Conditions with MPPT of Each PV Array. In Proceedings of the 2016 IEEE 2nd Annual Southern Power Electronics Conference (SPEC), Auckland, New Zealand, 5–8 December 2016; IEEE: Piscataway, NJ, USA, 2016; pp. 1–6. [CrossRef]
13. Acharya, A.B.; Ricco, M.; Sera, D.; Teodorescu, R.; Norum, L.E. Arm Power Control of the Modular Multilevel Converter in Photovoltaic Applications. *Energies* **2019**, *12*, 1620. [CrossRef]
14. Peftitsis, D.; Tolstoy, G.; Antonopoulos, A.; Rabkowski, J.; Lim, J.K.; Bakowski, M.; Ängquist, L.; Nee, H.P. High-Power Modular Multilevel Converters With SiC JFETs. *IEEE Trans. Power Electron.* **2011**, *27*, 28–36. [CrossRef]
15. Acharya, A.B.; Ricco, M.; Sera, D.; Teoderscu, R.; Norum, L.E. Performance Analysis of Medium-Voltage Grid Integration of PV Plant Using Modular Multilevel Converter. *IEEE Trans. Energy Convers.* **2019**, *34*, 1731–1740. [CrossRef]
16. Hagiwara, M.; Akagi, H. Control and Experiment of Pulsewidth-Modulated Modular Multilevel Converters. *IEEE Trans. Power Electron.* **2009**, *24*, 1737–1746. [CrossRef]

17. Yang, S.; Tang, Y.; Wang, P. Distributed Control for a Modular Multilevel Converter. *IEEE Trans. Power Electron.* **2017**, *33*, 5578–5591. [CrossRef]
18. Münch, P.; Görges, D.; Izák, M.; Liu, S. Integrated Current Control, Energy Control and Energy Balancing of Modular Multilevel Converters. In Proceedings of the IECON 2010-36th Annual Conference on IEEE Industrial Electronics Society, Glendale, AZ, USA, 7–10 November 2010; pp. 150–155. [CrossRef]
19. Soong, T.; Lehn, P.W. Internal Power Flow of a Modular Multilevel Converter With Distributed Energy Resources. *IEEE J. Emerg. Sel. Top. Power Electron.* **2014**, *2*, 1127–1138. [CrossRef]
20. Acharya, A.B.; Norum, L.E.; Sera, D. A Shadow Tolerant Configuration for PV Integration to Grid Using Modular Multilevel Converter. In Proceedings of the 2018 IEEE International Conference on Power Electronics, Drives and Energy Systems (PEDES), Chennai, India, 18–21 December 2018; pp. 1–6. [CrossRef]
21. IEEE. *IEEE Recommended Practice and Requirements for Harmonic Control in Electric Power Systems*; IEEE: New York, NY, USA, 2014; pp. 1–29. [CrossRef]
22. IEEE. *IEEE Standard for Interconnection and Interoperability of Distributed Energy Resources with Associated Electric Power Systems Interfaces*; IEEE: New York, NY, USA, 2018; pp. 1–138. [CrossRef]

Article

Energy Storage for 1500 V Photovoltaic Systems: A Comparative Reliability Analysis of DC- and AC-Coupling

Jinkui He [1], Yongheng Yang [1,*] and Dmitri Vinnikov [2,*]

[1] Department of Energy Technology, Aalborg University, 9220 Aalborg, Denmark; jhe@et.aau.dk

[2] Department of Electrical Power Engineering and Mechatronics, Tallinn University of Technology, 19086 Tallinn, Estonia

* Correspondence: yoy@et.aau.dk (Y.Y.); dmitri.vinnikov@taltech.ee (D.V.); Tel.: +45-9940-9766 (Y.Y.)

Received: 28 May 2020; Accepted: 19 June 2020; Published: 1 July 2020

Abstract: There is an increasing demand in integrating energy storage with photovoltaic (PV) systems to provide more smoothed power and enhance the grid-friendliness of solar PV systems. To integrate battery energy storage systems (BESS) to an utility-scale 1500 V PV system, one of the key design considerations is the basic architecture selection between DC- and AC-coupling. Hence, it is necessary to assess the reliability of the power conversion units, which are not only the key system components, but also represent the most reliability-critical parts, in order to ensure an efficient and reliable 1500 V PV-battery system. Thus, this paper investigates the BESS solutions of DC- and AC-coupled configurations for 1500 V PV systems with a comparative reliability analysis. The reliability analysis is carried out through a case study on a 160 kW/1500 V PV-system integrated DC- or AC-coupled BESS for PV power smoothing and ramp-rate regulation. In the analysis, all of the DC-DC and DC-AC power interfacing converters are taken into consideration along with component-, converter-, and system-level reliability evaluation. The results reveal that the reliability of the 1500 V PV inverter can be enhanced with the DC-coupled BESS, while seen from the system-level reliability (i.e., a PV-battery system), both of the DC- and AC-coupled BESSs will affect the overall system reliability, especially for the DC-coupled case. The findings can be added into the design phase of 1500 V PV systems in a way to further lower the cost of energy.

Keywords: energy storage; 1500 V photovoltaic (PV); reliability; cost-oriented design

---

1. Introduction

Solar energy installations have experienced rapid growth in the last decade [1], which brings both environmental and economic benefits. One of the driving forces behind this growth is that the solar industry keeps seeing innovation for both reducing the system cost and increasing the grid-integration performance [2]. The mainstream DC voltage has been increasing from 600 V or 1000 V (low voltage in relevant standards [3,4]) to 1500 V, which is the maximum voltage of the low voltage directive according to the IEC standards in order to reduce the cost of large-scale photovoltaic (PV) systems. By doing so, for a given capacity, the installation cost can be reduced to a large extent (fewer strings, connections, and less cabling) [5]. On the other hand, the high variability of the solar PV energy (due to the weather conditions) raises challenges for integrating these PV systems to the grid. In such a case, energy storage could be integrated to the PV systems for smoothing the output power of PV plants. Recently, this has also become a promising solution toward smart PV systems [6].

For 1500 V PV applications, several studies have shown that single-stage conversion PV systems (without DC-DC stage) outperform two-stage based PV systems with respect to size, efficiency, and cost-effectiveness [7–9]. For the same reason, the battery energy storage systems (BESS) for large-scale PV Plants are also based on single-stage conversion [10,11], i.e., one bi-directional DC-DC or DC-AC stage depending on the type of connection: DC-coupling or AC coupling. As illustrated in Figure 1a, in DC-coupling, the output of the BESS is connected to the DC side of the PV inverter, while, in AC-coupling, as shown in Figure 1b, the BESS is added to the PV system at its AC side. Both of the configurations have the potential to improve the grid-integration performance of 1500 V PV systems. Flexible power management is significantly enhanced in such systems.

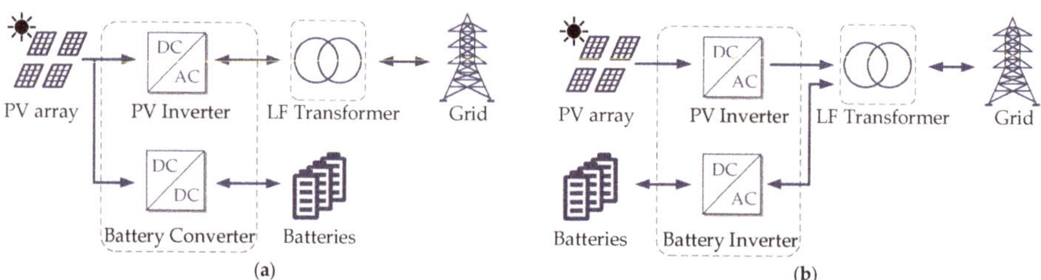

Figure 1. System diagram of the single-stage 1500 V PV system with integrated battery energy storage systems (LF: low-frequency transformer): (**a**) DC-coupled configuration and (**b**) AC-coupled configuration.

Many research efforts have been devoted to address the design and control of PV-battery systems. In [12,13], the common methods for power smoothing and ramp-rate reduction with BESS were compared in terms of power tracking performance and BESS capacity requirement, respectively. In [14], the coordinated control of a single-stage based PV system with a DC-coupled BESS was analyzed along with energy management. The methods to determine the optimal sizing of BESSs were developed in [15], where second-life Li-ion batteries were considered for a cost-effectiveness analysis. The benefits of the DC-coupled BESS for a large-scale PV plant were investigated in [16], which shows a higher efficiency and less energy losses (with oversized PV arrays) when compared to an AC-coupled configuration can be achieved. Moreover, the benefits of the AC coupling over DC coupling were investigated in [17], in terms of reduced integration challenges and increased design flexibility.

However, the prior-art studies did not fully cover the discussions with respect to the lifetime and reliability of the PV-battery systems. As the loading on the converters in the two configurations is different, the reliability performance will also vary, which, in turn, may affect the final design. When considering that, recently, the reliability analysis of the power converters in PV-battery systems has attracted increasing interest, such as those presented in [18,19]. More specifically, a single-phase DC-coupled PV-battery system was considered in [18], where the impact of different self-consumption control strategies on the PV inverter reliability was analyzed. This work was further strengthened in [19] by considering the reliability analysis of the remaining power converters (i.e., PV boost converter and battery converter), in which the system-level reliability was also investigated when considering both DC-coupled and AC-coupled BESS configurations. However, the above investigation was performed for residential PV systems. When it comes to large-scale PV applications, e.g., 1500 V PV-battery systems, different insights in reliability analysis may be offered, which, in turn, can provide further design considerations to enhance the system performance. Nevertheless, such an analysis has not been thoroughly and systematically discussed in the literature. Thus, it is necessary to explore the power converter reliability for large-scale PV-battery systems, following which, a proper design of these power converters can possibly be achieved.

With the above concerns, this paper investigates the BESS of DC- and AC-coupling for 1500 V PV systems with emphasis on the reliability comparison. The analysis is carried out through a case study instead of a new methodology preposition on a 160 kW/1500 V PV system, in which proper BESSs are designed for the DC- and AC-coupled configurations to smooth the PV power and limit the power ramp rate. The rest of this paper is organized, as follows. In Section 2, the system modeling is presented, which includes the mission profiles and basic components for the system under study.

In addition, the general control to enable the power smoothing and limit the power ramp rate is briefed. Subsequently, in Section 3, the reliability analysis is presented. First, a component-level reliability performance of all the power devices (i.e., IGBTs and diodes) within the two configurations is evaluated by estimating their lifetime under a real mission profile. Afterwards, the converter- and system-level reliability assessments based on the reliability block diagrams along with the Monte-Carlo simulations are carried out. Through this comparative reliability analysis, the most fragile part within the two systems can be identified; additionally, the overall system reliability can be improved by selecting an adequate BESS connection type, which is shown in Section 4. Following, the reliability benchmarking results are discussed further in Section 5. Finally, concluding remarks are given in Section 6.

2. System Modeling

In this paper, a single-stage 1500 V PV system is considered, as shown in Figure 1, where a BESS is deployed under the DC- or AC-coupled configuration for power smoothing. The reliability analysis that is given in this paper focuses on the power semiconductor devices. The power flow of the system can then be modeled, as shown in Figure 2, where T_a and SI are mission profile parameters, i.e., the ambient temperature and the solar irradiance, respectively; P_{pv} and V_{pv} respectively denote the power and voltage of the PV arrays; P_{inv_in} and P_{inv_out} represent the input and output power of the PV inverters; P_{bat_dc} and P_{bat_ac} are the input power of the battery converter and the battery inverter; V_g, V_{bat_dc}, and V_{bat_ac} represent the grid voltage, the battery voltage for the DC coupling, and the battery voltage for the AC coupling, correspondingly; and, T_j is the junction temperature of the power devices. The overall inputs are the mission profiles of the installation site that represent the system operating conditions [20] and the outputs are the thermal stress profiles of the power semiconductors in PV/battery converters for the next reliability analysis, as it can be observed in Figure 2. A description of each model for this mission profile translation is given in the following sections.

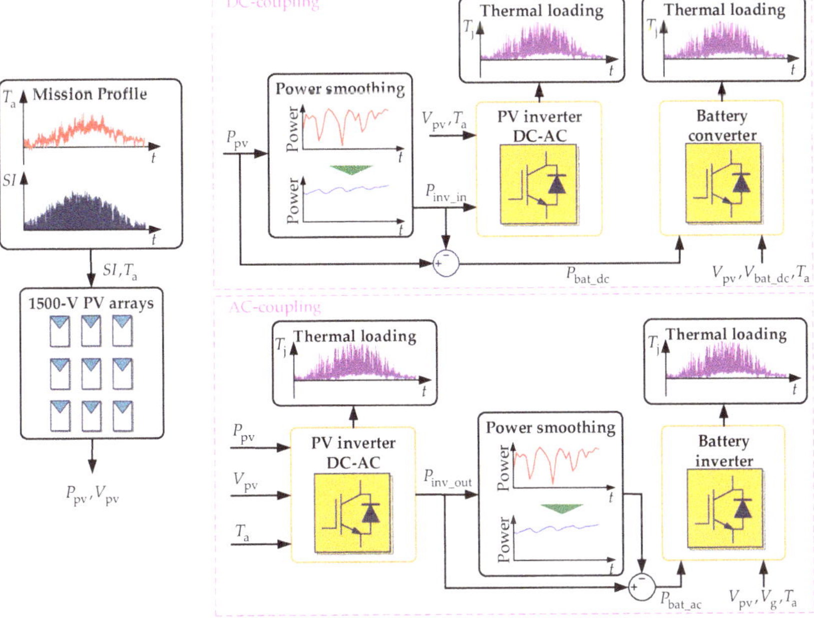

Figure 2. Schematic of the system modeling for the DC- and AC- coupling 1500 V PV systems.

2.1. Mission Profile

A one-year mission profile that is provided in [21] is used in this paper, which was recorded in the year of 2019 in Denmark with a sampling rate of 1 min./sample. It should be mentioned that the mission profile sampling rate could affect the reliability prediction, as discussed in [22], where a high sampling rate (e.g., 1 s to 1 min. per sample) is recommended to obtain as much information as possible in the mission profile translation process. In contrast, a lower sampling rate (e.g., 5 min. per sample) may give rise to a certain degree of uncertainty in the lifetime results. The mission profile is shown in Figure 3. As it can be observed in Figure 3, the solar irradiance and ambient temperature both vary in a wide range. Moreover, the frequent presence of clouds will inevitably affect the PV output power, where a BESS is highly expected to smooth the power fluctuations and limit the power ramp rate. This is to guarantee the system security and stability.

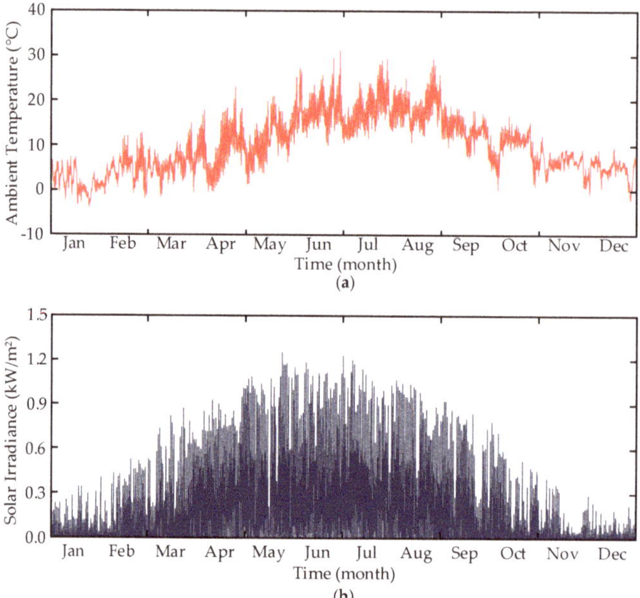

Figure 3. One-year mission profile: (**a**) ambient temperature and (**b**) solar irradiance.

2.2. PV Array and PV Inverter

In this study, it is assumed that the PV-battery system is rated at 160 kW. The JKM380M-72-V solar panel is selected to assemble the 1500 V PV arrays [23]. In this case, 432 solar panels (27 panels per string, 16 strings) are used to achieve the rated power around 160 kW and the maximum open-circuit voltage up to 1500 V. Subsequently, a three-level I-type inverter is employed for interfacing the PV arrays to the AC grid, as shown in Figure 4, where the overall control algorithms are also illustrated. It should be pointed out that when integrating a DC-coupled BESS to this PV system, an appropriate coordinated control of the PV system and the BESS should be considered [14]. The system specifications are given in Table 1. Regarding the power semiconductor devices, three 1200 V/300 A IGBT modules from Semikron are adopted [24] and, correspondingly, their heatsink sizing is designed to guarantee the maximum junction temperature below 125 °C during the rated operation with the ambient temperature being 50 °C.

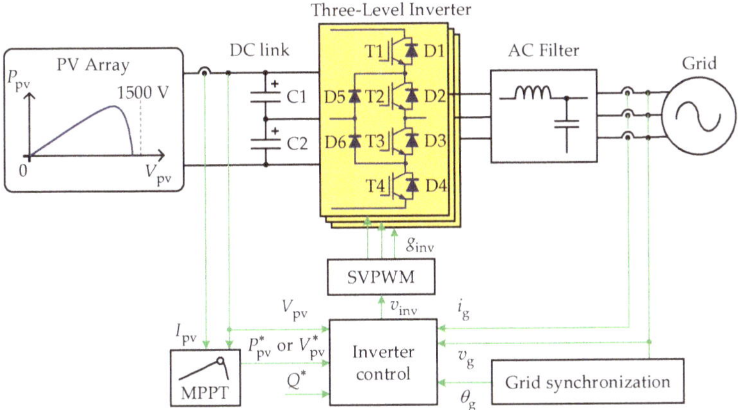

Figure 4. General configuration and control structure of the 1500 V PV system based on the three-level I-type topology: P_{pv}—PV power, V_{pv}—PV voltage, I_{pv}—PV current, P_{pv}^*—active power reference, V_{pv}^*—DC-link voltage reference, Q^*—reactive power reference, i_g—grid current, v_g—grid voltage, θ_g— phase angle of the grid voltage, v_{inv}—output voltage of the inverter, g_{inv}—gate signals, MPPT—maximum power point tracking, and SVPWM—space vector pulse width modulation.

Table 1. PV System Specifications.

Parameter	Value
Nominal power P_{nom}	160 kW
Power factor $\cos(\varphi)$	1.0
Grid line-to-line RMS [1] voltage V_{LL}	550 V
Grid frequency f_g	50 Hz
Switching frequency f_{sw}	5 kHz
Device type	SEMiX305MLI12E4 [24]
Heatsink thermal impedance per module $R_{th(s-a)}$	0.085 K/W

[1] Root Mean Square.

2.3. Power Smoothing Operation

The target of integrating battery energy storage systems for a PV system is to comply with certain requirements (decided by the transmission system operator) when injecting power into the grid. This paper focuses on the power smoothing and ramp-rate control with the consideration of their impact on the reliability of power semiconductor devices within the PV-battery system. Similar to the previous studies [13,15], in this work, it is assumed that the ramp-rate limit is 10% of the rated power per minute, which is defined by the Puerto Rico Electric Power Authority (PREPA) [25]. Subsequently, the ramp-rate compliant algorithm in [15] is applied to smooth the system output power, size the battery converter/inverter, and obtain the mission profile of the integrated battery system. The power smoothing results for the DC-coupled PV-battery system are shown in Figure 5. The smoothed power is less fluctuating when compared with the PV power before smoothing, as seen in Figure 5a. The difference, as shown in Figure 5b, is the power reference for the BESS. As seen in Figure 5b, in most days, the absorbed/injected power is within 100 kW. Following, Figure 6 presents the ramp rate results of the PV system with and without a 100 kW BESS. As observed in Figure 6, without the BESS, the output power of the PV system can experience even 80% rises or drops in a very short period (e.g., 1 min.), which may affect the overall supply and demand power balance, causing grid stability issues. In contrast, with the 100 kW BESS, the corresponding ramp rate is mainly distributed within the limit (i.e., 10% of the rated power per minute).

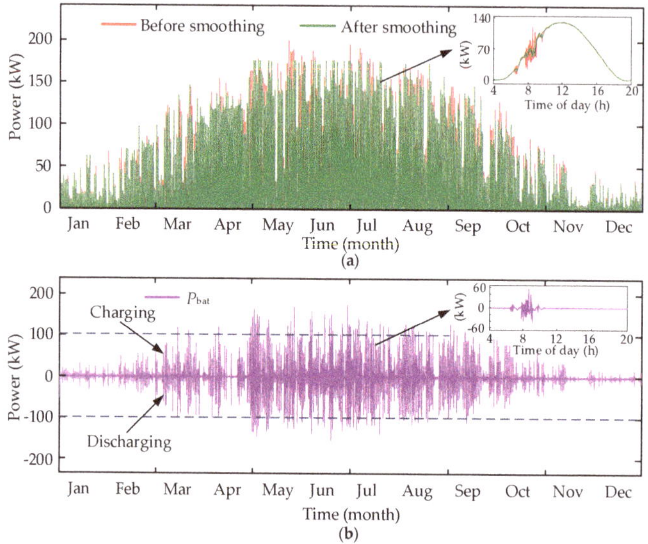

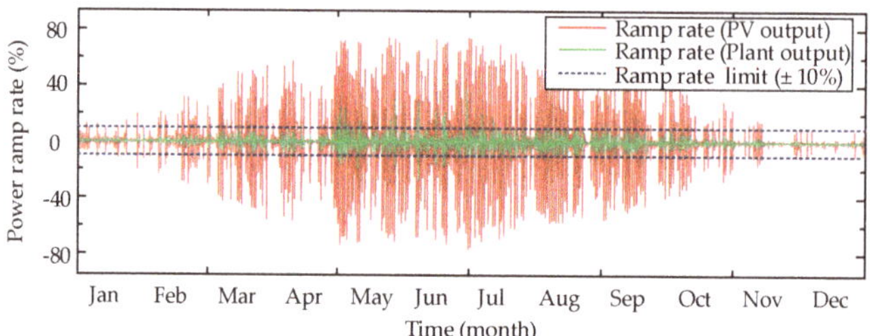

Figure 5. Power smoothing analysis: (**a**) input power of the PV inverter with and without power smoothing and (**b**) power reference for the DC-coupled BESS.

Figure 6. One-year power ramp-rate analysis: the ramp rate of PV output and the ramp rate of PV-BESS (plant) output when energy storage system is deployed for power smoothing.

2.4. Battery Energy Storage System

The reliability analysis that is given in this paper considers two typical energy storage systems shown in Figure 7. For the DC-coupled configuration, as shown in Figure 7a, a simple half bridge-based DC/DC topology is chosen to build an interleaved converter with three stages, interfacing the batteries to the DC side of the 1500 V PV system. While for the AC-coupled configuration, the same topology as for the PV inverter is employed as the interfacing unit between batteries and the point of common coupling (PCC), as shown in Figure 7b. Table 2 presents the designed ratings of the battery systems. Regarding the power modules, three 1200 V/200 A IGBT modules and three 1700 V/150 A IGBT modules from Semikorn are respectively used for the battery inverter and converter shown in Figure 7 [24]. Moreover, the corresponding heatsink parameter, heatsink-to-ambient thermal impedance per module $R_{\mathrm{th(s-a)}}$, is designed to ensure the junction temperature of the most stressed power devices is below 125 °C during the rated operation with the ambient temperature being 50 °C.

For the battery model, the two BESSs that are shown in Figure 7 are equipped with the same type of batteries as a commercial BESS for 1500 V PV applications [26], while the operating voltage of the two BESSs are different. In the case of the DC-coupled configuration, the voltage of the PV system at the maximum power point (MPP) varies between 1000 V and 1300 V under the Denmark mission profile (see Figure 3). The battery voltage range is 670 V to 870 V, which is lower than the minimum MPP voltage to ensure the converter charges the batteries in the buck mode, and in the boost mode, the batteries are discharged. On the other hand, for the AC-coupled configuration, the battery operating voltage is 860 V to 1120 V, ensuring the battery can be discharged when considering a 10% variation of the grid voltage. For the sake of simplicity, it is assumed that the battery voltage is at the upper and lower limit during the charging and discharging mode, respectively. Detailed sizing of the battery storage is out of the scope of this paper and, thus, the capacity of the battery systems is assumed to be sufficient during operation. Table 2 summarizes the battery specifications.

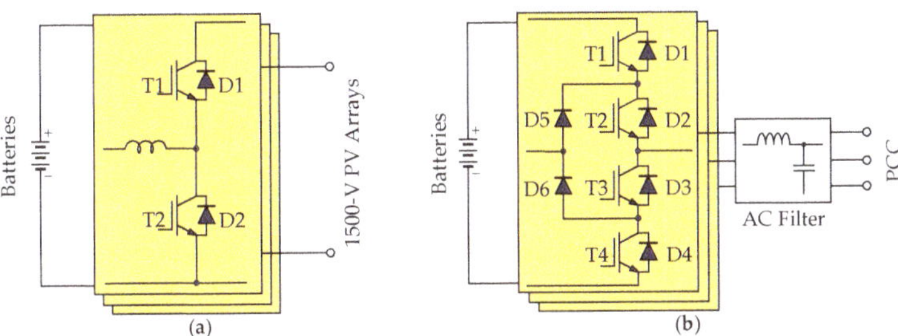

Figure 7. Battery energy storage system: (**a**) for the DC-coupled configuration and (**b**) for the AC-coupled configuration (PCC: the point of common coupling).

Table 2. BESS Specifications.

Battery Converter	
Nominal power P_{nom}	100 kW
Switching frequency f_{sw}	5 kHz
Heatsink thermal impedance per module $R_{th(s-a)}$	0.131 K/W
Device type	SKM150GB17E4G
Battery Inverter	
Nominal power P_{nom}	100 kW
Switching frequency f_{sw}	5 kHz
Heatsink thermal impedance per module $R_{th(s-a)}$	0.148 K/W
Device type	SEMiX205MLI12E4
Battery	
Battery type	Samsung SDI, 3.68 V/94 Ah
Nominal cell voltage/Operating voltage P_{nom}	3.68 V/3.2–4.15 V
Number of cells per rack for batteryconverter (operating voltage)	210 (670–870 V)
Number of cells per rack for battery inverter (operating voltage)	270 (860–1120 V)

2.5. Mission Profile Translation

With the above operation conditions, the thermal loading of all the power semiconductor devices (i.e., the IGBTs and diodes in the PV-battery systems) during one-year operation is investigated. Based on the input power of the PV and battery converters, the thermal loading of the power devices is determined from the loss and thermal models of the components, and this translation is normally achieved through two-dimensional (2-D) look-up tables to process the long-term simulations. More detailed analysis and steps regarding the mission profile translation have been discussed in [27].

Figure 8 presents the resultant junction temperature of the IGBT T1 (see Figure 4) in the PV inverter integrated with the AC- or DC-coupled BESS. It can be seen in Figure 8 that the DC-coupled BESS can considerably reduce the thermal stress of the IGBT T1 during the power smoothing process. This is also true for the remaining devices (e.g., IGBT T2, diode D1, D2, and D5), which is not presented here, as their temperature profiles are similar to the presented one, and they will be considered in the lifetime and reliability analysis in the following sections. Figure 9 shows the temperature profiles of the power devices within the AC- and DC-coupled BESS converters. The power devices in the battery inverter have a similar temperature profile, as shown in Figure 9a, while they present a significant difference, as shown in Figure 9b, for the battery converter. The variety of these temperature profiles will cause cyclic thermo-mechanical stresses among the different materials inside the power devices, which finally lead to cumulative bond-wire and solder-joint fatigue after a certain number of cycles. The evaluation of the damage will be discussed in the next section. Nevertheless, as seen from the translated thermal loading profiles on the power converters, the coupling configurations will affect the reliability performance of the power converters and, thus, the entire PV-battery system.

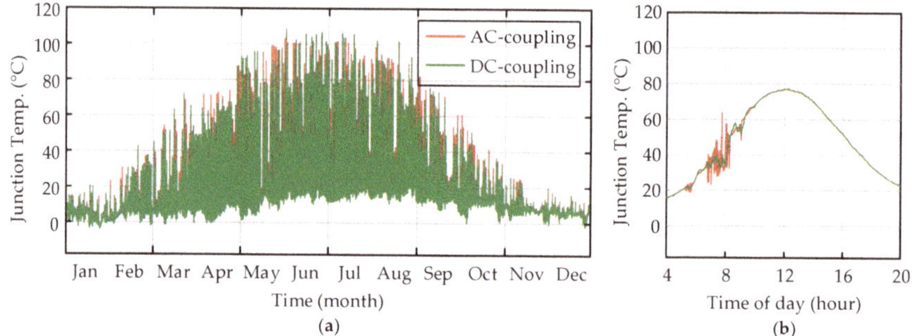

Figure 8. Junction temperature of the IGBT T1 (see Figure 4) in the PV inverter integrated with the AC- or DC-coupled BESS: (**a**) one-year operation and (**b**) one-day operation.

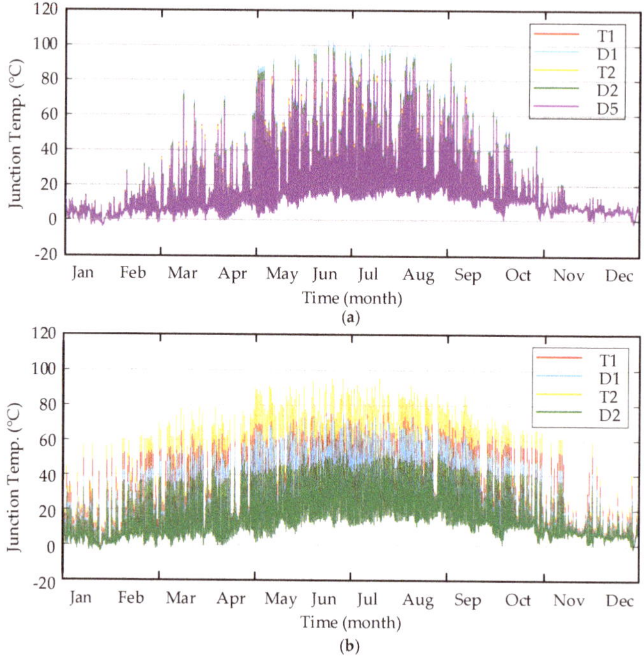

Figure 9. Junction temperature of the power devices within the battery converters (referring to Figure 7): (**a**) in the battery inverter and (**b**) in the battery converter.

3. Component Reliability Analysis

Normally, the power device reliability is expressed in terms of lifetime consumption (*LC*), which indicates how much lifetime has been consumed since the beginning of th operation. In this paper, the *LC* is obtained according to the Miner's rule [28], which is expressed as

$$LC = \sum_i \frac{n_i}{(N_f)_i} \tag{1}$$

where n_i is the number of cycles under certain stress conditions and $(N_f)_i$ is the corresponding number of cycles to failure with the same stress conditions that is dependent on the lifetime model. Notably, the *LC* calculation according to (1) assumes that various stress cycle events are independent, and the caused damage can be linearly accumulated. The device is considered to reach its end of life when the *LC* exceeds one (or 100%).

In this paper, the number of cycles to failure N_f is evaluated based on the Bayerer model [29]. This model is an empirical model describing N_f in relation to certain stress conditions (i.e., the minimum junction temperature $T_{j(min)}$ and cycle amplitude ΔT_j), which is expressed as

$$N_f = A \cdot (\Delta T_j)^{-\beta_1} \cdot \exp\left(\frac{\beta_2}{T_{j(min)} + 273}\right) \cdot t_{on}^{\beta_3} \cdot I^{\beta_4} \cdot V^{\beta_5} \cdot D^{\beta_6} \tag{2}$$

in which the impact of the heating time t_{on}, current per wire bond I, blocking voltage V, and bond wire diameter D are also considered according to individual power laws. The model parameters and coefficients are summarized in Table 3. More discussions on this lifetime model are provided in [29].

Table 3. Parameters and coefficient of the Bayerer Model [29].

Parameter and Coefficient	Value/Test Condition
Technology factor A	9.34×10^{14}
Minimum junction temperature $T_{j(min)}$	$20\,°C \leq T_{j(min)} \leq 120\,°C$
Cycle amplitude ΔT_j	$45\,K \leq \Delta T_j \leq 150\,K$
Heating time t_{on}	$1\,s \leq t_{on} \leq 15\,s$
Current per bond wire I	3 A to 23 A
Voltage class/100 V	6 V to 33 V
Bond wire diameter D	75 μm to 500 μm
Coefficient β_1–β_6	{−4.416, 1285, −0.463, −0.716, −0.761, −0.5}

Additionally, the number of cycles n_i in (1) is not directly available, since the temperature profiles obtained in the previous section, as shown in Figure 8; Figure 9, are irregular. In order to obtain the number of cycles n_i at a certain stress condition (i.e., the cycle amplitude, mean junction temperature, and cycle period t_{cyc}), a rainflow counting analysis [30] is performed, categorizing the irregular thermal cycles into several regular cycles. Subsequently, the number of cycles n_i can be obtained.

Applying the above calculation, the one-year *LC* of all the IGBTs and diodes in the PV-battery systems can be obtained. Figure 10 summarizes the results. For the PV inverters, as expected, the DC-coupled configuration alleviates the loading on the PV inverter. It can be seen in Figure 10a,b that the IGBTs and diodes in the PV inverter with the DC-coupled BESS have much lower *LC* than the corresponding ones in the PV inverter with the AC-coupled BESS. Additionally, in both cases, the clamping diodes (e.g., D5 referring to Figure 4) are the most stressed devices (i.e., with the highest *LC*). This can be considered when designing the power converters, e.g., to select a high reliability diode. Regarding the power converters for the batteries, as can be seen in Figure 10c,d, several power devices in both converters have higher *LC* when compared with the devices in the PV inverters. For both the DC- and AC-coupled configurations, it can be expected that the converter-level reliability of the battery converters will be lower than that of the PV inverters. Although the battery inverter consumes less life under the mission profile, it has more power components, when compared to the battery converter. The overall lifetime may be different.

So far, the component reliability of all the power interfacing converters within the DC- and AC-coupled PV-battery systems has been analyzed, with which the most fragile component can be identified. Furthermore, the corresponding *LC* results provide the basis for the next converter- and system-level reliability analysis.

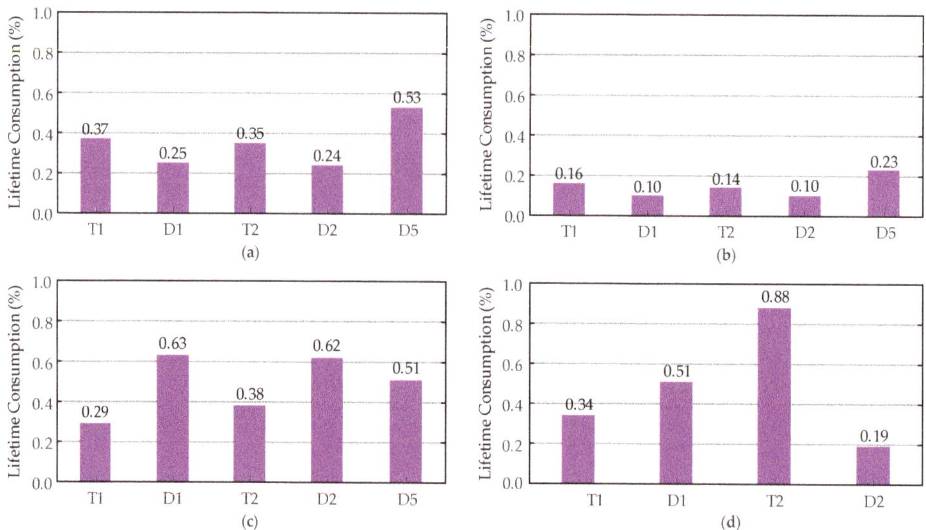

Figure 10. One-year *LC* results of the power devices in different power converters: (**a**) PV inverter in the AC-coupled configuration, (**b**) PV inverter in the DC-coupled configuration, (**c**) battery inverter in the AC-coupled configuration, and (**d**) battery converter in the DC-coupled configuration.

4. System-Level Reliability Benchmarking

With the static damage obtained in the previous section, the lifetime of the power devices can be obtained as certain fixed values. This is far from reality, since the power device lifetime could present variations due to the uncertainties in device parameters and experienced stresses. Therefore, the lifetime prediction should consider these uncertainties and, thus, provide statistical lifetime values. In this section, a statistical approach that is based on the Monte-Carlo analysis is applied [31], in which the variations of the model parameters in (2) and the thermal stresses are introduced with 5% variations to represent the uncertainties. Notably, to assist the analysis, the dynamic stress parameters (i.e., $T_{j(min)}$, ΔT_j, and t_{on}) are normally converted into equivalent static ones (i.e., $T'_{j(min)}$, $\Delta T'_j$, and t'_{on}), which can produce the same one-year *LC* when applying them to the *LC* calculation process [32]. By doing so, the system-level reliability can be predicted.

The Monte-Carlo simulations are conducted when considering a population of 10,000 samples, following which the obtained lifetime data for a certain device are fitted with the Weibull distribution as [33]

$$f(x) = \frac{\beta}{\eta}\left(\frac{x}{\eta}\right)^{\beta-1} e^{-\left(\frac{x}{\eta}\right)^\beta}, \quad F(x) = 1 - e^{-\left(\frac{x}{\eta}\right)^\beta} \tag{3}$$

where $f(x)$ and $F(x)$, respectively, represent the probability density function (PDF) of the Weibull distribution and the cumulative density function (CDF, also referred to as the unreliability function) with x, η, and β being the operation time, the scale parameter, and the shape parameter, respectively.

Subsequently, the reliability assessment of the PV-battery system follows the steps from the component level, the converter level, to the system level, and the corresponding lifetime values are obtained in terms of B_{10} lifetime, which represents the total operation time when 10% of the populations will fail. For the component-level reliability analysis, the B_{10} lifetime of each power device can be obtained from the corresponding $F(x)$ curve. While investigating the converter-level and system-level reliability, it can be performed by using the Reliability Block Diagram (RBD), which describes the reliability interaction between each device and subsystem in the entire system. Figure 11a shows the RBD of the considered PV-battery systems. For the converter-level RBD, if any of the IGBTs or diodes fails, it is considered that the converter cannot function. Thus, the series-connected RBD is considered for these power converters, as shown in Figure 11a. Subsequently, the unreliability function for the converters $F_{con}(x)$ can be calculated as

$$F_{con}(x) = 1 - \prod_{i=1}^{n}\left(1 - F_{comp(i)}(x)\right) \tag{4}$$

in which $F_{comp(i)}(x)$ represents the unreliability function of the ith device in the converters. Regarding the system-level RBD, as shown in Figure 11b, the series connection of the converter-level RBDs is considered for both the DC- and AC-coupled system configuration. Thus, the system unreliability is calculated as

$$F_{sys}(x) = 1 - (1 - F_{pv}(x))(1 - F_{bat}(x)) \tag{5}$$

where $F_{pv}(x)$ and $F_{bat}(x)$ represent the unreliability functions of the PV converter and the battery converter, respectively.

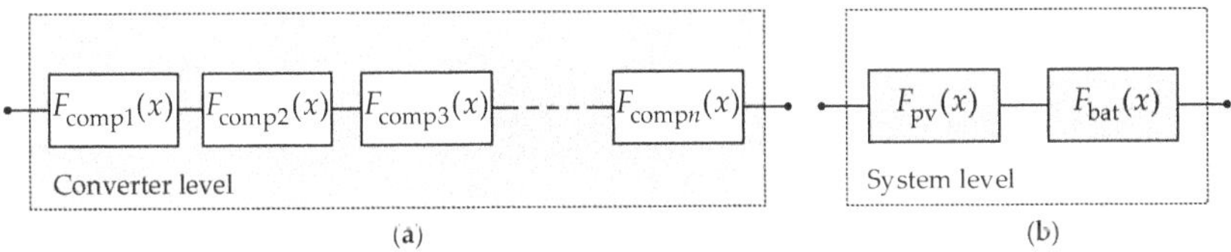

Figure 11. Series connection of the reliability block diagram: (**a**) converter level and (**b**) system level, where $F_{compi}(x)$ represents the unreliability function of the ith device in the converter, and $F_{pv}(x)$ and $F_{bat}(x)$ represent the unreliability functions of the PV and battery converters, respectively.

The converter-level unreliability functions of the power converters within the two considered PV-battery systems are shown in Figure 12, along with the corresponding component-level functions of their power devices. The B_{10} lifetime results are in accordance with the LC comparison in the previous section. For the PV inverters, as shown in Figure 12a,b, the reliability of the two inverters is dominated by the reliability of the clamping diodes (e.g., D5 referring to Figure 4) and, consequently, the B_{10} lifetime of the PV inverter with the DC-coupled BESS is 101.6 years, which is even higher than the twice of the B_{10} lifetime of the PV inverter with the AC-coupled BESS. Notably, the results also imply that the two inverters have excessive design margins for the considered Denmark mission profile. Regarding the BESS converters, for the battery inverter, as shown in Figure 12c, instead of the clamping diodes, the outer diodes (e.g., D1 referring to Figure 7b) will have the lowest B_{10} lifetime due to its bidirectional operation, and the inverter B_{10} lifetime is 77.1 years. As observed in Figure 12d, the reliability of the battery converter for the DC-coupled BESS is mainly affected by the lower IGBTs (e.g., T2 referring to Figure 7a) and its B_{10} lifetime is 68.8 years. Notably, both the battery inverter and the battery converter are less reliable than the corresponding PV inverters, which will seriously affect the overall reliability of the PV-battery systems.

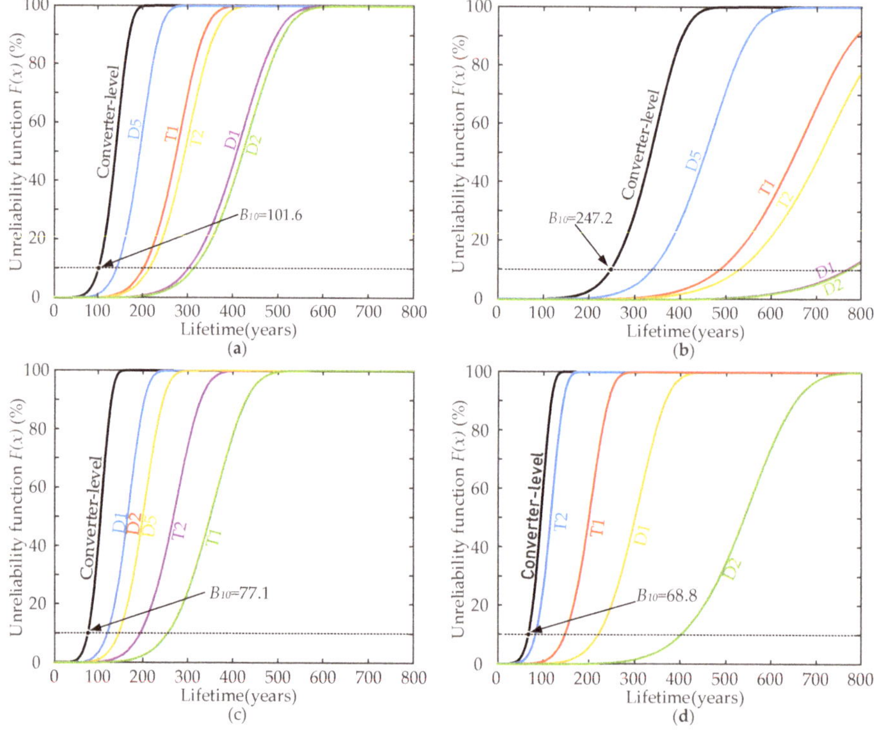

Figure 12. Unreliability functions of the different power converters: (**a**) PV inverter with the AC-coupled BESS, (**b**) PV inverter with the DC-coupled BESS, (**c**) battery inverter for the AC-coupled BESS, and (**d**) battery converter for the DC-coupled BESS.

Figure 13 (the solid lines) shows the system-level unreliability functions of the considered PV-battery systems, where the corresponding converter-level unreliability functions are also given (the dashed lines). The B_{10} lifetime of the DC-coupled configuration is 68.8 years, which is six years shorter than that of the AC-coupled configuration. As expected, both the reliability of the DC-and AC-coupled configuration are dominated by the reliability of their battery converters, especially for the DC-coupled case. This is expected, as the battery is used to balance and smooth the fluctuating power from the PV array. Although the integrated DC-coupled BESS can enhance the reliability of the PV inverter to a large extent, the much less reliable battery converter will limit the overall reliability performance. Hence, for the case study in this paper, the AC-coupled configuration is a better option, featuring a more balanced and higher reliability performance. It should be mentioned that the redundancy of the interleaved battery converter is not considered in the above comparison for comparing the two configurations under the full power smoothing capability. Obviously, it is expected that the reliability of the DC-coupled configuration can be improved with a proper redundancy design of the battery converter. For instance, adding one more paralleled bidirectional converter stage to the battery converter (see Figure 7a), which can achieve a three-out-of-four redundancy [34]. By doing so, the reliability of the battery converter will be improved considerably, as shown in Figure 13b, achieving higher system-level reliability than the AC-coupled configuration in Figure 13a, while its hardware cost might probably still less than the three-level battery inverter for the AC-coupled configuration.

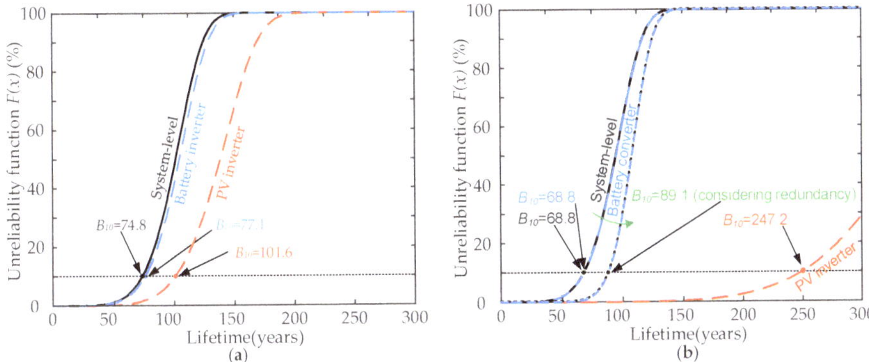

Figure 13. Unreliability function of the DC- and AC-coupled PV-battery system: (**a**) AC-coupled configuration and (**b**) DC-coupled configuration.

5. Discussion

The presented comparative reliability analysis has provided information on the differences of the two configurations for 1500 V PV systems in terms of reliability. This section discusses the possible reliability-affecting factors, as well as some opportunities that could be further explored for improving their reliability performance in practice.

From the case study, it is concluded that, as compared with the AC-coupled configuration, the DC-coupled configuration has a shorter lifetime period and lower reliability. This is in agreement with the hypothesis in Section 1, but this comparison result is different from the conclusion in [19]. A comparative reliability evaluation has been performed for residential PV-battery systems in [19], where it was shown that the DC-coupled configuration was more reliable. The main reason behind this difference is the increased DC voltage of the 1500 V PV system, which has a considerable impact on the lifetime and reliability of the battery converter for DC-coupling, as observed in the above analysis. In such a case, instead of using the bidirectional converter that is based on 1700-V half-bridge power modules, a three-level bidirectional converter might be a better option [35], which can be applied to this case with the series connection of two 1200-V half-bridge power modules. In all, it is indicated that the reliability performance is dependent on applications.

On the other hand, the reliability improvements could also be achieved through power devices advancements. For instance, the Silicon Carbide (SiC) devices offer superior features, like low switching losses, fast switching speed, high voltage blocking capability, and high allowable operating temperatures [36]. These features are very suitable for the power converters in both configurations. At the same time, when compared with the IGBT modules, the SiC modules with improved packaging technologies follow different failure mechanisms or modes. New lifetime models are needed to properly analyze their reliability [37].

It should be noted that the passive components (e.g., the DC-link capacitors) and gate drivers can also affect the lifetime and reliability of these power converters, which will be the future work.

6. Conclusions

In this paper, the battery energy storage for 1500 V PV systems was investigated with a comparative reliability analysis of two configurations, i.e., DC-coupling and AC-coupling. The presented analysis can be used to assess how different configurations affect the reliability of the power interfacing units and, thus, justify the selection of configurations for the 1500 V PV-battery systems. With the component-, converter-, and system-level reliability analysis, the most fragile part in each level can be identified, which will contribute to the design-for-reliability of the PV-battery systems with a predictable reliability performance. The presented case study has shown that the AC-coupled configuration has slightly better reliability performance, whereas the DC-coupled configuration can become more reliable with a proper redundancy design. In all, the exploration in this paper has demonstrated that it is of importance to design for reliability at the system level. Moreover, this can be further extended to examine the overall system cost of 1500 V PV systems during the life-cycle and, then, it will be beneficial to the reduction of the cost of PV energy through design.

Author Contributions: Conceptualization, J.H. and Y.Y.; methodology, software, validation, formal analysis, investigation, resources, data curation, writing—original draft preparation, visualization, J.H.; writing—review and editing, supervision, project administration, funding acquisition, Y.Y. and D.V. All authors have read and agreed to the published version of the manuscript.

Funding: This research was funded by the Novo Nordisk Fonden through the Interdisciplinary Synergy Programme (Award Ref. No.: NNF18OC0034952).

Conflicts of Interest: The authors declare no conflict of interest.

References

1. Rechargenews. 'Exponential' Global Solar Growth to Continue with 142GW Added in 2020. Available online: https://www.rechargenews.com/solar/exponential-global-solar-growth-to-continue-with-142gw-added-in-2020/2-1-733672 (accessed on 20 May 2020).
2. Kouro, S.; Leon, J.I.; Vinnikov, D.; Franquelo, L.G. Grid-Connected Photovoltaic Systems: An Overview of Recent Research and Emerging PV Converter Technology. *IEEE Ind. Electron. Mag.* **2014**, *9*, 47–61. [CrossRef] [CrossRef]
3. International Electrical Testing Association (NETA). ANSI/NETA ECS-2020: Standard for Electrical Commissioning Specifications for Electrical Power Equipment and Systems. Available online: https://www.netaworld.org/standards/ansi-neta-ecs (accessed on 10 June 2020).
4. Electrical Installation Wiki. Electrical Regulations and Standards. Available online: https://www.electrical-installation.org/enwiki/Electrical_regulations_and_standards (accessed on 10 June 2020).
5. Inzunza, R.; Okuyama, R.; Tanaka, T.; Kinoshita, M. Development of a 1500VDC photovoltaic inverter for utility-scale PV power plants. In Proceedings of the IEEE 2nd International Future Energy Electronics Conference (IFEEC), Taipei, Taiwan, 1–4 November 2015; pp. 1–4.
6. Mansiri, K.; Sukchai, S.; Sirisamphanwong, C. Fuzzy Control for Smart PV-Battery System Management to Stabilize Grid Voltage of 22 kV Distribution System in Thailand. *Energies* **2019**, *11*, 1730. [CrossRef]
7. Serban, E.; Ordonez, M.; Pondiche, C. DC-bus voltage range extension in 1500 V photovoltaic inverters. *IEEE J. Emerg. Sel. Top. Power Electron.* **2015**, *3*, 901–917. [CrossRef] [CrossRef]
8. Stevanovi'c, B.; Serrano, D.; Vasi'c, M.; Alou, P.; Oliver, J.A.; Cobos, J.A. Highly efficient, full ZVS, hybrid, multilevel DC/DC topology for two-stage grid-connected 1500-V PV system with employed 900-V SiC devices. *IEEE J. Emerg. Sel. Top. Power Electron.* **2019**, *7*, 811–832. [CrossRef] [CrossRef]
9. Yang, Y.; Kim, A.K.; Blaabjerg, F.; Sangwongwanich, A. *Advances in Grid-Connected Photovoltaic Power Conversion Systems*; Woodhead Publishing: Cambridge, UK, 2019; pp. 37–38.
10. Ingeteam. Battery Inverter up to 1.64 MVA with 1500 V Technology. Available online: https://www.ingeteam.com/en-us/sectors/photovoltaic-energy/p15_24_288_45/storage-b-series-1500v-1640kva.aspx (accessed on 20 May 2020).
11. SMA. SMA Energy Systems for Large-Scale Operations. Available online: https://files.sma.de/dl/2485/ENERGYSYSTEMLARGE-KDE1917.pdf (accessed on 20 May 2020).
12. Atif, A.; Khalid, M. Saviztky—Golay Filtering for Solar Power Smoothing and Ramp Rate Reduction Based on Controlled Battery Energy Storage. *IEEE Access* **2020**, *8*, 33806–33817. [CrossRef] [CrossRef]
13. Martins, J.; Spataru, S.; Sera, D.; Stroe, D.-I.; Lashab, A. Comparative Study of Ramp-Rate Control Algorithms for PV with Energy Storage Systems. *Energies* **2019**, *12*, 1342. [CrossRef]
14. Myneni, H.; Kumar, G.S. Energy Management and Control of Single-Stage Grid Connected Solar PV and BES System. *IEEE Trans. Sustain. Energy* **2019**. [CrossRef] [CrossRef]
15. Saez-de-Ibarra, A.; Martinez-Laserna, E.; Stroe, D.; Swierczynski, M.; Rodriguez, P. Sizing Study of Second Life Li-ion Batteries for Enhancing Renewable Energy Grid Integration. *IEEE Trans. Ind. Appl.* **2016**, *52*, 4999–5008. [CrossRef] [CrossRef]

16. Akeyo, O.M.; Rallabandi, V.; Jewell, N.; Ionel, D.M. The Design and Analysis of Large Solar PV Farm Configurations With DC-Connected Battery Systems. *IEEE Trans. Ind. Appl.* **2020**, *56*, 2903–2912. [CrossRef] [CrossRef]

17. Eaton. White Paper WP083023EN: AC vs. DC Coupling in Utility-Scale Solar Plus Storage Projects. April 2016. Available online: https://www.eaton.com/ecm/idcplg?IdcService=GET_FILE&allowInterrupt=1&RevisionSelectionMethod=LatestReleased&Rendition=Primary&dDocName=WP083023EN (accessed on 20 May 2020).

18. Sangwongwanich, A.; Angenendt, G.; Zurmuhlen, S.; Yang, Y.; Sera, D.; Sauer, D.U.; Blaabjerg, F. Enhancing PV inverter reliability with battery system control strategy. *CPSS Trans. Power Electron. Appl.* **2018**, *3*, 93–101. [CrossRef] [CrossRef]

19. Sandelic, M.; Sangwongwanich, A.; Blaabjerg, F. Reliability Evaluation of PV Systems with Integrated Battery Energy Storage Systems: DC-Coupled and AC-Coupled Configurations. *Electronics* **2019**, *8*, 1059. [CrossRef] [CrossRef]

20. Sangwongwanich, A.; Yang, Y.; Sera, D.; Blaabjerg, F. Mission Profile-Oriented Control for Reliability and Lifetime of Photovoltaic Inverters. *IEEE Trans. Ind. Appl.* **2020**, *56*, 601–610. [CrossRef] [CrossRef]

21. Technical University of Denmark. DTU Climate Station Data. Available online: http://climatestationdata.byg.dtu.dk/ (accessed on 20 May 2020).

22. Vernica, I.; Wang, H.; Blaabjerg, F. Impact of Long-Term Mission Profile Sampling Rate on the Reliability Evaluation of Power Electronics in Photovoltaic Applications. In Proceedings of the 2018 IEEE Energy Conversion Congress and Exposition (ECCE), Portland, OR, USA, 23–27 September 2018; pp. 4078–4085.

23. Jinko. JKM380M-72-V. 2020. Available online: https://www.http://www.jinkosolar.com/ (accessed on 20 May 2020).

24. Semikron, Nuremberg, Germany. IGBT Modules. Available online: https://www.semikron.com/products/product-classes/igbt-modules.html (accessed on 20 May 2020).

25. Gevorgian, V.; Booth, S. *Review of PREPA Technical Requirements for Interconnecting Wind and Solar Generation;* National Renewable Energy Lab. (NREL): Golden, CO, USA, 2013. [CrossRef]

26. Sungrow. Energy Storage: ST1370KWH-1000. Available online: https://www.sungrowpower.com/en/products/storage-system/turnkey-solution/st1370kwh-1000 (accessed on 20 May 2020).

27. Yang, Y.; Wang, H.; Blaabjerg, F.; Ma, K. Mission profile based multi-disciplinary analysis of power modules in single-phase transformerless photovoltaic inverters. In Proceedings of the 15th European Conference on Power Electronics and Applications (EPE), Lille, France, 2–6 September 2013; pp. 1–10.

28. Miner, M. Cumulative Damage in Fatigue. *J. Appl. Mech.* **1945**, *12*, 159–164.

29. Bayerer, R.; Herrmann, T.; Licht, T.; Lutz, J.; Feller, M. Model for Power Cycling lifetime of IGBT Modules—Various Factors Influencing Lifetime. In Proceedings of the 5th International Conference on Integrated Power Electronics Systems, Nuremberg, Germany, 11–13 March 2008; pp. 1–6.

30. Huang, H.; Mawby, P. A. A lifetime estimation technique for voltage source inverters. *IEEE Trans. Power Electron.* **2013**, *28*, 4113–4119. [CrossRef] [CrossRef]

31. Yang, Y.; Sangwongwanich, A.; Blaabjerg, F. Design for reliability of power electronics for grid-connected photovoltaic systems. *CPSS Trans. Power Electron. Appl.* **2016**, *1*, 92–103. [CrossRef] [CrossRef]

32. Reigosa, P.D.; Wang, H.; Yang, Y.; Blaabjerg, F. Prediction of Bond Wire Fatigue of IGBTs in a PV Inverter Under a Long-Term Operation. *IEEE Trans. Power Electron.* **2016**, *31*, 7171–7182. [CrossRef]

33. ZVEI. How to Measure Lifetime for Robustnesss Validation—Step by Step. Berlin, Germany, rev. 1.9. 2012. Available online: https://www.zvei.org/en/press-media/publications/how-to-measure-lifetime-for-robustness-validation-step-by-step/ (accessed on 20 May 2020).

34. Zhang, Y.; Wang, H.; Wang, Z.; Blaabjerg, F.; Saeedifard, M. Mission Profile-Based System-Level Reliability Prediction Method for Modular Multilevel Converters. *IEEE Trans. Power Electron.* **2020**, *35*, 6916–6930. [CrossRef] [CrossRef]

35. Jin, K.; Yang, M.; Ruan, X.; Xu, M. Three-Level Bidirectional Converter for Fuel-Cell/Battery Hybrid Power System. *IEEE Trans. Ind. Electron.* **2010**, *57*, 1976–1986. [CrossRef] [CrossRef]

36. Zhang, L.; Yuan, X.; Wu, X.; Shi, C.; Zhang, J.; Zhang, Y. Performance Evaluation of High-Power SiC MOSFET Modules in Comparison to Si IGBT Modules. *IEEE Trans. Power Electron.* **2019**, *34*, 1181–1196. [CrossRef] [CrossRef]
37. Ni, Z.; Lyu, X.; Yadav, O.P.; Singh, B.N.; Zheng, S.; Cao, D. Overview of Real-Time Lifetime Prediction and Extension for SiC Power Converters. *IEEE Trans. Power Electron.* **2020**, *35*, 7765–7794. [CrossRef] [CrossRef]

Article

Step-Up Series Resonant DC–DC Converter with Bidirectional-Switch-Based Boost Rectifier for Wide Input Voltage Range Photovoltaic Applications

Abualkasim Bakeer, Andrii Chub * and Dmitri Vinnikov *

Department of Electrical Power Engineering and Mechatronics, Power Electronics Group, Tallinn University of Technology, 19086 Tallinn, Estonia; abbake@taltech.ee
* Correspondence: andrii.chub@taltech.ee (A.C.); dmitri.vinnikov@taltech.ee (D.V.)

Received: 25 June 2020; Accepted: 17 July 2020; Published: 21 July 2020

Abstract: This paper proposes a high gain DC–DC converter based on the series resonant converter (SRC) for photovoltaic (PV) applications. This study considers low power applications, where the resonant inductance is usually relatively small to reduce the cost of the converter realization, which results in low-quality factor values. On the other hand, these SRCs can be controlled at a fixed switching frequency. The proposed topology utilizes a bidirectional switch (AC switch) to regulate the input voltage in a wide range. This study shows that the existing topology with a bidirectional switch has a limited input voltage regulation range. To avoid this issue, the resonant tank is rearranged in the proposed converter to the resonance capacitor before the bidirectional switch. By this rearrangement, the dependence of the DC voltage gain on the duty cycle is changed, so the proposed converter requires a smaller duty cycle than that of the existing counterpart at the same gain. Theoretical analysis shows that the input voltage regulation range is extended to the region of high DC voltage gain values at the maximum input current. Contrary to the existing counterpart, the proposed converter can be realized with a wide range of the resonant inductance values without compromising the input voltage regulation range. Nevertheless, the proposed converter maintains advantages of the SRC, such as zero voltage switching (ZVS) turn-on of the primary-side semiconductor switches. In addition, the output-side diodes are turned off at zero current. The proposed converter is analyzed and compared with the existing counterpart theoretically and experimentally. A 300 W experimental prototype is used to validate the theoretical analysis of the proposed converter. The peak efficiency of the converter is 96.5%.

Keywords: photovoltaic (PV); DC–DC converter; series resonance converter; wide range converter; bidirectional switch; conversion efficiency

---

1. Introduction

Nowadays, with climate change around the world being evident, electrification is considered as a viable solution for the energy transition [1]. Therefore, power electronic converters face numerous new applications [2]. Among different emerging applications, there is a demand for high-performance DC–DC converters suitable for the integration of low-voltage energy sources and battery energy storage in DC microgrids [3]. In some cases, like photovoltaic (PV) module-level power electronic applications, both high-voltage step-up and the wide input voltage range regulation capability are required to interface individual PV modules that can supply their maximum power at very different voltages due to shading effects [4]. Therefore, the associated DC–DC interface converter has to regulate the input voltage in a wide range while providing high efficiency to draw the maximum available power from a PV module.

Usually, high-voltage step-up applications require galvanic isolation as a high-frequency transformer to step up voltage efficiently. Several DC–DC converters have been proposed to solve the voltage variation [5–7]. These topologies vary in their structure, complexity, and other aspects. The isolated buck-boost converters were justified as a suitable solution for high step-up wide-range applications. Usually, they have active switches at both sides of the converter to implement voltage buck and boost functionalities on different converter sides. Among these topologies, the series resonant converters (SRCs) have demonstrated high performance in target PV applications. They provide soft-switching of semiconductor components and good utilization of the isolation transformer [8,9]. The SRC topology is similar to the LLC converter topology that is investigated in many industrial applications [10,11]. However, the ratio between the magnetizing and the resonant inductances is several times higher in the SRC compared to the LLC converter. In general, the resonant converter applies the frequency modulation to control the DC voltage gain. However, this study targets low-power compact SRCs that use a small (low-cost) resonant inductor in the resonant tank. Even though such implementation results in low values of the quality factor, their DC voltage gain can be controlled using the pulse width modulation (PWM), which simplifies the converter design.

In the galvanically isolated buck-boost SRCs, the input voltage buck functionality is usually implemented by PWM [12] or phase-shifted modulation (PSM) [13] of the front-end inverter. These modulation methods have been already verified in numerous studies that date back as far as 1988 [12]. From the recent reports, it could be concluded that the highest efficiency is achieved for the input voltage buck operation by using PSM and hybrid PSM methods [14]. Currently, much attention is given to the input voltage boost implementation in the SRC [8]. The implementation of a boost rectifier usually achieves this [15]. As this study targets high-voltage step-up applications, the boost rectifiers are based on the voltage-doubler rectifier (VDR) to minimize the transformer turns ratio. Typical boost VDR is based on replacing diodes with the metal oxide semiconductor field-effect transistors (MOSFETs) and their control with short pulses [16] or double-pulse modulation [17]. Power losses in the boost VDR could be reduced if only one diode is replaced with a MOSFET, which, however, results in higher peak current of the resonant inductor and thus can compromise its size [9]. On the other hand, the implementation of a four-quadrant bidirectional switch in parallel to the transformer secondary winding allows for a significant reduction of the switch voltage stress and thus switching losses. At the same time, it keeps the positive and negative magnitudes of the resonant current balanced. Due to these advantages, topology in [18] is considered in this study. Comprehensive analysis shows that this topology cannot operate in a wide range of voltages and power. Therefore, a new converter is proposed to extend the input voltage regulation range by rearranging positions of the resonant tank elements. The proposed converter topology is based on the topology in [18], where the position of the resonance capacitor is moved to be placed between the bidirectional switch and the transformer secondary winding. The proposed converter is feasible in a wide range of the resonant inductance values without affecting the input voltage regulation range, which is not feasible for the baseline topology from [18]. This paper proposes a new SRC topology with a modified boost VDR and verifies it in the voltage range suitable for the module-level PV applications. The main hypothesis is that it is possible to extend regulation voltage and power range of the baseline topology by rearranging the resonant capacitor position. There are three main contributions: identification and experimental verification of the limits of the input voltage and power regulation range in the converter [18], synthesis of the converter with improved input voltage and power regulation range, and derivation of its steady-state mathematical model that is verified experimentally.

The rest of the paper starts with Section 2 that describes the proposed topology and provides its comprehensive analysis. Section 3 presents a comparison between the proposed and the baseline SRC topology. The results of experimental verification are discussed in Section 4. Finally, Section 5 draws the conclusion.

2. Topology Description and Modulation

2.1. Topology Description

The configurations of the proposed and conventional topologies are shown in Figure 1a,b, respectively. The proposed topology employs the voltage-doubler rectifier in the output side. The resonance capacitor (C_r) is placed before the bidirectional switch in the proposed topology. At the same time, it is integrated as part of the voltage-doubler rectifier in Figure 1b. In the proposed SRC topology, the capacitors C_1 and C_2 have much larger capacitance than C_r to keep the resonance frequency (f_r) constant according to (1). The value of resonance inductance L_r equals either the value of the isolation transformer leakage inductance (L_{lk}) or the sum of the transformer leakage inductance and the external inductance (L_{ext}) if the leakage inductance is low. The cost and size of the converter can be reduced by utilizing only the transformer leakage. The average voltage of the resonant capacitor (V_{Cr}) equals zero due to the symmetry of the VDR during the switching period as will be explained later. Contrary to the topology [18], the average voltage across the resonant capacitors ($C_r/2$) equals half of the output voltage (i.e., $V_{OUT}/2$). The bidirectional switch comprises the two MOSFETs Q_1 and Q_2. The conventional voltage-source full-bridge inverter is employed at the input side. The transistors of the input-side bridge are driven with complementary pulses of nearly 0.5 duty cycle, considering a small dead time between the control signals in the same leg. The input-side full-bridge inverter feeds the isolating transformer *TX* with a balanced rectangular voltage that features positive and negative magnitudes equal to the input voltage. The magnetizing inductance of the transformer (L_m) provides auxiliary circulating current that assists the soft switching of the primary-side transistors. The magnetizing current can recharge their parasitic output capacitance during the short dead times. The isolating transformer *TX* steps up the voltage fed by the input bridge inverter by the turns ratio *n*.

$$f_r = \frac{1}{2\pi \sqrt{L_r C_r}} \tag{1}$$

where, L_r is the resonant inductance.

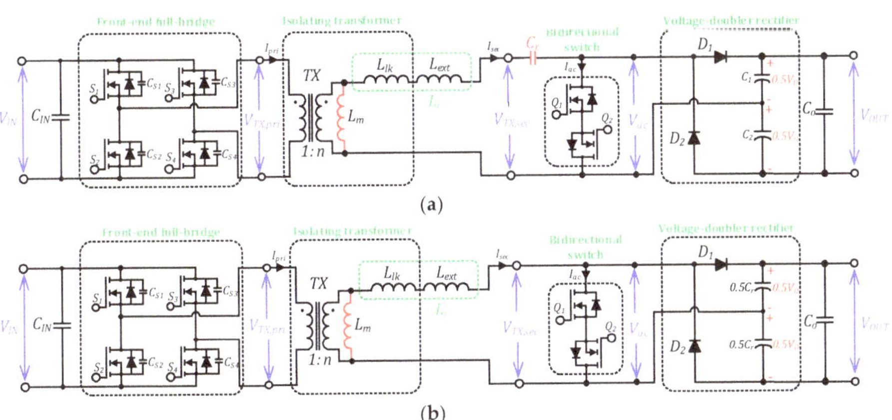

Figure 1. Converter topology of (**a**) the proposed SRC based on the modified VDR with a bidirectional switch and (**b**) the SRC presented in [18].

2.2. PWM Schemes for the Boost VDR

The switches Q_1 and Q_2 form the bidirectional switch. They are used to short-circuit the transformer output winding; so, the resonant inductor can increase its energy serving as a boost inductor of the AC boost converter. The two switches are connected in the back-to-back configuration. The bidirectional switch allows its current to flow in both directions, while it can block both voltage polarities. Two PWM schemes could be employed to generate the gating signal for the switches Q_1 and Q_2, as shown in Figure 2. First, only one of the switches is turned on at each half-cycle and forms a path for the current through the body diode of the other switch, as shown in Figure 2a. For example, during the positive voltage half-wave across the transformer secondary winding, the transistor Q_1 is turned on, which results in the conduction of the body diode of the transistor Q_2. In the other PWM scheme from Figure 2b, the switches are controlled with overlapped signals of equal duty cycle, which are shifted regarding the control signals of the input-side switches. Therefore, the body diodes are not used at all, and synchronous rectification is implemented to reduce the conduction losses. In both cases, the switching frequency of the switches Q_1 and Q_2 is the same as that of the primary-side switches. The voltage boosting mode occurs during two equal time intervals with the cumulative duty cycle D_b, which are separated in time by half of the switching period T_{SW}.

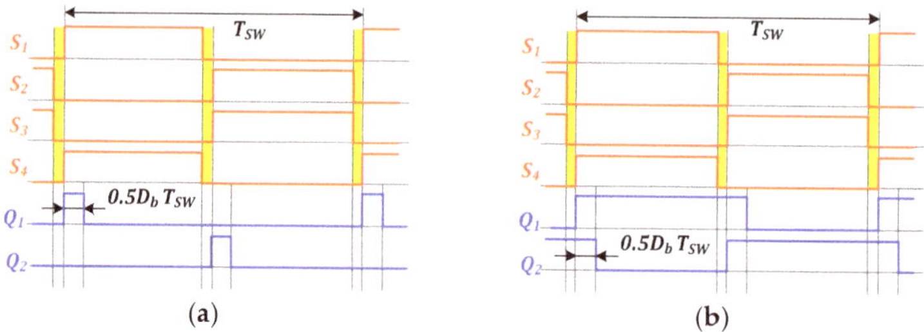

Figure 2. Possible PWM schemes for the bidirectional switch: (**a**) simple boost PWM and (**b**) phase-shifted PWM with overlapping signals.

The peak-to-peak ripple of the capacitor voltage is affected by the output power level (P_{OUT}), as given in (2). It is worth mentioning that the proposed topology can operate with the PWM technique from Figure 2b in a limited range of power and voltage. Abnormal operation occurs when the maximum capacitor voltage is larger than the voltage of the transformer secondary winding (i.e., $\Delta V_{Cr}/2 > n \cdot V_{IN}$). As shown in Figure 3, when a positive voltage feeds the secondary winding of the transformer, the current begins to flow in the reverse direction after discharging the stored energy. This reverse current increases the conduction losses of the converter, which results in the deterioration of the system efficiency and reduction of the DC voltage gain.

$$\Delta V_{Cr} = \frac{P_{OUT}T_{SW}}{2nV_{IN}C_r},$$

(2)

where T_{SW} is the switching period, n is the transformer turns ratio, and V_{IN} is the input voltage.

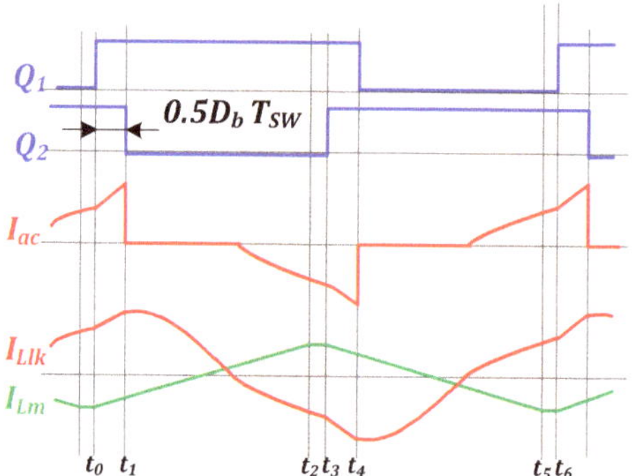

Figure 3. Abnormal operation of the proposed SRC with a boost VDR in the case of PWM scheme from Figure 2b when $\Delta V_{Cr}/2 > nV_{IN}$.

3. Steady-State Analysis and Comparison

3.1. Description of the Operating Principle

The main voltage and current waveforms of the proposed topology and the state-plane trajectory of the state variables are given in Figures 4 and 5. The resonant current ($i_{LIk}(t)$) is multiplied by the resonant impedance Z_r defined in (3) to have the same units of the axes. The steady-state analysis was performed based on the following assumptions:

1. The output voltage (V_{OUT}) is ripple-free due to the high value of the output capacitance (C_O).
2. The output capacitances (C_1, C_2, C_O) are much larger than the resonant capacitance (C_r).
3. The PWM scheme in Figure 2a is applied to the switches Q_1, Q_2.
4. The system is lossless.

$$Z_r = \sqrt{\frac{L_{lk}}{C_r}}. \tag{3}$$

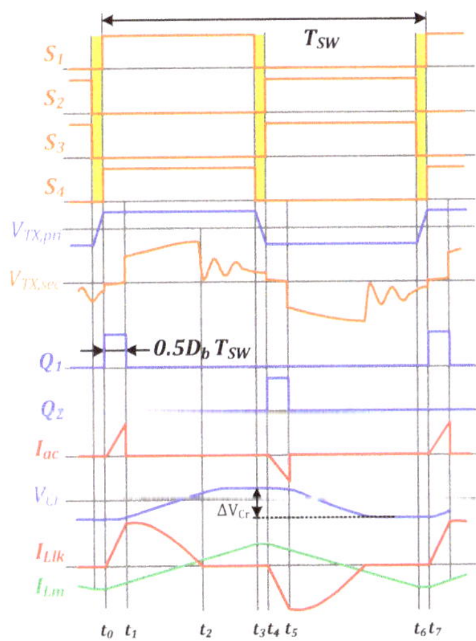

Figure 4. Sketch of idealized voltage and current steady-state waveforms of the proposed converter.

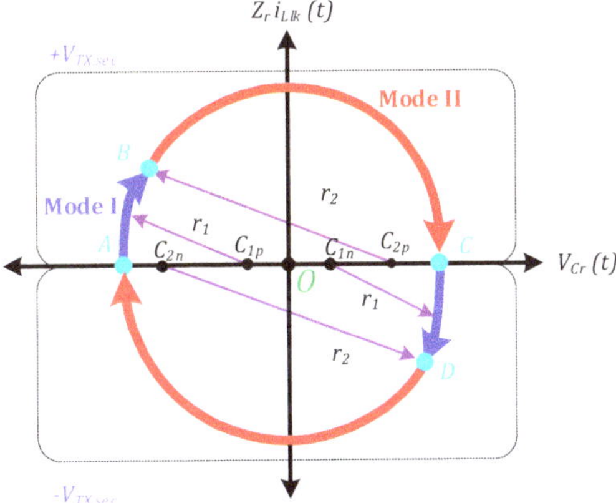

Figure 5. The state-plane trajectory of the resonance tank of the proposed SRC.

3.2. Modes of Operation

Mode I *[$t_0 < t \leq t_1$]*: This time interval corresponds to the first half of the voltage boosting mode with the duty cycle of $D_b/2$. A positive voltage is applied to the transformer primary winding by turning on the switches S_1 and S_4 and the secondary winding voltage equals nV_{IN}. For better understanding, the equivalent circuit of this interval is given in Figure 6a. The resonance capacitor voltage has a minimum value of $-\Delta V_{Cr}/2$ at the time instant t_0. Also, the initial resonance current is nearly zero prior to turning on the switch Q_1 that will be turned on at nearly zero current switching (ZCS) conditions. The voltage of the resonant capacitor assists the secondary winding voltage to accelerate the current charging of the resonant inductor, similar to the conventional boost converter. On the state-plane, the voltage of the resonant capacitor moves from point A to point B during this time interval with roughly sinusoidal shape, as shown in Figure 5. The state variable equations for the resonance current and voltage can be expressed in time domain as in Equations (4) and (5), respectively.

$$i_{LIk}(t) = \frac{r_1}{Z_r}sin(\pi - \omega_r(t - t_0)), \tag{4}$$

$$v_{Cr}(t) = nV_{IN} + r_1cos(\pi - \omega_r(t - t_0)), \tag{5}$$

$$r_1 = nV_{IN} + \frac{\Delta V_{Cr}}{2}, \tag{6}$$

where r_1 refers to the radius of the trajectory arc segment with center at $(nV_{IN}; 0)$, $\omega_r = 2\pi f_r$ is the angular resonance frequency in rad/s.

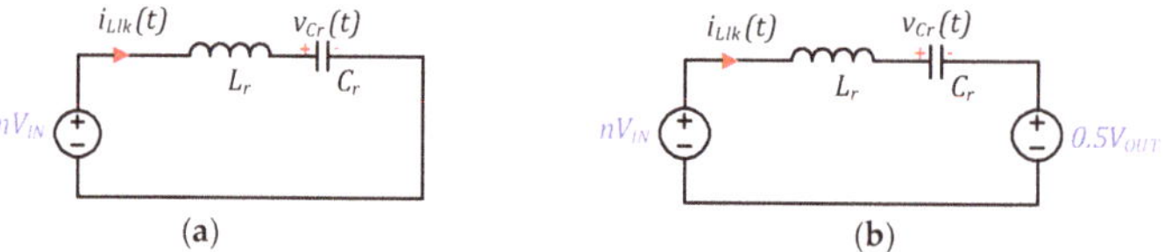

Figure 6. Equivalent circuit of the resonant tank operation during: (**a**) Mode I and (**b**) Mode II.

Mode II *[$t_1 < t \leq t_2$]*: The switch Q_1 is switched off and the bidirectional switch current I_{ac} drops to zero. The stored energy in the resonant inductor is releasing directly to the load, as shown in Figure 6b. The diode D_1 begins to conduct as it has a forward-biased state. The current of the resonant inductance is still flowing in the same direction with reversed voltage polarity, that is, it equals $(V_{OUT}/2 - nV_{IN})$ at the instant t_1. During this interval, the resonant inductor resonates with the resonant capacitor, and therefore, the resonance capacitor voltage moves from point B to point C along the trajectory curve. The length of this path is represented by the angle β (in rad). The capacitor voltage reaches its maximum value at $\Delta V_{Cr}/2$, and the resonant current reaches zero at the instant t_2. Equations (7)–(9) describe the converter operation in this mode before the resonant current drops to zero.

$$i_{LIk}(t) = \frac{r_2}{Z_r}sin(\beta - \omega_r(t - t_1)), \tag{7}$$

$$v_{Cr}(t) = nV_{IN} - \frac{V_{OUT}}{2} + r_2cos(\beta - \omega_r(t - t_1)), \tag{8}$$

$$r_2 = -nV_{IN} + \frac{V_{OUT}}{2} + \frac{\Delta V_{Cr}}{2}, \tag{9}$$

where r_2 refers to the radius of the arc trajectory segment with center at $((nV_{IN} - V_{OUT}/2); 0)$.

Mode III *[$t_2 < t \leq t_3$]*: The resonant current equals zero as all the stored energy is released into the load in the previous mode, and the diode D_1 is turned off at ZCS. Consequently, the converter enters the discontinuous conduction mode (DCM) and no energy is transferred to the load during this mode. The capacitor voltage remains constant at its maximum value at $\Delta V_{Cr}/2$ until the end of this period at $T_{SW}/2$.

Mode IV [$t_3 < t \leq t_4$]: This refers to the dead-time interval and the primary switches (S_1, S_4) are turned off at t_3. The magnetizing current denotes the circulating current referred to the primary winding to charge/discharge the parasitic output capacitance of switches (S_1, S_4) and (S_2, S_3), respectively. Therefore, the voltage across the switches (S_2, S_3) equals zero before the time instant t_4. The values of the dead time and the magnetizing inductance are interdependent and must ensure full discharging of the switch parasitic capacitances. This allows the primary switches to be turned on at zero voltage switching (ZVS) conditions.

Mode V [$t_4 < t \leq t_0$]: This represents the negative half-cycle of the switching period. The converter operates similar to that during time interval [$t_0; t_4$].

3.3. DC Voltage Gain Derivation

The segments of the state-plane trajectory of the resonant tank that correspond to the positive and negative half-cycles are symmetric. Therefore, only the trajectory segment A-B-C is considered to derive the DC voltage gain expression for the proposed converter. The general expressions for the circle radius in Modes I and II are given by Equations (10) and (11), respectively. The two circles intersect at point B, which results in (12).

$$r_1^2 = (v_{Cr} - nV_{IN})^2 + (Z_r i_{Llk})^2, \tag{10}$$

$$r_2^2 = \left(v_{Cr} - nV_{IN} + \frac{V_{OUT}}{2}\right)^2 + (Z_r i_{Llk})^2, \tag{11}$$

$$(v_{Cr}(t_1) - nV_{IN})^2 + (Z_r i_{Llk}(t_1))^2 - r_1^2 = \left(v_{Cr}(t_1) - nV_{IN} + \frac{V_{OUT}}{2}\right)^2 + (Z_r i_{Llk}(t_1))^2 - r_2^2. \tag{12}$$

The resonant inductor current and resonant capacitor voltage at the instant t_1 can be given as:

$$i_{Llk}(t_1) = \frac{r_1}{Z_r} sin(\omega_r t_1), \tag{13}$$

$$v_{Cr}(t_1) = nV_{IN} + r_1 cos(\omega_r t_1) \tag{14}$$

The resonance path angle β is derived from Equations (7) and (13) as follows:

$$\beta = \pi - sin^{-1}\left(\frac{r_1}{r_2} sin(\omega_r t_1)\right). \tag{15}$$

Then, by substituting $t_1 = D_b T_{SW}/2$ in Equations (13)–(15), the cumulative duty cycle of the converter can be expressed as:

$$D_b = \frac{2cos^{-1}\left(\dfrac{\frac{T_{SW}P_{OUT}}{4C_r}(4-G_n)+nV_{IN}V_{OUT}}{nV_{IN}V_{OUT}+\frac{T_{SW}P_{OUT}}{4C_r}G_n}\right)}{\omega_r T_{SW}}, \tag{16}$$

where $G_n = \frac{V_{OUT}}{nV_{IN}}$ is the normalized DC voltage gain.

It follows from Equation (16) that the duty cycle depends not only on the level of the converter output power but also on the resonance tank parameters.

For the topology in [18], the duty cycle D_b is given in Equation (17).

$$D_b = \frac{2L_{lk}\sqrt{\frac{2P_{OUT}T_{SW}}{V_{OUT}C_r}(V_{OUT} - nV_{IN})}}{Z_r nV_{IN}T_{SW}}. \tag{17}$$

3.4. Comparison of DC Voltage Gain and Input Operating Range

This section provides a comparison between the proposed and the baseline [18] topologies. The main feature of the proposed topology is the input voltage regulation in a wide voltage and power range. The superiority over the baseline SRC topology is achieved as the proposed converter can provide much higher DC voltage gain at the same duty cycle D_b, as shown in Figure 7a. Moreover, the proposed converter is much less sensitive to the value of the resonant inductor, while the baseline SRC topology shows a strong dependence of the available voltage and power regulation range on the resonant inductance value. Based on Equation (16), the operating range limit is shown in Figure 7b where it is compared to the target operating range defined by the maximum input current $I_{INm} = 20$ A, the maximum input power $P_{INm} = 300$ W, and the maximum input voltage $V_{INm} = 30$ V, while the minimum input voltage is limited to 10 V to limit the converter power loss. The given target operating range is typical for module-level PV applications, as the interface converter should be capable of operating under partial shading, when the global maximum power point can occur at voltage as low as 10 V. The area highlighted with yellow color shows the region where the baseline SRC topology from [18] cannot operate as the limiting line is drawn theoretically for a critical case of $D_b = 0.8$. In ractice, the duty cycle value is always below unity $D_b < 1$, which means that experimental regulation range of the baseline topology will be even more limited than that in Figure 7b.

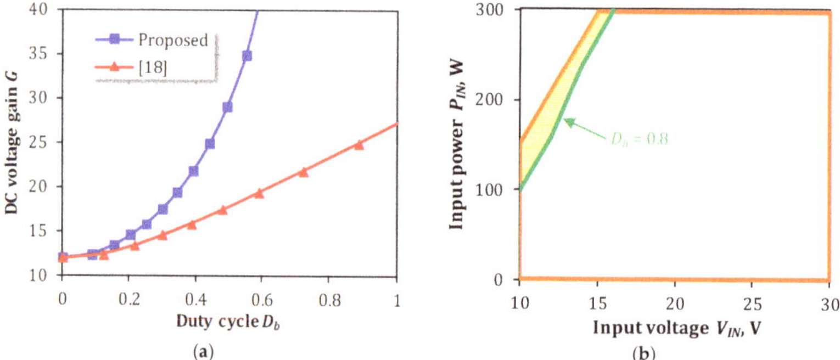

Figure 7. Comparison between the proposed topology and [18] in terms of (a) the DC voltage at $P_{IN} = 300$ W and (b) target and feasible input voltage and power ranges for $L_r = 100$ µH.

4. Experimental Results

4.1. Description of the Experimental Prototype

A 300 W prototype was built to demonstrate the feasibility of the proposed converter. The parameters and components used in the setup are listed in Table 1. The converter is designed for the target operating range shown in Figure 7b, which suits the PV applications and is similar to the previous studies [18–20]. Only generic Si MOSFETs are used in the input side to reduce the converter cost. The use of SiC devices at the output side is unavoidable due to high switching frequency.

The isolating transformer was built on an ETD39 core of 3C95 ferrite material. The transformer turns ratio $n = 6$ yields the output voltage of 350 V according to (18), which results in the boost mode at the input voltage below 30 V. The value of the output voltage is suitable for the integration with residential DC microgrids. The number of primary and secondary turns equals 9 and 54, respectively. This design yields the maximum flux density of the core of 60 mT in the worst case. Therefore, the core losses can be minimized. To reduce the skin effect and proximity losses in the transformer, 90×0.2 and 90×0.1 litz wires were used for the primary and secondary windings, respectively. They were interleaved to reduce the leakage inductance and insulated by the Kapton polyimide tape to minimize the capacitance between the layers. An external inductor was connected in series with the secondary winding of the transformer to increase the resonance inductance. The resonance frequency of the converter was aimed close to the switching frequency to ensure the ZVS of the primary-side MOSFETs [9]. The dead-time period between S_1, S_2 or S_3, S_4 equaled 190 ns. The PWM control signals were generated using the low-cost microcontroller STM32F334. The system efficiency was measured by a Yokogawa WT1800 precision power analyzer.

$$n = \frac{V_{OUT}}{2V_{INm}}. \tag{18}$$

Table 1. Setup parameters and components.

Parameter	Symbol	Value
Input voltage range	V_{IN}	10:30 V
Input-side capacitor	C_{IN}	150 µF
Transformer leakage inductance	L_{lk}	4 µH
Transformer magnetizing inductance	L_m	1.3 mH
External inductor	L_{ext}	92.5 µH
Resonance capacitor	C_r	30 nF, metal film
Voltage-doubler capacitors	C_1, C_2	150 µF, electrolytic
Output-side capacitors	C_o	150 µF
Output voltage	V_{OUT}	350 V
Switching frequency	F_s	95 kHz
Components	**Symbol**	**Part Number**
Primary-side switches	S_1, S_2, S_3, S_4	FDMS86180
Bidirectional switch	Q_1, Q_2	SCT2120AF
Output-side diodes	D_1, D_2	C3D02060E

4.2. Steady-State Waveforms

The steady-state experimental waveforms of the proposed converter at the maximum input voltage are given in Figure 8a–c. The operating power was 300 W and the bidirectional switch was turned off. The primary switches $S_{1,2,3,4}$ were driven with a nearly 0.5 duty cycle, so the primary winding voltage of the transformer had the square shape in Figure 8a. The proposed converter operated in the DC transformer (DCX) mode. The peak-to-peak voltage of the transformer primary was 60 V and the input current equaled 10 A. The primary current was a periodic sine wave with a peak value of 8 A. The oscillations appeared in the primary winding voltage and current due to the parasitic capacitances of the semiconductors. In Figure 8b, the bidirectional switch had zero current and the peak-to-peak voltage of 350 V. The secondary winding voltage had the same shape of the primary winding voltage, but the magnitude was multiplied by n (i.e., 180 V). The peak value of the secondary winding current was 2.4 A. The theoretical and measured peak-to-peak of the resonance capacitor voltage was 174 and 175, respectively. These values are well-matched, which proves the theoretical analysis. The output voltage was constant at 350 V and C_1 had half of the output voltage (i.e., 175 V). The voltage swing of the resonant capacitor equaled 300 V, which was slightly higher than the calculated value of 280 V.

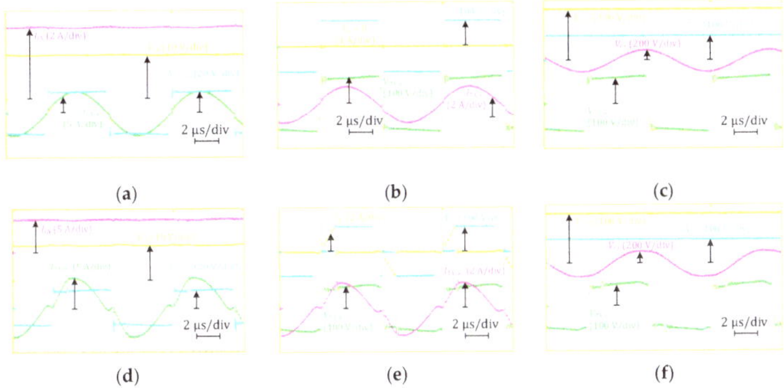

Figure 8. Steady-state experimental waveforms of the proposed converter operating at (**a–c**) $P_{IN} = 300$ W, $V_{IN} = 30$ V, $D_b = 0$, and (**d–f**) $P_{IN} = 300$ W, $V_{IN} = 25$ V, $D_b = 0.215$.

For the converter operation in the boost mode, the steady-state voltage and current waveform are given in Figure 8d–f. The operating power was $P_{IN} = 300$ W and the input voltage equaled $V_{IN} = 25$ V. The cumulative duty cycle of the bidirectional switch equaled $D_b = 0.215$. The drawn current from the supply equaled 12 A. The average voltage of the resonance capacitor was zero and the peak-to-peak ripple voltage equaled roughly 230 V. The output voltage was constant at $V_{OUT} = 350$ V. Small parasitic oscillations occurred due to hard-switching of the output transistors.

The state-plane trajectory of the converter in the DCX and the boost modes are presented in Figure 9. In the case of the DCX mode, the state-plane trajectory has a circular shape because the voltages and currents of the resonant tank are virtually sinusoidal. The radius of each curve corresponds to ΔV_{Cr}.

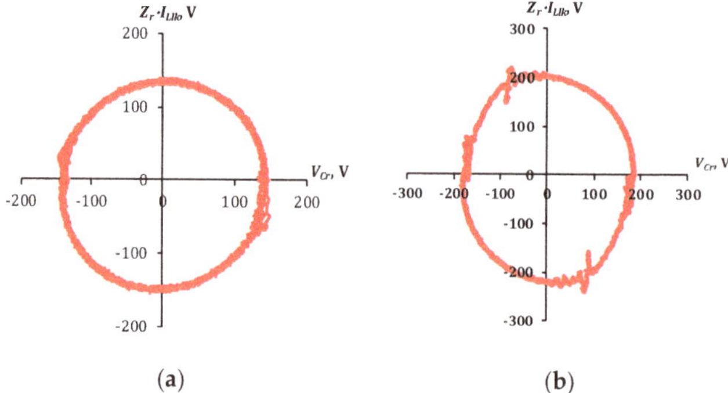

(a)   (b)

Figure 9. Experimental state-plane trajectory for the proposed converter at $P_{IN} = 300$ W corresponding to (**a**) the DCX mode and (**b**) the boost mode.

4.3. Performance Verification

The theoretical and measured conversion gain of the converter as a function in the duty cycle is plotted in Figure 10. The DC voltage gain of the converter equals 12 at $D_b = 0$. Small differences between the theoretical and measured DC voltage gain values occurred due to the influence of power losses. Therefore, at the input power of $P_{IN} = 200$ W, deviations between the theoretical and measured DC voltage gain values are more significant because the converter experiences higher losses. Evidently, the input power level affects the duty cycle at the same DC voltage gain, for example, at the gain of $G = 20$: $D_b = 0.22$ at $P_{IN} = 50$ W and $D_b = 0.32$ at $P_{IN} = 200$ W.

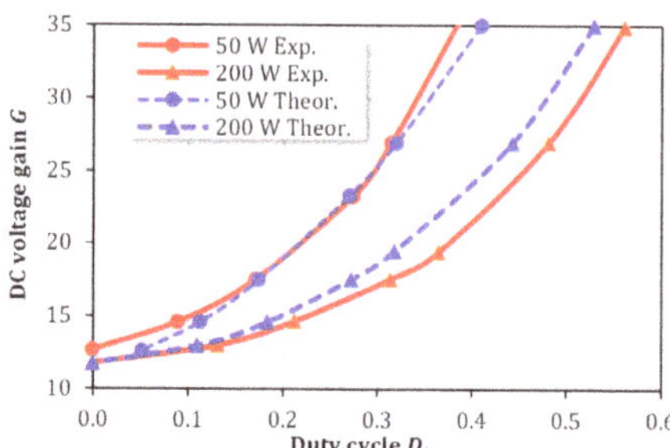

Figure 10. Calculated and measured DC voltage gain of the converter versus the duty cycle at two input power levels.

The input voltage regulation range was measured for the proposed and the baseline topologies. As it was predicted, the baseline converter from [18] features limited input voltage and power regulation range, which is more restricted than was predicted theoretically. The area highlighted with yellow color in Figure 11 describes the input voltage and power range, where the proposed converter can operate and thus achieves superiority over the baseline converter from [18]. Experimental study of the baseline converter shows that it cannot regulate the voltage at duty cycles above some critical value, which drops almost linearly from 0.7 at V_{IN} = 10 V to 0.46 at V_{IN} = 22 V, which corresponds to the limit shown as the green line in Figure 11.

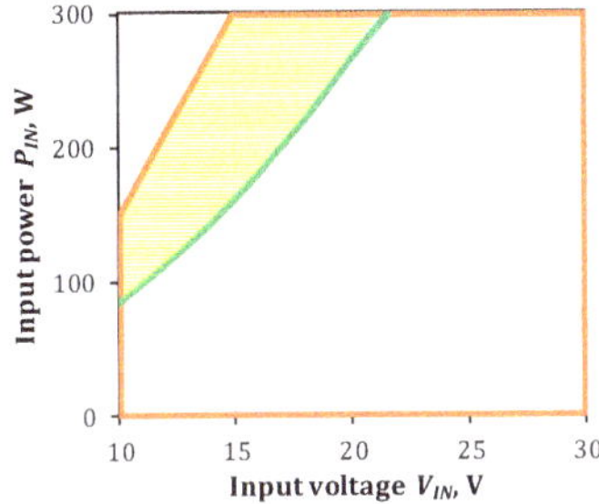

Figure 11. Comparison of the experimental limit of the input voltage and power regulation range in relation to the target operating range for the proposed and the baseline SRC topologies.

Next, it is essential to examine the efficiency of the converter with different input voltages and power levels. The efficiency measured across the wide input voltage range is shown in Figure 12a. The maximum efficiency of the proposed converter equaled 96% and 96.5% at P_{IN} = 50 W and P_{IN} = 200 W, respectively. These values were achieved when the input voltage was at its maximum level (D_b = 0) and the bidirectional switch was turned off. The converter is similar to the traditional SRC and the transformer current had the sinusoidal shape. Accordingly, a full soft-switching was achieved in the converter semiconductors. When the input voltage decreased, the converter activated PWM of the bidirectional switch to boost the transformer voltage up to 350 V. With D_b > 0, the converter lost the full soft-switching feature and the converter efficiency decreased together with the input voltage, causing also higher conduction losses. In addition, the efficiency was affected by the operating power due to the change in the quality factor of the resonant tank. The efficiency was also measured versus the input power for different input voltages, as shown in Figure 12b. The efficiency curves are flat across the wide input power range for the input voltages of over 20 V.

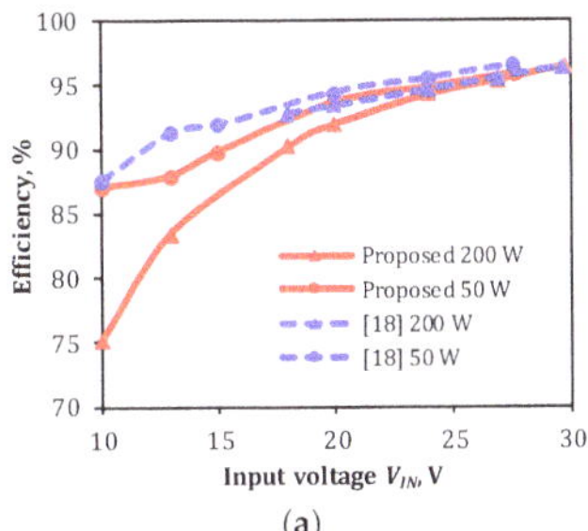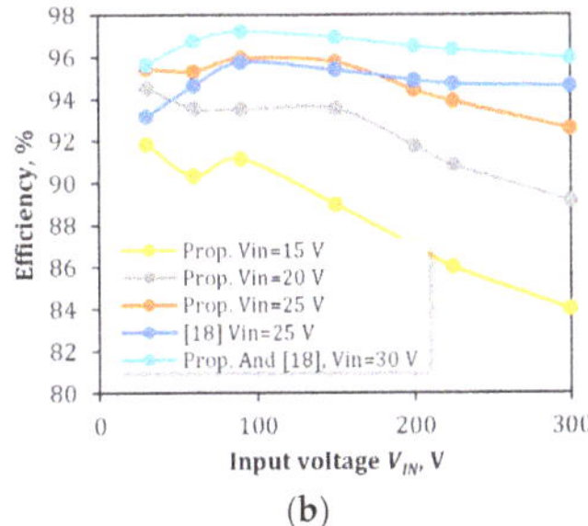

Figure 12. Comparison of the measured efficiency of the proposed and the baseline converters versus (**a**) the input voltage at two power levels and (**b**) the input power at different input voltage levels.

Figure 12 also includes several efficiency curves for the baseline converter from [18] to compare it to the proposed converter. It should be noted that the proposed converter features slightly lower efficiencies at higher power, which results from the higher voltage stress of the bidirectional switch in the proposed topology, but higher efficiency at light load, as shown in Figure 12b for the input voltage V_{IN} = 25 V. However, this is an acceptable drawback, considering that the proposed converter can operate in the most critical operating points, that is, at low input voltages and high input currents (cf. Figure 11). For example, the converter from [18] cannot achieve the input power of above 200 W at the input voltage below 18 V, as shown in Figure 12a.

An application-specific study was performed to verify the capability of the proposed converter to perform maximum power point tracking (MPPT). A simplified control approach based on the hill-climbing MPPT algorithm with direct perturbations of the duty cycle was used similar to [21]. For the given converter, operation with a PV module resulted in a reduction in the input voltage when the duty cycle D_b was increased. Hence, the direct MPPT can be implemented, avoiding a PI controller. The test was performed using the solar array simulator (SAS) Agilent E4360A as the input power source and electronic DC load Chroma 63204 in the constant voltage mode. The MPPT routine shown in Figure 13 corresponds to the converter operation with a 48-cell monocrystalline-Si PV module Sharp NQ-R258H. When the converter was first connected to the SAS, the current spike appeared due to the charging of the internal capacitances. When the output voltage reached the reference value of 350 V, the DC load started operation and the converter drew a minimum current from the SAS needed to keep the DC load running. After that, the MPPT started with a delay of 0.6 s. The converter reached the maximum power point of the corresponding PV module in 2.5 s. It should be noted that the resonant capacitor experienced an offset of around 200 V caused by start-up transients. Nevertheless, this offset disappeared after several seconds of operation. More importantly, it could be seen that the voltage ripple of the resonant capacitor depended tightly on the input power.

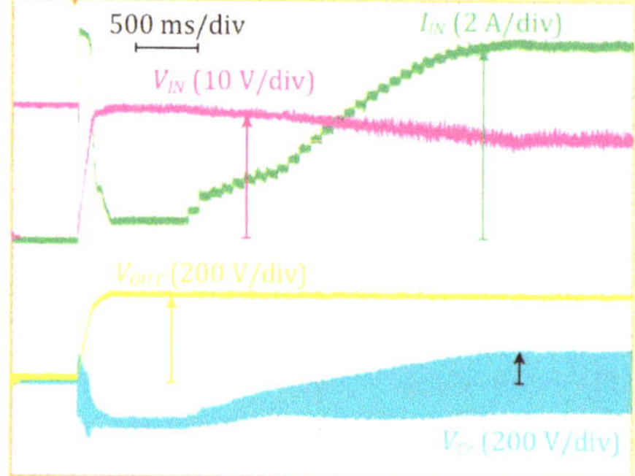

Figure 13. Experimental MPPT routine.

5. Conclusions

This paper presents a new series resonant DC–DC converter with voltage boost capability achieved by using the output-side boost voltage-doubler rectifier. Contrary to the baseline topology, the proposed converter contains the resonant capacitor between the transformer secondary winding and the bidirectional switch. As a result, the proposed converter can operate in the range of low voltages and high currents where the baseline topology cannot operate at all. However, the proposed converter experiences higher voltage stress of the bidirectional switch than that of the baseline counterpart due to the influence of the resonant capacitor voltage ripple, which results in efficiency reduction. Nevertheless, the proposed converter expends the input voltage and power operating range and achieves a peak efficiency of 96.5%. The proposed converter was justified for the module-level PV applications as it is capable of performing the maximum power point tracking and covers a wide voltage range of possible maximum power points that could be observed under shading operation. The main implementation challenge of the proposed converter is related to the design of the isolating transformer. In cost-sensitive applications, it is advisable to integrate the resonant inductor into the transformer, which, however, could result in high proximity losses in the transformer windings. Therefore, future research will be focused on the optimization of the passive components of the converter.

Author Contributions: Conceptualization, A.B. and A.C.; methodology, A.B.; software, A.B.; validation, A.C. and A.B.; formal analysis, A.C. and A.B.; investigation, A.B.; resources, D.V.; data curation, A.C.; writing—original draft preparation, A.C. and A.B.; writing—review and editing, A.C. and D.V.; visualization, A.C. and A.B.; supervision, A.C.; project administration, A.C.; funding acquisition, A.C. All authors have read and agreed to the published version of the manuscript.

Funding: This research was supported in part by the Estonian Research Council grant PSG206, and in part by the Estonian Centre of Excellence in Zero Energy and Resource Efficient Smart Buildings and Districts, ZEBE, grant 2014-2020.4.01.15-0016 funded by the European Regional Development Fund.

Conflicts of Interest: The authors declare no conflict of interest.

Nomenclature

PV	Photovoltaic
SRC	Series resonance converter
ZVS	Zero voltage switching
LLC	Inductor-inductor-capacitor resonant converter
PWM	Pulse width modulation
PSM	Phase-shifted modulation
VDR	Voltage-doubler rectifier
MOSFET	Metal oxide semiconductor field-effect transistor
C_r	Resonant capacitance (F)
f_r	Resonant frequency (Hz)
L_r	Resonant inductance (H)
L_{lk}	Leakage inductance (H)
L_{ext}	External inductance (H)
V_{Cr}	The average voltage of the resonant capacitor (V)
V_{OUT}	Output voltage (V)
L_m	The magnetizing inductance of the transformer (H)
n	Turns ratio of the transformer
D_b	Cumulative boosting duty cycle
T_{SW}	Switching period (s)
P_{OUT}	Output power (W)
ΔV_{Cr}	The peak-to-peak ripple of the resonant capacitor voltage
V_{IN}	Input voltage (V)
C_O	Output capacitance (F)
Z_r	Resonant impedance (Ω)
ZCS	Zero current switching
ω_r	Angular resonant frequency (rad/s)
β	Length of the resonant path (rad)
DCM	Discontinuous conduction mode
G_n	Normalized DC voltage gain
P_{INm}	Maximum input power (W)
V_{INm}	Maximum input voltage (V)
I_{INm}	Maximum input current (A)
P_{IN}	Input power (W)
MPPT	Maximum power point tracking
SAS	Solar array simulator

References

1. Dennis, K. Environmentally beneficial electrification: Electricity as the end-use option. *Electr. J.* **2015**, *28*, 100–112. [CrossRef]
2. Zhang, G.; Li, Z.; Zhang, B.; Halang, W.A. Power electronics converters: Past, present and future. *Renew. Sustain. Energy Rev.* **2018**, *81*, 2028–2044. [CrossRef]
3. Batarseh, I.; Alluhaybi, K. Emerging opportunities in distributed power electronics and battery integration: Setting the stage for an energy storage revolution. *IEEE Power Electron. Mag.* **2020**, *7*, 22–32. [CrossRef]

4. Ravyts, S.; Van de Sande, W.; Vecchia, M.D.; den Broeck, G.V.; Duraij, M.; Martinez, W.; Daenen, M.; Driesen, J. Practical considerations for designing reliable DC/DC converters, applied to a BIPV case. *Energies* **2020**, *13*, 834. [CrossRef]

5. Cha, H.; Peng, F.Z.; Yoo, D. Z-Source resonant DC-DC converter for wide input voltage and load variation. In Proceedings of the 2010 International Power Electronics Conference—ECCE ASIA, Sapporo, Japan, 21–24 June 2010. [CrossRef]

6. LaBella, T.; York, B.; Hutchens, C.; Lai, J.-S. Dead time optimization through loss analysis of an active-clamp flyback converter utilizing gan devices. In Proceedings of the 2012 IEEE Energy Conversion Congress and Exposition (ECCE), Raleigh, NC, USA, 15–20 September 2012. [CrossRef]

7. Wang, Y.; Liu, R.; Han, F.; Yang, L.; Meng, Z. Soft-Switching DC–DC converter with controllable resonant tank featuring high efficiency and wide voltage gain range. *IET Power Electron.* **2020**, *13*, 495–504. [CrossRef]

8. Chub, A.; Vinnikov, D.; Lai, J.-S. Input voltage range extension methods in the series-resonant dc-dc converters. In Proceedings of the 15th Brazilian Power Electronics Conference and 5th IEEE Southern Power Electronics Conference (COBEP/SPEC'2019), Santos, Brazil, 1–4 December 2019. [CrossRef]

9. Kim, J.-W.; Park, M.-H.; Han, J.-K.; Lee, M.; Lai, J.-S. PWM resonant converter with asymmetric modulation for ZVS active voltage doubler rectifier and forced half resonance in PV application. *IEEE Trans. Power Electron.* **2020**, *35*, 508–521. [CrossRef]

10. Fei, C.; Lee, F.C.; Li, Q. High-Efficiency high-power-density LLC converter with an integrated planar matrix transformer for high-output current applications. *IEEE Trans. Ind. Electron.* **2017**, *64*, 9072–9082. [CrossRef]

11. Nabih, A.; Ahmed, M.; Li, Q.; Lee, F.C. Transient control and soft start-up for 1 MHz LLC converter with wide input voltage range using simplified optimal trajectory control. *IEEE J. Emerg. Sel. Top. Power Electron.* **2020**. [CrossRef]

12. Vandelac, J.-P.; Ziogas, P.D. A DC to DC PWM series resonant converter operated at resonant frequency. *IEEE Trans. Ind. Electron.* **1988**, *35*, 451–460. [CrossRef]

13. Kim, E.-H.; Kwon, B.-H. Zero-Voltage- and zero-current-switching full-bridge converter with secondary resonance. *IEEE Trans. Ind. Electron.* **2010**, *57*, 1017–1025. [CrossRef]

14. Pahlevani, M.; Pan, S.; Jain, P. A hybrid phase-shift modulation technique for DC/DC converters with a wide range of operating conditions. *IEEE Trans. Ind. Electron.* **2016**, *63*, 7498–7510. [CrossRef]

15. Wu, H.; Zhang, J.; Qin, X.; Mu, T.; Xing, Y. Secondary-Side-Regulated soft-switching full-bridge three-port converter based on bridgeless boost rectifier and bidirectional converter for multiple energy interface. *IEEE Trans. Power Electron.* **2015**, *31*, 1017–1025. [CrossRef]

16. Son, S.; Montes, O.A.; Junyent-Ferre, A.; Kim, M. High step-up resonant DC/DC converter with balanced capacitor voltage for distributed generation systems. *IEEE Trans. Power Electron.* **2019**, *34*, 4375–4387. [CrossRef]

17. Zhao, X.; Chen, C.-W.; Lai, J.-S. A high-efficiency active-boost-rectifier-based converter with a novel double-pulse duty cycle modulation for PV to DC microgrid applications. *IEEE Trans. Power Electron.* **2019**, *34*, 7462–7473. [CrossRef]

18. LaBella, T.; Yu, W.; Lai, J.-S.; Senesky, M.; Anderson, D. A Bidirectional-Switch-Based wide-input range high-efficiency isolated resonant converter for photovoltaic applications. *IEEE Trans. Power Electron.* **2014**, *29*, 3473–3484. [CrossRef]

19. Zhao, X.; Zhang, L.; Born, R.; Lai, J.-S. A high-efficiency hybrid resonant converter with wide-input regulation for photovoltaic applications. *IEEE Trans. Ind. Electron.* **2017**, *64*, 3684–3695. [CrossRef]

20. Vinnikov, D.; Chub, A.; Liivik, E.; Roasto, I. High-Performance quasi-z-source series resonant DC–DC converter for photovoltaic module-level power electronics applications. *IEEE Trans. Power Electron.* **2017**, *32*, 3634–3650. [CrossRef]

21. Chub, A.; Zinchenko, D.; Vinnikov, D.; Blinov, A. Three-Port flyback converter for photovoltaic module integration in bipolar dc microgrids. In Proceedings of the 2020 IEEE International Conference on Industrial Technology (ICIT), Buenos Aires, Argentina, 26–28 February 2020. [CrossRef]

MDPI

St. Alban-Anlage 66

4052 Basel

Switzerland

Tel. +41 61 683 77 34

Fax +41 61 302 89 18

www.mdpi.com

Energies Editorial Office

E-mail: energies@mdpi.com

www.mdpi.com/journal/energies